Mechanical Alloying

Mechanical Alloying

Energy Storage, Protective Coatings, and Medical Applications

Third Edition

Prof. Dr. M. Sherif El-Eskandarany
Energy and Building Research Center, Nanotechnology and Advanced Materials Program, Kuwait Institute for Scientific Research, Kuwait

ELSEVIER

William Andrew
Applied Science Publishers

William Andrew is an imprint of Elsevier
The Boulevard, Langford Lane, Kidlington, Oxford, OX5 1GB, United Kingdom
50 Hampshire Street, 5th Floor, Cambridge, MA 02139, United States

Library of Congress Cataloging-in-Publication Data
A catalog record for this book is available from the Library of Congress

British Library Cataloguing-in-Publication Data
A catalogue record for this book is available from the British Library

ISBN: 978-0-12-818180-5

For information on all William Andrew publications
visit our website at https://www.elsevier.com/books-and-journals

Publisher: Matthew Deans
Acquisitions Editor: Christina Gifford
Editorial Project Manager: Peter Adamson
Production Project Manager: Kamesh Ramajogi
Designer: Mark Rogers

Typeset by Thomson Digital

To Egypt, Japan, and Kuwait
How do I say thank you when "Thank You" isn't enough?

To My Lovely Wife Mitsuko El-Eskandarany
I owe a debt of gratitude that I cannot entirely repay!

Writing a book is harder than I thought and more rewarding than I could have imagined. None of this would have been possible without the Upper Manager of the Kuwait Institute of Scientific Research (Dr. Samira A.S. Omar) and the Executive Director of the Energy and Building Research Center (Dr. Osamah Alsayegh), as well as my kind colleagues of Nanotechnology and Advanced Materials Program, Energy and Building Research Center, Kuwait Institute of Scientific Research.

Thank you very much for your kind support and continuous encouragement.

I am eternally grateful to Ms. Christina Gifford, Acquisitions Editor-Elsevier, and Dr. Peter W. Adamson, Editorial Project Manager-Elsevier, for their very kind support and for their helpful advices.

Contents

About the author xi
Preface xiii
Acknowledgment xv

1 Introduction 1
1.1 Advanced materials 1
1.2 Strategies used for fabrications of advanced materials 1
1.3 Mechanically assisted approach 4
1.4 Thermal approach 8
References 10

2 Characterizations of mechanically alloyed powders 13
2.1 Introduction 13
2.2 Examples of characterization techniques 13

3 The history and necessity of mechanical alloying 19
3.1 History of story of mechanical alloying 19
3.2 Fabrications of ODS alloys 21
3.3 Fabrications of other advanced materials 27
3.4 Mechanical alloying, mechanical grinding, mechanical milling, and mechanical disordering 30
3.5 Types of ball mills 31
3.6 Mechanism of mechanical alloying 47
3.7 Necessity of mechanical alloying 48
References 49

4 Controlling the powder-milling process 55
4.1 Factors affecting the MA/MD/MM 55
References 91

5 Ball milling as a superior nanotechnological fabrication's tool 93
5.1 Introduction 93
5.2 Nanocrystalline materials 94
5.3 Formation of nanocrystalline materials by ball milling technique 99
5.4 Selected examples 101
5.5 Effect of ball milling on the structure of carbon nanotubes 102
5.6 Pressing and sintering of powders materials 105

5.7 Consolidation of nanocrystalline powders 108
5.8 Spark plasma sintering for consolidation of ball-milled nanocrystalline powders 110
5.9 Fabrication of nanodiamonds and carbon nanotubes by milling 114
References 125

6 Mechanochimical process for fabrication of 3D nanomaterials 129
6.1 Introduction 129
6.2 Reduction of Cu_2O with Ti by room temperature rod milling 129
6.3 Properties 130
6.4 Mechanism of MSSR 136
6.5 Fabrication of nanocrystalline WC and nanocomposite WC-MgO refractory materials by MSSR method 138
6.6 c-BN 147
6.7 NbN 150
References 152

7 Fabrication of nanocrystalline refractory materials 153
7.1 Introduction 153
7.2 Preparation challenges and difficulties 154
7.3 Synthesizing and properties of mechanically solid-state reacted tic powders 155
7.4 Other carbides produced by mechanical alloying 165
References 190

8 Fabrication of and consolidation of hard nanocomposite materials 193
8.1 Introduction and background 193
8.2 Fabrications methods of particulate MMNCs 195
8.3 WC-based nanocomposites 207
8.4 Fabrication of metal matrix/carbon nanotubes nanocomposites by mechanical alloying 225
References 227

9 Solid-state hydrogen storage nanomaterials for fuel cell applications 229
9.1 Introduction 229
9.2 Hydrogen energy 231
9.3 Solid-state hydrogen storage 235
9.4 Magnesium hydride as an example of solid-state hydrogen storage material 237
References 259

10 Mechanically induced-catalyzation for improving the behavior of MgH_2 263
10.1 Introduction 264
10.2 Scenarios for improving the behavior of MgH_2 264

10.3 Combination of cold rolling and ball milling for improving the kinetics behavior of MgH_2 powders 286
References 289

11 Implementation of MgH_2-based nanocomposite for fuel cell applications 293
11.1 Introduction 293
11.2 Hydrogen reactors 294
References 308

12 Utilization of ball-milled powders for surface protective coating 309
12.1 Introduction 309
12.2 Thermal spraying 310
References 333

13 Mechanically induced solid-state amorphization 335
13.1 Introduction 335
13.2 Fabrication of amorphous alloys by mechanical alloying process 336
13.3 Crystal-to-glass transition 339
13.4 Mechanism of amorphization by mechanical alloying process 342
13.5 The glass-forming range 361
13.6 Amorphization via mechanical alloying when ΔH^{for}= Zero; mechanical solid-state amorphization of $Fe_{50}W_{50}$ binary system 366
13.7 Special systems and applications 373
13.8 Difference between mechanical alloying and mechanical disordering in the amorphization reaction of $Al_{50}Ta_{50}$ in a rod mill 377
13.9 Mechanically induced cyclic crystalline-amorphous transformations during mechanical alloying 385
13.10 Consolidation of multicomponent metallic glassy alloy powders into full-dense bulk materials 396
13.11 Recent studies 411
References 412

14 Mechanical alloying for preparing nanocrystalline high-entropy alloys 417
14.1 Introduction 417
14.2 Preparations of nanocrystalline HEAs by mechanical alloying 421
References 428

15 Biomedical applications of mechanically alloyed powders 431
15.1 Introduction 431
15.2 Metallic biomaterials 432
15.3 Mechanical alloying for fabrication of metallic biomaterials 434
References 439

Index 441

About the author

M. Sherif El-Eskandarany, a full Professor of Materials Science and Nanotechnology, gained his Master and Doctor Degrees at Tohoku University, Japan. He worked as a Professor at Institute for Materials Research, Tohoku University, Japan, and Professor at Faculty of Engineering, Al-Azhar University, Egypt. Until 2007, he worked as First-Under-Secretary of Egyptian Minster of Higher Education and Scientific Research, and the former Vice-President of The Academy of Scientific Research and Technology of Egypt. He has joined Kuwait Institute for Scientific Research to work as Senior Research Scientist in 2007. Since then, he works as Senior Research Scientist and Program Manager of Nanotechnology and Advanced Materials. He is the founder of Nanotechnology and Advanced Materials of KISR and the Project Leader of Establishing Nanotechnology Center in Kuwait. In 2018, he promoted to Principle Research Scientist. He has published more than 280 peer-reviewed papers in high-cited international scientific journals in the field of materials science, nanoscience, and nanotechnology, and more than 250 papers in the proceedings of several international conferences. He awarded six patents from the United States Patent and Trademark Office in the area of nanomaterials, protective coating, and hydrogen storage nanocomposites. He is the author of six scientific books and received many national and international awards, two of them given by the His Excellency the Former Egyptian President and the other one given by His Highness The Prince of Kuwait.

Preface

The mechanical alloying (MA) process, using ball-milling and/or rod-milling techniques, has received much attention as a powerful tool for the fabrication of several advanced materials including equilibrium, nonequilibrium (e.g., amorphous, quasicrystals, nanodiamonds, carbon nanotubes, nanocrystalline powders, etc.), and nanocomposite materials. In addition, it has been employed for reducing some metallic oxides by milling the oxide powders with metallic reducing agents at room temperature. The MA is a unique process in that a solid-state reaction takes place between the fresh powder surfaces of the reactant materials at room temperature. Consequently, it has been employed to produce alloys and compounds that are difficult or impossible to be obtained by the conventional melting and casting techniques.

This book is intended primarily to serve as an introduction to the MA process, including a general description of the process, starting material requirements, the equipment, characterizations of the milled powders, consolidation techniques and utilizations of the mechanically alloyed/milled powder for hydrogen storage, surface protective coating, and medical applications.

This book contains several examples of selected advanced materials that have been fabricated by MA during the last 3 decades. This book aims at the scientists on materials, nanoscience, nanotechnology, powder metallurgy, and metallurgists in industry. Researchers, senior undergraduate, and graduate students may also benefit from this book.

Chapter 1 gives a brief introduction about the approaches used for preparations of advanced materials through mechanically and thermally assisted approaches. Chapter 2 presents a brief introduction related to the characterization techniques used to investigate the general properties of ball-milled materials. A brief history usage of ball milling for fabrications of wide spectrum of advanced and new materials within the last 45 years is presented in Chapter 3. In this chapter, the types of ball mills used in MA are presented. Chapter 4 elaborates and discusses in more details the factors affecting the MA, mechanical milling, and mechanical disordering processes.

Chapter 5 presents the applications of ball-milling technique in nanotechnology for fabrications of nanocrystalline materials through top-down approach. This chapter gives several typical examples of the nanocrystalline systems fabricated by ball milling and how they can be consolidated into bulk nanocrystalline materials, using advanced consolidation techniques. The mechanism of the mechanochemical process, using high-energy ball milling, is discussed in Chapter 6. In Chapter 7, we shall discuss the mechanism of synthesizing nanocrystalline metal carbides, using high-energy ball-milling technique. Chapter 7 gives some typical examples of employing of the ball-milling method for producing useful metal carbides and nitrides nanopowders.

Employing of mechanical mixing for fabrications of nanocomposite powders is discussed in Chapter 8, where reactive ball-milling technique for synthesizing metal hydrides at room temperature is discussed in details in Chapters 9. The effect of nanocatalysts on improving the hydrogenation/dehydrogenation behaviors of metal hydrides is discussed in Chapter 10. Chapter 11 discusses the implementation of MgH_2-based nanocomposite for fuel cell applications.

Chapter 12 is dedicated to present and discuss the solid-state amorphization reactions for fabrications of amorphous and metallic glassy powders that are elaborated and discussed in Chapter 12. We shall discuss one of the most vital applications of mechanically alloyed and ball-milled powders in the area of surface protective coating, using advanced spraying technology in Chapter 13, where Chapter 14 will present the milling role for fabrication of different types of high-entropy alloys. Finally, the application of MA for fabrications of biomaterials used in biomedical sector will be shown and discussed in Chapter 15.

I hope that this book will be useful for the readers and can give a helpful introduction for engineers and researchers who are going to start their projects on materials production through MA.

M. Sherif El-Eskandarany
October 2019
Kuwait

Acknowledgment

The continual financial supports and encouragements given by the Kuwait Government and Kuwait Foundation for the Advancement of Sciences are deeply appreciated.

Introduction

1

However, there are many approaches and techniques used for producing the advanced materials, mechanical alloying has been receiving great attentions and considerations as a unique process for synthesizing of new advanced materials families that cannot be obtained by any other techniques. Nanostructured materials, nanoparticles, nanocomposites, carbon nanotubes, and amorphous and metallic glassy alloys are some of those new engineering materials that can be successfully obtained by such a room-temperature way of fabrication (Fig. 1.1).

1.1 Advanced materials

Life in the current 21st century cannot be dependent on limited groups of materials; instead, it is dependent on unlimited families of advanced materials. Laptop computers, digital cameras, smart cell phones, nanosensors, microwave ovens, computerized cars, bio-microelectromechanics (bio-MEMS), thin films photovoltaics, and many other intelligent devices and instruments used in many sectors require special type of materials that are enjoying superior properties. In spite of the traditional categories of materials (metals and metal alloys, ceramics, polymer, and composites) that do not match well with the whole modern industries requirements, a newcomer so-called "advanced materials" has found an important space in the functional classifications of the materials. However, the advanced materials can be defined in numerous ways based on their properties and usages; we can defined them as those materials that show advances over the traditional materials and used for manufacturing of high-tech products. Thus, the advanced materials refer to all new materials and developing to the existing materials to obtain superior unique and performance in one or more properties. Amorphous and metallic glasses, nanomaterials and nanocomposites, biomaterials, semiconductors, and smart or the so-called intelligent materials are some types of the advanced materials used in different sectors (Fig. 1.2).

1.2 Strategies used for fabrications of advanced materials

However, there are numerous standard fabrications methods that have been used for producing and fabrications of the traditional materials; including hydrometallurgy, pyrometallurgy, and powder metallurgy, the advanced materials cannot be prepared easily by any of them. Within the last 6 decades, materials scientists have developed

Mechanical Alloying. http://dx.doi.org/10.1016/B978-0-12-818180-5.00001-7

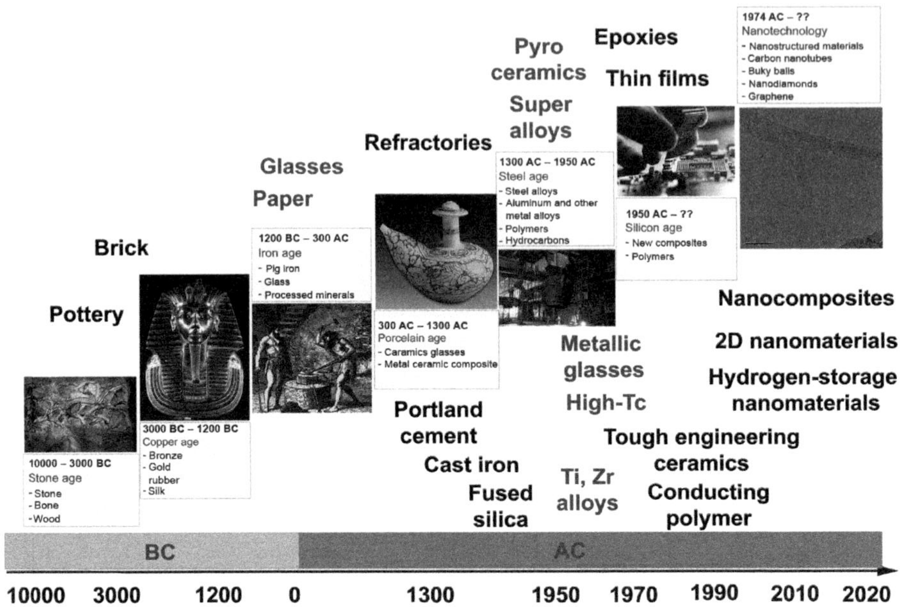

Figure 1.1 Schematic illustration, displaying the human civilizations, starting from Stone Age to Nanotechnology Age.

several synthesizing approaches and methods used for synthesizing of new materials families, the so-called advanced or "high-tech" materials that show unusual and superior chemical, physical, and mechanical properties.

These new routes of materials processing and fabrications have led to control the materials subatomic structure and tailoring the materials with desired and predetermined structure. Tailoring the materials structure through controlling their atomic arrangements (e.g., long-range ordered or short-range ordered) affects the whole materials properties to attain high-performance characterizations. Moreover, controlling the morphological and microscopic (shape and size) characterizations of the materials leads to significant changes in their properties and behaviors. It can be concluded that the way in which a material is produced (materials processing and fabrications) affects the atomic arrangements and microscopic properties of it and this will not only lead to develop the whole properties of the product but it also affect its performance and future applications, as schematically presented in Fig. 1.3, which show the interrelationship between preparations and processing, structure, properties, and performance.

The recent investigations made by the materials scientists within the last few decades enable us to prepare wide types of advanced materials through new approaches of preparations. In general, the strategies used for fabrication of the advanced materials may be classified as: (1) mechanically assisted approach, (2) mechanically induced solid-state reaction (MISSR) approach, (3) thermally assisted approach, (4) high-energy-assisted approach, (5) chemically assisted approach, (6) lithographic approach, (7) vapor deposition approach, and (8) liquid-phase fabrication approach.

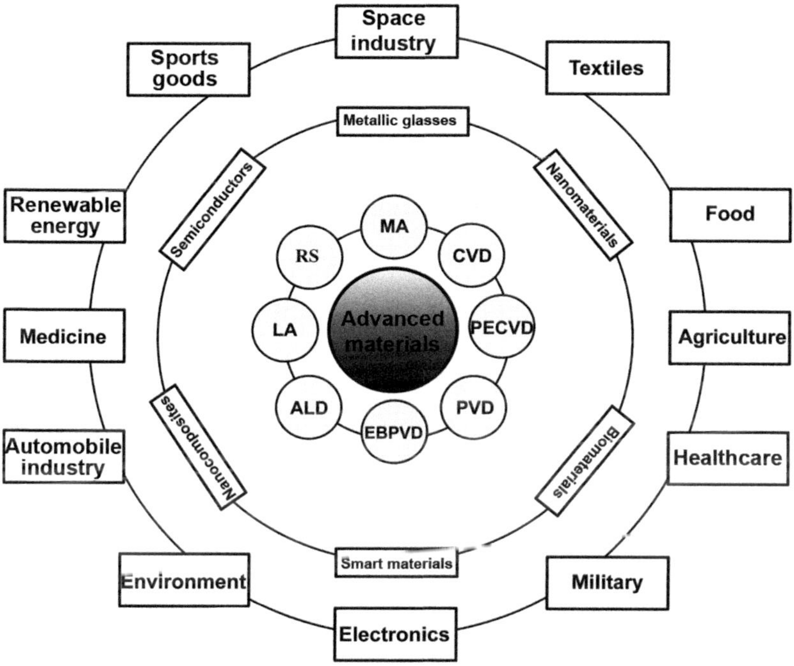

Figure 1.2 Advanced materials, such as metallic glasses, nanomaterials, biomaterials, smart materials, nanocomposites, semiconductors, etc., that are used for different industrial, medical, electronic, and many other sectors are prepared by wide variety of materials processing.

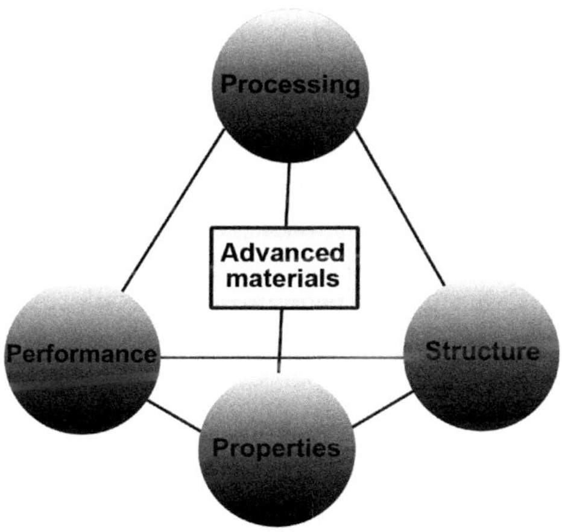

Figure 1.3 Schematic presentation of processing/structure/properties/performance interrelationship of advanced materials.

Some examples of new materials preparation methods such as ball milling, rapid solidifications (RSs), atomization, sputtering, chemical vapor deposition (CVD), electron beam physical vapor deposition (EBPVD), arc discharge, laser ablation (LA), photolithography, nanoimprint lithography (NIL), solgel, atomic force microscope (AFM) nanostencil, plasma enhanced chemical vapor deposition (PECVD), physical vapor deposition (PVD), and atomic layer deposition (ALD), which have been used for producing wide varieties of advanced materials, are illustrated in Fig. 1.2.

1.3 Mechanically assisted approach

1.3.1 Powder metallurgy

Powder metallurgy (P/M) can be simply defined as the cost-effective process of manufacturing of components and tools starting from metallic, ceramic, or composite powders. In fact, P/M is not a new process in our world of materials science; it dated back to 3000 BC, when the Egyptian employed it for preparing iron powder from "sponge iron" to make their tools.[1] At that time high-temperature furnaces were not available yet. Since then, P/M has been considered as a practical process that can be successfully used for net-shape or near-net-shape forming of high melting point metals, metal oxides, and cemented carbides without any needs of melting and casting castings processes. The development of this process has led today to produce high-quality iron powder by grinding and then ball milling of the sponge iron into fine particles and followed by heating the as-milled iron powders in hydrogen to remove the oxides. Modern P/M technology has started in the 2nd decade of the last century with a glorious achievement at that time, when a mass production of qualified tungsten carbide powders could be produced in industrial scale. During the period of time between 1920s and 1940s, the worldwide interest in P/M technology was monotonically increased, especially after the mass production of porous bronze brushes for bearings in 1920s.[2] During the World War II and till 1960s, a wide varieties of new composites, ferrous- and nonferrous-based materials were developed. Within the last 50 years, glorious progress in the area of powder consolidation has been achieved and new powder pressing techniques, such as cold/hot isostatic pressing, spark plasma sintering, shock-wave consolidation, induction hot pressing, etc., have been introduced. Accordingly, P/M has been drastically grown due to its capability of producing large-scale volume of powders and consolidated precision and complex near-net-shaped dense components. Fig. 1.4 presents a flow sheet of a typical powder metallurgical process. Ball milling and gas atomization are typical examples of P/M techniques used for fabrication of advanced materials powders with fine particle size.

1.3.2 Ball milling

With regard to the scope of this book we are much focused on the fabrication of advanced materials by ball-milling technique. Ball-milling technique via mechanical alloying (MA),[3] mechanical disordering (MD),[4] reactive ball milling (RBM),[5]

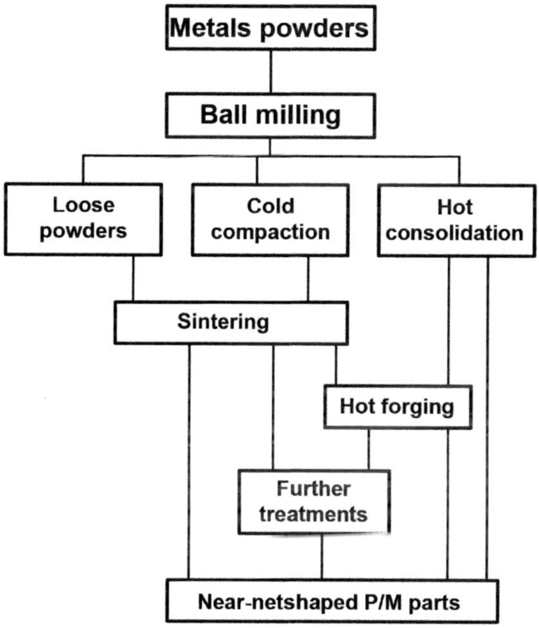

Figure 1.4 Flowchart diagram shows the mechanical methods of powder production and their consequence consolidations.

MISSR,[6,7] mechanically induced solid-state mixing (MISSM),[8] thermally assisted solid-state amorphization (TASSA),[9] electric discharge-assisted mechanical milling,[10] and mechanochemistry approaches has been considered as one of the most powerful routes for producing wide variety of advanced materials.[11,12]

Fundamentally, the term *milling* may be referred to as the breaking down of relatively coarse materials to the ultimate fineness. In the mineral processing (so often called mineral dressing[13]), the as-mined ore or the so-called run-of-mine, which contains the valuable mineral, contains also a significant volume fractions of invaluable raw materials known as "gangue minerals." In mineral dressing point of view, milling is a typical comminuting process that aims to separate the grains of valuable minerals from the "gangue," using rotating cylindrical steel vessels, which contain a charge of loose balls that is enjoying free motion inside the mill. Apart from the ore comminution, milling is also used for preparing materials for many industrial applications, such as milling of quartz to fine powder (under 70 μm in diameter), milling of talc to produce body powder, milling of iron ore for preparation of pellets, preparation of ultrafine powders for pharmacological applications,[14] and many others.

1.3.3 Mechanical alloying

The MA[3] is an MISSR that takes place between the diffusion couples, powders of the reactant material, in a room-temperature reactor known as ball mill.[15] Over the last 5 decades, ball milling has evolved from being a standard technique in mineral dressing and powder metallurgy, used primarily for particle size reduction and powder blending, to its present status as an important method for the preparation of either materials with enhanced physical and mechanical properties or, indeed, new phases, or new engineering materials. It is one of the most powerful nanotechnology tools for preparing a wide range of nanocrystalline, nanoparticles, and nanocomposites materials via the so-called top-down approach.[16] Accordingly, the term MA is becoming increasingly common in the materials science, metallurgy, and nanotechnology literatures.[17,18] So far, the MA process, using ball-milling[19] and/or rod-milling techniques,[20,21] has received much attention as a powerful tool for fabrication of several advanced materials (Fig. 1.5).

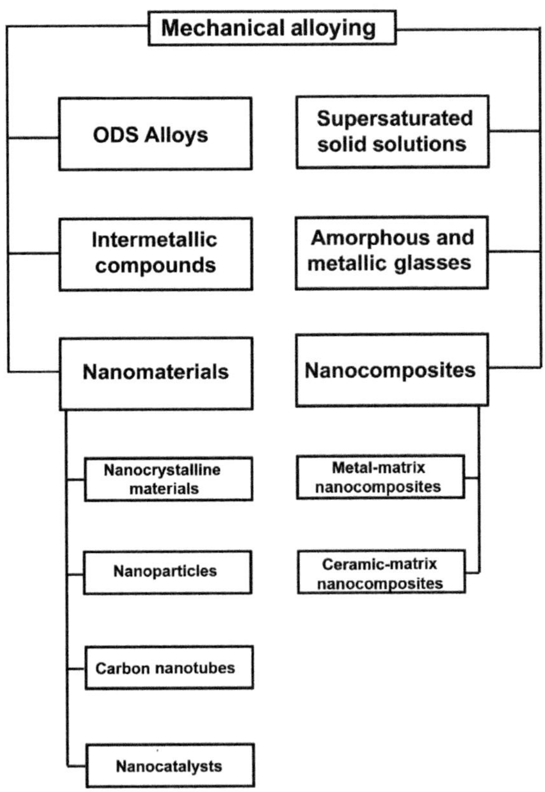

Figure 1.5 Mechanical alloying is a pioneer process for preparing a wide variety of advanced materials at room temperature. Fabrications of industrial scale of ODS alloys, intermetallic compounds, and amorphous and metallic glassy alloys, as well as nanocrystalline and nanocomposite materials at room temperature are some advantages of this process.

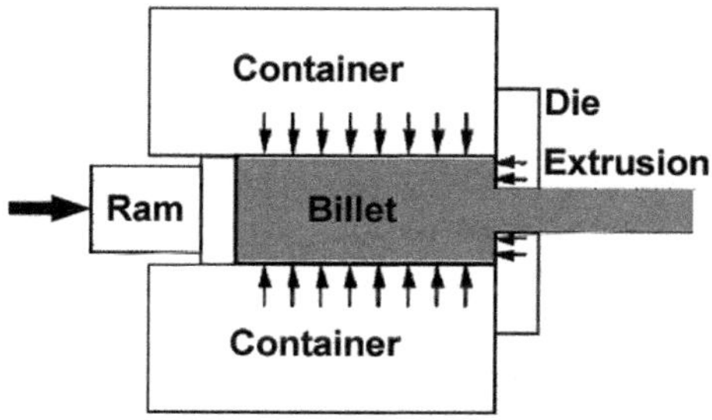

Figure 1.6 Schematic presentation shows the extrusion procedure, which is a typical plastic deformation process by which a block/billet/consolidated powder bulk material of metal is reduced in cross-section with forcing it to flow through a die orifice under high pressure (see the text).

1.3.4 Severe plastic deformation

Extrusion is a typical plastic deformation process by which a block/billet/consolidated powder bulk material of metal is reduced in cross-section by forcing it to flow through a die orifice under high pressure (Fig. 1.6). However, extrusion process is a well-known mechanical method used for more than 100 years to minimize materials thickness and refine structure; it can find wide applications for fabrications of nanostructured bulk materials, such as bulk nanostructured Cu, as was demonstrated by Champion et al.[22] In their experiments, the loose copper powders obtained from cyro-melting process were first consolidated into bulk material, using cold isostatic pressing and then sintered in hydrogen gas atmosphere. The as-sintered bulk Cu specimen were encapsulated in a copper sheath and densified up to 100% by hydrostatic extrusion. The as-extruded Cu sample that has nanocrystalline-grained structure with an average size of about 60 nm in diameter exhibits high compressive performance with a Hall–Petch yield stress.[22]

It has been demonstrated by McFadden et al.[23] that superplasticity behavior of metals, which is defined as the ability of a material to sustain large plastic deformation, is of industrial interest, as it forms the basis of a fabrication method that can be used to produce components having complex shapes from materials that are hard to machine, such as metal matrix composites and intermetallics. They presented unique results related to the low-temperature superplasticity in nanocrystalline nickel metal, nanocrystalline aluminum alloy (1420-Al), and nanocrystalline nickel aluminide (Ni_3Al). They investigated that the nanocrystalline nickel was found to be superplastic at low temperature (470°C), which corresponds to 0.36 melting temperature (T_m), m being the lowest normalized superplastic temperature reported for any crystalline material. They also found that nanocrystalline Ni_3Al is enjoying superplastic behavior at a temperature of 450°C, being below the superplastic temperature in the microcrystalline regime.[23]

In 2001, Baró et al.[24] addressed an important review article that presented the results of experimental and theoretical studies of diffusion and related phenomena, such as grain growth, creeps, superplasticity, in bulk nanostructured materials upon subjecting to severe plastic deformation conducting by equal channel angular pressing and high pressure torsion processes. They described these processes as unique approach for synthesizing of nanostructured materials for not only scientific interests but also potentials of industrial applications. They believe that bulk nanostructured metals and alloys would possess several unique physical and mechanical properties, which differ from those of their coarse-grained counterparts, upon subjecting to high density of severe deformation-induced nonhomogeneities in the interfacial structures as well as the presence of high-density ensembles of lattice defects.[24]

1.4 Thermal approach

1.4.1 Rapid solidification

RS is a typical melt-quenching technique that can be classified into three categories: the so-called (1) spinning, (2) droplet, and (3) surface melting.

Melt-spinning approach: Melt spinning is a well-known process used for producing amorphous metal and metallic glassy alloys in the form of thin strips (called ribbons). This approach, which was introduced by Pol Duwez and his group in 1960,[25] has led to synthesizing of several hundred binary, ternary, and multicomponent amorphous and metallic glassy systems during the last 5 decades.[26–29] In the melt-spinning process, a molten metal alloy system is subjected to very rapid cooling rate reaches to 10^6 Ks^{-1}. A detailed comprehensive review article on the production of amorphous and metallic glasses by melt-spinning technique was published elsewhere.[30]

In the typical melt-spinning process, certain amount (5–100 g) of small pieces of alloy is placed inside a crucible (usually made of quarts glass or sometimes made of boron nitride for high-temperature use) surrounded by induction coil, as illustrated in Fig. 1.7. Applying high current leads to raise the temperature of the alloy inside the crucible; accordingly, it melts. Then the molten metallic drops are ejected by Ar-pressurization through a fine nozzle onto a fast-rotating copper wheel, which usually rotates at 5000–7000 rpm. Such high rotation rate offers the RS rates (10^5–10^6 Ks^{-1}) that are required to freeze the atoms of the liquid phase (molten metal) into solid amorphous solid. It is worth mentioned that melt spinning of molten metallic system does not only lead to the formation of amorphous and metallic glassy alloys, but also lead to the formation of other nonequilibrium phases, such as nanocrystalline,[31] quasi-crystalline phases,[32,33] and supersaturated solid solution.[34]

In addition to the melt-spinning approach, there are a number of other RS methods, such as water quenching, high pressure casting, copper mold casting, arc-melting method, and some others. More detailed information on the formation of bulk metallic glassy alloys has been recently published by Suryanarayana and Inoue.[29]

1.4.2 Droplet method: gas/water atomization

The simple definition of atomization is the conversion of a molten metal into aerosol particles by forcing through nozzle at high pressure. Gas atomization, which

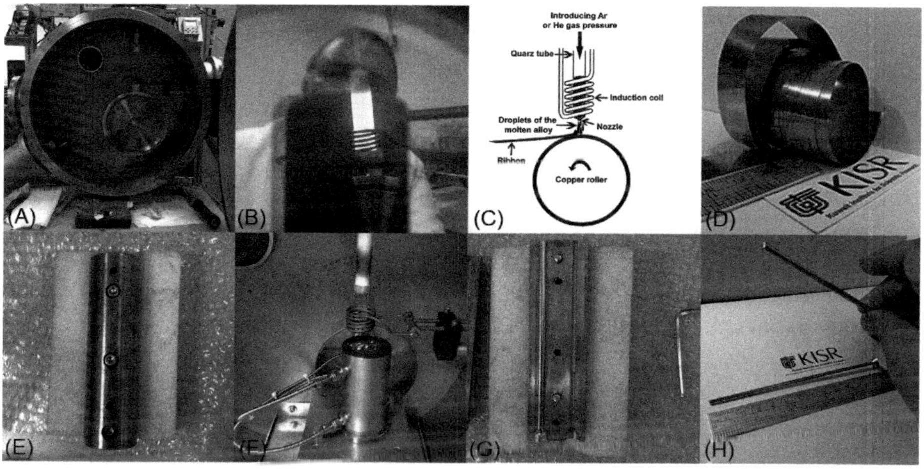

Figure 1.7 Schematic presentation illustrates the fundamental of the melt-spinning (A–C) and casting (E–G) processes used mainly for preparing ribbons- (D) and rod-like (H) metallic glassy materials. The samples were prepared by the author, using a melt-spinner equipment, PA 500, Edmund Bühler, Germany, housed in the Nanotechnology Laboratory, Energy and Building Research Center (EBRC), Kuwait Institute for Scientific Research (KISR).

is achieved in atomizing media of nitrogen, argon, or air, is a process dedicated to synthesis of high-quality metal powders with controlled particle sizes.[35] The fundamental principle in this fabrication approach is that a pendant drop or free falling melt-stream tends to break up into droplets as a result of surface tension. This process can be used to fabricate either ordered or disordered material powders. During the gas atomization process, the molten alloy is atomized thanks to inert gas jets into fine metal droplets that cool down during their fall in the atomizing tower. The end product of the metallic powders obtained by gas-atomization possess has perfectly spherical shape with relatively smooth surfaces combined with a high cleanliness level. The fine powder of the end product can be then consolidated into bulk nanostructured complex shapes. Gas and water atomization has been recently used for preparing metallic glassy alloy powders, such as $Al_{82}Ni_{10}Y_8$,[36] $Al_{84}Gd_6Ni_7Co_3$,[37] Fe-based magnetic,[38] and metallic glassy matrix of Zr-based alloys reinforced by ZrC powders.[39]

1.4.3 Thermal plasma processing

Thermal plasma processing is one of the most powerful technologies used for synthesizing of wide varieties of advanced materials, including nanocrystalline metallic[40] and ceramics[41] materials that possess unique and unusual properties. This process is characterized by the applications of high temperature being, in the range between 1,700°C and about 30,000°C.[42] The thermal plasma approach, which is a single-step process, offers a high-quenching rate in the range between 10^5 and 10^7 Ks^{-1}.

1.4.4 Vapor deposition

Vapor deposition techniques have been used for producing different types of materials, including fibers, nanotubes, powders, thin films, multilayer coatings, and graded composition deposits for many of years. Different categories of advanced materials, including nanocrystalline and amorphous alloys, have been produced in the form of thin films, with thickness in the range between few nanometers and thousands of nanometers.[43] There are two strategies of producing thin films via vapor deposition technique, known as: (1) PVD, and (2) CVD.

PVD process: PVD and EBPVD processes are atomistic deposition processes in which material is vaporized from a solid or liquid source in the form of atoms or molecules, transported in the form of a vapor through a vacuum or low pressure gaseous (or plasma) environment to the substrate where it condenses.[43] There are many ways for film depositions via PVD process, such as sputtering deposition, arc vapor deposition, and ion plating.

CVD: CVD process, which belongs to the class of vapor-transfer processes, is considered to be one the most powerful tools for producing all types of advanced materials. It is a versatile process capable to synthesize of coatings, powders, fibers, nanotubes, and monolithic components. This process can be also used to produce most of metal and metal alloys and their compounds, such as carbides nitrides and oxides. In addition, it has been used to prepare semiconductors, including carbon and silicon, as well as nonmetal systems such as metal oxides. CVD can be simply defined as the process that enables the deposition of a solid on a heated surface from a chemical reaction in the vapor phase.[44] Recently, CVD is used in conjunction with PVD processes and led to generation of new system such as PECVD and activated sputtering.

References

1. Powder Metal Technologies and Applications, Handbook: Vol. 7, *ASM International* (1990)
2. Upadhyaya, G.S., Powder Metallurgy Technology, *Cambridge International Science Publishing* (2002)
3. Benjamin, J. S., *Metall. Trans.*,1:2943 (1970)
4. El-Eskandarany, M. Sherif, Aoki, K., and Suzuki, K., *J. Alloys Compounds*, 177:229 (1991)
5. El-Eskandarany, M. Sherif, Aoki, K., Sumiyama, K., and Suzuki, K., *Mater. Sci. Forum*, 88:801 (1992)
6. El-Eskandarany, M. Sherif, *Materials Transactions, JIM*, 36:182 (1995)
7. Schaffer, G.B., and McCormick, P.G., *Metallurgical Transaction A*, 21:2789 (1990)
8. El-Eskandarany, M. Sherif, *J. of Alloys Comp.*, 279:263 (1998)
9. El-Eskandarany, M. Sherif, Aoki, K., and Suzuki, K., *J. Appl. Phys.*, 71:2924 (1992)
10. Calka, A., and Wexler, D., *Nature*, 419:147 (2002)
11. El-Eskandarany, M. Sherif, *Mechanical Alloying for Fabrication of Advanced Engineering Materials*, William Andrew Publishing, Inc, New York, the U.S.A, (2001)
12. Suryanarayana, C., *Mechanical Alloying and Milling*, Marcel Dekker, New York, the U.S.A, (2004)

13. Wills, Barry A., and Napier-Munn, Tim, *Will's Mineral Processing Technology: An Introduction to the Practical Aspects of Ore Treatment and Mineral*, 7th, Edition, Elsevier Science & Technology Books (2006)
14. Remington: *The Science and Practice of Pharmacy*, 21st Edition, Pharmaceutical Press (2011)
15. El-Eskandarany, M. Sherif, Saida, J., and Inoue, A., *Acta mater.*, 51:1481 (2003)
16. El-Eskandarany, M. Sherif, *Journal of Nanoparticles*, 2:14 (2009)
17. Campbell, S. J., and Kaczmarek, W. A., *Mössbauer Spectroscopy Applied to Materials and Magnetism,* (G. J. Long and F. Grandjean, eds.) Plenum Press, New York, 2:273 (1996)
18. Binns, Chris, *Introduction to Nanoscience and Nanotechnology*, Wiley, New York (2010)
19. Benjamin, J. S., *Scientific American*, 234:48 (1976)
20. El-Eskandarany, M. S., Aoki, K., and Suzuki, K., *J. Less-Common Met.*, 167:113 (1990)
21. Smeulders, D.E., et al., J. Materials Science, 40:655 (2005)
22. Champion, Y., Guerin-Mailly, S., Bonnentien, J.-L., and Langlois, P., *Scripta mater.*, 44:1609 (2001)
23. McFadden, S. X., Mishra, R. S., Valiev, R. Z., Zhilyaev, A. P., and Mukherjee, A. K., Nature, 398:684 (1999)
24. Baró, M. D. et al., *Rev.Adv.Mater.Sci.*, 2:1 (2001)
25. Klement, W., Willens, R. H., Duwez, Pol, *Nature*, 187:869 (1960).
26. Johnson, W.L., *MRS Bull*, 24:42 (1999)
27. Inoue, A., Wang, X.M., and Zhang, W., Rev.Adv.Mater.Sci., 18:19 (2008)
28. Qin, C., Zhao, W., and Inoue, A., *International Journal of Molecular Sciences*, 12:2275 (2011)
29. Suryanarayana, C., and Inoue, A., *Bulk Metallic Glasses*, CRC Press, Taylor& Francis Group, New York, The U.S.A. (2011)
30. Suryanarayana, C., Rapid solidification. *In Processing of Metals and Alloys*, ed. R.W. Cahn. Vol. 15 of Materials Science and Technology—A Comprehensive Treatment, pp. 57–110. Weinheim, Germany: VCH (1991)
31. Zhang, Yang-huan, Li, Bao-wei, Ren, Hui-pin, Ding, Xiao-xia, Liu, Xiao-gang, and Chen, Le-le, *Journal of Alloys and Compounds*, 509:2808 (2011)
32. Shechtman, D., Blech, I., Gratias, D., and Cahn, J.W., *Phys. Rev. Lett.*, 53:1951 (1984)
33. Antonione, C., Battezzati, L., and Marino, F., *Journal of the Less Common Metals*, 145:421 (1988)
34. Dong, Z.F., Lu, K., and Hu, Z.Q., *Nanostructured Materials*, 11:351 (1999)
35. Jones, H.. In: Suryanarayana, C., ed. *Nonequilibrium Processing of Materials.* Oxford, UK: Pergamon, pp. 23–45 (1999)
36. Liu, Yong, et al., *Intermetallics*, 13:393 (2005)
37. Surreddi, K.B., Scudino, S., Sakaliyska, M., Prashanth, K.G., Sordelet, D.J., and Eckert, J., *Journal of Alloys and Compounds*, 491:137 (2010)
38. Liu, Yong, Niu, Sen, Li, Fei, Zhu, Yitian, and He, Yuehui, Powder Technology, 213:36 (2011)
39. Kawamura, Y., Mano, H, and Inoue, A., *Scripta Materialia*, 43:1119 (2000)
40. Cinca, N., and Guilemany, J.M., *Intermetallics*, 24:60 (2012)
41. Szépvölgyi, János, Mohai, Ilona, Károly, Zoltán, Gál, Loránd, 28: 895 (2008)
42. Ananthapadmanabhan, P. V., In: Suryanarayana, C., ed. *Nonequilibrium Processing of Materials.* Oxford, UK: Pergamon, pp. 121–150 (1999)
43. Mattox, Donald M., *Handbook of Physical Vapor Deposition (PVD) processing*, Noyes Publications, Westwood, New Jersey, U.S.A. (1998)
44. Pierson, Hugh. O., Handbook of Chemical Vapor Deposition (CVD), Noyes Publications, Westwood, New Jersey, U.S.A. (1999)

Characterizations of mechanically alloyed powders

2

2.1 Introduction

The area of materials characterization is huge and grown decade by another. At the moment, there is a wide spectrum of sophisticated equipment, used to offer reliable information of qualitative and quantitative information on chemistry, structure, morphology, composition (Fig. 2.1), as well as physical and mechanical properties. Numerous methods, approaches, and devices varied from traditional to sophisticated instruments can be used for mechanically alloyed powder characterizations based on the facilities availability and the properties that required to be investigated.

2.2 Examples of characterization techniques

2.2.1 Photon probe methods

X-ray diffraction (XRD): XRD is the most common method used to investigate the general crystal structure of powder, solid, and liquid materials. This important technique is used to identify crystalline/noncrystalline phases of materials. XRD diffraction of any materials can be used for quantitative phase analysis, and to obtain lattice constant (a_o), crystallite size, and lattice strains. Modern XRD devices (Fig. 2.2) provide the possibility of small angle X-ray scattering (SAXS) measurements. SAXS is mainly used for surface analysis and measuring nanoparticles in the range between 1 and 100 nm in diameter.

Fourier-transform infrared spectroscopy (FT-IR): FT-IR (Fig. 2.3) is a spectroscopic chemical analytical technique, used to measure the infrared intensity against wavelength of light. This technique is commonly used to identify organic, polymeric, and, in some cases, inorganic materials. Reference to the wave number, infrared light can be categorized as far infrared (4–400 cm^{-1}), mid infrared (400–4,000 cm^{-1}), and near infrared (4,000–14,000 cm^{-1}).

Inductively coupled plasma mass spectrometry (ICP-MS): ICP-MS is the most precious technique used for elemental analysis of the materials. This technique is highly sensitive and capable of multielement trace analysis and ultra trace analysis, often at the parts-per-trillion level. Testing for trace elements can be performed on a range of materials from super alloys to high purity materials.

X-ray photoelectron spectroscopy (XPS): XPS is a sophisticated technique, used for analyzing the surface chemistry of materials. This technique is able to measure the elemental composition, empirical formula, and chemical state and electronic state of the elements within a material. XPS spectra are obtained by irradiating a solid surface

Mechanical Alloying. http://dx.doi.org/10.1016/B978-0-12-818180-5.00002-9

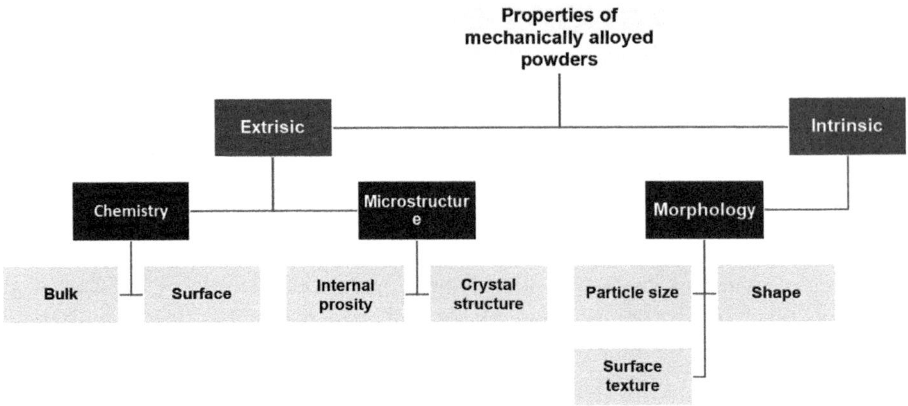

Figure 2.1 The general extrinsic and intrinsic properties of mechanically alloyed powders can be measured by different techniques on regard to their morphological, chemical, and microstructure properties.

Figure 2.2 9kW-SmartLab-XRD/SAXS (Rigaku-Japan) equipment housed in the Nanotechnology Laboratory, Energy and Building Research Center, Kuwait Institute for Scientific Research.

Figure 2.3 FT-IR (FTIR-8400S, Shimadzu, Japan) equipment housed in the Nanotechnology Laboratory, Energy and Building Research Center, Kuwait Institute for Scientific Research.

with a beam of X-rays while simultaneously measuring the kinetic energy and electrons that are emitted from the top 1–10 nm of the material being analyzed.

Energy dispersive X-ray fluorescence (XRF): XRF (Fig. 2.4) technique is nondestructive approach used to investigate the elemental constituents within a material. This technique offers a simple methodology based on the principle that individual atoms can emit X-ray photons of a characteristic energy or wavelength. Either the energy or wavelengths corresponds to a specific element within the sample and allows it to be identified. Based on the total counts of the characteristic energy or wavelength, the quantification of the element can also be determined. XRF has wide range of applications in different areas.

2.2.2 Photon probe methods

Field emission scanning electron microscope/EDS (FE-SEM/EDS): FE-SEM/EDS (Fig. 2.5) is used to investigate the morphology (e.g., particle sizes and shapes), metallographic details, imperfections, and topology of nanocrystalline powders and bulk materials. This technique is also used to investigate the elemental compositions within the materials in submicron scale.

Figure 2.4 XRF (S8 TIGER 4 kW) system provided by Bruker-Germany and housed in the Nanotechnology Laboratory, Energy and Building Research Center, Kuwait Institute for Scientific Research.

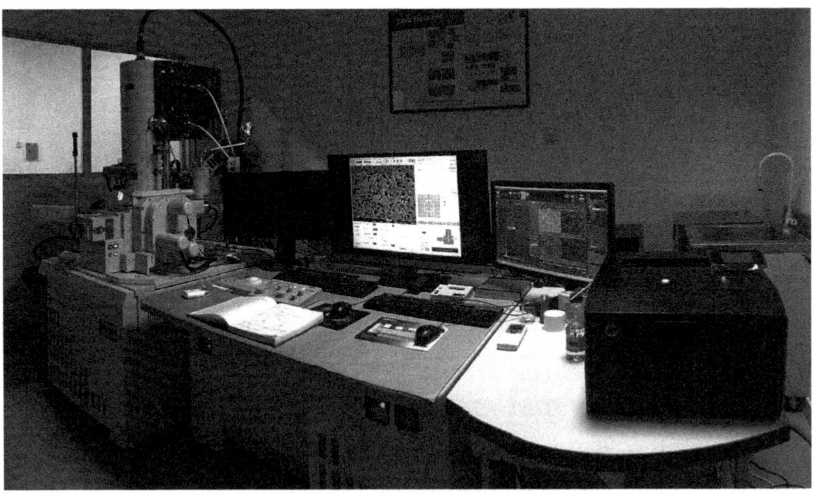

Figure 2.5 FESEM/EDS unit (JSM-7800F, JEOL-Japan) housed in the Nanotechnology Laboratory, Energy and Building Research Center, Kuwait Institute for Scientific Research.

Figure 2.6 200 kV FE-HRTEM/EDS/STEM (JEOL-Japan) equipment housed in the Nanotechnology Laboratory, Energy and Building Research Center, Kuwait Institute for Scientific Research.

Field emission high-resolution transmission electron microscope/energy dispersive X-ray spectroscopy (FE-HRTEM/EDS): This sophisticated technique (Fig. 2.6) is used to investigate the local structure (2–5 nm) of materials, particle and grain shape, and sizes, as well as nanoscaled elemental analysis.

2.2.3 Scanning probe methods

Atomic force microscopy (AFM): AFM (Fig. 2.7) is used to investigate the topology and surface structure of nanopowders and nanocrystalline bulk materials.

2.2.4 Thermodynamic methods

Differential scanning calorimeter (DSC) and thermogravimetry/differential thermal analysis (TG/DTA): DSC and TG/DTA (Fig. 2.8) are used to investigate the thermal stability, glass transition temperature (T_g), crystallization (T_x), melting

Figure 2.7 AFM system (Agilent 5600LS, USA) housed in the Nanotechnology Laboratory, Energy and Building Research Center, Kuwait Institute for Scientific Research.

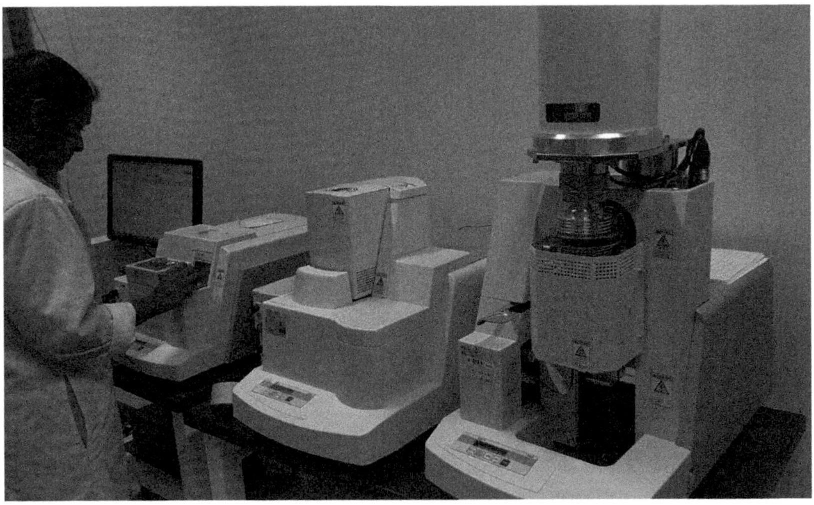

Figure 2.8 DSC/TG/DTA/TMA unit (Shimadzu Thermal Analysis System) housed in the Nanotechnology Laboratory, Energy and Building Research Center, Kuwait Institute for Scientific Research.

(T_m), and phase transformation temperatures. TG is used to investigate the mass loss versus temperatures.

Thermal mechanical analysis (TMA): This technique is used to investigate some of mechanical properties for bulk materials versus temperatures.

The history and necessity of mechanical alloying

3

The capability of any societies along the human history on developing and instigating of new materials that fit their needs has led to the advancement of their performance and ranking in the worldwide. The gap differences on the "level of life," indexed by the progress made on health, education, industry, economic, culture, etc., between a country to country and region to another are always attributed to the man's ability for developing materials and manufacturing equipment and devises used for materials fabrications and characterizations. However, there are many approaches and techniques used for producing the advanced materials, mechanical alloying has been receiving great attentions and considerations as a unique process for synthesizing of new advanced materials families that cannot be obtained by any other techniques. Nanostructured materials, nanoparticles, nanocomposites, carbon nanotubes, and amorphous and metallic glassy alloys are some of those new engineering materials that can be successfully obtained by such a room temperature way of fabrication. The photo shows the STEM-BF image of a Cu-based alloy nanoparticle prepared by mechanical alloying (Fig. 3.1) (After El-Eskandarany, unpublished work, 2014).

3.1 History of story of mechanical alloying

It is well established that melting and casting techniques, using of fine inert particles or the so-called filamentary reinforcements with high temperature strengths combined with soft conventional matrix metals, can produce composites which outperform superalloys.[1] The use of such inert fine particles to improve mechanical properties of metals at elevated temperatures was first exploited in 1910 by Coolidge[2] in thoriated tungsten to increase creep resistance. It has been reported that the first dispersion-strengthened material designed as a structural load-bearing system was sintered aluminum particles (SAP) in 1952.[3] Liquid metallurgy, using melting technique, has also been employed successfully for producing of ThO_2-dispersed nickel.[4]

There are four existed techniques known as (1) simple mixing, (2) ignition surface coating, (3) internal oxidation, and (4) selective reduction are available to combine oxide dispersion strengthening and solid solution strengthening in alloy systems containing relatively nonreactive elements.[5] In fact, these techniques were found to be unsuitable and mismatching for production of oxide dispersion-strengthened precipitation-hardened nickel-base superalloys[6] due to the reactiveness of the Al, Ti, and Cr, which are existed in the alloy.

In 1960s, Benjamin (The God Father of MA) has introduced a unique process, known as MA process, at The International Nickel Company (INCO) as part of a program to produce a material combining oxide dispersion strengthening with gamma prime

Mechanical Alloying. http://dx.doi.org/10.1016/B978-0-12-818180-5.00003-0

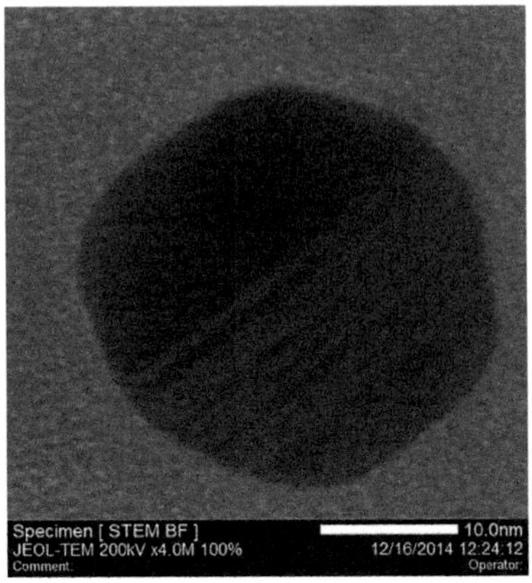

Figure 3.1 STEM-BF image of a Cu-based alloy nanoparticle prepared by mechanical alloying. (*After El-Eskandarany, unpublished work, 2014.*)

precipitation hardening in a nickel-based superalloys intended for gas turbine applications.[6] It is worth to be mentioned here that the term "mechanical alloying" was first phrased by Ewan C. MacQueen in a US patent owned by INCO.[7] It should be emphasized that the original MA process was the by-product of research into different subjects. In the early 1960s, INCO had developed a process for manufacturing graphite aluminum alloys by injection of nickel-coated graphite particles into a molten bath by argon sparging. A modification of the same technique was tried to inoculate nickel-based alloys with dispersion of nickel-coated, fine refractory particles. The reason for nickel coating was to render the normally unwetted oxide particles wettable by a nickel-chromium alloy. Early experiments used metal-coated zirconium oxide purchased from an outside vendor. The examinations of these materials revealed no differences between the inoculated materials and uninoculated alloys. Examinations of the inoculants revealed that they were zirconia-coated nickel rather than nickel-coated zirconia. Attention was directed to ball milling as a means of coating oxide particles with nickel. Ball milling had been used for the coating of tungsten carbide with cobalt for well over 70 years.[8] Small amounts of nickel-coated thoria and zirconia were successfully produced in a small high-speed shaker mill. This process, in particular, was used to coat oxides with metals that could not be applied by chemical process due to their reactivity. Since the apparatus employed, a small high-energy ball mill, could produce only 1 cm^3 of powder per single milling run, these powders were used only for studies of the rate of rejection of oxide powders from molten alloys. Compacts of composite powders were partially melted in an arc melter, sectioned, and examined metallographically.

In mid-1966, attention was turned to the ball-milling process that had been used to make metal powders for wetting studies as a means of making the alloy itself by powder metallurgy. The reason was attributed to the capability of this process to coat hard phases (e.g., WC or ZrO_2) with a soft phase (Co or Ni).[9] This choice is attributed to the fact that the ball-milling process could be employed to coat hard phases, such as tungsten carbide or zirconium oxide with soft phases such as cobalt or nickel. It was also considered that:

- Both welding and fracturing occur during ball milling of metallic powders.
- High-energy ball mills can greatly increase the rate of the grinding and fracture processes.
- Virtually any composition could be manufactured using a mixture of elemental and readily available powders instead of having to rely on atomized pre-alloy powders, which are relatively expensive.
- The thermodynamic activity of reactive gamma prime forming elements, such as aluminum and titanium, could be reduced in orders of magnitude by combining them with less reactive metals, such as nickel, in intermetallic compounds.

These key ideas led to the concept of establishing a kneading action which would refine the internal structure of powders while maintaining their overall particle size at a relatively coarse level, preventing pyrophoricity.[8]

3.2 Fabrications of ODS alloys

As was pointed out in the previous section, Benjamin[6] introduced a pioneering development on ball-milling technique for producing ODS alloys that were used for high temperature structural applications, such as jet engine parts. This unique method could be successfully used for preparing fine, uniform dispersions of oxide particles (Al_2O_3, Y_2O_3, ThO_2) in nickel-base superalloys. It is worth noting that these materials cannot be obtained by the conventional powder metallurgy method. During the 1970s, research programs concerned the nature and mechanism of the MA process itself and the design of special equipment for carrying out the process. At that time, MA was well known as a process for the fabrication of several ODS alloys.[9] During the 1970s, research programs concerned the nature and mechanism of the MA process itself and the design of special equipment for carrying out the process. At that time, MA was well known as a process for the fabrication of several ODS alloys.[10–18]

Table 3.1 presents selected metal matrix composites and ODS alloys produced by MA technique and their applications.[19,20]

Table 3.1 Typical examples of selected ODS alloys produced by mechanical alloying process and their applications.[19,20]

Alloy	Applications
ODS Ni-based alloys	High temperature applications in engine, air and aerospace industries, glass processing, chemical plants and reactors
Dispersion hardened Al-based alloys, such as Al/Al_2O_3 composites	Engine industry, high absorbing materials, optical devices, high temperature batteries, compressors, fans
Dispersion hardened Cu-based alloys	Contact materials, electromechanical components, springs
Cu-W alloys	Spot welding electrodes, contact materials, carrier for electronic components
Dispersion hardened Ti-based alloys	Air and aerospace industries, medical applications

3.2.1 ODS Ni-base superalloys and Fe-base high-temperature alloys

The major commercial dispersion-strengthened Ni-base superalloys produced to date by mechanical alloying process are INCONEL alloys MA 754, MA 758, and MA 6000. The compositions for some types of these alloys are presented in Table 3.2.[14,19–21] The mechanically alloyed powders of MA 6000 (Table 3.2), which are age hardened to incorporate gamma prime strengthening with dispersion strengthening, are used in gas turbine engine blades (Fig. 3.2).[22]

Table 3.2 Elemental chemical composition of Ni- and Fe-based ODS superalloys produced by mechanical alloying process (in wt.%).[14,19–21]

Alloy	Ni	Fe	Cr	Co	W	Mo	Ta	Al	Ti	Y_2O_3
MA 754	Bal.	20	0.3	—	—	—	—	0.5	—	0.6
MA 758	Bal.	30	0.3	—	—	—	—	0.5	0.5	0.6
MA 6000	Bal.	15	4.5	—	2	—	2	0.5	4	1.1
MA 760	Bal.	20	6	—	2	—	2	—	3.5	1.1
MA 757	Bal.	16	4	—	—	—	—	0.5	—	0.6
MA 957	—	Bal.	14	—	—	0.3	—	—	1	0.25
MA 956	—	Bal.	20	—	—	—	—	4.5	0.5	0.5
PM 1000	Bal.	3	20	—	—	—	—	0.3	0.5	0.6
PM 1500	Bal.	3	30					0.3	0.5	0.6
PM 2000	—	Bal.	20					5.5	0.5	0.5

Figure 3.2 MA superalloy components of a swirler assembly for a power station, made of INCOLOY alloy MA 956.[22]

3.2.1.1 INCONEL MA 754

In 1980, Weber[23] prepared a large amount of mechanically alloyed ODS superalloy (INCONEL MA 754). It is essentially a Ni-20% Cr alloy strengthened by about 1 vol % Y_2O_3. The Ni, Cr, and Y_2O_3 powders were milled until a homogeneous Ni-20% Cr alloy was formed in which the Y_2O_3 particles were uniformly distributed. The fabricated alloy powder was then consolidated by hot extrusion which was followed by hot rolling. A recrystallization step, often directional, followed consolidation that resulted in elongated, high-aspect-ratio grains that were very stable owing to the inert oxide pinning. After the directional crystallization, the grains had typical dimensions of ~500–700 μm parallel to the working direction and ~15 μm perpendicular to this direction. The typical BFI of the fabricated MA 754 is displayed in Fig. 3.3, which shows the oxide distribution in the metallic matrix. Benjamin et al.[11] have suggested that fine particles are a uniform dispersion of stable yttrium aluminates formed by the reaction between the added Y_2O_3, excess oxygen in the powder, and the aluminum added to getter oxygen. In Fig. 3.4, the larger particles are titanium carbonitrides. The dispersoids are typically 14 nm in diameter with an average spacing of 0.2 μm. The 1093°C stress rupture properties of INCONEL MA 754 are compared to those of other alloys in Fig. 3.4. The MA 754 alloy, like other ODS materials, has a very fine, flat, log stress-log rupture life slope compared to conventional alloys. The strength of MA 754, about 100 MPa for 100 h life, is somewhat higher than both of the other ODS alloys and several times greater than conventional materials, MAR-M alloy 509 and alloy 80A. Thus, while MA 754 alloy is comparable to TD (thoria dispersed) NiCr, it has a nonradioactive dispersoid and high strength, so it is suitable for applications such as gas turbine vanes.

3.2.1.2 INCONEL MA 6000

The iron-based INCOLOY alloy MA 956 contains about 20% Cr, 4.5% Al, 0.5% Ti, and about 0.5% Y_2O_3. It can be used at operating temperatures of over 1300°C in a

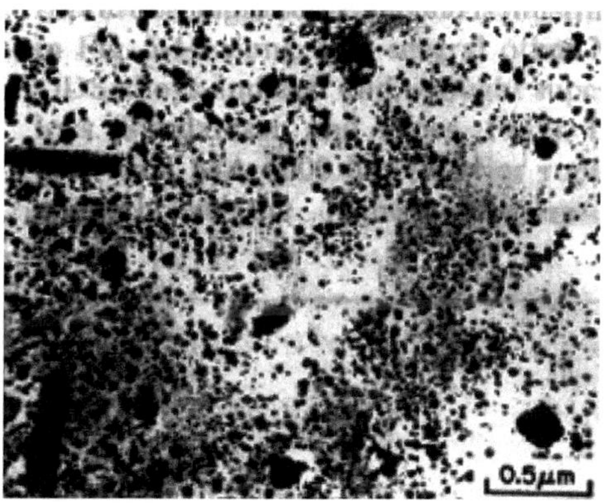

Figure 3.3 BFI micrograph of INCONEL MA 754 showing uniform distribution of the fine primary dispersion, the presence of coarser carbonitrides and microtwins. *(After Weber.)*[23]

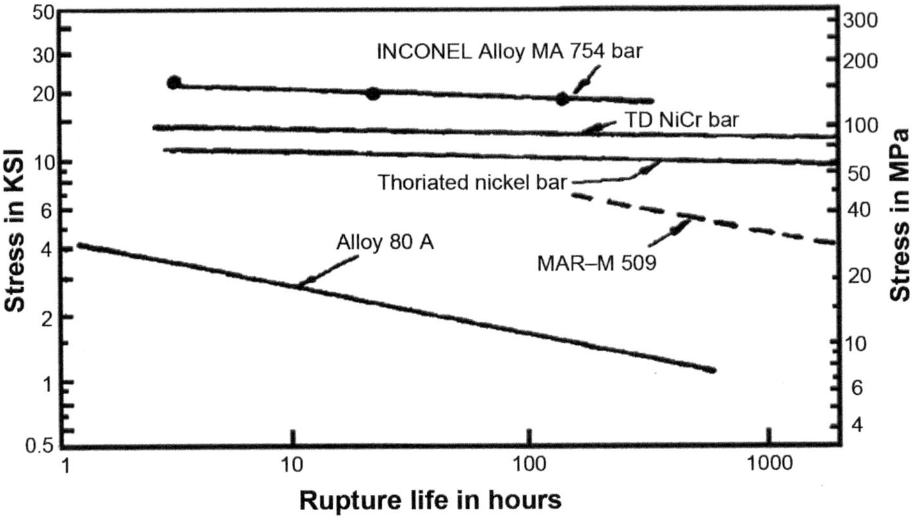

Figure 3.4 Stress rupture properties of MA 754 at 1093°C compared to those of other alloys. *(After Weber.)*[23]

corrosive atmosphere.[24] INCONEL MA 6000 combines precipitation strengthening (gamma prime, γ' precipitates) from its Al, Ti, and Ta content for intermediate temperature strength with oxide dispersion strengthening from the Y_2O_3 addition for strength and stability at very high temperatures. It contains about 25% of γ' precipitates. The dispersoid dimensions are 30 nm average diameter and 0.1 μm average spacing. As in MA 754, the Y_2O_3 reacts with oxygen and aluminum to form uniform dispersions of yttrium aluminates, for example, yttrium-aluminum garnet $5Al_2O_3\,3Y_2O_3$ (YAG). The YAG dispersoid appears to be very stable.

Negligible coarsening is observed for stress rupture tests at ~750°C and only small coarsening for temperatures of 950–980°C at rupture lives beyond 10^4 h. These changes cause no serious loss in the load-bearing capability of the alloy, for practical applications. Fig. 3.5 shows the 1000-h creep rupture strength for MA 6000 alloy compared with several other MA ODS alloys. It is clear that MA 6000 has superior rupture strength at the highest temperatures (less than 900°C) and comparable strength to MAR-M 200 at intermediate temperatures. MA 6000 alloy has been used to a small extent, so far, in gas turbine engine blades.

3.2.1.3 INCONEL MA 956

MA 956 also has excellent fabricability. It can be cold-worked and can be joined by several welding techniques. MA 956 sheets can be bent more than 150 degree around a diameter equal to twice the sheet thickness.[23] As we have mentioned earlier, the first ODS material designed as a structural system was sintered Al powder (SAP) as developed by Irmann.[3] Sintered Al powder (SAP) displayed a greater strength than pure Al and no changes were observed after extended heating near the melting point. In 1957, Benjamin et al.[13] developed further SAP. Bars of SAP were made by a mixture of

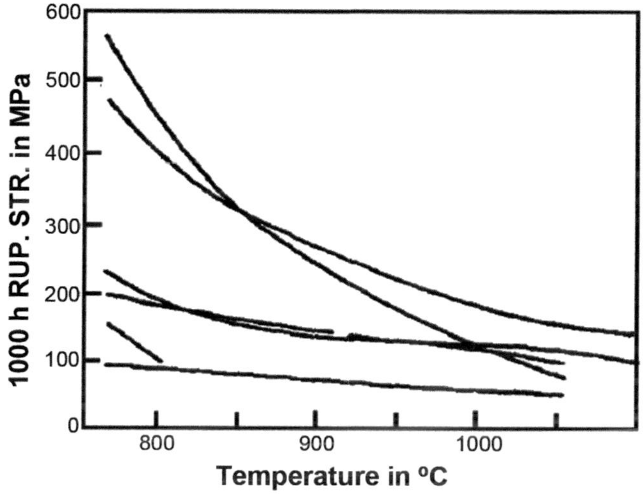

Figure 3.5 Creep rupture at 1000 h for MA alloys compared with cast and wrought superalloys: (1) MA 956, (2) MA 764, (3) MA 753, (4) MA 6000, (5) MAR-M 200, and (6) $Ni_{80}Al_{20}$. *(After Sundaresan et al.)*[24]

1%–10% vol % Al_2O_3 powder in pure Al powder. The Al_2O_3 particles in SAP exhibit a wide distribution of sizes (~10 nm–1 μm). In order to maximize the strength of Al-Al_2O_3 ODS alloys it was desired to obtain a finer and more uniform distribution of the Al_2O_3 dispersion in the Al matrix.

Fig. 3.6 shows the variation of ultimate tensile strength as a function of the combined volume fraction of Al_2O_3 and carbon and compares it to some SAP materials of comparable strength. The correlation between the ultimate tensile strength of MA

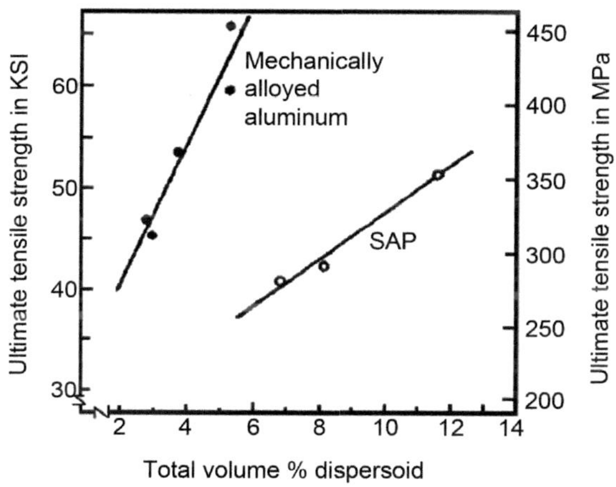

Figure 3.6 Room temperature tensile strength versus dispersoid content for SAP and MA Al. *(After Benjamin et al.)*[13]

Al and Al_2O_3 content alone is not as good; indicating that certain carbon dispersion contributes to the strengthening.[13] High tensile strengths of the heats were obtained with low dispersoid levels (less than 5.4 vol %). Sintered Al powder (SAP) products containing this content of Al_2O_3 dispersoid typically show tensile strengths at room temperature of around 241 MPa.[25] Strengths of 310–448 MPa are only obtained in SPA products with more than about 7 vol % Al_2O_3.[26]

It was found necessary to use a process control agent, such as a stearic acid or methanol, to prevent excessive welding.[13] The decomposition of the process control agent during mechanical alloying and subsequent powder compaction by hot pressing and extrusion resulted in significant contamination by carbon. The carbon was believed to present as Al_4C_3 dispersoids. Thus, dispersion strengthening in these materials comes from both Al_2O_3 and Al_4C_3 dispersoids.

Conventional Al–Mg alloys have inherently good corrosion characteristics and dispersion strengthening of such alloys through MA was seen to be of technological value. It was upon this basis that the MA process was extended to produce a dispersion-strengthened Al-4 vol % Mg alloy by Benjamin et al. in 1981.[15] A mixture of commercial Al and Mg powders is charged into a high-energy ball mill (Szegvari attritor)[27] filled with small steel grinding balls. The powder particles are trapped between colliding grinding balls and severe plastic deformation results. The surfaces of the starting Al and Mg powders are covered by adsorbed gases, hydrated oxides, and other thin amorphous compounds, such as carbonates. As displayed in Fig. 3.7, flattening during MA processing increases the surface-to-volume ratio of the powders and ruptures the surface films, exposing atomically clean metal.

As shown in Fig. 3.8A, cold welds are forced between powder particles, building up composite particles and entrapping fragments of the surface films within the composites. Adding a control agent during MA process leads to increased oxygen and carbon contents of the powders. This pickup occurs by the formation of new adsorbed surface layers (Fig. 3.8B). Further processing of the composite metal particles (Fig. 3.8C)

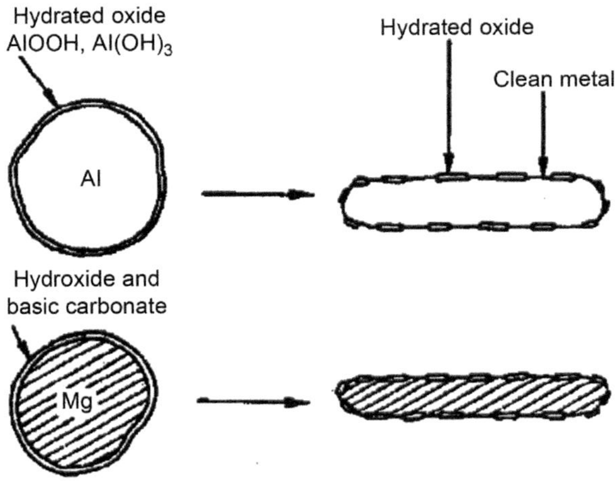

Figure 3.7 Rupture of surface contamination films by plastic deformation during MA process of Al and Mg powders. *(After Benjamin et al.)*[15]

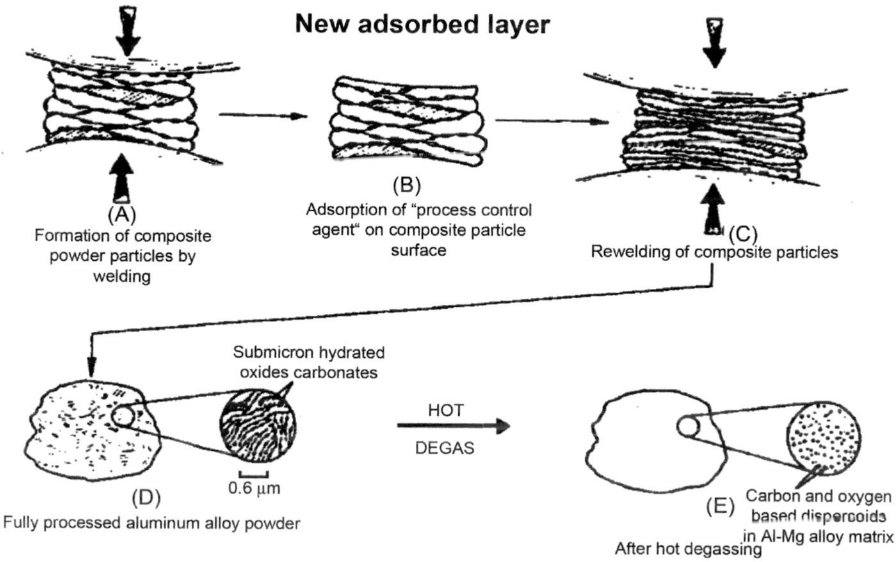

Figure 3.8 Schematic diagram of the MA process applied to Al–Mg alloys. *(After Benjamin et al.)*[15]

caused further welding and plastic deformation. This thins the lamellae of the original starting powders, and causes further fragments within the composite particles. The internal homogeneity of the composite particles is improved by this thinning of the internal lamellae. In addition to welding, which tends to increase average particle size, fracturing occurs, which tends to decrease average particle size. The fracturing is assisted to some extent by the increased carbon and oxygen content, due to the occlusion of a process control agent

With further processing, the internal structure of the powders continues to be refined. Local composition, from particle to particle and within the particle, approaches the average composition of the desired alloy. The end point of the MA process is reached when the surface film fragments are randomly distributed throughout the powder particles and metallic alloying has been completed on a submicron scale.

3.3 Fabrications of other advanced materials

Apart from the fabrication of ODS alloys by the ball-milling technique, for their subsequent beneficiation, White[28] observed the formation of an amorphous phase by ball-milling elemental Nb and Sn powders at room temperature. In 1983, Koch et al.[29] reported the first novel technique for formation of $Ni_{60}Nb_{40}$ amorphous alloy by high-energy ball milling of elemental Ni and Nb powders. Since then, the MA method has been successfully employed for the formation of a large number of amorphous alloys. This technique leads to the formation of several alloys that cannot be prepared by liquid metallurgy, such as Al–Ta[30] and Al–Nb[31] binary systems.

Moreover, MA was also employed for formation of nanocrystalline single phase solid solutions of Fe–Cu[32,33] alloys. Moreover, the formation of supersaturated solid solutions in Zr–Al[34,35] and immiscible Ni–Ag[36] via MA process have been reported during 1990s. Extended solid solutions far beyond the thermodynamic equilibrium have been reported upon mechanical milling for alloys with negative enthalpies of mixing by Schwaz in 1998.[37] More recently, solid solubility extension of Mg in Al has observed for the binary Al–Mg system.[38]

An attractive application of the ball-milling technique has been demonstrated by El-Eskandarany et al. for preparing metal nitrides (e.g., Fe_4N,[39] TiN,[40] AlTaN,[41] and NbN[42]) by milling the elemental powder under nitrogen gas flow. The investigation of this method, which is called reactive ball milling (RBM),[39] has been employed for preparing different metal hydrides with high hydrogen storage capacity.[43–49] A very recent important review article on the preparations of hydrogen storage materials by mechanochemical process has been addressed by Hout et al.[50]

Within the last 2 decades, the ball-milling technique has been proposed for formation of nanocrystalline materials at room temperature, including pure metals,[51,52] metal alloys,[53–59] intermetallic powders,[60,21] metal carbides,[61–65] and metal oxide nanoparticles.[66–69] Moreover, mechanical alloying was employed for preparations of nanocatalysts such as Fe–Ni,[70] Ni–Mo,[71] Raney–Ni,[72] and Pt–Co alloys[73] powders that possess high-surface area. More recently, a hydrogenation catalyst including a base material coated with a catalytic metal, which was prepared by MA, has been proposed for the remediation of contaminated compounds.[74]

In 1998, El-Eskandarany[75] prepared homogeneous nanocomposite Al/SiC_p materials by milling the elemental powders of Al and β-SiC in a high-energy ball mill. It should be noted that this composite material is difficult to obtain by the conventional liquid metallurgy method due to the poor wettability between molten Al (or Al alloys) and the reinforcement material of SiC. In addition, the liquid metallurgy method usually leads to an undesirable reaction between SiC and molten Al, producing brittle phases of Al_4C_3 and Si. Since then, ball-milling technique has been employed for preparing wide range of nanocomposites, including metal matrix and ceramics matrix composites.[76–80] The powders of as-produced nanocrystalline and nanocomposite that were consolidated successfully by different powder pressing techniques into fully dense nanocrystalline compacts show unusual unique physical and mechanical properties.[81–85] More recently, single-wall carbon nanotubes (SWCNTs) have been prepared by ball milling of graphite powders at room temperature.[86–90]

In addition, it has been employed for reducing some metallic oxides by milling the oxide powders with metallic reducing agents at room temperature.[91–95]

Furthermore, the MA method is the most powerful tool for preparing wide varieties of nonequilibrium materials[96]; such as amorphous[29,97–100] and metallic glassy[101–106] alloy powders with wide range of amorphization[97,107] and excellent glass forming ability[108] with wide supercooled liquid region.[109] Consequently, it is also used to produce alloys and compounds that are difficult or impossible to be obtained by conventional melting and casting techniques.[110,111]

Fig. 3.9 shows a schematic time-line presentation of the development of mechanical alloying process during the last 6 decades.

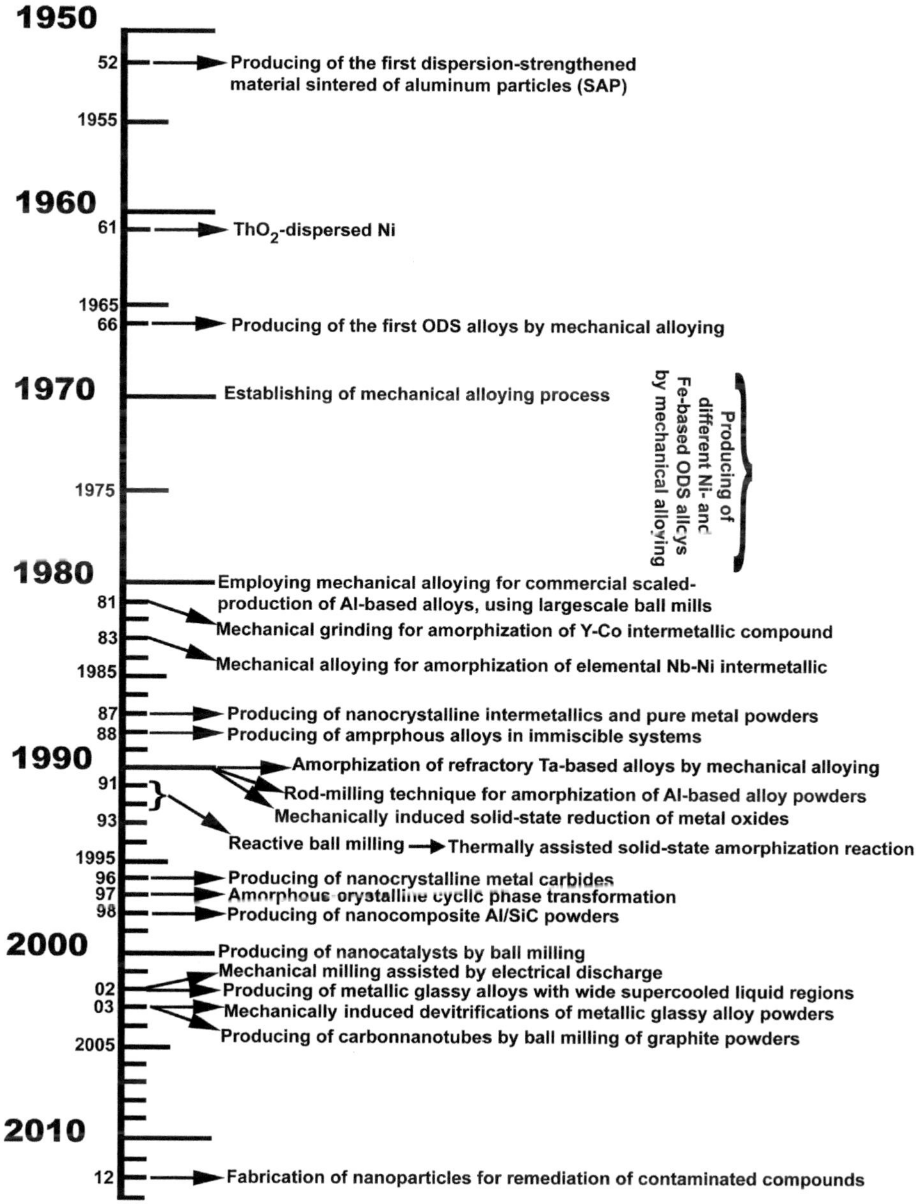

Figure 3.9 Schematic time-line presentation of the development of mechanical alloying process during the last 6 decades.

3.4 Mechanical alloying, mechanical grinding, mechanical milling, and mechanical disordering

As was mentioned in Chapter 1, the traditional objectives of ball milling are particle size reduction (breaking down the minerals until every particle is either fully mineral or fully gangue), mixing and blending, and particle shaping. In this book, however, we will focus only on the applications of milling (ball milling and rod milling) for producing of engineering materials via solid-state reaction, solid-state reduction, mechanically induced phase transformation, and mechanically induced solid-state mixing. According to the free energy changes that take place on the starting materials upon subjecting the powders to high-energy milling process, the milling strategy can be classified into two categories known as (1) *mechanical alloying*[6,8] and (2) *mechanical grinding* (MG).[112]

Benjamin[6] has defined the MA process as a method for producing composite metal powders with a controlled fine microstructure. It occurs by the repeated fracturing and rewelding of a mixture of powder particles in a highly energetic ball mill. As originally carried out, the process requires at least one fairly ductile metal to act as a host or binder.[13] Since the pioneering investigation of Koch et al.,[29] MA has been considered as a powerful approach method for achieving a solid-state reaction between two (or more) elemental metal powders (diffusion couples) at an ambient temperature, using ball-milling method. The end product obtained after the compilation of MA process is usually a single homogeneous powder of metastable phase, such as amorphous, nanoscaled materials, quasicrystals, big cubes, etc.

The term *mechanical milling* (MM), which was proposed in 1990,[113] has been used to express about the grain size reduction of the material powders that are achieved upon introducing heavy lattice imperfections (e.g., lattice and point defects) during the milling process. MM may be accompanied with grain size reduction and/or crystal-metastable phase transformations. In 1991, the term MG was replaced by new terminology called *mechanical disordering* (MD).[114] MD has been mainly used as a more suitable terminology, which reflects the mechanism of mechanically induced crystal-to-metastable phase transformations in much better understanding.

Fig. 3.10 shows a schematic free-energy (ΔG) diagram for the phases involved in the MA and MD processes, starting from: (1) mixture of two elemental powders (A + B), and (2) an intermetallic phase of AB that are subjected to MA and MD processes, respectively. Obviously presented in Fig. 3.10 the MA and MD processes are reactions going in thermodynamically opposite directions of each other. As mentioned before, the MA process synthesizes amorphous alloy powders by reacting the starting elemental crystalline powders (point 1 in Fig. 3.10). The initial state is just a mixture of crystals of pure A and B so that the free energy of this state (point 1 in Fig. 3.10) is located along the straight line that is joining the free energies of the pure elemental powders of A and B. Point 2 is the free energy of the single amorphous phase (a-AB) that is formed by milling the elemental powders of A and B for a certain period of time. Point 3 is the free energy of the single crystalline intermetallic alloy with the same composition. The mechanical solid-state amorphization reaction takes place at

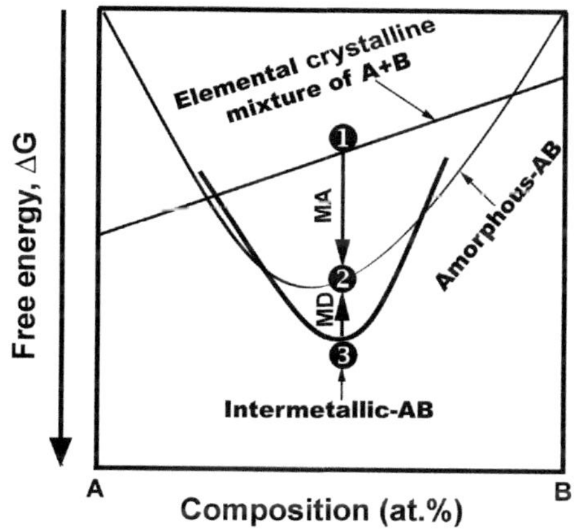

Figure 3.10 Schematic free-energy (ΔG) diagram for the phases involved in the MA and MD processes, starting from: (1) mixture of two elemental powders (A + B), and (2) an intermetallic phase of AB that are subjected to MA and MD processes, respectively. *(After Susuki, K.)*[117]

a constant temperature that is sufficiently high enough to allow for the mechanical interdiffusion of A in B and/or B in A, but too low for promoting the nucleation and/or growth of the crystalline intermetallic alloy. This kinetic selection of the reaction path is possible in binary systems in which the elements have largely different diffusivities in each other and in the amorphous phase.[115–117]

The major process in MA for producing quality powders of alloys and compounds with well-controlled microstructure and morphology is the repeated welding, fracture, and rewelding of the reactant mixed powders. Several types of mills have been employed for such purpose. The MA process can be successfully performed in both high-energy mills (attritor-type ball mill, planetary-type ball mill, centrifugal-type ball mill, and vibratory-type ball mill) and low-energy tumbling mills (e.g., ball and rod mills). In the MD process, however, crystalline alloy or compound powders are transformed into the amorphous solid state by relaxing the short-range order without compositional changes.[117]

3.5 Types of ball mills

However, there are many types of ball mills, such as drum ball mills, jet ball mills, bead mills, horizontal rotary ball mills, vibration ball mills, and planetary ball mills; they may be grouped or classified into two types according to their rotation speed as follows: (1) *high-energy ball mills*, and (2) *low-energy ball mills*. In fact, choosing the right ball mill depends on the milling objectives. For example, the characteristics properties of those ball mills used for reduction of the particle size of the starting materials via top-down approach (MM process), or for mechanically induced solid-state mixing for fabrications of composite and nanocomposite powders, may differ widely from those mills used for achieving mechanically induced solid-state reaction between the starting reactant materials of elemental powders (MA process), or for tackling dramatic phase transformation changes on the structure of the starting materials (MD). Nowadays, most of the ball mill manufactures introduce their equipment to be utilized successfully in the aforementioned processes so that a single ball mill can cover all proposes of the milling process. In this section, brief explanations of some common ball mills types will be described, taking some well-known commercially available equipment as typical examples.

3.5.1 High-energy ball mills

3.5.1.1 Attritor or attrition ball mill

Szegvari introduced this type of mill to the industry in 1922 in order to quickly attain fine sulfur dispersion for use in the vulcanization of rubber.[18] Attrition ball mills are powerful mill and can be utilized successfully utilized in the three major areas of MM, MA, and MD. Commercial types of attritors are available from Union Process (www.unionprocess.com) and Zoz GmbH (www.zoz.de). The attritor, which always refers to a simple and effective processing, is so called as a "stirred ball mill."[118] Attrition ball-milling process is simple and effective process. In attrition, the material is comminuted by means of free moving beads, which are set in motion by a stirrer. The grinding effect depends upon the stirrer speed, on the stirrer and the chamber geometry. Attritors are used for grinding materials with considerably different properties, for example, soft pigment, but also very hard ceramics.

The illustration of this mill, which is known as Szegvari attritor grinding mill, is shown in Fig. 3.11A and B. In this mill, the starting materials are charged into a stationary tank with the milling media (balls). The milling tools (the vials and balls) can be made of carbon steel, stainless steel, chrome steel, tungsten carbide, and ceramic. The milling procedure takes place by the stirring action, the so-called agitation, of an agitator, which has a vertical rotating central shaft with horizontal arms called impellers (Fig. 3.11B) and run at speed in speed between 75 and 500 rpm. Some high-speed attrition mills can be operated at a much higher speed in the range between 400 and 2000 rpm. This causes the media to exert both impact (Fig. 3.11C) and shearing (Fig. 3.11D) forces on the material.[118] The capacity (volume) of the attritor used for the milling process ranges from 3.8×10^{-3} m^3 to 3.8×10^{-3} m^3. These mills are designed to work under vacuum or in the presence of an inert gas, such as Ar or He. RBM can be also achieved successfully upon attrition the metallic powders under nitrogen or hydrogen gas flow to synthesize metal nitrides and hydrides, respectively. To minimize the temperature rise that usually occurs upon high-energy milling, the milling system here can be cooled down during the process by introducing a continuous water flow, as shown in Fig. 3.11B.

Kimura and his coworker[119] at the National Defense Academy of Japan developed an attritor ball mill with a rotation speed of about 500 rpm. In addition, they equipped the mill with several devices to control and measure the applied torque during the MA process. They could minimize the oxygen contamination content during the MA

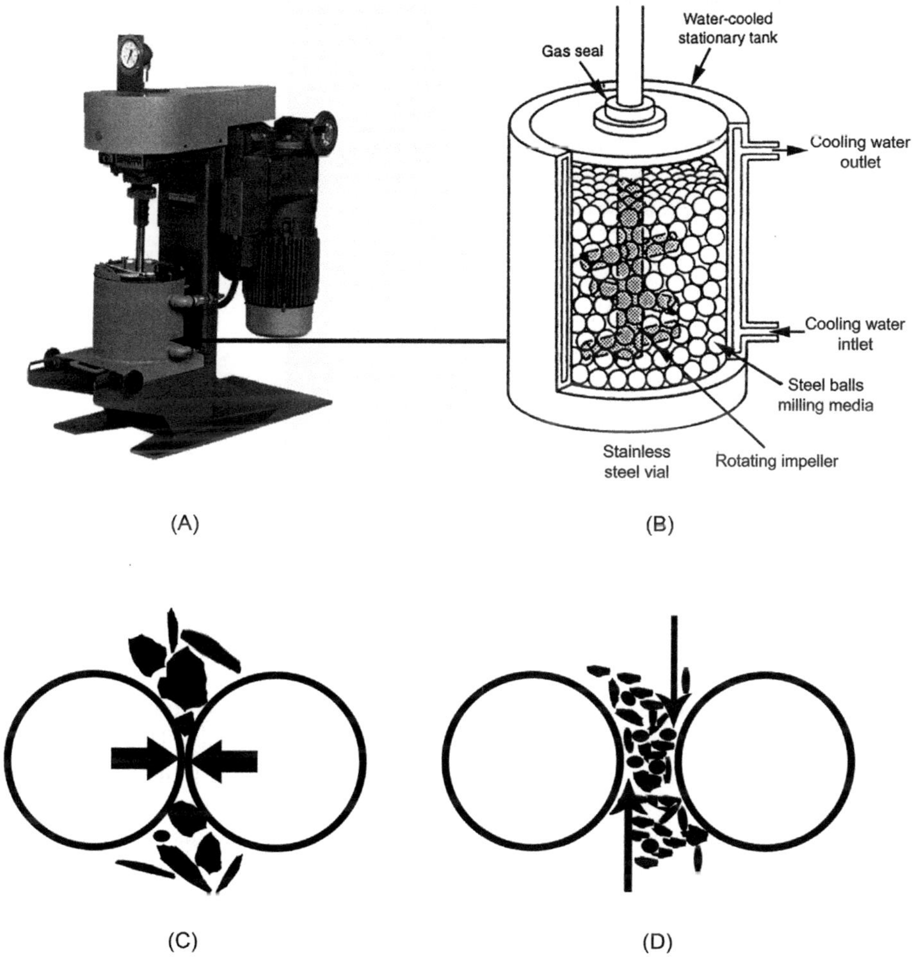

Figure 3.11 In the Szegvari attritor ball mill, the ball charge is activated by impellers radiating from a rotating vertical shaft which rotates at speeds up to 250 rpm.

experiments by continuous evacuation (using rotary and diffusion pumps) of the vial and introducing a continuous flow of an argon gas. In addition, the milling temperature could be controlled by flushing the outermost shell of the vial with current water. They have proposed this milling tool for the synthesizing of several amorphous alloy powders.

3.5.1.2 Shaker mills

Spex (SPEX CertPrep, Inc., Metuchen, NJ, USA[120]) is one example of the shaker mills. The company offers two models: the so-called Spex 8000M (Fig. 3.12A) and Spex 8000D (Fig. 3.12B). Whereas the first model holds a single vial in an arm which

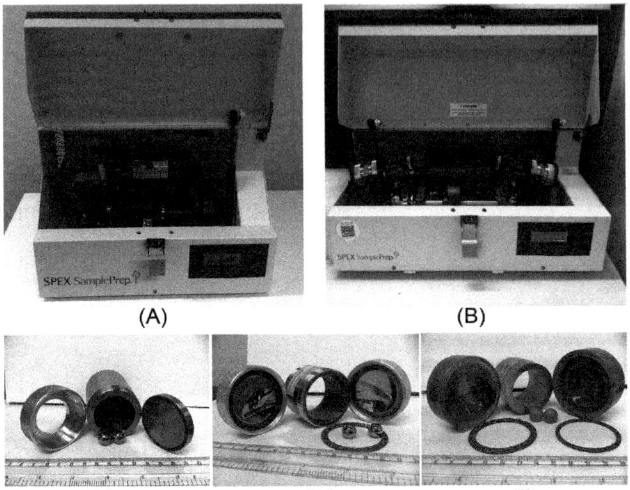

Figure 3.12 Photos of (A) SPEX 8000M and (B) SPEX 8000D high-energy ball mill. The milling tools (vials and balls) used of this type of ball mills are either made of (C) steel, (D) WC, or (E) agate steel. The equipment is housed in the Nanotechnology Laboratory, Energy and Building Research Center (EBRC), Kuwait Institute for Scientific Research (KISR).

oscillates along several axes at a very high frequency, the second model holds two vials with two arms, as shown in see Fig. 3.12A and B, respectively. This high-energy ball mill agitates the charge of powder (usually not more than 10 g) and balls in three mutually perpendicular directions at approximately 1200 rpm. The motion of the balls in a vibratory mill is very complicated, since both direct collisions with the top and bottom of the vial as well as glancing collisions with the sides occur. In this mill, the back-and-forth shaking fashion motion is combined with lateral movements of the ends of the vial. At each swing of the vial, the milling media (balls) impact against the sample powder and the end of the vial. Due to the amplitude (about 5 cm) and speed (~1200 rpm) of the clamp motion, the ball rotates at high velocities, as high as 5 m/s. Accordingly, the force generated by balls' impacts is extraordinary great. SPEX has a wide variety of optional milling tools (vials and balls) materials such as tungsten carbide, hardened steel, stainless steel, agate, alumina, zirconia, silicon nitride, plastic, and methacrylate. Fig. 3.12C–E presents some typical examples of SPEX milling tools made of tungsten carbide, hardened steel, and agate, which are shown together with their gaskets.

Another type of the vibratory ball mill, which is used at the van der Waals–Zeeman Laboratory, consists of a stainless steel vial with a hardened steel bottom, and a single hardened steel ball of 6 cm in diameter (Fig. 3.13). The vial of this mill is evacuated during milling to a pressure of 10^{-6} Torr, in order to avoid reactions with a gas atmosphere. Subsequently, this mill is suitable for mechanical alloying of some special systems that are highly reactive with the surrounding atmosphere, such as rare earth elements.

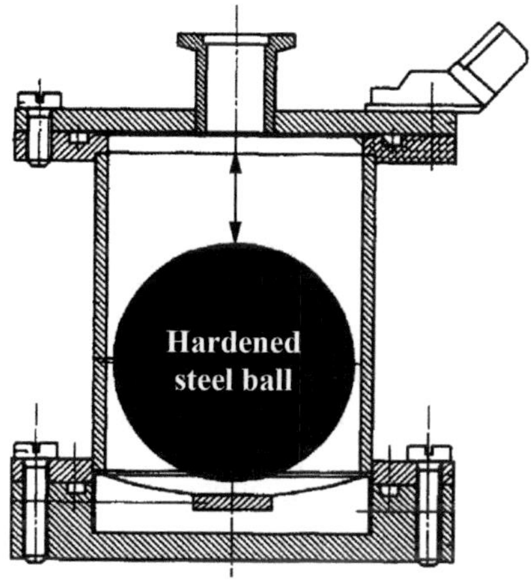

Figure 3.13 Schematic illustration of the vibratory ball mill, which is used at the van der Waals–Zeeman Laboratory.

3.5.1.3 RETSCH mixer mills MM 200 and MM 400

Retsch Mixer Mills (Retsch GmbH, Retsch-Allee 1-5, 42781 Haan, Germany)[121] are laboratory scale shaker mills developed for mechanical alloying, mechanical disordering, homogenizing, and mixing of small sample amounts quickly and efficiently. The MM 200 (Fig. 3.14) model can be used for wet and dry milling, using variety of milling vials made of stainless steel, WC, agate, and ZrO_2. Likewise SPEX 8000D, the MM 400 can mountain two vials (1.5–50 mL vol) to mill two individual samples

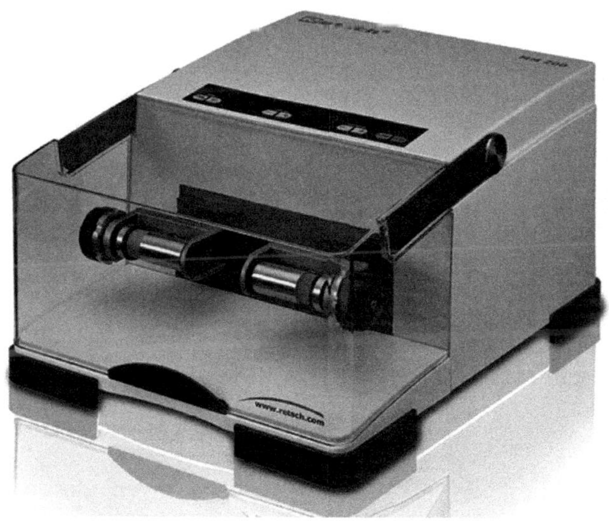

Figure 3.14 A photo of Retsch Mixer Mill (MM 200) high-energy ball mill.[121]

from 2 to 20 mL at the same time with a frequency ranging between 3 and 30 Hz (180–1800 min^{-1}).

3.5.1.4 Super Misuni

Super Misuni (Nissin Giken Co., Ltd, Sitama, Japan[122]) is another example of a vibratory ball mill (Fig. 3.15) that is enjoying many interesting features, making it unique device for preparing rather large amount (15–20 g) of many types of advanced materials under controlled milling conditions with milling speed reaches to 1600 rpm. The milling process on this mill can be performed under vacuum or at gas flow. Moreover, the milling vials can be continuously cooled by continuous water flow. One merit of Super Misuni is the ability to perform the milling process under a variety of temperature, ranging from −190°C up to 300°C.

SPEX has produced a series type of mills, the so-called Freezer/Mill, with different models (6770, 6870, and more recently 6970EFM (Fig. 3.16)). Freezer/Mills, which allow cryogenic milling of the samples at a temperature of −196°C using liquid nitrogen, are desirable mills used for milling wide variety of temperature sensitive materials, including soft metallic powders, polymers, metallic glassy alloys, ceramics, and many others. Whereas the 6770 model deals with small amount of samples (0.1–4 g), the two other models (6870 and 6970EFM) are capable to mill large amount of samples, as large as 50 and 100 g, respectively, using enclosed liquid nitrogen auto-fill system.

Different type of RETSCH Mixer Mill is CryoMill (Fig. 3.17), which is designed for cryogenic milling at a low temperature (−196°C), using liquid nitrogen. It features an integrated cooling system, which continually cools the milling vial with the liquid nitrogen before and during the grinding process. The inertia of the milling media

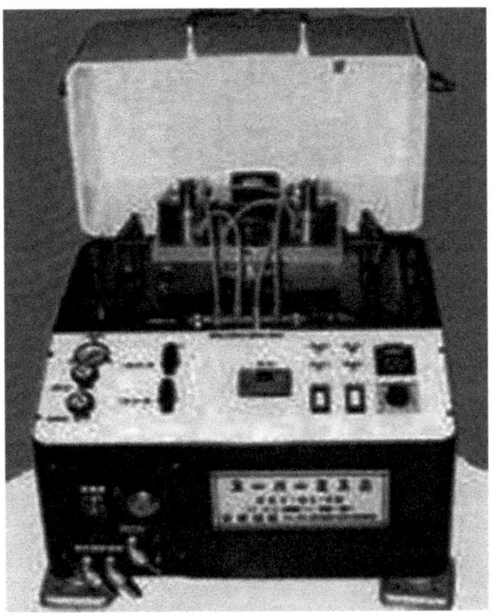

Figure 3.15 A photo of Super Misuni (Nissin Giken Co., Ltd, Sitama, Japan) high-energy ball mill.[122]

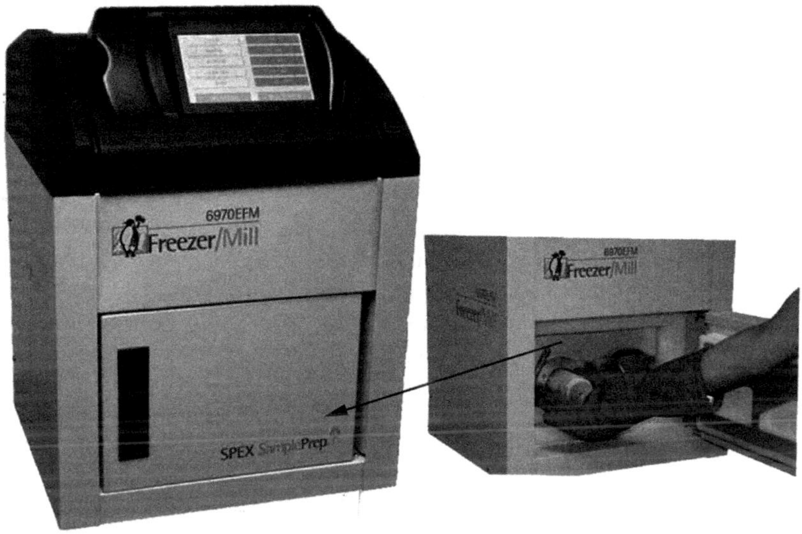

Figure 3.16 SPEX Freezer/Mill-6970EFM.[120]

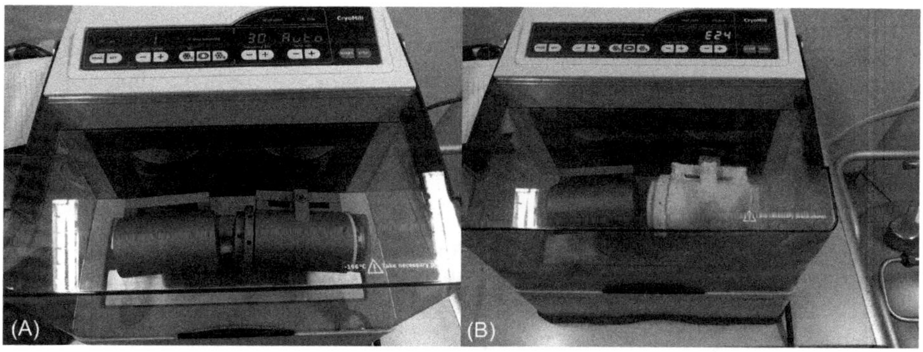

Figure 3.17 A photo of Retsch Mixer CryoMill. This type of high-energy ball mill can be used at (A) ambient temperature mode or (B) under flow of a liquid nitrogen. The equipment is housed in the Nanotechnology Laboratory, Energy and Building Research Center (EBRC), Kuwait Institute for Scientific Research (KISR).

(balls) is this mill causes them to impact with high energy on the sample powders at the rounded ends of the milling vial and pulverize them with a vibrational frequency ranging between 3 and 25 Hz (180–1500 min^{-1}).

3.5.1.5 Planetary ball mills

The Planetary Ball Mills are the most popular mills used in MM, MA, and MD scientific researches for synthesizing almost all of the materials presented in Fig. 1.2. In this type of mills, the milling media have considerably high energy, because milling stock and balls come off the inner wall of the vial (milling bowl or vial) and the effective centrifugal force reaches up to 20 times gravitational acceleration.

The centrifugal forces caused by the rotation of the supporting disc and autonomous turning of the vial act on the milling charge (balls and powders). Since the turning directions of the supporting disc and the vial are opposite, the centrifugal forces alternately are synchronized and opposite. Therefore, the milling media and the charged powders alternatively roll on the inner wall of the vial are lifted and thrown off across the bowl at high speed, as schematically presented in Fig. 3.18.

However, there are some companies in the worldwide manufacture and sale numbers of planetary-type ball mills, Fritsch GmbH (www.fritsch-milling.com) and Retsch (http://www.retsch.com) are considered to be the oldest and principle companies in this area.

Fritsch produces different types of planetary ball mills with different capacities and rotation speed. Perhaps, Fritsch Pulverisette P5 (Fig. 3.19A) and Fritsch Pulverisette P6 (Fig. 3.19B) are the most popular models of Fritsch-planetary ball mills. A variety of vials and balls made of different materials with different capacities, starting from 80 up to 500 mL, are available for the Fritsch Pulverisette planetary ball mills; these include tempered steel, stainless steel, tungsten carbide, agate, sintered corundum, silicon nitride, and zirconium oxide. Fig. 3.20 presents 80 mL-tempered steel vial (A) and 500 mL-agate vials (B) together with their milling media that made of the same materials.

More recently and in Year 2011, Fritsch GmbH (http://www.fritsch-milling.com) introduced a new high-speed and versatile planetary ball milled called Planetary Micro Mill PULVERISETTE 7 (Fig. 3.21). The company clams this new ball mill will be helpful to enable extreme high-energy ball milling at rotational speed reaches

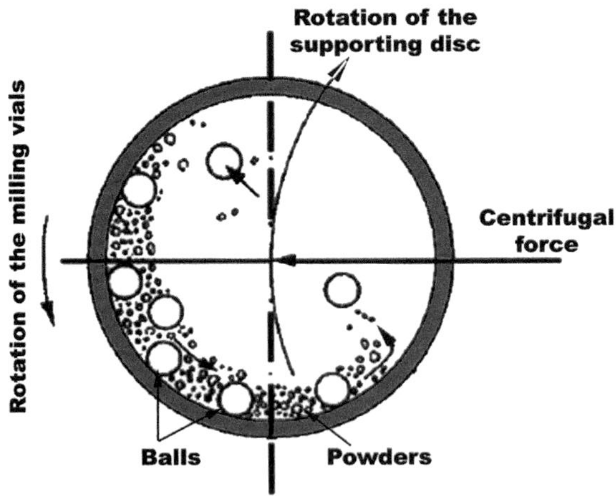

Figure 3.18 Schematic drawing of a high-energy planetary ball mill. The w_d and w_v are the angular velocities of the disc and the vials, respectively.

Figure 3.19 Photos of Fritsch planetary-type high-energy ball mill of (A) Pulverisette P5 and (B) Pulverisette P6. The equipment is housed in the Nanotechnology Laboratory, Energy and Building Research Center (EBRC), Kuwait Institute for Scientific Research (KISR).

Figure 3.20 Photos of the vials used for Fritsch planetary ball mills with capacity of (A) 80 mL and (B) 500 mL. The vials and the balls shown in (A) and (B) are made of tempered steel agate materials, respectively (Nanotechnology Laboratory, Energy and Building Research Center (EBRC), Kuwait Institute for Scientific Research (KISR)).

Figure 3.21 Photo of pulverisette-7-premium line equipment and its stainless steel vial produced by Fritsch, Germany (http://www.fritsch-milling.com).

to 1100 rpm. This allows the new mill to achieve sensational centrifugal accelerations up to 95 times Earth gravity. They also mentioned that the energy application resulted from this new machine is about 150% greater than the classic planetary mills. Accordingly, it is expected that this new milling machine will enable the researchers to get their milled powders in short ball milling time with fine powder particle sizes that can reach to be less than 1 μm in diameter. The vials available for this new type of mill have sizes as 20, 45, and 80 mL. Both the vials and balls can be made of the same materials, which are used to manufacture of the large vials used for the classic Fritsch planetary ball mills that shown earlier of this text.

Retsch also produced a number of capable high-energy planetary ball mills with different capacities (http://www.retsch.com/products/milling/planetary-ball-mills/), namely Planetary Ball Mill PM 100 (Fig. 3.22A), Planetary Ball Mill PM 100 CM, Planetary Ball Mill PM 200, and Planetary Ball Mill PM 400 (Fig. 3.22B). Likewise Fritsch, Retsch offers high-quality ball milling vials with different capacities (12, 25, 50, 50, 125, 250, and 500 mL) and balls of different diameters (5–40 mm), as exemplified in Fig. 3.23. These milling tools can be made of hardened steel as well as other different materials such as carbides, nitrides, and oxides.

Both Fritsch and Retsch companies have offered special types of vials that allow monitoring and measure the gas pressure and temperature inside the vial during the high-energy planetary ball milling process. Moreover, these vials allow milling the powders under inert (e.g., argon or helium) or reactive gas (e.g., hydrogen or nitrogen)

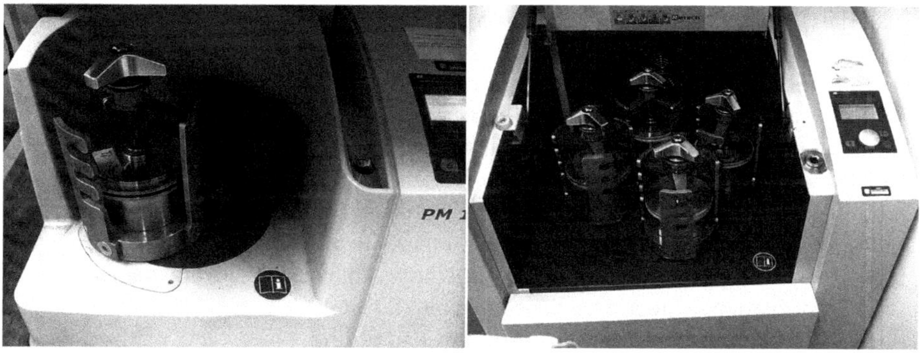

Figure 3.22 Photos of Retsch planetary-type high-energy ball mill of (A) PM 100 and (B) PM 400. The equipment is housed in the Nanotechnology Laboratory, Energy and Building Research Center (EBRC), Kuwait Institute for Scientific Research (KISR).

with a maximum gas pressure of 500 kPa (5 bar). It is worth to be mentioned here that such a development made on the vial's design allows the users and researchers to monitor the progress tackled during the mechanical alloying and mechanical disordering processes by following up the phase transformations and heat realizing upon RBM, where the interaction of the gas used with the freshly created surfaces of the powders during milling (adsorption, absorption, desorption, and decomposition) can be monitored. Furthermore, the data of the temperature and pressure driven upon using this system is very helpful when the ball mills are used for the formation of stable (e.g., intermetallic compounds) and metastable (e.g., amorphous and nanocrystalline

Figure 3.23 Photos of the vials used for Retsch planetary ball mills with capacity of (A) 80 mL, (B) 250 mL, and (C) 500 mL. The vials and the balls shown are made of tempered steel (Nanotechnology Laboratory, Energy and Building Research Center (EBRC), Kuwait Institute for Scientific Research (KISR)).

materials) phases. In addition, measuring the vial temperature during blank (without samples) high-energy ball mill can be used as an indication to realize the effects of friction, impact, and conversion processes.

More recently, evico-magnetics (www.evico-magnetics.de) has manufactured an extraordinary high-pressure milling vial with gas-temperature monitoring (GTM) system. Likewise, both system produced by Fritsch and Retsch, and the developed system produced by evico-magnetics allow RBM but at very high gas pressure that can reach to 15,000 kPa (150 bar). In addition, it allows in-situ monitoring of temperature and of pressure by incorporating GTM. The vial, which can be used with any planetary mills, is made of hardened steel with capacity up to 220 mL. The manufacture offers also two-channel system for simultaneous use of two milling vials.

3.5.1.6 The uni-ball mill

The uni-ball mill, which is a universal ball mill type, was introduced by Calka and Radlinski in 1991.[123] Fig. 3.24 displays a schematic presentation of this type of mill. From the presented illustration, one may notify that uni-ball mill is not just a conventional ball mill since the ball movements can be confined to the vertical plan by the

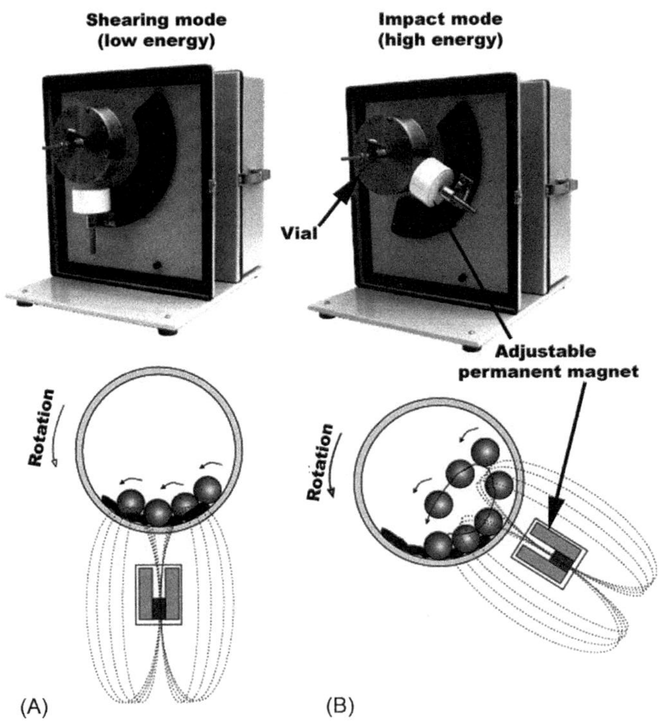

Figure 3.24 Photos and schematic presentations of the uni-ball mill was introduced by Calka and Radlinski in 1991[124] in the (A) low-energy and (B) high-energy modes.

cell walls and severely controlled by an external magnetic field that is generated by adjustable permanent magnet placed nearby the vials. Changing the magnet's position leads to affect the mode of the movement of the ferromagnetic balls from shearing mode (low energy mode) to impact mode (high-energy mode), as presented in Fig. 3.24A and B, respectively.

In this type of mills, the intensity and direction of the field can be extremely controlled. By adjusting the spatial dimensions of the magnetic field, the ball trajectories, the impact energy, and the shearing energy can be varied. In this design of the mill, permanent magnets are used. There are three general patterns of the ball movement that can be achieved using this device. When the magnet is positioned below the mill, the magnetic field holds the balls in the bottom part of the chamber rotating with a frequency ω_c as shown in Fig. 3.25.[124] Friction causes the balls to rotate in the same direction with the frequency $\omega_b = \omega_c R/r$, where R is the radius of the vial and r is the radius of the ball. Periodically, the outer ball on the right-hand side gets released completes the most of the circle being pushed against the vial wall by the centrifugal force, and hits the left-most ball at the bottom.[124] The powders are worked by both of impact and shear stresses. Two useful variations of this case could be achieved, as illustrated in Fig. 2.25B and C. By slowing down the milling rotation speed, a situation

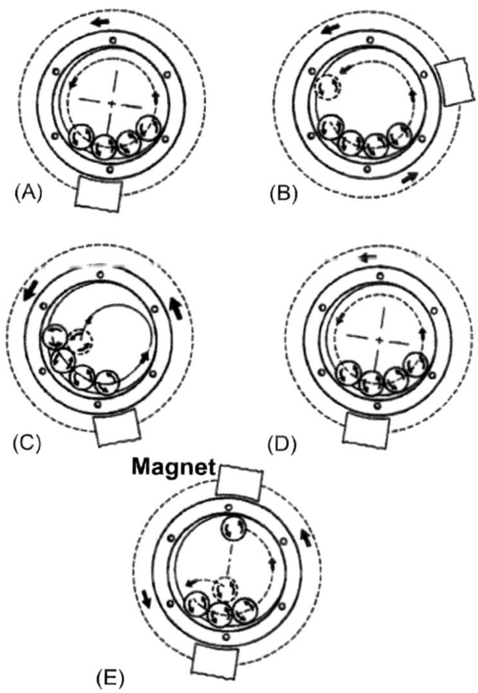

Figure 3.25 Schematic presentations show the different patterns of the ball movement inside uni-ball mill that resulted upon the position of the magnet[123] of the uni-ball mill, which was introduced by Calka and Radlinski in 1991 in the (A) low-energy and (B) high-energy modes.

may achieved when the ball released from the bottom is not fully pinned to the wall by the centrifugal force and can either hit one of the bottom balls or the opposite vial wall.

The balls may be confined to the bottom part of the chamber for all time either by increasing the intensity of the magnet or by decreasing the rotational speed, as shown in Fig. 3.25D.[123] In this case the balls both rotate and oscillate around the equilibrium position at the bottom and the powder is worked mostly by shearing. In Fig. 3.25E, the ball movement caused by the centrifugal force can be halted in two opposite positions, at the lowest, and the highest point inside the vial. The ball trapped by the magnetic attraction in the upper position rotates with the rotation of the mill vial and can be released to fall vertically on top of one of the bottom balls. It can be summarized that the adjustment of the magnet position leads to change the distance between the milling media and the magnet that controls and changes the milling energy from a low rotation frequency to a high rotation frequency. Accordingly, one can select between the two modes based on his desired impact energy and the type of the ball-milled system (e.g., miscible or immiscible, glass forming ability, etc.). It should be mentioned here that this type of mill can be performed at an intermediate mode (combination between low and high-energy modes) by simple adjusting the position of the magnet.

3.5.2 Low-energy tumbling mill

Tumbling mills are defined as cylindrically shaped shells, which rotate about a horizontal axis. Loads of balls or rods are charged into the mill to act as milling media. The powder particles of the reactant materials meet the abrasive and/or impacting force which reduces the particle size and enhances the solid-state reaction between the elemental powders.

3.5.2.1 Tumbler ball mill

The tumbler ball mills date back to 1876[125] and are characterized by the use of balls (made of iron, steel, or tungsten carbide) as milling media. The capacities of these mills are governed by several variables (ratio of mill length to diameter, speed of mill, size of balls, particle size, etc.) that should be adjusted and balanced.

In this type of mills, the useful kinetic energy can be applied to the powder particles of the reactant materials[125] (Fig. 3.26) by:

- Collision between the balls and the powders
- Pressure loading of powders pinned between milling media or between the milling media and the liner
- Impact of the falling milling media
- Shear and abrasion caused by dragging of particles between moving milling media
- Shock wave transmitted through crop load by falling milling media

The tumbler ball mills have been successfully used for preparing several kinds of mechanically alloyed powders[126] However, this kind of low-energy mill may lead to an increase in the required milling time for a complete MA process; it produces homogeneous and uniform powders.[108] In addition, it is cheaper than those of the high-energy mills and can be self-made with lower costs. Moreover, tumbling mills are operated simply with low maintenance requirements.

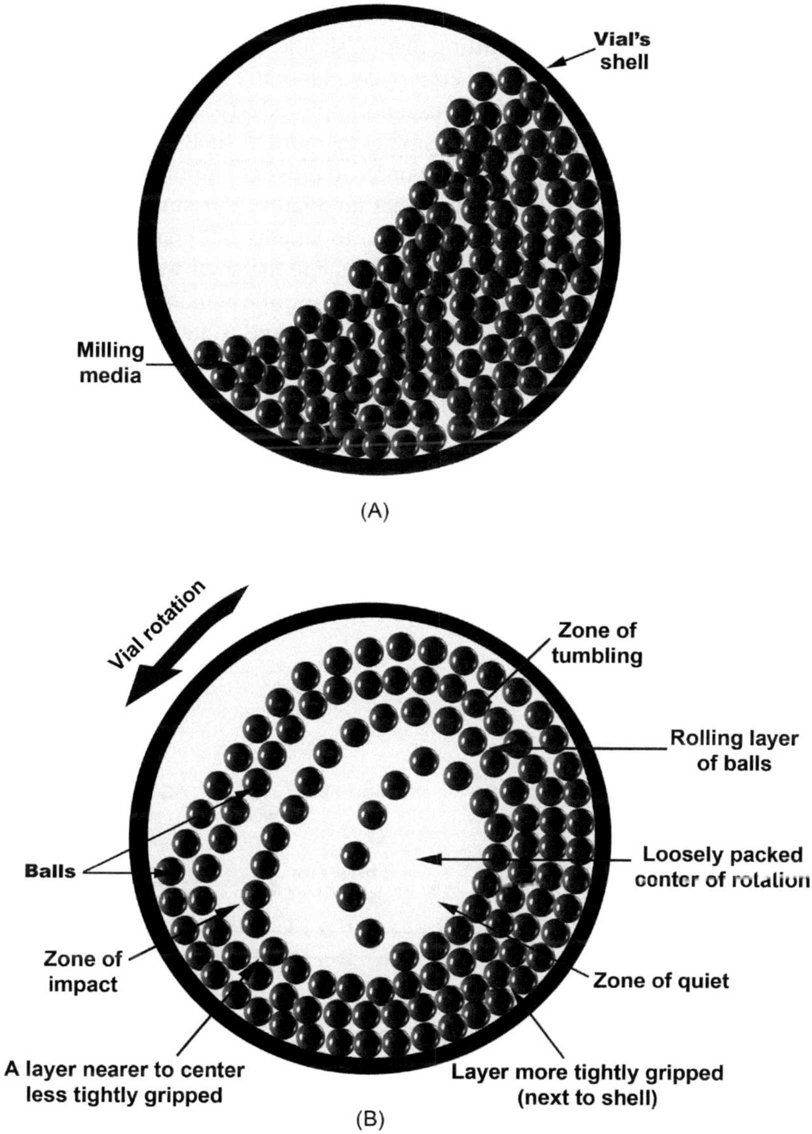

Figure 3.26 Schematic presentations show the ball positions and movement inside the vial of a tumbler mall mill at the (A) static and (B) dynamic modes.

3.5.2.2 Tumbler rod mill

Our current knowledge of the materials which are fabricated by MA has shown that almost all ball-milled alloy powders are contaminated with iron, when stainless steel balls and vial are used; this is a natural consequence of the collision between milling media. Therefore, MA method is faced with a serious problem which has impeded progress. Since the ball–powder–ball collision in a tumbling, planetary, or vibrating mill can be the main source of iron contamination, different kinds of mills, in which there is no collision between the milling media, should be used.

In 1990, El-Eskandarany et al.[111] employed a laboratory scale rod mill for preparing a large amount (30 g) of homogeneous amorphous $Al_{30}Ta_{70}$ powder. In their experiments, they made a stainless steel (SUS 304) cylindrical shell and used 10 stainless steel (SUS 304) rods as milling media. In order to prevent jamming of the rods inside the shell, the shell has been designed so that its length (250 mm) is greater than its diameter (120 mm) and the rods have been cut to lengths (200 mm) less than the full length of the shell. The movement of the rods inside the shell was directly observed through a thick and transparent plastic plate sealing the window of the shell. This observation has shown that the milling occurs by the line contact of rod–powder–rod extending over the full length of the shell. The results have shown that a single phase of amorphous Al_xTM_{100-x} (TM; Ti, Zr, Hf, Nb, and Ta) powders with low iron contamination content can be formed via rod-milling technique.[48–50] They reported that rod-milling technique leads to the formation of high-thermal stable, homogeneous, and low iron-contaminated amorphous alloys.

Fig. 3.27 shows the concentration content of iron concentration in mechanically alloyed $Al_{30}Ta_{70}$ as a function of (*A*) rod-milling and (*B*) ball-milling times, as well

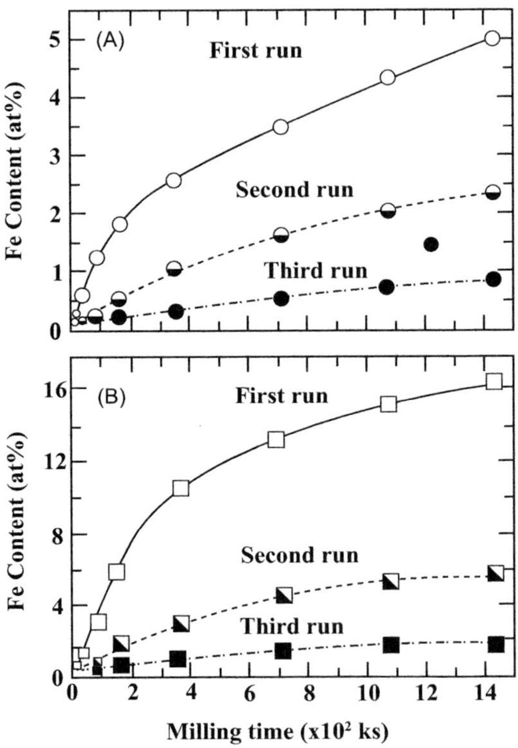

Figure 3.27 Iron contamination content in mechanically alloyed $Al_{30}Ta_{70}$ powders as a function of milling repetition of milling process: (A) rod-milling time and (B) ball-milling time. *(After El-Eskandarany et al.)*[111]

as the effect of repeating the MA process. At the beginning of the first milling run, the milling media were used in the absence of mechanically alloyed powder coatings. After this run, the milling media which had been coated by the powders were used again in the second and the third milling runs. A drastic decrease in iron content from the first to the third milling runs was observed. It also shows that the amount of iron contamination in the rod-milled powders is lower than in the ball-milled powders.

In the ball-milling (BM) process the starting elemental powders usually agglomerate at the early stage of milling to form powder particles of greater diameters, as large as several hundred microns, and this is followed by continuous disintegration until the particle size is less than a few microns. As shown in Fig. 3.28, the rod milling (RM) leads to a similar behavior for the variation in powder diameters. In RM, however, the average diameter of the agglomerate powders is very small and the subsequent disintegration into fine powders proceeds at a high rate to provide a narrow size distribution.

3.6 Mechanism of mechanical alloying

As previously mentioned, the main process which takes place in a mill during the MA method to produce quality powders with controlled microstructure is the repeated welding, fracturing, and rewelding of a mixture of powders of the diffusion couples. It is critical to establish a *balance* between fracturing and cold welding in order to mechanically alloy successfully. Two techniques are proposed by Gilman and Benjamin[22] to reduce cold welding and promote fracturing. The first technique is to modify the surface of the deforming particles by addition of a suitable processing control agent (PCA) (wet milling) that impedes the clean metal-to-metal contact necessary for cold welding. The second technique is to modify the deformation mode of the powder particles so that they fracture before they are able to deform to the large compressive

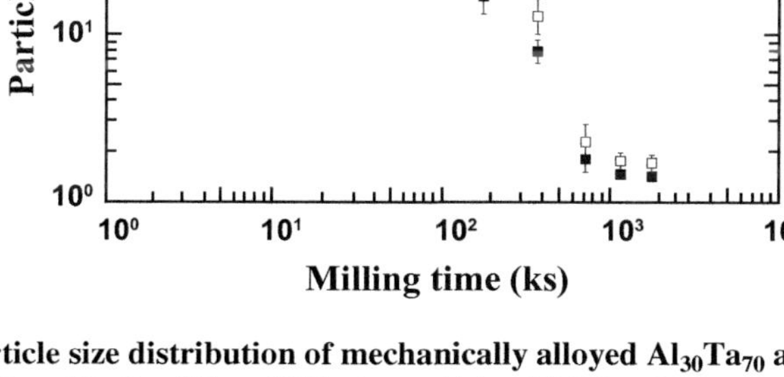

Figure 3.28 Particle size distribution of mechanically alloyed $Al_{30}Ta_{70}$ as a function of rod-milling (RM) and ball-milling (BM) times. *(After El-Eskandarany et al.)*[111]

strains necessary for flattening and cold welding. Cooling the mill chamber is an approach to accelerate the fracture and establishment of steady-state processing (effect of milling temperature).[28]

We should emphasize that milling the powders of certain metals which cold-weld easily (e.g., Ti, Zr, Al, Pb, Zn, Ag, etc.) with an organic agent (PCA)[127] may lead to an undesired reaction between the PCA and the milled powders, specifically those pure metals of the 4*f* and 5*f* elements.

3.6.1 Ball–powder–ball collision

The starting material powders that are mechanically alloyed can be two (or more) metallic powders, powders of intermetallic compound(s) or dispersoid powders. The MA process starts by blending the two (or more) individual powder constituents in order to obtain the final or the so-called end product after certain hours of milling (dry or wet). The morphology of the powders is modified when they are subjected to ball collisions (Fig. 3.29). It is worth noting that the effects of collisions on the milled powders depend on the type of the constituent particles.

It has been shown that the initial ball–powder–ball collision causes the ductile metal powders to flatten and work harder when they are cold-welded and heavily mechanically deformed. Therefore, they flatten, overlap, and atomically clean metal interfaces. They are brought into intimate contact, forming layered structure of composite particles consisting of various combinations of the starting ingredients, as schematically shown in Fig. 3.29. Further milling results in cold welding and deformation of the layered particles and a refined microstructure is obtained. Due to the initially low hardness of the starting elemental powders, the lamellar spacing of the agglomerated particles is quickly reduced upon further milling. Increasing the MA time increases the hardness and this leads to fracturing of the agglomerated powders into smaller particles. Further milling time leads to an interdiffusion reaction that takes place at the clean or fresh surfaces of the intimate layers in the powder particles, to form an alloy.

3.7 Necessity of mechanical alloying

Mechanical alloying is a unique process for the formation of several alloys and compounds that are difficult or impossible to be produced by the conventional melting and casting technique. For example, Al–Ta binary system (Fig. 3.30) shows a remarkable

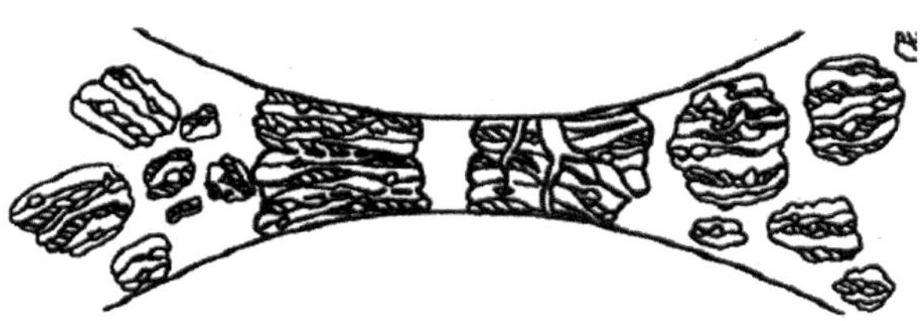

Figure 3.29 Ball–powder–ball collision of powder mixture during mechanical alloying. *(After Gilman et al.)*[18]

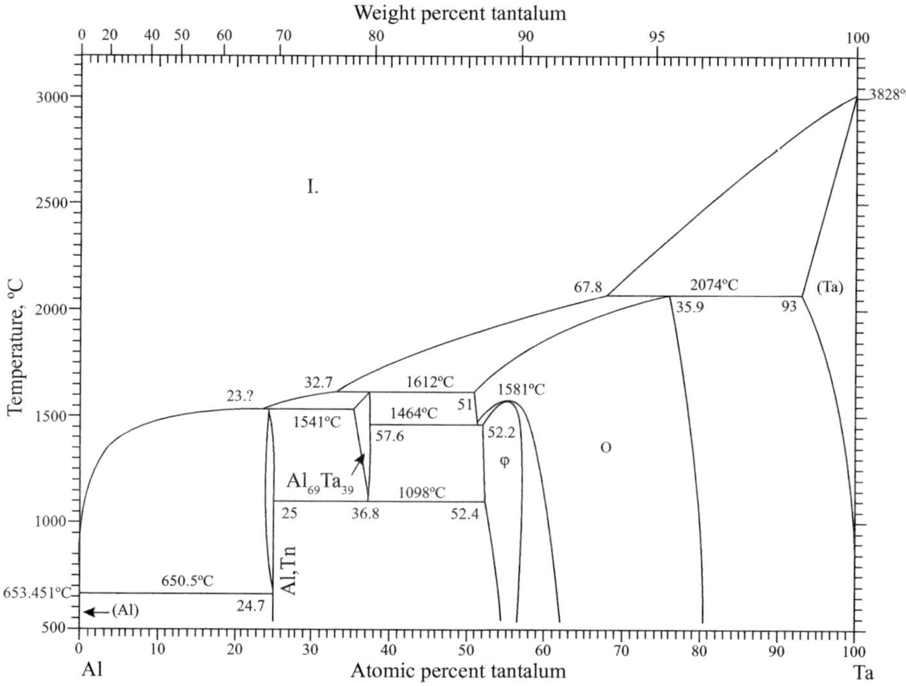

Figure 3.30 Phase relation of Al–Ta binary system. *(After Witusiewicz et al.)*[128]

gap difference between the melting points of Al (661°C) and Ta (3020°C). This gap difference restricts the production of such promising advanced materials that are used as capacitors in industrial applications. The MA method leads to the fabrication of such new amorphous material with a wide range of formation.

References

1. Greco, V. P., *Review and update on electrocomposites*, US Army Armament Research Development and Engineering Center, Watervliet, New York, The USA (1992) "Approved for public release; distribution unlimited.
2. Coolidge, W.D., *Proc. Am. Inst. Elec. Eng.*, 961 (1910)
3. Irmann, R., *Metallurgia*, 46:125 (1952)
4. Alexander, G. B., U.S. Patent 2, 972, 529 (1961)
5. Soni, P.R., *Mechanical Alloying; Fundamentals and Applications*, Cambridge International Science Publishing, Cambridge, UK (2001)
6. Benjamin, J.S., *Met. Trans.*, 1:2943 (1970)
7. Benjamin, J.S., *MPR*, 45:122 (1990)
8. Benjamin, J. S., *Sci. Forum*, 88–90:1 (1992)
9. Zbiral, J., Jangg, and Korb, G., *Sci. Forum*, 88–90:19 (1992)

10. Wright, L. G., and Wilcox, B. A., *Metall. Trans.*, 5:957 (1974)
11. Benjamin, J. S., and Volin, T. E., *Metall. Trans.*, 5:1929 (1974)
12. Gessinger, G. H., *Metall. Trans. A*, 7:1203 (1976)
13. Benjamin, J. S., and Bomford, M. J., *Metall. Trans. A*, 8:1301 (1977)
14. Gilman, G., and Nix, W. D., *Metall. Trans. A*, 12:813 (1981)
15. Benjamin, J. S., and Schelleng, R. D., *Metall. Trans. A*, 12:1827 (1981)
16. Benn, R. C., Benjamin, J. S., and Austin, C. M., *High Temperature Alloys: Theory and Design*, Warrendale, PA, TMS-AIME (1984)
17. Kang, S. K., and Benn, R. C., *Metall. Trans. A*, 18:747 (1987)
18. Gilman, P. S., and Benjamin, J. S., *Ann. Rev. Mater. Sci.*, 13:279 (1983)
19. Carreño-Morelli, E., et al. Mater. Sci. Eng. A, 251:48 (1998)
20. Hort, Norbert, and Kainer, Karl Ulrich, Chapter 10: Powder Metallurgically Manufactured Metal Matrix Composites, In Kainer, Karl U., Metal Matrix Composites, WILEY-VCH Verlag GmbH & Co. KGaA, Weinheim (2006) pp. 243-276
21. Suryanarayana, C., *Prog. Mater. Sci., 46*:1 (2001)
22. Sundaresan, R., and Froes, F.H., *J. of Metals*, Aug. 1987, pp. 22-27.
23. Weber, J. H., *The 1980s-Payoff Decade for Advanced Materials*, 25:752, Azusa, CA, S.A.M.P.E. (1980)
24. Sundaresan, R., and Froes, F. H., *Modern Developments in Powder Metallurgy*, 21:439 (1988)
25. Bloch, E. A., Met. Rev., 6:193 (1961)
26. Hansen, N., Metall. Trans., A1:545 (1970)
27. U.S. Patent No. 2,764,359, Szegvari, A. (1956)
28. White, R. L., *The Use of Mechanical Alloying in the Manufacture of Multifilamentary Superconductor Wire*, Ph.D. Thesis, Stanford University (1979)
29. Koch, C. C., Cavin, O. B., McKamey, C. G., and Scarbourgh, J. O., *Appl. Phys. Lett.*, 43:1017 (1983)
30. El-Eskandarany, M. S., Aoki, K., and Suzuki, K., *J. Alloys Comp.*, 186:15 (1992)
31. El-Eskandarany, M. S., Aoki, K., and Suzuki, K., *Scripta Metall.*, 25:1695 (1991)
32. Yavari, A. R., and Desré, P. J., in: Ordering and Disordering in Alloys, (A. R. Yavari, ed.), p. 414, Elsevier (1992)
33. Shingu, P. H., Ishihara, K. N., and Kuyama, *J., Proc. of the 34th Japan Congress on Materials Research*, p. 19, The Society of Materials Science, Kyoto, Japan (1991)
34. Fecht, H. J., Han, G., Fu, Z., and Johnson, W. L., *J. Appl. Phys.*, 67:1744 (1990)
35. Xu, J., Herr, U., Klassen, T., and Averback, R. S., *J. Appl. Phys.*, 79: 3935 (1996)
36. Biswas, A., Dey, G. K., Haq, A. J., Bose, D. K., and Banerjee, S., *J. Mater. Res.*, 11:599 (1996)
37. Schwarz, R. B., *Materials Science Forum*, 269:269 (1998)
38. Scudino, S., Sakaliyska, M., Surreddi, K.B., and Eckert, J., *Journal of Alloys and Compounds*, 483:2 (2009)
39. El-Eskandarany, M. Sherif, Aoki, K., Sumiyama, K., and Suzuki, K., Mater. Sci. Forum, 88:801 (1992)
40. El-Eskandarany, M. S., Sumiyama, K., Aoki, K., and Suzuki, K., J. Mat. Res., 7:888 (1992)
41. El-Eskandarany, M. S., Aoki, K., and Suzuki, K., *Appl. Phys. Lett.*, 60:1562 (1992)
42. El-Eskandarany, M. S., Sumiyama, K., Aoki, K., Masumoto, T., and Suzuki, K., *J. Mat. Res.*, 9:2891 (1994)
43. Harris JH, Curtin WA, and Schultz L, *J Mater Res*, 3:872 (1988)
44. Aoki K., Memezawa A., and Masumoto T., *J Mater Res*, 8:307 (1993)

45. El-Eskandarany, M. S., and Ahmed, H. A., Sumiyama, K., and Suzuki, K., *J. Alloys Comp.*, 218:36 (1995)
46. Aoyagi, H., Aoki, K., and Masumoto, T., *J. Alloys Comp.*, 231:804 (1995)
47. Grigorova, E., Khristov, M., Khrussanova, M., and Peshev, P., *J Mater Sci*, 43:5336 (2008)
48. Kalisvaart, W.P. et al., *International Journal of Hydrogen Energy*, 5:2091 (2010)
49. Minella CB, Rongeat C, Domenech-Ferrer R, Lindemann I, Dunsch L, Sorbie N, et al., *Faraday Discuss*, 151:253 (2011)
50. Huot, J., Ravnsbæk, D.B., Zhang, J., Cuevas, F., Latroche, M., and Jensen, T.R., *Progress in Materials Science*, 58:30 (2013)
51. Hellstern, E., Fecht, H.J., Fu, Z., and Johnson, W.L., *J. Appl. Phys.*, 65:305 (1989)
52. Fecht, H.J., Hellstern, E., Fu, Z., and Johnson, W.L., Metall. Trans. A, 21:2333 (1990)
53. Eckert, J., Holzer, J. C., Krill, C. E. III, and Johnson, W. L., *J. Mater. Res.*, 7:1751 (1992)
54. Abe, Y. R., and Johnson, W. L., *Mater. Sci. Forum*, 88–90:513 (1992)
55. Kuyama, J., Ishihara, K. N., and Shingu, P. H., *Mater. Sci. Forum*, 88–90:521 (1992)
56. Morris, M. A., and Morris, D. G., *Mater. Sci. Forum*, 88–90:529 (1992)
57. Zhang, X., Wang, H., Kassem, M., Narayan, J., and Koch, C.C., *J. Mater. Res.*, 16:3485 (2001)
58. Zhang, X., Wang, H., Kassem, M., Narayan, J., and Koch, C.C., *Scr. Mater.*, 46:661 (2002)
59. Zhang, X., Wang, H., Scattergood, R.O., Narayan, J., and Koch, C.C., *Mater. Sci. Eng.* A, 344:175 (2003)
60. Suryanarayana, C., *Intermetallics*, 3:15 (1995)
61. El-Eskandarany, M. S., *J. Alloys Comp.*, 305: 225 (2000)
62. Yamamoto, Takeshi A., Kondou, Takayuki, Kodera, Yasuhiro, Ishii, Takashi, Ohyanagi, Manshi, and Zuhair A. Munir, *Journal of Materials Engineering and Performance*. 14:460 (2005)
63. El-Eskandarany, M. Sherif, Sumiyama, K., and K. Suzuki, *J. Mat. Res.*, 10:659 (1995)
64. Lohse, B. H., Calka, A.E., and Wexler, D., *J. Mater Sci.* 40:655 (2007)
65. El-Eskandarany, M. Sherif, El-Mahdy, Amir A., Ahmed, H.A., and Amer, A.A., *J. of Alloys Comp.*, 312:315 (2000)
66. Güth, Konrad, Lyubina, Julia, Gebel, Bernhard, Schultz, Ludwig and Gutfleisch, Oliver, *Journal of Magnetism and Magnetic Materials*, 324:2731 (2012)
67. Salah, Numan, et al., *Int J Nanomedicine*, 6:863 (2011)
68. Suwanboon, S. and Amornpitoksuk, P., *Procedia Engineering*, 32:821 (2012)
69. Khayati, G.R., E. Nourafkan, G. Karimi, and Moradgholi, J. *Advanced Powder Technology*, 24:301 (2013)
70. Zeifert, B.H. et al., *Materials Letters*, 43:244 (2000)
71. Estrada-Guel, Ivanovich, et al., *Microscopy and Microanalysis*, 9:612 (2003)
72. Zeifert, Beatriz, *Rev.Add.Mater.Sci.*, 18:632 (2008)
73. Farhat, Z., *and* Alfantazi, A., *Materials Science and Engineering* A, 476:169 (2008)
74. Quinn, Jacqueline W., Clausen, Christian A., Geiger, Cherie L., and Aitken, Brian S., *US Patent*, US 8,288,307 B2 (2012)
75. El-Eskandarany, M. S., *J. Alloys Comp.*, 279:263 (1998)
76. Ying, D.Y., and Zhang, D.L., *Materials Science and Engineering* A, 286:152 (2000)
77. Li, Jianlin Li, Fei, and Hu, Keao, *Journal of Materials Processing Technology*, 147:236 (2004)
78. Heidarpour, A., Karimzadeh, F., and Enayati, M.H., *J. Alloys and Compounds*, 477:692 (2009)
79. Wang, Huihua, Wu, Wenyuan, Sun, Shucheng, Bian, Xue, and Tu, Ganfeng, *Ceramics International*, 37:2689 (2011)

80. El-Eskandarany, M. S., Omori, M., Konno, T. J., Sumiyama, K., Hirai, T., and Suzuki, K., *Metall. Trans. A*, 29:1973 (1998)
81. Rawers, J., *Nanostruct. Mater.*, 11:513 (1999)
82. Korth, G.E., and Williamson, R.L., *Met. Mater. Trans.* A, 26:2571 (1995)
83. Xu, C., Furukawa, M., Horita, Z., and Langdon, T.G., *Adv. Eng. Mater.*, 5:359 (2003)
84. El-Eskandarany, M. Sherif, *J. Alloys and Compounds*, 391:228 (2005)
85. Khayati, Gholam Reza, and Janghorban, Kamal, *Advanced Powder Technology*, 23:808 (2012)
86. Jazi, E.H., Borhani, G., and Farsani, R.E. *Micro & Nano Letters, IET.*, 7:448 (2012)
87. Yadav, Thakur Prasad, Yadav, Ram Manohar, and Singh, Dinesh Pratap, *Nanoscience and Nanotechnology*, 2:22 (2012)
88. Smart, S.K., Ren, W.C., Cheng, H.M., Lu, G.Q., and Martin, D.J., *Int. J. Nanotechnol.*, 4:618 (2007)
89. Liu F., Zhang X., Cheng J., Tu J., Kong F., Huang W., and Chen C., *Carbon*, 41:2527 (2003)
90. Zhao, Weifeng, Fang, Ming, Wu, Furong, Wu, Hang, Wang, Liwei, and Chen, Guohua, *Journal of Materials Chemistry*, 20:5817 (2010)
91. Schaffer, G. B., and McCormick, P. G., *Metall. Trans. A*, 21:2789 (1990)
92. Matteazzi, P., and Le Caër, G., *Mater. Sci. Eng. A*, 149:135 (1991)
93. El-Eskandarany, M. S., *Materials Transactions, JIM*, 36:182 (1995)
94. El-Eskandarany, M. S., *J. of Alloys Comp.*, 296:175 (2000)
95. El-Eskandarany, M. S., El-Bahnasawy, H.N., Ahmed, H. A., and Eissa, N.A., *J. of Alloys Comp.*, 314:286 (2001)
96. Schwarz, R.B., Petrich, P.R, and Saw, C.K., *J. Non-Cryst. Solids*, 76:281 (1985)
97. El-Eskandarany, M. Sherif, Itoh, F., Aoki, K., and Suzuki, K., *J. Non-Cryst. Solids*, 118: 729 (1990)
98. Seidel, M., Eckert, J., and Schultz, L., *J. Appl. Phys.*, 77:5446 (1995)
99. Seidel, M., Eckert, J., Bacher, I., Reibold, M., and Schultz, L., Acta Mater., 48: 3657 (2000)
100. Zhang, L.C., and Xu, J., *Materials Letters*, 56:615 (2002)
101. El-Eskandarany, M. Sherif, and Inoue, A., *Met. Trans.*, 33A:135 (2002)
102. El-Eskandarany, M. Sherif, Saida, J., and A. Inoue, *J. Mater. Res.* 18:250 (2003)
103. El-Eskandarany, M. Sherif, and Inoue, A., *Physical Review B*, 75:224109 (2007)
104. Lee, P.Y., Liu, W.C., Lin, C.K., and Huang, J.C., *Materials Science and Engineering: A*, 449:1095 (2007)
105. Li, Qiang, et al., *Journal of Materials Processing Technology*, 209:3285 (2009)
106. Suryanarayana, C., and Sharama, S., *Funct. Mater. Lett.*, 02:147 (2009)
107. El-Eskandarany, M. Sherif, *J. of Alloys Comp.*, 284:295 (1999)
108. El-Eskandarany, M. Sherif, Zhang, Wei, and Inoue, A., *Materials Transactions, JIM*, 43:1422 (2002)
109. El-Eskandarany, M. Sherif, and Inoue, A., *Materials Transactions, JIM* 43:770 (2002)
110. El-Eskandarany, M. S., Aoki, K., and Suzuki, K., *J. Less-Common Met.*, 167:113 (1990)
111. Suryanarayana, C., *Rev. Adv. Mater. Sci.*, 18:203 (2008)
112. Ermakov, A. E., Yurchikov, E. E., and Barinov, V. A., *Phys. Met. Metall.*, 52:50 (1981)
113. Jang, J. S. C., and Koch, C. C., *J. Mater. Res.*, 5:498 (1990)
114. El-Eskandarany, M. S., Aoki, K., and Suzuki, K., *J. Alloys Comp.*, 177:229 (1991)
115. Schwarz, R. B., and Johnson, W. L., *Phys. Rev. Lett.*, 51:415 (1983)
116. Schwarz, R. B., and Petrich, R. R., *J. Less-Common Met.*, 140:171 (1988)
117. Suzuki, K., J. *Non-Cryst. Solids*, 112:23 (1989)

118. Dry Grinding Attritor, www.unionprocess.com (2014)
119. Kimura, H., Kimura, M., and Ban, T., *Proceeding of the 2nd International Conference on Rapidly Solidified Materials*, San Diego, CA, published by ASM International, p. 172 (Mar. 7–8, 1988)
120. http://www.spexsampleprep.com/products_by_category.aspx?cat=3
121. http://www.rctsch.com (2014)
122. http://www.aaamachine.com/products/other/pdf/Super_Misuni_NEV-MA8_brochure.pdf (2014)
123. Calka, A., and Radlinski, A.P., *Mat. Sci. and Eng.*, A134: 1356 (1991)
124. Ajaol, Tawfk Taher, *The Development and Characterization of a Ball Mill for Mechanical Alloying*, MSc. Thesis, Queen's University, Kingston, Ontario, Canada (1999)
125. Processing Handbook, American Institute of Mining, Metallurgical and Petroleum Engineers, Inc., (1985)
126. Taggart, A. F., Handbook of Mineral Dressing: Ores and Industrial Minerals, John Wiley & Sons Inc., New York (1927)
127. Rairden, J. R., and Habesch, E. M., *Thin Solid Films*, 83:353 (1981)
128. Witusiewicz, V.T., Bondar, A. A., Hecht, U., Zollinger, J., Petyukh, V.M., Fomichov, O.S., Voblikov, V.M., and Rex, S., *Intermetallics*, 18: 92 (2010)

Controlling the powder-milling process

4

4.1 Factors affecting the MA/MD/MM

Likewise any other processes used for synthesizing and fabrications of materials, the powder-milling process via mechanical alloying (MA), mechanical disordering (MD), or mechanical milling (MM) techniques is also affected by several factors that are playing very important roles in the fabrication of homogeneous materials.[1] It is well known that the properties of the milled powders of the final product, such as the particle size distribution, the degree of disorder, or amorphization, and the final stoichiometry, depend on the milling conditions and, as such, the more complete the control and monitoring of the milling conditions, the better end product is obtained.[1–3] Fig. 4.1 shows a schematic presentation of some of these chief factors that are summarized as follows[3]:

- Type of mills (e.g., high-energy mills and low-energy mills)
- Shape of the milling vials
- The materials of milling tools (e.g., ceramics, stainless steel, and tungsten carbide)
- Milling speed (the speed at which the outer layer of the charge in the vial will centrifuge)
- Milling time
- Types of milling media (e.g., balls or rods)
- Milling atmosphere (e.g., air, nitrogen, and an inert gas)
- Milling environment (e.g., dry milling or wet milling)
- Milling media-to-powder weight ratio
- Milling temperature

4.1.1 *Types of ball mills*

However, there are many types of ball mills, such as drum ball mills, jet ball mills, bead mills, roller ball mills, vibration ball mills, and planetary ball mills; they can be grouped or classified into two types according to their rotation speed as follows: (1) *high-energy ball mills*, and (2) *low-energy ball mills*. Table 4.1 presents characteristics comparison between three types of ball mills (attritors, vibratory mills, planetary ball mills and roller mills) that are intensively used on MA, MD, and MM techniques.

In fact, choosing the right ball mill depends on the objectives of the process and the sort of materials (hard, brittle, ductile, etc.) that will be subjecting to the ball milling process. For example, the characteristics properties of those ball mills used for reduction of the particle size of the starting materials via top-down approach or the so-called mechanical milling (MM process), or for mechanically induced solid-state mixing for fabrications of composite and nanocomposite powders, may differ widely from those mills used for achieving mechanically induced solid-state reaction (MISSR) between

Mechanical Alloying. http://dx.doi.org/10.1016/B978-0-12-818180-5.00004-2

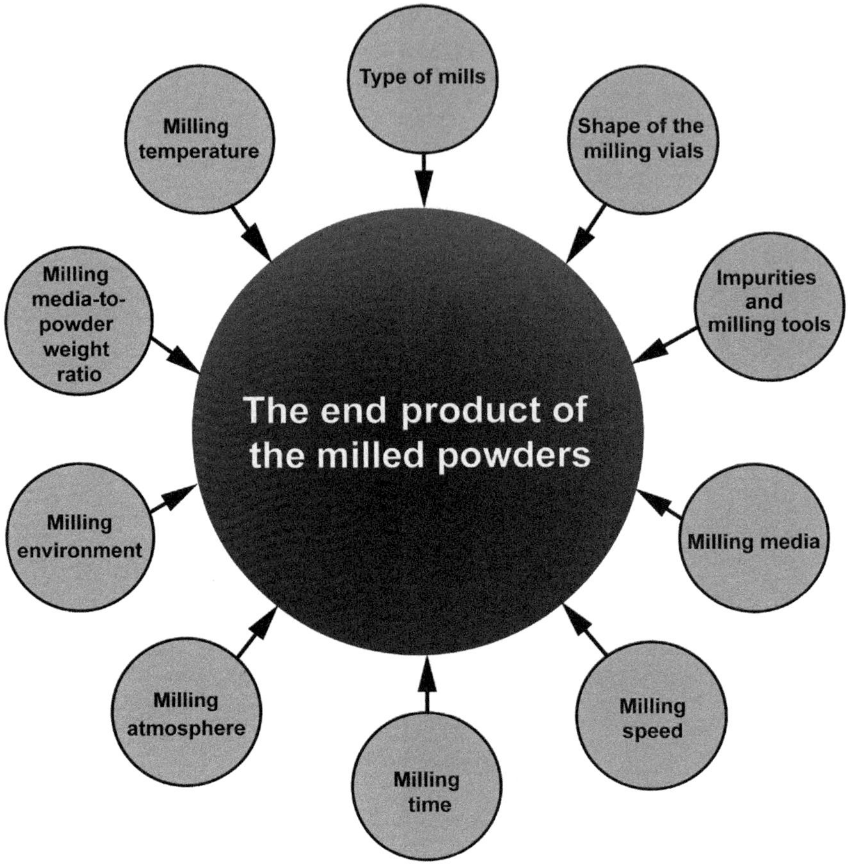

Figure 4.1 Schematic presentation of the main factors that affect the MA, MD, and MM processes.

the starting reactant materials of elemental powders (MA process), or for tackling dramatic phase transformation changes on the structure of the starting materials (MD). Nowadays, most of the ball mill manufactures introduce their equipment to be utilized successfully in all of the aforementioned processes so that single ball mill equipment can cover all proposes of the milling process.

Martinez-Sanchez et al.[4] have pointed out that employing of high energy ball mills leads not only to contaminate the milled amorphous powders with significant volume fractions of impurities that come from milling media that move at high velocity, but it also affects the stability and crystallization properties of the formed amorphous phase. They have proved that the properties of the formed amorphous phase ($Mo_{53}Ni_{47}$) powder depend on the type of the ball mill equipment (SPEX 8000D Mixer/Mill and Zoz Simoloter mill) used in their important investigations. This was indicated by the high contamination content of oxygen on the amorphous powders prepared by SPEX 8000D Mixer/Mill, when compared with the corresponded amorphous powders

Table 4.1 Characteristics of different types of ball/rod mills used for MA, MD, and MM processes.

	Attritors	Vibratory ball mills	Planetary ball mills	Roller mills
				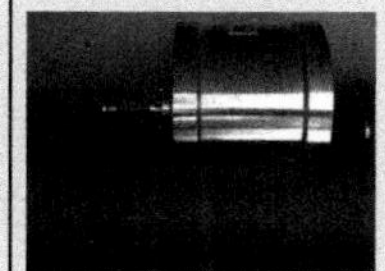
Ball velocity (m/s)	4.5–5.1	≤3.9	0.24–11.24	< 5
Kinetic energy (10^{-3} J/hit)	<10	120	0.4–884	0–190
Shock frequency (Hz)	>1000	15–200	4.5–100	0–2.4
Power (W/g/ball or rod)	<0.001	0.005–0.24	0–1.604	0–0.1

prepared by Zoz Simoloter mill. Accordingly, they have attributed the poor stabilities, indexed by the crystallization temperature of the amorphous phase formed by SPEX 8000D Mixer/Mill to the presence of foreign matter (impurities).

4.1.2 Shape of the milling vials

Another important issue that affects the ball milling process is the shape of the milling vials. The internal shape of the vial can be flat-ended type or concave-ended type. This effect was investigated in 1992 by Harringa et al.,[5] when they employed SPEX 8000 Mixer/Mill for preparing of doped $Si_{80}Ge_{20}$ alloy powders, using both flat- and round-ended vials made of hardened tool steel. The milled powders obtained from milling in both vials were characterized by XRD technique after different ball milling times. Fig. 4.2 shows the XRD patterns view of the (111) line for the samples milled for 3 and 6 h in both vials (flat- and round-ended). When they notified that the sample, which was milled for 3 h in the flat-ended steel vial, shows considerably more alloying than that one which obtained after the same milling time but using the round-ended vial, indicated by both the line shape and intensity. They pointed out that the rate of milling of the samples milled in the round-ended steel vial alloy is a much slower than when comparing with those samples milled in the flat-ended vial. Increasing the ball milling time to 6 h, single phase of $Si_{80}Ge_{20}$ alloy is obtained when using flat-ended. Contrary to this, the sample milled for the same ball milling time with using

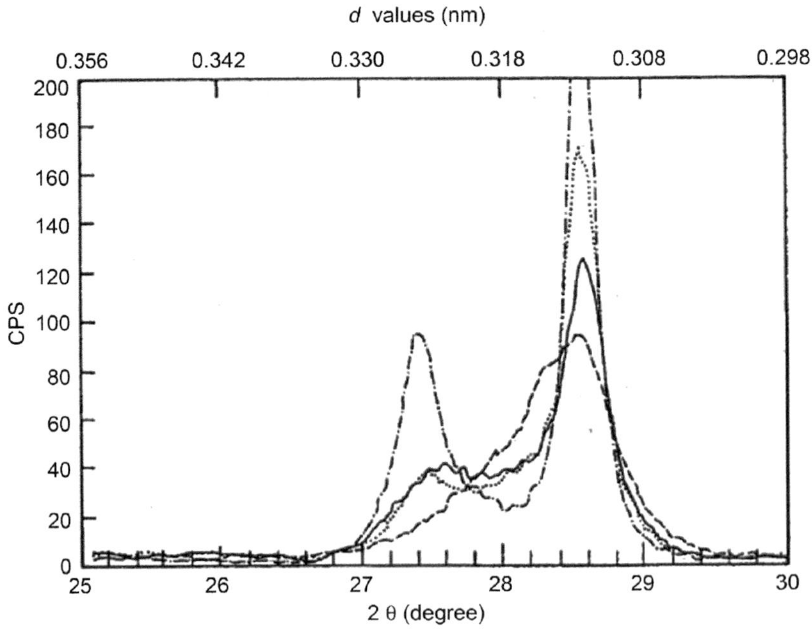

Figure 4.2 Diffraction line (111) of $Si_{80}Ge_{20}$ samples obtained after milling for: (-.-) 3 and (...) 6 h in a round-ended vial, and ( ) 3 and (---) 6 h in a flat-ended vial. [5]

rounded-end vial is suffering from the existence of unprocessed Ge crystals. Further increasing of the milling times (15 h) improved the quality of the powder obtained in the rounded-end vial to have the similar structure of the powder obtained after 9 h of milling via flat-ended vial. It was pointed out that with using the round-ended vial the balls may roll around the end of the vial rather than hitting it and this leads to the accumulation of unprocessed powders in that region.

4.1.3 Impurities and the milling tools

In spite of the obvious benefits of using MA, MD, and MM for preparations of a wide spectrum of advanced materials, the contamination of the milled powders that usually happened upon high-energy ball milling of the reactant powders is considered to be the most serious problems of the ball milling process. The source of contaminations can be classified into the following categories:

- *Organic contaminations*: It is usually introduced to the powders upon milling with liquid lubricants of some organic compounds known as process control agent (PCA). The PCA plays a very important role on milling of ductile materials that always subjected to severe plastic deformation and cold working that always lead to agglomerate the powders that tend to stuck on the surface of the milling tolls (balls and the internal surfaces of the vials) without breaking the powder particles and/or alloying. This phenomenon of powder agglomeration (Fig. 4.3) is sometimes known as negative milling since the powders, instead to be reduced in sizes, tend to gain larger sizes due to the excessive cold working. These lubricants, which

Figure 4.3 Photo of $Ni_{33}Al_{67}$ powders after planetary dry ball milling for 25 h at rotation speed of 250 rpm in an argon gas atmosphere, using Cr-tempered steel milling tools and in the absence of PCA. (Unpublished work of the author)

act as particle-agglomeration-inhibitor-surfactants, added to the powders during milling. Accordingly, the surface-tension of the powder particles, which their surfaces absorb the surfactant, tend to have lower values that permit perfect alloying between the reactant material powders without agglomerations. However, there is a long list of liquid organic PCA that can be used effectively (1–5 wt.% of the powder charge) such as ethyl alcohol, ethanol, etc., but they can be also in the solid form such as oxalic acid, boric acid, and alumina. When using PCA, the as-milled powders of the end product should be subjected to drying and surface activation processes before any further process such as powder consolidations. In general, most of these lubricants used as PCA are not stable against temperature since they have low melting and boiling points. As a result, they so often decomposed during excessive high-energy ball milling and interact with the fresh surfaces of the powders to form undesired interstitial compounds, such as carbides, nitrides, etc., which change the expected properties of the end product. It is thought that the existence of such compound contaminations has no harmful effect on the properties since they are homogeneously dispersed in the matrix of the milled material powders. Hence, they can improve some of the mechanical properties of the end product, as it is pointed out by Frazier et al.[6]

- In addition to the contamination by the PAC and in many cases, the milling tools used are not perfectly dried after washing and cleaning process. Accordingly, the powders milled with such humid surface of the milling tools lead the materials suffer from the existence of high moisture content that should be removed by drying in ovens at 120–150°C. It would be necessary to mention here that all the as-received and as-milled powders should be stored in vacuum-silica gel dissectors (Fig. 4.4).
- *Gas contamination* (e.g., oxygen, nitrogen): It is introduced to the milled powders due to the lack of sealing the milling vial that contains the powder charge. Milling of metallic powder on air atmosphere or using unsealed milling vials usually leads to the formation

Figure 4.4 The powders of as-ball-milled samples are sealed under argon gas atmosphere in Pyrex glass bottles and stored in vacuum-silica gel dissectors. The photo taken from KISR-Nanotechnology Lab, Kuwait.

Figure 4.5 Photos taken from KISR-Nanotechnology Lab, Kuwait, for the author while handling the ball-milled powders inside a glove box (mBRAUN Glove Box, Dieselstr. 31, D-85748 Garching-Germany; http://www.mbraun.com) under a helium gas atmosphere.

of significant volume fractions of high-temperature phases of metal oxide that cannot be decomposed simply upon power drying. In order to minimize the gas contamination content that usually happened upon exposure of the powders in the open air, the powders should be handled, mixed, balanced, and sealed into the vials together with the balls inside a glove box filled with an inert gas atmosphere of argon or helium (Fig. 4.5). It is worth to be mentioned here that some materials like Ti, Zr, and Hf are very sensitive to nitrogen gas. Hence, significant mole fractions of metal nitride would be existed in the final product of the milled powders. Thus, nitrogen gas would not be recommended for using as an inert gas media.

- Solid contamination: powder contaminations with foreign solid materials may have one (or more) of the following reasons:
    - Materials of the milling tools are introduced to the powders during the ball milling process as a result of frictions between the milling media (balls) or between the ball and the internal wall of the milling vial. It is well known that the magnitude of contamination depends on many parameters such as the milling medium, milling time, milling speed, and also the difference in strength and hardness between the powders and the milling tools. When using, for example, stainless, tool steel, Cr-tempered steel, and WC milling tools, fine chips of these materials are found to be existed in the milled powders. To minimize the effect of these foreign materials on changing the properties of the milled samples, it is recommended to consider the following issues:

- When milling hard material powders like, for example, Nb, Ta, or W it is not suitable to use the soft stainless steel alloy of SUS304. The fine milled powder particles on the mill charge will act as abrasion media of the milling tools, thus a significant volume fractions of the SUS304 (Fe, Cr, Ni) would be peeled off from their surfaces and mixed and in many cases alloyed with the milled powders during the ball milling process. It was found that 33 at.% of iron could be detected in the end product of pure W powders that were milled in steel milling tools for 50 h, using SPEX Mixer/Mill.[7] In most cases, these metallic alloy contaminations with their constitution elements can react with most of the metallic system and form undesired phases with poor properties. Suryanarayana[7] pointed out that using steel as milling medium, a large amount of iron (20 at.%) was found in WC/Co composite upon milling for 310 h, using SPEX Mixer/Mill. Thus, using hard tool steel or WC always recommended to be used as milling tools in these cases.
- Secondly, and more efficient is to use milling tools made of the same or similar materials of the milled powders. As an example, milling tools made of WC or WC/Co are always considered to be used when mixing of WC/Co powders to fabricate composite materials.
- It is better to utilize specific milling tools for one system rather than using the same milling tools for milling different systems. If it is necessary to change from system to another, a perfect cleaning procedure should be tackled in order to insure the removal of all the powders of the previous system, especially in the dead zone of the vials at the corners and the internal walls. It is better in all cases to use new set of milling media for new powder system.
- Existing of silica sand and dust in the as-milled sample powders are common problems that should be avoided by maintaining the MA lab in a proper clean conditions and to be isolated by simple closed doors from the outside. Moreover, the milling tools should be cleaned and kept inside vacuum-silica gel dissectors in order to protect them from any expected contamination by the dust. Such storing conditions offer prefect protection against corrosion for the steel-based milling tools.
- The type of the ball mill used should be considered as an important factor. It is believed that roller ball milling leads to bring more contamination to the powders when compared with the roller rod mills.[8] This can be understood from the difference in the mechanisms between the two categories of mills, as previously discussed in Chapter 3.

4.1.4 Milling media

Milling media possess significant influencing performance on the ball milling process in terms of the milling energy and size of the end product. The sort of materials used for manufacturing of the milling media (balls and vials) is important but crucial due to the impact of the milling balls on the inner walls of the milling vials. Selection of the milling media materials including shapes and sizes depends upon several factors, some of them are interrelated. In general, when a material is to be ball-milled there are certain critical factors, which have to be taken into account. In general, the milling media should fulfill two major requirements: (1) they should have large surface area to provide suitable contact with the material being milled; and (2) they should be as heavy as possible to have sufficient energy required for size reduction of the powder particles. Some of these principle factors that govern the selections of the milling media are discussed as follows.

- *Hardness*: The hardness of material powders is considered to be the most important characteristic to realize when deciding on what type of milling media to choose. The harder the milling media, the better the milling efficiency. Using milling media made of hard materials such as hardened steel, tungsten carbide, agate, and zirconia leads to maximize the milling efficiency and thus minimizes the milling time required to get fine and homogenous powder particles. However, using such hard milling media may lead also to contaminate the milled powders with foreign materials that come from the ball wear during the milling process. Table 4.2 presents the Vickers hardness values of selected materials used for manufacturing of the milling media.
- *Specific gravity*: The specific gravity of the milling media also plays a very important role on the milling process. It is well known that balls with high density and large diameters would give excellent results due to the expected generation of high impact forces applied on the milled powders. In general, the balls should enjoy more dense values when compared with the ball-milled powders. The densities' values of some selected materials used for producing the milling media are listed in Table 4.2.
- *Brittleness*: Another important factor that governs the selection of the milling media is the brittleness of the milled powders. Brittleness is the degree to which a material will easily break. Almost all of ceramics, including metal oxides and metal carbides, are brittle. Breaking down the brittle materials can be performed successfully by selecting the proper type of the milling media based on their properties that are presented in Table 4.2. In general, materials that are enjoying ductility such as metals and some metal alloys cannot easily be ball-milled.
- *Balls sizes*: The efficiency of ball milling depends on the surface area of the milling media that used in the process. The milling process is affected by number of contact points between the balls and the powder particles. Hence, the angle of nip presents a very important factor so that the ball sizes must be carefully chosen in relation to the largest and hardest particles of the feed powders. In general, the capacity of a ball mill increases by decreasing the ball diameters. Moreover, the rate of ball-to-ball contacts per unit time increases with decreasing the ball diameters because the number of balls in the mill increases. It is predicted that using different ball sizes can lead to higher collision energy that will help on improving the milling process.[9] Thus, balls should be as small as possible and the charge of the balls should

Table 4.2 Selected properties for some different milling media used for ball milling.

Materials	Vickers hardness (GPa)	Density (g/cm^3)	Abrasion resistance	Suitable to be used with
Agate, SiO_2	12.6	2.65	Good	Soft to medium-hard samples
Corundum, Al_2O_3	20	3.8	Fairly good	Medium-hard, fibrous samples
Silicon nitride, Si_3N_4	17.5	3.25	Excellent	Abrasive samples
Zirconium oxide, ZrO_2	12.5	5.9	Very good	Abrasive, fibrous samples
Stainless steel (Fe, Cr, Ni)	2.4	7.8	Fairly good	Medium-hard, brittle samples
Tempered steel (Fe, Cr)	3.29-3.99	7.9	Good	Hard, brittle samples
Tungsten carbide composites (WC/Co)	15	14.9	Very good	Hard, abrasive samples

be graded such that the largest balls are just heavy enough to grind the largest and hardest particles in the feed, whereas the small balls are responsible for powder refining.

The dynamics of the balls of different sizes during attrition of Cu-15 vol.% Nb powders have been studied by Cook and Courtney, using cinematographic techniques.[10] In their experiments, an attritor with a clear tank was specially designed to study media dynamics. A high-speed video camera was used to record the motion of the balls with different diameters inside the attrition, which its impeller rotated at different speeds. The metallographic examinations of the samples obtained after ball milling for 2–16 h using balls with 7.64-mm diameter and a mixture of 7.64-mm diameter and 4.76-mm balls are shown in Figs. 4.6 and 4.7, respectively.[10]

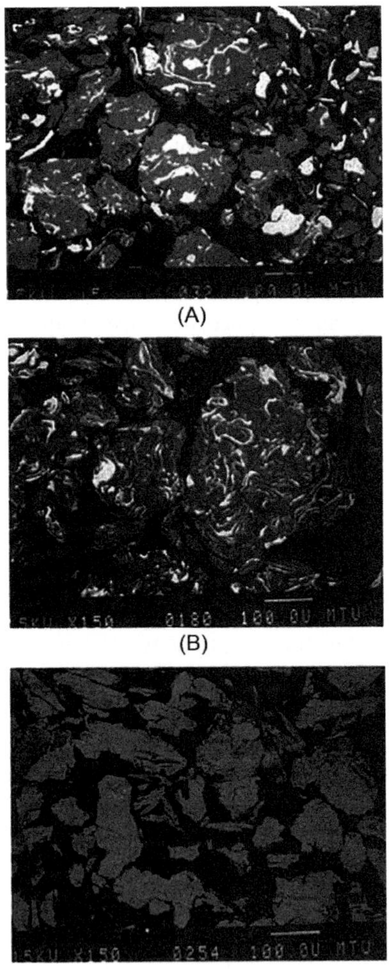

Figure 4.6 Micrographs of Cu-15 vol.% Nb powders milled with 7.64-mm-diameter balls after attrition milling time of (A) 2 h, (B) 4 h, and (C) 16 h.[10]

Obviously shown, after 16 h of ball milling using a mixture of 7.64-mm diameter and 4.76-mm balls (Fig. 4.7C) the powder has finer particle size in comparison to that of Fig. 4.6C that indicates that a greater particle fracture rate can be obtained when milling with differently sized balls. Thus, relative particle fracture rates are increased during the fracture dominant stage when differently sized balls were employed.[10]

Recently, Vaezi et al.[11] have reached to the same conclusion but with a different binary system (Cu-50% Fe) when they ball-milled the starting material powders with high-energy planetary type, using different ball sizes. They conclude that using a mixture of 5-mm and 10-mm balls leads to enhance the milling kinetics.

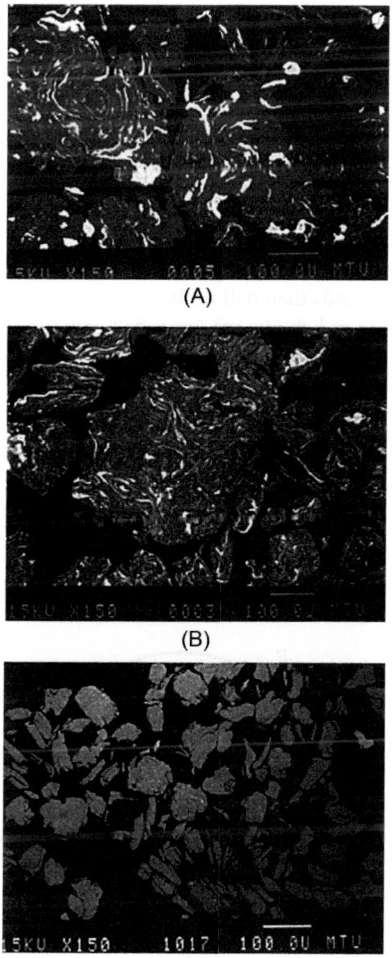

(A) (B) (C)

Figure 4.7 Micrographs of Cu-15 vol.% Nb powders milled with 7.64-mm-diameter balls mixed with 4.76-mm balls after attrition milling time of (A) 2 h, (B) 4 h, and (C) 16 h.[10]

Shape of the milling media: As discussed in Chapter 3, the shape of the milling media can be balls, rods, or barrels. In general, impact, shear stresses and their combination present the extreme forces required for powder refining conducted in ball-milling or rod-milling systems. In the tumbling and ball mills and roller mills, the rotation and friction of the vial shell to transfer energy to the milling media should be used. The energy that is required to break the particles in the mill comes from the rotational energy that is supplied by the rotator drums, which are directly connected to the drive motor. Thus, the internal frictional forces and ball-powder-ball collisions lead the media to rise before gravity forces make them fall. Based on the motion of the milling media (balls), the ball mill can be classified into three zones: (1) shearing zone, where the balls are lifted due to the rotated vial action and frictional forces, (2) cataracting zone, where the balls are falling due to the gravity, and (3) inactive zone, where the balls are not moving (Fig. 4.8). Each impact event is considered to deliver a finite amount of energy to the charge which in turn is distributed unequally to each particle that is in the neighborhood of the impacting media particles and which can therefore receive a fraction of the energy that is dissipated in the impact event.

However, balls have a greater surface area per unit weight than rods that make them better suited for fine particle size reduction; rods have several advantages, which can be summarized as follows:

- Milling rods do not require cascading as do ball charges, thus enabling rod mills to be operated at lower peripheral speeds than ball mills,
- The existed void spaces in rod charge are less when compared with the ball charge,
- Rods offer great milling contact between the powder charges per surface,
- The action within the rods causes the energy of rods to be directed to the largest sized particles of the powders, and
- Since there is no collision between the rods during the rod milling process, the volume fractions of contamination that usually introduced by the milling tools to the powders are less.

In the recent years, a different type of milling media, the so-called Cylpebs, which was developed by with different shapes have appeared in the market, such as

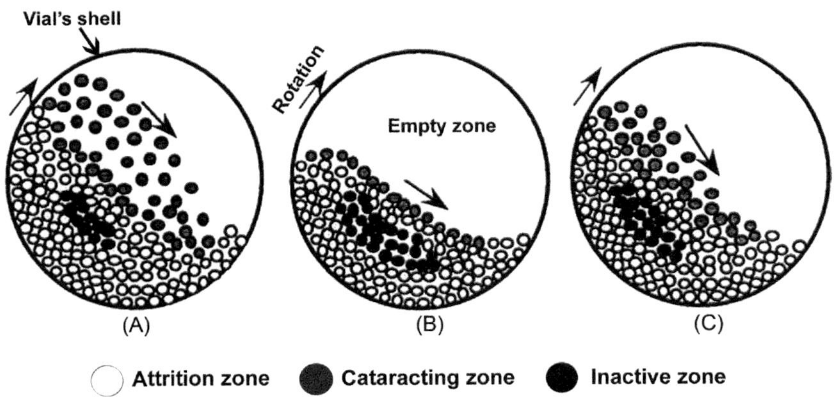

Figure 4.8 Schematic presentation of the ball motions in a tumbling mill showing (A) attrition, (B) cataracting, and (C) inactive zones.

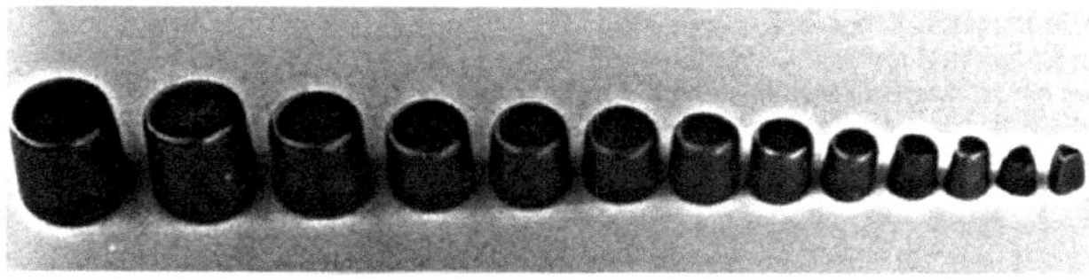

Figure 4.9 Doering Cylpebs are slightly tapered cylindrical grinding media with length equaling diameter, and all the edges being radiuses. The largest Cylpebs available in the market are of the size 85 mm × 85 mm, and the smallest ones are 8 mm × 8 mm. Because of their geometry, Cylpebs have greater surface area than balls of the same mass.
Source: www.donhad.com.au.

developed by Doering International, Powerpebs by the Donhad, and Millpebs by the Wheelabrator Allevard Enterprise, are available in the market. These milling media have barrel- or cylindrical-like shapes of length equaling diameter, and all the edges being radiuses (Fig. 4.9). It is claimed by the manufacture (Doering International; www.donhad.com.au) that for a given charge volume, the bulk density of Cylpebs is 9% greater than steel balls, 12% greater than cast balls. Moreover, the surface area of Cylpebs is 14.5% greater than balls of equal weight. The density and surface area combination deliver 25% greater grinding capacity in the mill charge. The grinding performance of the Cylpebs should then be correspondingly higher compared with the steel balls.[12] Because of their cylindrical geometry, Cylpebs have greater surface area and higher bulk density compared with balls of similar mass and size, as presented in Table 4.3. The milling mechanism with using Cylpebs is considered to be a combination between ball milling and rod milling. Beside the point contact action similar to the balls, Cylpebs have other milling actions resulted from line contact along the cylindrical section and area contact between the end faces on the powder particles, being similar to the mechanism of rods. The line contact and area contact increase tendency for milling to take place preferentially on the larger powder particles, as schematically shown in Fig. 4.10. Once the large particles are caught on the line or between the face areas, this prevents the smaller particles being broken further, which is similar to the rod mill practice.[12]

It is worth to be mentioned here that the milling vial should not be completely filled with the powder and milling medic charges. There should be enough space that allows the milling media to move around freely in the vial that provide alloying and particle size reduction. Suryanarayana[13] has suggested that 50% of the vial volume should be left free, as schematically presented in Fig. 4.11.

4.1.5 Milling speed

Milling speed or the rate of circulation is defined as the number of circulations of the mill a particle makes per mill revolution. Milling speed is one of the most important milling parameters that controls the milling process because it provides the mill with the energy required for (1) powder particles and grain-size refining (top-down approach), (2) solid-state reaction, (3) solid-state reduction, (4) solid-state mixing, and (5) phase transformations. Selection of the proper milling speed means controlling

Table 4.3 Comparative data of some physical properties for selected Cylpebs and balls used for powder milling.[12]

Cylpebs ($\gamma = 7.65$ kg/L)					Balls ($\gamma = 7.85$ kg/L)				
Dimension (mm)	Mass (g)	Surface area (cm^2)	Specific surface area (cm^2/g)	Bulk density (t/m^3)	Dimension (mm)	Mass (g)	Surface area (cm^2)	Specific surface area (cm^2/g)	Bulk density (t/m^3)
$\phi 20 \times 20$	48.0	19.0	0.40	4.9	$\phi 20$	33.0	12.5	0.38	4.5
$\phi 16 \times 16$	23.5	12.0	0.51	4.9	$\phi 15$	13.0	7.0	0.54	4.5
$\phi 12 \times 12$	11.1	6.8	0.55	4.9	$\phi 10$	4.5	3.1	0.69	4.5

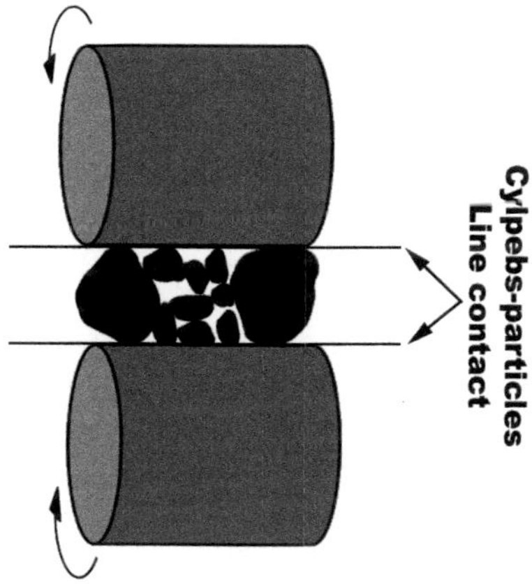

Figure 4.10 Schematic presentation showing the milling action of linear contact of Cylpebs during the milling process.

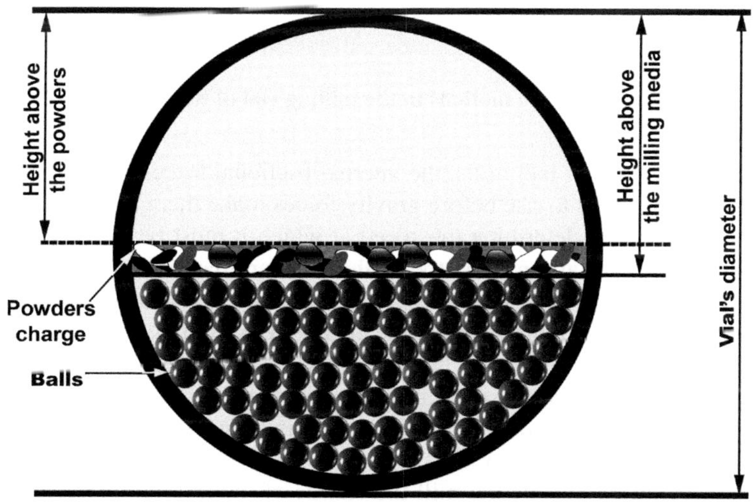

Figure 4.11 Schematic presentation of the cross-sectional view for a milling vial filled with powders charge and balls. The volume of the total charge (powders + balls) should be in the range between 40% and 50% of the total volume of the vial.

the energy input required achieving the aforementioned process, based on the goal of the experiments. However, one may realize that the faster the mill rotates, the higher would be the kinetic energy ($E = 1/2\ mV^2$) input into the powders, but there are different difficulties usually resulted upon applying extraordinary milling speed. Milling the powders at high speeds leads to increasing the contamination by the materials of the milling tools due to the high rate of wear happened. Moreover, pinning of balls to the inner walls of the vial and increasing of the vial temperature limit the maximum speed that can be used.[14]

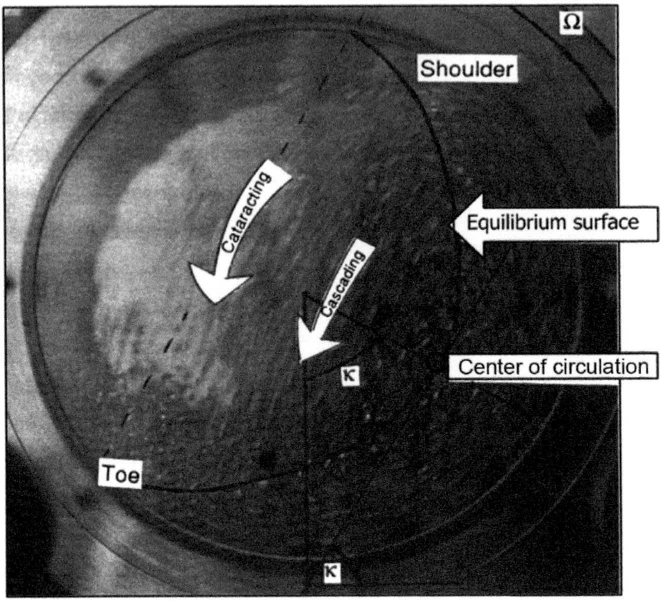

Figure 4.12 Balls and particle motions in the milling vial of tumbling mill.[16]

In the tumbling/roller ball mills, the internal frictional forces and ball-powder-ball collisions lead the balls to rise before gravity forces make them fall (Fig. 4.12). Thus, it is very important to determine the speed at which it must be run in order that the descending balls will fall onto the toe of the powder charge, and not onto the inner walls of the vial. When the ball mill rotates at low speeds, the balls tend to roll down to the toe of the mill and this cascading leads to finer grinding, with increased slimes powder production and increased liner wear. At higher speeds the milling media is projected clear of the charge to describe a series of parabolas before landing on the toe of the charge. This cataracting leads to comminution by impact and a coarser end product with reduced liner wear.[15]

In traveling around inside the mill the charge of the milling media and the powders follow a path that has two parts: (1) the lifting section near to the inner walls of the vial is circular while (2) the drop back to the toe of the mill charge is parabolic (Fig. 4.13A). Assume a ball, which is lifted up the vial of a mill of radius R, revolving at N rpm. The ball abandons its circular path for a parabolic path at point P (Fig. 4.13B), when the weight of the balls is just balanced by the centrifugal force, that is, when $mV^2/R = mg\cos\alpha$[15] (m is the mass of the balls (kg), V is the linear velocity of the balls, and g is the acceleration due to gravity).

Taking into account several parameters such as the diameters of the balls (d) and the vial (D), the critical speed of a tumbling ball mill (N_c) can be estimated from the following equation[15]:

$$N_C = \frac{42.3}{\sqrt{D-d}}\,\text{rev}\,\text{min}^{-1}$$

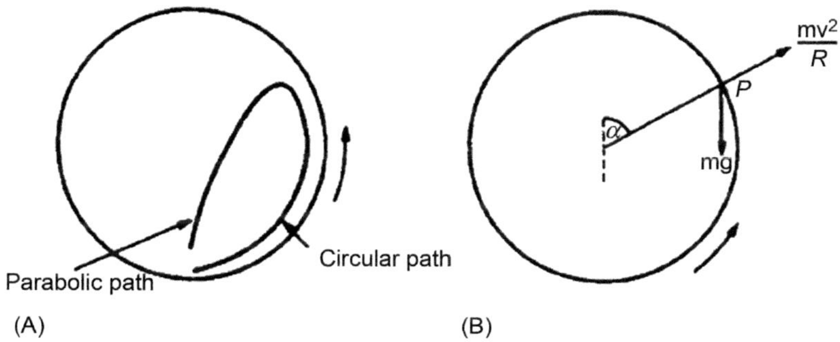

Figure 4.13 (A) Trajectory of milling media in tumbling mill, and (B) forces acting on the milling media.[15]

It is worth to be mentioned here that if the mill speed is 100% of critical and the charge centrifuging, the concepts of a shoulder and toe are no longer meaningful for the particles in contact with the shell. The head will be the highest point in the mill and the shoulder will not exist. The concepts can then be applied to the outermost layer of charge that is not centrifuging. If the mill speed is sufficiently low to prevent a cataracting stream developing, the departure shoulder and the head will be close together, and the impact toe and bulk toe will be one region.[16]

For planetary ball mills, Abdellaoui and Gaffet[17,18] have proposed interesting model to compute the collision velocity and kinetic energy released from the ball to the powders, as presented in the following equations:

$$\left\|\overline{V_c}\right\|^2 = (R\Omega)^2 + (r - r_b)^2\,\omega^2\left(1 + \frac{2\omega}{\Omega}\right)$$

$$E_k = \frac{1}{2}m\left\|\overline{V_c}\right\|^2$$

where R is the distance between the disc center and the vial center, r is the vial radius, r_b is the ball radius, Ω is the disc rotation speed, ω is the vial rotation speed, and m is the ball mass.[18]

In 2004, Hashimoto et al.[19] pointed out to the importance of theoretical investigations regarding effect of the rotation and revolution speeds and their directions on the impact energy of balls and grinding characteristics of the product in the mill. They have studied experimentally and theoretically the effect of the rotation-to-revolution speed ratio on the grinding performance using a gibbsite sample ground by three kinds of planetary mills. They calculated the specific impact energy of the balls [E_W] from the relative impact velocity between two colliding balls and/or a ball colliding against the inner wall of the vial mill [v_j] as presented in the following equation[19]:

$$E_W = \sum_{j=1}^{n} \frac{1}{2W} m V_j^2$$

where m is the mass of a ball, n denotes the number of collision of a ball against other balls and/or the internal wall of the vial within a second, and W is the weight of the sample charged into the vial. Based on the earlier presented equation, it shows the grinding rate, K_P, increases with increasing the revolution speed of the disk, N_r (Fig. 4.14) due to its proportion to the kinetic energy of balls (Fig. 4.15).

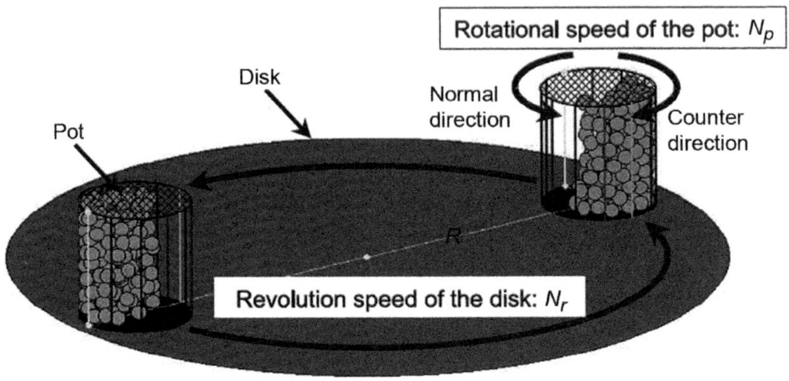

Figure 4.14 Schematic configuration of a planetary ball mill.[19]

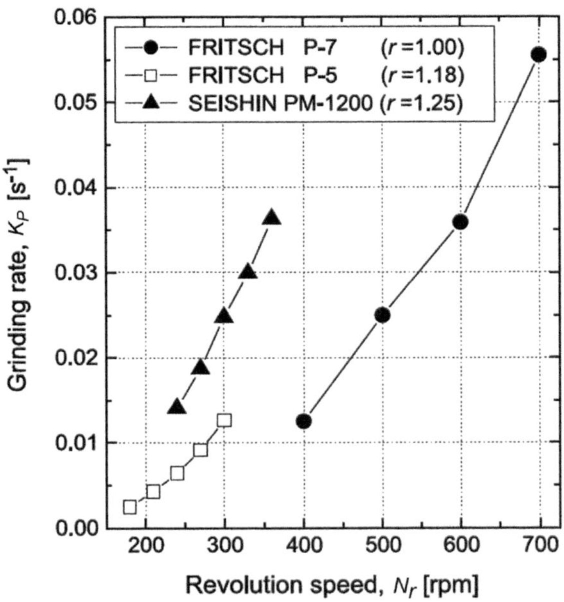

Figure 4.15 Relation between grinding rate (K_p) and revolution speed for the samples milled in different planetary ball mills (P-7, P-5, and SEISHIN PM-1200). The rotational-to-revolution speed ratio (N_p/N_r), which is denoted by r, is shown for each mill.[19]

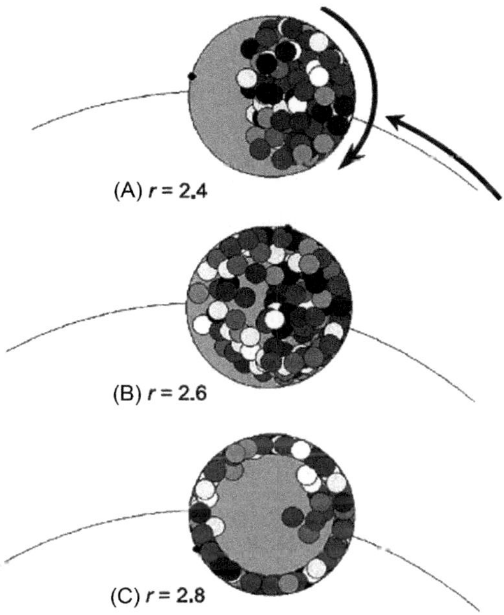

Figure 4.16 Snapshots of the balls motion during milling at different speed ratios.[19]

According to the snapshot of the balls motion during milling at different speed ratios (r), the ball motion shifts from cascading ($(a)r = 2.4$), into cataracting ($(b)r = 2.6$) followed by rolling ($(c)r = 2.8$) with increasing r, as presented in Fig. 4.16.[19] These results lead to realize that the balls' motion is always in rolling when r is considerably a large value due to the centrifugal force caused by the rotation of the pot. Thus, rolling motion is ineffective for grinding performance due to a reduction considerably in collision of balls each other or against the inner wall of the vial.[19]

In 1992, Calka et al.[20] pointed out to the formation of different nanostructure phases of vanadium carbide upon ball milling of elemental of 70V-30C and 50V-50C powders, using different types of ball mills (high-energy, medium-energy, and low-energy). Based on their results, a single phase of intermetallic compounds of V_2C or VC with grain sizes of 10 nm in diameter was obtained when the powders milled in a high-energy ball mill. In contrast to this, different types of nanostructures were obtained upon milling the same starting materials with similar compositions but in a medium energy ball mill.[20] In this case, the end product, which was a mixture of V_2C or V crystals plus C amorphous phase, transformed into intermetallic V_2C or VC upon annealing the powders at 600°C for 2 h. The starting powders, milled with a low-energy ball mill, led to the formation of a mixture composited of nanocrystalline phases of unprocessed V and amorphous carbon, obtained after 1000 h of milling (end product). Again, these products were transformed into V_2C or VC upon powder annealing at 600°C for 2 h.[20]

In 1999, El-Eskandarany et al.[21] investigated the effect of milling speed on crystalline solid-state cyclic phase transformations on Al–Zr binary system upon ball milling their elemental powders with a nominal composition of $Al_{50}Z_{50}$, using planetary ball mill (Fritsch P5) operated at different milling speed of 90, 180, 270, and 360 rpm. Based on their XRD results (Fig. 4.17), when the powders were milled at a rate of 360 rpm, a single phase of fcc-ZrAl solid solution was formed after 12 h. This metastable phase no longer appears upon further milling to 24 h and transformed into a single amorphous, which is transformed into orthorhombic–Zr–Al intermetallic phase. They observed cyclic phase transformation between the obtained intermetallic phase and the amorphous phase upon increasing the MA time at such high milling rate (Fig. 4.17). Contrary to this, the cyclic phase transformation between the formed two phases were absent for the samples milled at lower milling rates (90–270 rpm) using the same experimental conditions and length of time, as indexed by the XRD shown in Fig. 4.18. This cyclic phase transformation, which is affected by the milling speed, is not unique for Al–Zr binary system, but it has been observed in different systems metallic systems.[22–24]

It has been reported by Bab et al.[25] that the gas–solid reaction kinetics for HfN system, which was taken place in a horizontal oscillatory ball mill (Retsch MM2), depends on the milling speed. They notified that the number of nitrogen molecules absorbed by metallic Hf to form a single phase of cubic-HfN nanocrystalline powders increases (Fig. 4.19B) as the milling speed increased to be in the range between 25 and 29 Hz. Below this milling speed (21 Hz) the end product was mixture of unprocessed hcp-Hf plus cubic-HfN, as presented in Fig. 4.20, which shows the XRD patterns of

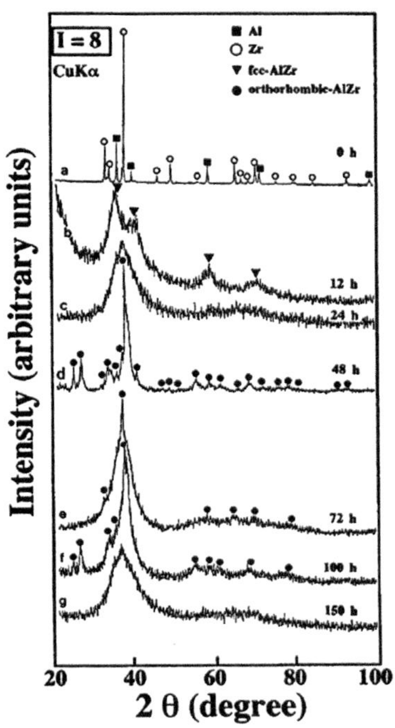

Figure 4.17 XRD patterns of the $Al_{50}Zr_{50}$ powder that ball-milled at a milling speed of 360 rpm after (A) 0 h, (B) 12 h, (C) 24 h, (D) 48 h, (E) 72 h, (F) 100 h, and (G) 150 h of the MA time.[21]

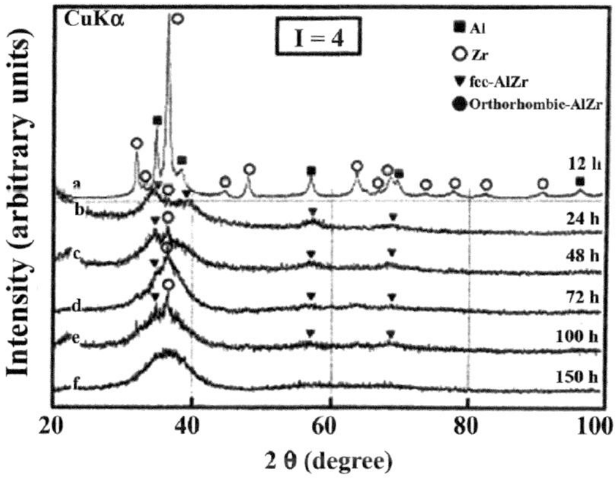

Figure 4.18 XRD patterns of the $Al_{50}Zr_{50}$ powder that ball-milled at a milling speed of 180 rpm after (A) 12 h, (B) 24 h, (C) 48 h, (D) 72 h, (E) 100 h, and (F) 150 h.[21]

as-ball-milled Hf powders in nitrogen atmosphere for 8 h of ball milling times at different milling speeds and gas pressures.

Rochman et al.[26] investigated the effect of milling speed on the structure and mechanical properties of amorphous and supersaturated solid solution of Fe-7 mass% C-6 mass% Mn prepared by planetary ball milling their elemental powders. Based on their results, the solubility of carbon atoms into a-Fe is dependent on the milling speed with the same milling time (100 h), as presented in Fig. 4.21. From the figure one can say that increasing the milling speed to 120 rpm leads to increase the carbon content in a-Fe to the level of 1.3%. Contrary to this, the carbon contents of those other samples milled for 100 h at 80 and 100 rpm were far below than this value, as presented in Fig. 4.21. Moreover, they have notified that milling the system at high milling speed (120 rpm) led to form fine powder particles with narrow particle size distribution.[26] When the powders of Fe-7C-6Mn consolidated when the powders were HIPed at 853 K, the hardness of the HIPed compact obtained from the powders milled at 100 rpm is very low and similar to that of the HIPed compact obtained from as-received iron powders (around HV = 100). However, in the case of HIPing at 1053 K, the hardness increases up to 498. The hardness values of the consolidated samples that obtained by HIPing the powders milled at 120 rpm show extraordinary high values (781 for HIP at 853 K and 1070 for HIP at 1053 K).

More recently, it has been shown by Ebrahimi-Purkani et al.[27] that the formation of Mg_2Ni intermetallic compound can be obtained only for those samples that were planetary ball milled for 60 h at milling speed of 600 rpm.

4.1.6 Milling time

One advantage of MA, MD, and MD methods over any other methods is the possibility to monitor the progress of the solid-state reaction, solid-state mixing, phase

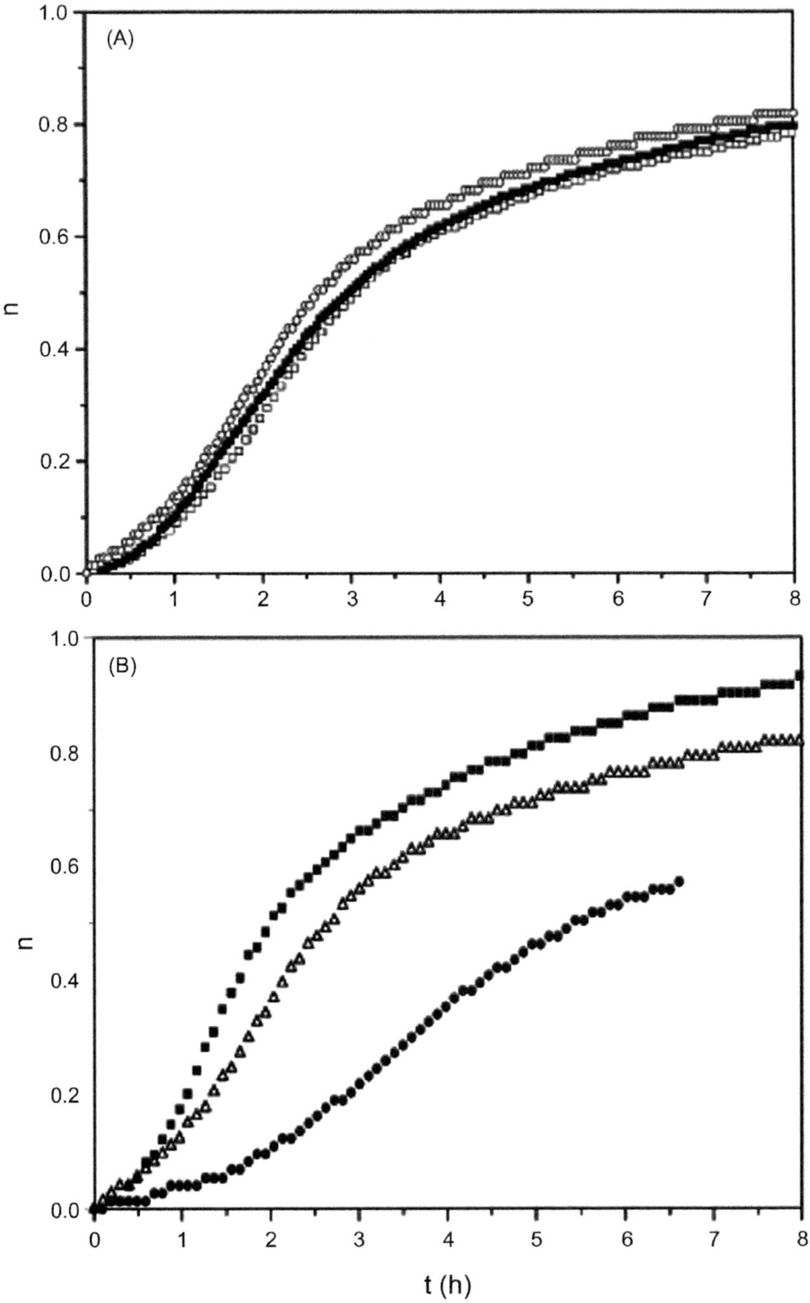

Figure 4.19 Number of absorbed nitrogen atoms per metal atom versus milling time: (A) at pressures of 0.15 (□), 0.20 (■), and 0.30 MPa () for a milling frequency of 25 Hz and (B) at milling frequencies of 21 (•), 25 (Δ), and 29 Hz (■) for a pressure of 0.30 MPa.[25]

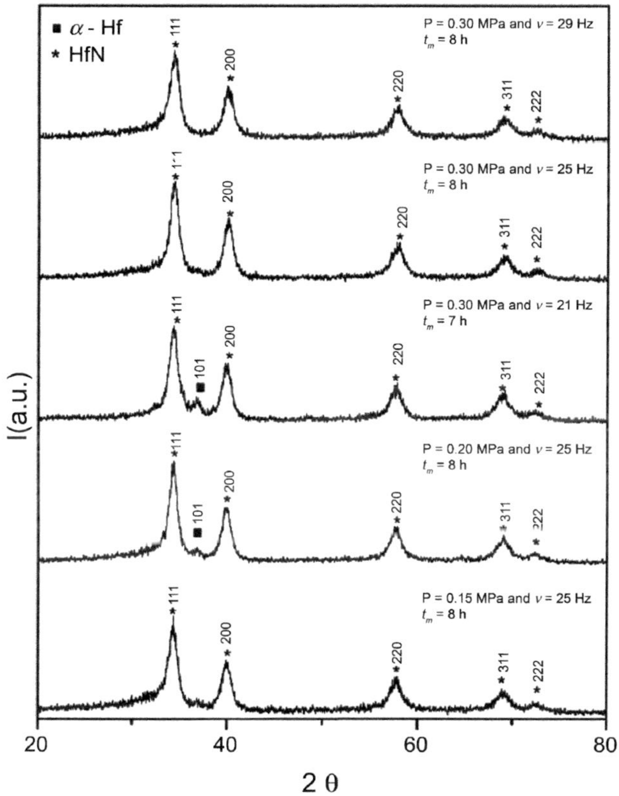

Figure 4.20 XRD diffraction patterns for ball-milled Hf powders milled under nitrogen gas atmosphere under different milling conditions.[25]

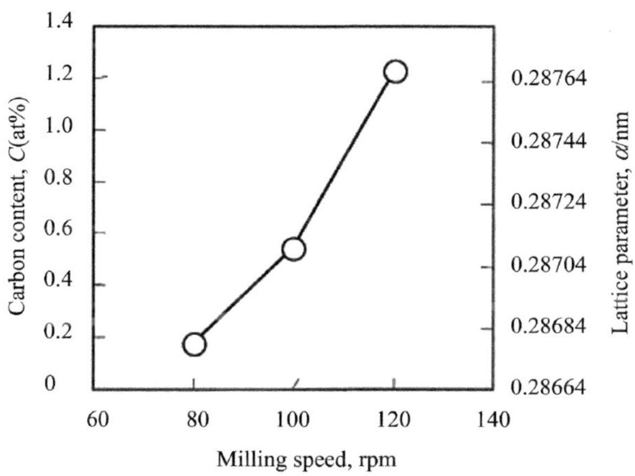

Figure 4.21 Change in the carbon content and lattice parameter of α-Fe as a function of milling speed.[26]

transformations, and the degree of powder mechanical mixing that taken place upon ball milling the starting materials for different milling times. Within the last 30 years, almost all the published scientific research articles have focused on studying the effect of the ball milling time on the structural, morphological, thermal stability, chemical, physical, and mechanical properties of the ball-milled materials. It is worth to be mentioned here that the ball milling time required to prepare the same material is dependent on the ball mill used. High energy ball mills have high kinetic energy that allows the solid-state reaction, phase transformations, and particle size reduction to be achieved in shorter time, when compared with the low energy ball mills that always require longer milling.

Several characterizations techniques and devices, such as X-ray diffraction (XRD), transmission electron microscope (TEM), scanning electron microscope (SEM), differential scanning calorimeter (DSC), etc., have been employed to monitor the progress in the MISSR via ball milling process. There is no more fast and easy way to get quick information on the milling progress more than XRD that gives a full picture on the mean structural changes upon powders ball milling for different length of time (Fig. 4.18). This technique can be also used to calculate the change on the lattice parameters, a_o (Fig. 4.22), crystallite size, and the lattice strains (Fig. 4.23) upon powder ball milling for different ball milling time. The development on the local structural changes of the powders milled for different ball milling time can be monitored beyond the atomic level by TEM and high-resolution TEM (HRTEM) that provides the opportunity for local analysis with a regional diameter of less than 5 nm (Fig. 4.24).

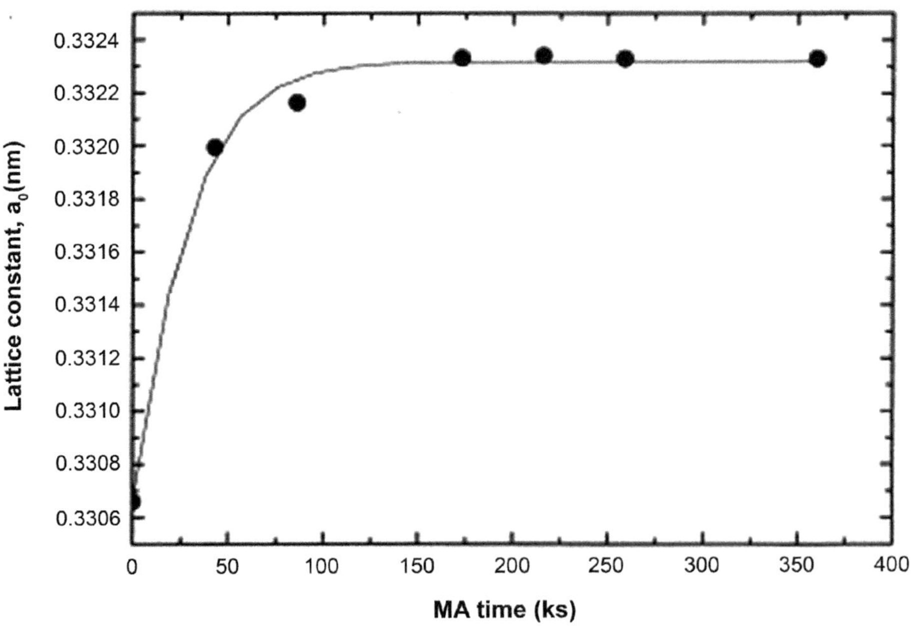

Figure 4.22 Dependence of the lattice constant, *a* for bcc-Nb on MA time during ball milling elemental $Nb_{50}Zr_{10}Al_{10}Ni_{10}Cu_{20}$ powders.[28]

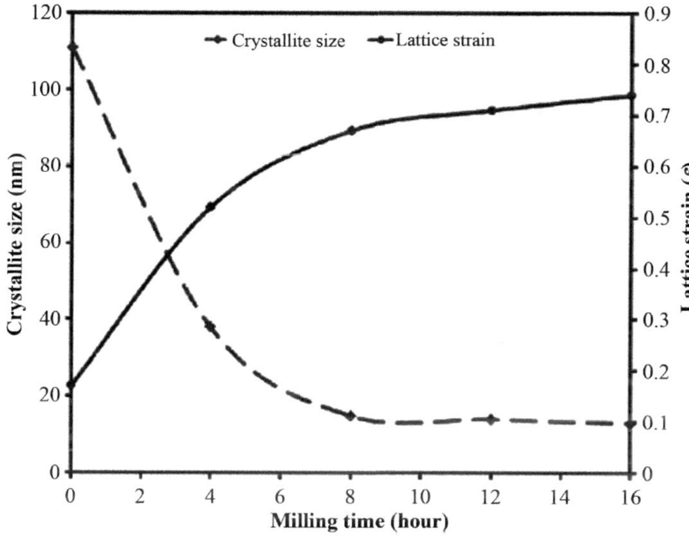

Figure 4.23 Variations of crystallite size and lattice strain of ZnO powder particles versus milling times.[29]

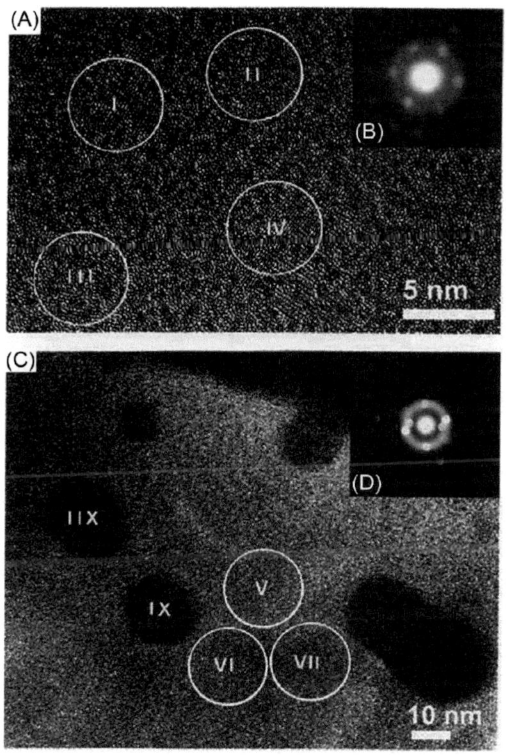

Figure 4.24 HRTEM image (A and C) and the corresponding NBDP (B and D) of mechanically alloyed $Zr_{65}Al_{7.5}Ni_{10}Cu_{12.5}Pd_5$ powders after 259 and 432 ks of ball milling, respectively.[30]

In spite of the melt spinning technique that leads to the formation of amorphous and metallic glassy alloys through a single-stage step, MA offers an excellent opportunity to monitor the stages that the powders passed from its starting long-range ordered to short-range ordered structure. Such amorphization reaction that promoted in a linear fashion with increasing the ball milling time can be followed up with using DSC method, which gives several useful information, such as glass transition temperature (T_g), crystallization temperature (T_x), super cooled liquid region (ΔT_x), and enthalpy change of crystallizations (ΔH_x), as exemplified in Fig. 4.25.

There is no more fundamental method to monitor the morphological changes, which are taken place in the ball-milled powders upon milling for different milling times, more than SEM technique. The metallographic developments of the ball-milled powder particles can be followed up by light optical microscope and SEM techniques that can give primary judgment on the formed phase, as exemplified in Figs. 4.26 and 4.27.

4.1.7 Milling atmosphere

We mean by milling atmosphere used here the gas atmosphere inside the milling vial, where the MA, MD, or MM take place under it. The ball/rod milling process can

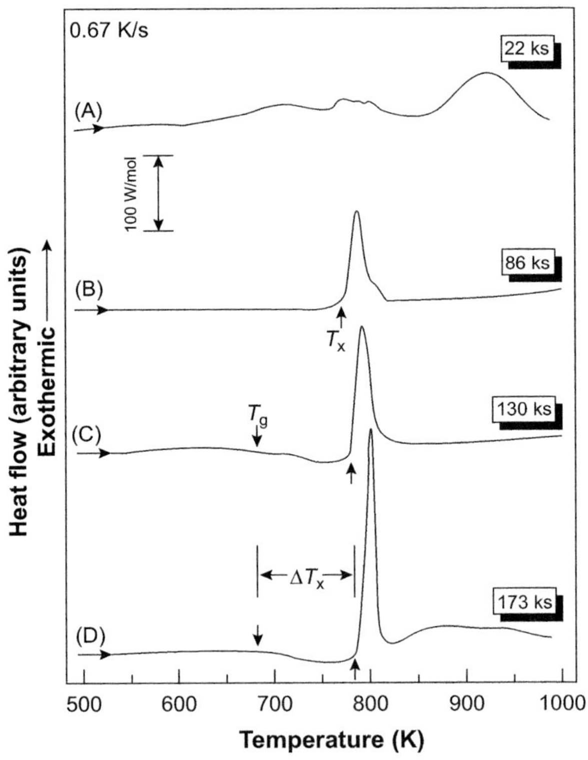

Figure 4.25 DSC thermograms of mechanically alloyed $Zr_{65}Al_{7.5}Ni_{10}Cu_{12.5}Pd_5$ powders after (A) 22 ks, (B) 86 ks, (C) 130 ks, and (D) 173 ks of ball milling.[30]

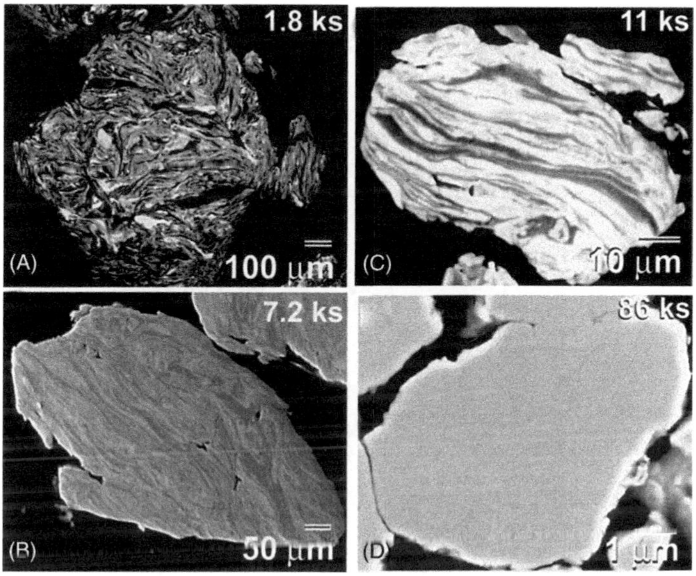

Figure 4.26 SEM micrographs of the cross-sectional view for mechanically alloyed $Co_{75}Ti_{25}$ powders after ball milling for (A) 1.8, (B) 7.2, (C) 11, and (D) 86 ks. [31]

be operated under normal atmosphere (air), vacuum, or under inert gas atmosphere such as argon and helium gas. In fact, the milling atmosphere is considered as one of the most important factors that governs the milling process. The oxygen existed in the air atmosphere is considered to be the most serious problems of MA process since it leads to contaminate the milled powders and changes the properties of the end product. Koch et al.[32] reported that ball milling $Ni_{60}Nb_{40}$ powders in air or under helium atmospheres lead to the formation of an amorphous phase; however, they have different thermal stabilities. It was shown in their experiments that the crystallization temperature of the prepared amorphous alloy in air was lower than that one prepared in helium atmosphere, implying the oxygen role on decreasing the thermal stability of the amorphous alloys.

Consequence of the increased oxygen content: In contrast to this, it has been suggested by Yavari et al.[33] that the amorphization reaction, which takes place in a ball mill between immiscible A-B binary couples with $\Delta H_{mix} < 0$, is attributed to the presence of some oxygen (~5 at.%) and the ternary mixtures are found to be miscible with $\Delta H_{mix} < 0$.

Wang et al.[34] mechanically alloyed elemental $Ni_{60}Ti_{40}$ powders under different milling atmosphere of argon, nitrogen, and oxygen gases. An amorphous phase was obtained when the powders were ball-milled under argon or nitrogen gas atmosphere for 20 h. No distinguishable structural changes could be detected when the two species of the amorphous alloys were further ball-milled in argon and nitrogen gases for 12 h. However, in an oxygen atmosphere, the amorphous phase was crystallized

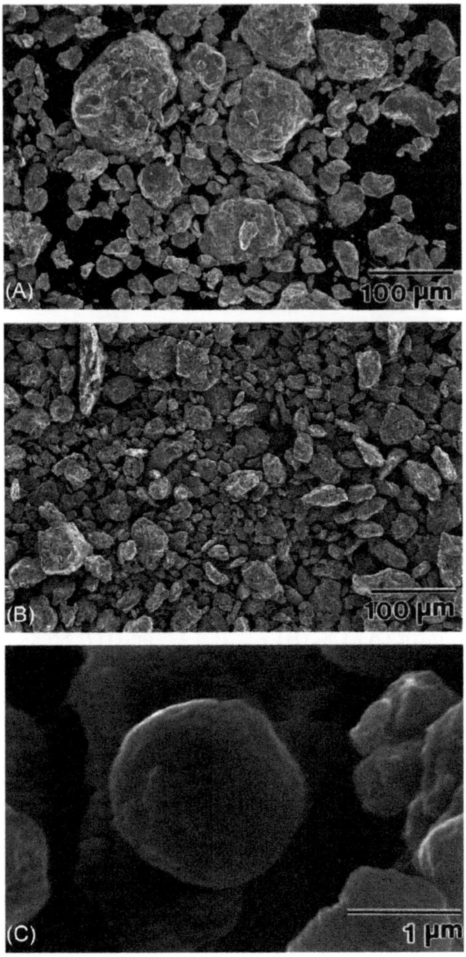

Figure 4.27 SEM micrograph of $Nb_{50}Zr_{10}Al_{10}Ni_{10}Cu_{20}$ powders after (A) 43 ks, (B) 86 ks, and (C) 720 ks of MA.[28]

into intermetallic compounds and was accompanied by the occurrence of titanium oxides in the initial stage. They reported that further milling of the powders in an oxygen atmosphere, the formed intermetallic compounds were decomposed and only the nickel saturation solution and titanium oxides remained due to the heat released during the oxidation.

It has been shown that very fine powders have relatively large surface areas and, thus, are highly reactive not just with oxygen, but also with other gases, such as nitrogen[35] or hydrogen.[36] This so-called reactive ball milling[32] will be discussed later in another chapter of this book.

4.1.8 Milling environment

Milling environment is another important factor that controls the ball milling process and the properties of the end product. The milling environment used here refers to whether the ball milling process takes place in wet or dry mode. Wet milling technique usually employs a lubricant agent or organic surfactants, generally known as PCA in order to overcome the excessive cold-welding effect, which usually happened upon dry milling of metal powder systems, especially for low-melting temperature and ductile metals such as Sn-based, Mg-based, and Al-based powder materials. These PCAs, which are used to controlling the cold-welding effect and to enhance efficiency of the ball milling process,[37] are mostly organic surfactants such as methanol, ethanol, and stearic acid. During the ball milling process, the PCAs are adsorbed on the surface of the milled powders and act as powder particles agglomeration inhibitors. Adding such PCAs to the metal powder charge will modify and lubricate the clean surfaces of the deformed particles created during the ball milling process and hence impedes these new deformed surfaces from the cold welding. Most of the authors, who are suggesting usage of wet milling mode, have added an amount of PCAs in the range between few drops up to 5 wt.%.[38–42]

In general, wet ball milling of solid powders is more intensive than dry milling. The great effectiveness of wet milling is attributed to the saturation of unsaturated bonds of the powder surfaces by the PCAs used. Such unsaturated bonds are developed during the disintegration of the powder particles during the ball milling process. Thus, the presence of the PCAs layers between the grains can promote the powder disintegrations.

It has been shown by Shaw et al.[41] that adding stearic acid and methanol as PCAs can effectively prevent excessive cold welding of Al particles during ball milling of elemental $Al_{93}Fe_3Ti_2Cr_2$ powders. In spite of the obvious advantages obtained with using PCAs, they also notified that using such PCAs, especially with high concentration, led to decreases the rate of the grain-size refinement (Fig. 4.28), the lattice-stain increment (Fig. 4.29), and the formation of Al-based solid solutions (Fig. 4.30).[41] On metallographic point of view, it is shown that the microstructural evolution of Al alloy powder particles is affected by the addition of PCAs (stearic acid and methanol) so that concentration of the PCA added to the ball-milled $Al_{93}Fe_3Ti_2Cr_2$ powders, the more Fe, Ti, and Cr particles remain undissolved, as indicated by the unprocessed elemental metal particles in Fig. 4.31.[41]

4.1.9 Ball-to-powder weight ratio

Whereas most of the ball-milled systems usually prepared with using ball-to-powder weight ratio (W_b:W_p) in the range between 10:1 and 20:1, the effect W_b:W_p on the amorphization reaction of $Al_{50}Ta_{50}$ alloy powders in a low-energy ball mill was investigated in 1991 by El-Eskandarany et al.[42] They have used 90, 30, 20, 10, and 3 g of powders to obtain W_b:W_p ratios of 12:1, 36:1, 54:1, 108:1, and 324:1, respectively.

The XRD patterns of mechanically alloyed $Al_{50}Ta_{50}$ powders as ball-milled for 1440 ks (400 h) as a function of the W_b:W_p ratio is presented in Fig. 4.32. Single

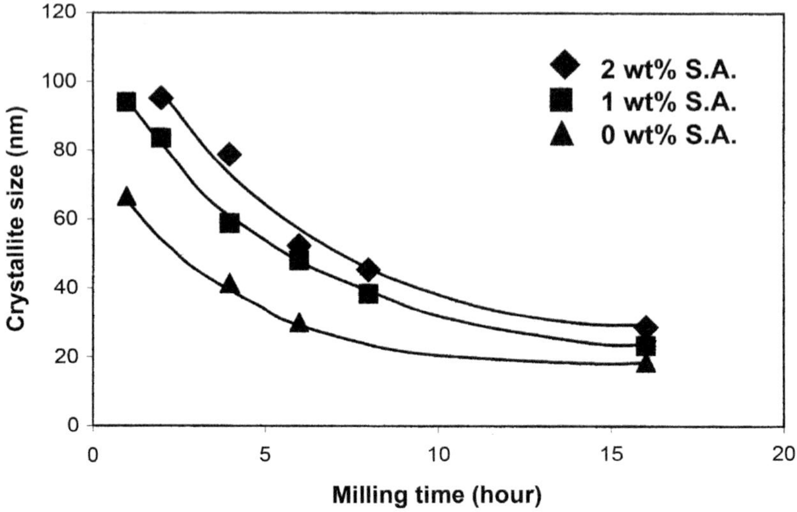

Figure 4.28 Calculated crystallite sizes of fcc-Al in the Al alloy powder mixtures with and without stearic acid as a function of milling time.[41]

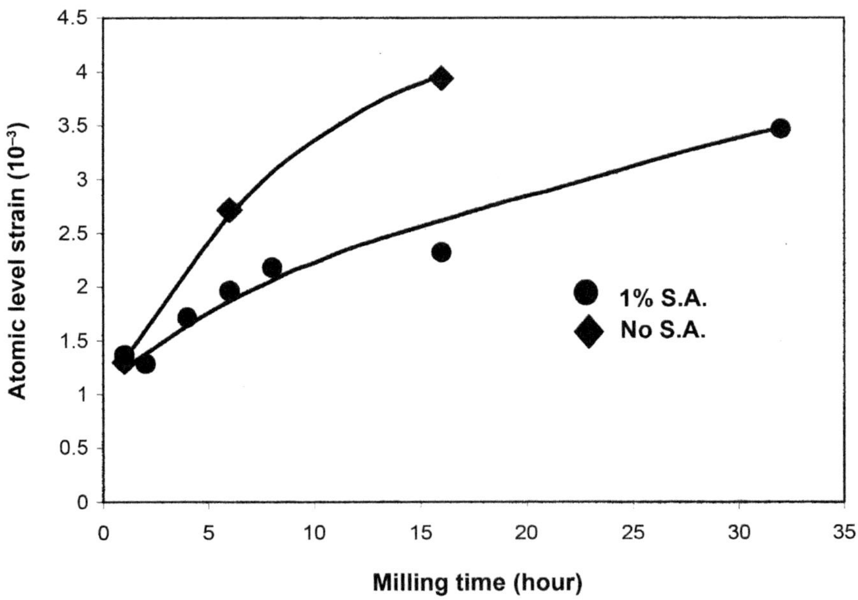

Figure 4.29 The atomic level strain of fcc-Al in the Al alloy powder mixtures without and with 1 wt pct stearic acid as a function of milling time.[41]

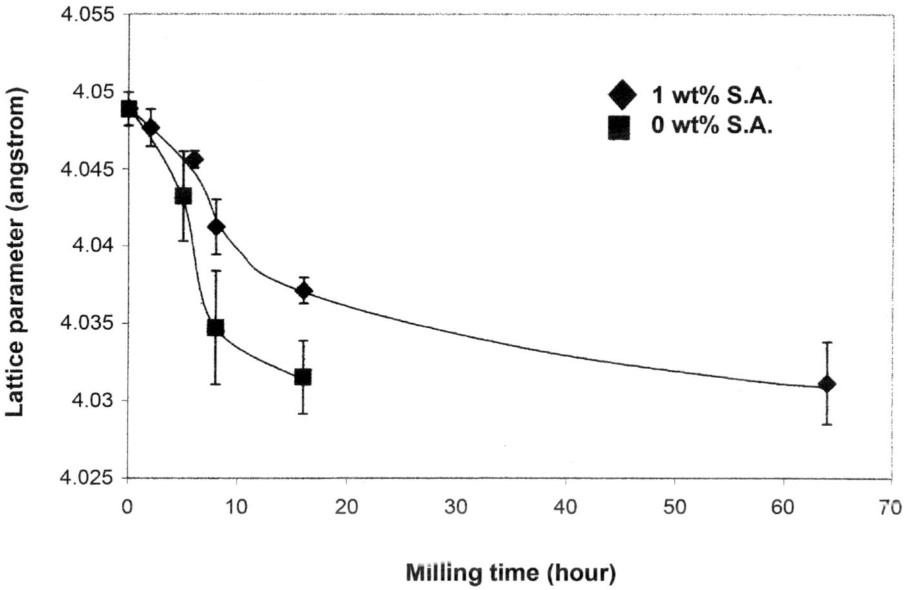

Figure 4.30 The lattice parameter of fcc-Al in the Al alloy powder mixtures without and with 1 wt pct stearic acid as a function of milling time.[41]

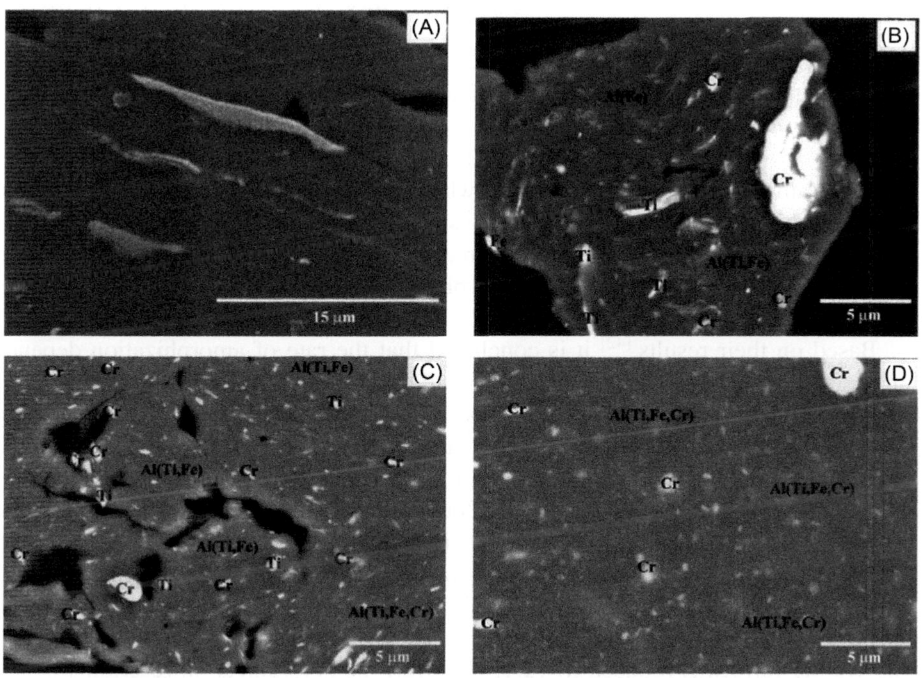

Figure 4.31 SEM backscatter electron images of the microstructure of Al alloys with 1 wt pct stearic acid milled for (A) 2 h, (B) 8 h, (C) 16 h, and (D) 64 h. The letters in the micrographs indicate the major composition of that particular location detected using EDS.[41]

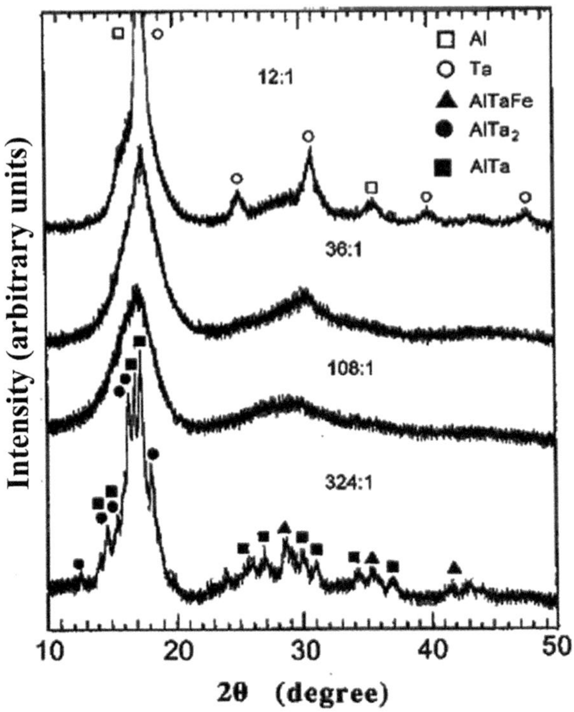

Figure 4.32 The X-ray diffraction patterns (XRD) of $Al_{50}Ta_{50}$ alloy powders as ball-milled for 1440 ks (400 h) as a function of the ball-to-powder weight ratio, W_b:W_p. *(After El-Eskandarany et al.)*[42]

phase of amorphous alloys is obtained when ratios 36:1 and 108:1 were used. The Bragg peaks of elemental Al and Ta crystals still appear when the W_b:W_p ratio is 12:1, indicating that the amorphization reaction is not completed. In contrast, when the W_b:W_p ratio is 324:1, the amorphous phase coexists with the crystalline phases of AlTa, $AlTa_2$, and AlTaFe.

Based on their results,[42] it is concluded that the rate of amorphization depends strongly on the kinetic energy of the ball mill charge and this depends on the number of opportunities for the powder particles to be reacted and interdiffused. Increasing the W_b:W_p ratio accelerates the rate of amorphization, which is explained by the increase in the kinetic energy of the ball mill charge per unit mass of powders. It has been shown in this study that the volume fraction of the amorphous phase in the mechanically alloyed ball-milled powders increases during the early stage of milling, 86–173 ks (48 h) with increasing W_b:W_p ratio. It is noted that further increasing this weight ratio leads to the formation of crystalline phases and this might be related to the high kinetic energy of the ball mill charge which is transformed into heat. When the W_b:W_p ratio was reduced to 12:1, however, the amorphization reaction was not completed. This indicates that the kinetic energy of the mill charge is insufficient for complete transition from the crystalline to the amorphous phase.

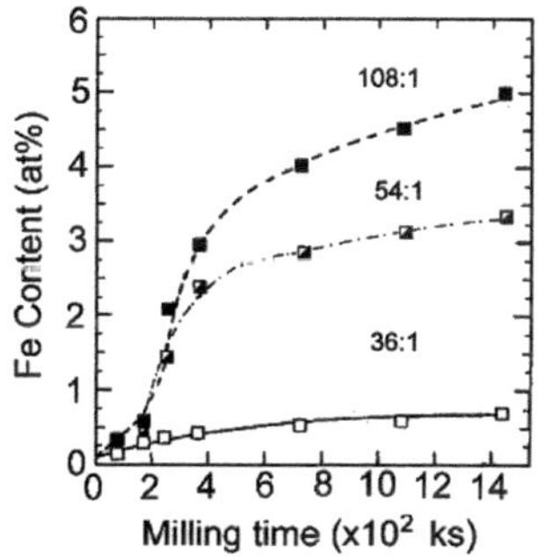

Figure 4.33 Iron contamination content in mechanically alloyed $Al_{50}Ta_{50}$ powders as a function of the ball milling time and ball-to-powder weight ratio, W_b:W_p. *(After El-Eskandarany et al.)*[42]

It is worth noting that powder particles reached the minimum of extreme fineness when using a high W_b:W_p ratio. One disadvantage of using such a high weight ratio is being the high concentration of iron contamination which is introduced to the milled powders during the MA process, as presented in Fig. 4.33.

Romankova et al.[43] have applied the vibration ball milling for coating of stainless steel balls during milling of Ti–Al powders. They examined metallographically the development of the Ti–Al coating structure after milling for 60 min as a function of the ball-to-powder weight ratio for 6 mm balls (Fig. 4.34).

The results showed that the milling energy increased with increasing the number of balls. When the weight ratio was 3:1, the substrate could be covered with a thin Al

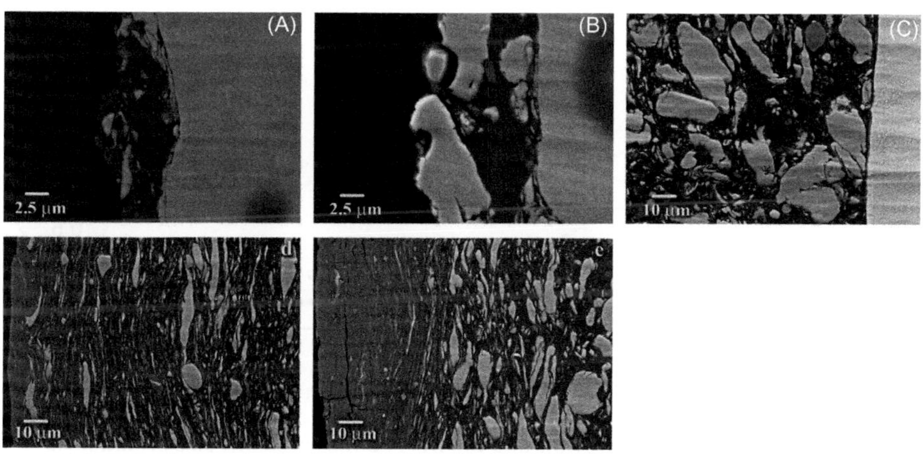

Figure 4.34 Formation of Ti–Al coatings as a function of the ball-to-powder weight ratio W_b:W_p for 6 mm balls: (A) 3:1, (B) 4:1, (C) 6:1, (D) 8:1, and (E) 14:1.[43]

layer (Fig. 4.34A). For this case, only small Ti particles were embedded into the Al matrix. It should be noted that the substrate underwent plastic deformation under the ball impacts and its surface became slightly bent. When the weight ratio was increased to 4:1, the energy was sufficient to embed larger Ti particles in the Al layer than at ratio 3:1 (Fig. 4.34B). Al bound these Ti particles to the substrate. They notified that, at the 4:1 ratio, the growth for the Ti–Al coating across the substrate was clustered; this resulted in a hillock-like morphology and increased the surface roughness. Upon further increasing the ball-to-powder weight ratio from 6:1 to 14:1, the coating roughness gradually decreased. They also reported that the lamellar structure was refined when the ball-to-powder weight ratio was 14:1, as presented in Fig. 4.34E.

More recently, Waje et al.[44] have studied the effect of the ball-to-powder weight ratio (BPR) on the crystallite size of ball-milled $CoFe_2O_4$ nanoparticles, using XRD (Fig. 4.35). From their results it can be seen that the particle size decreases linearly from 15.3 to 11.4 nm when used BPR of 8:1 and 30:1, respectively.

4.1.10 Milling temperature

The ball milling temperature of both MA and MD is very important parameter that affects the processing and the structure of the end product. Considering the schematic

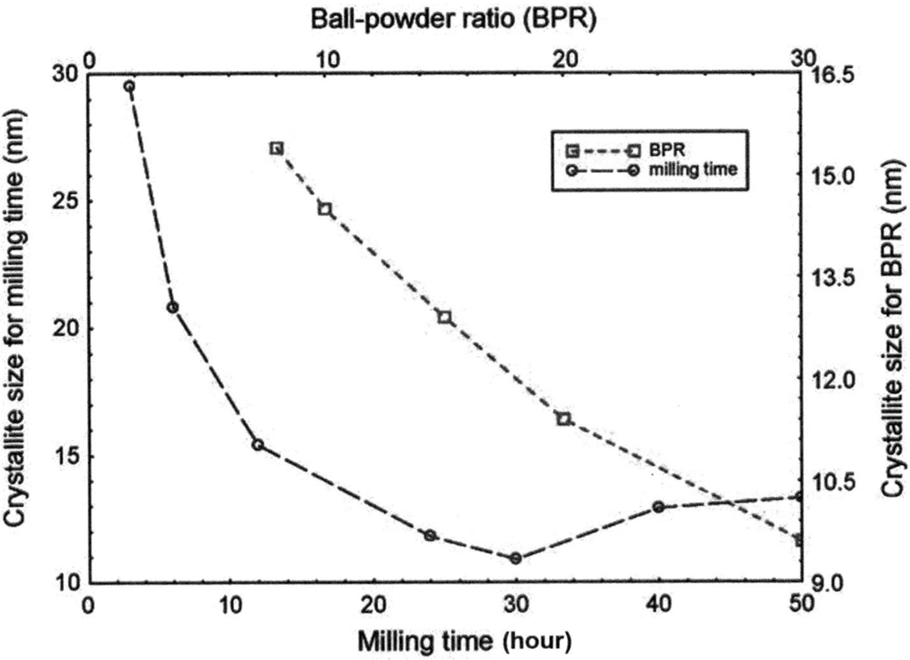

Figure 4.35 Patterns of milling time and the BPR on the crystallite size evolution of $CoFe_2O_4$.[44]

presentation displayed in the last chapter (Fig. 3.9) one can realize that the MA via ball milling is a mixture of elemental A and B powders (point 1 in the aforementioned figure) processed at ambient temperature that can be enhanced if the powder is milled at higher temperatures. This is because we need to increase the negative value of the free energy to obtain the end product of the reactant phase (e.g., an amorphous AB, point 2). On other words, increasing the milling temperature during MA should assist the MISSR between the diffusion couples, A and B. Accordingly, the MISSR kinetics would be fast, leading to the formation of the end product in a shorter ball milling time. Contrary to this, when we start ball milling with an intermetallic compound-AB, for example (point 3 in Fig. 2.9), this requires to reduce the free energy from large negatives value to less values (MD process) in order to get a metastable phase without any compositional changes, like an amorphous phase that is denoted in Fig. 3.9 by point 2. Thus, ball milling at high temperature would not be helpful in this case since it stabilizes the existence of AB compound, where an amorphous AB can hardly be obtained. This means that ball milling of an intermetallic compound at lower temperatures (using liquid nitrogen, e.g.) will assist the long-range ordered (intermetallic phase) to short-range ordered phase transformations. We should emphasize that metastable phases are obtained if the competing stable phases are suppressed either by the milling action or by the interdiffusion pathway. In particular, intermetallic phases are often destabilized by disordering upon milling with respect to disordered solid solutions or amorphous phases.

The effect of milling temperature on the amorphization of mechanically alloyed $Cu_{30}Ta_{70}$ has been investigated in 1991 by Lee et al.[45] In their experiments, they ball-milled the elemental powders of $Cu_{30}Ta_{70}$ at two different controlled temperatures of 37°C and 200°C for 100 h under the same milling conditions. They realized that ball milling at 200°C leads to the formation of an amorphous phase coexisted with lower volume fractions of unprocessed bcc-Ta crystals (Fig. 4.36). This implies the role of

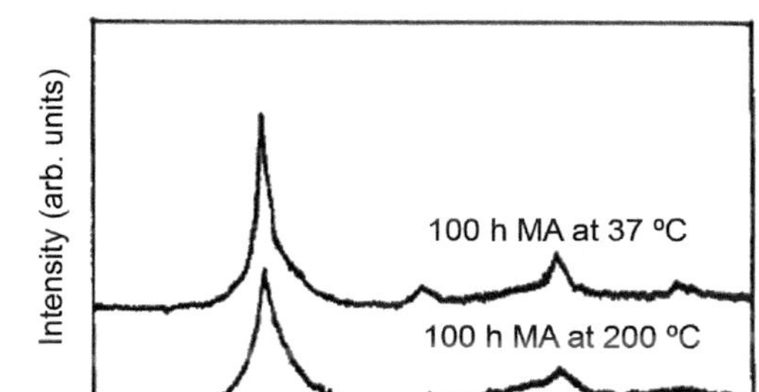

Figure 4.36 X-ray diffraction spectra for a mixture of $Cu_{30}Ta_{70}$ powders subjected to MA for 100 h.[45]

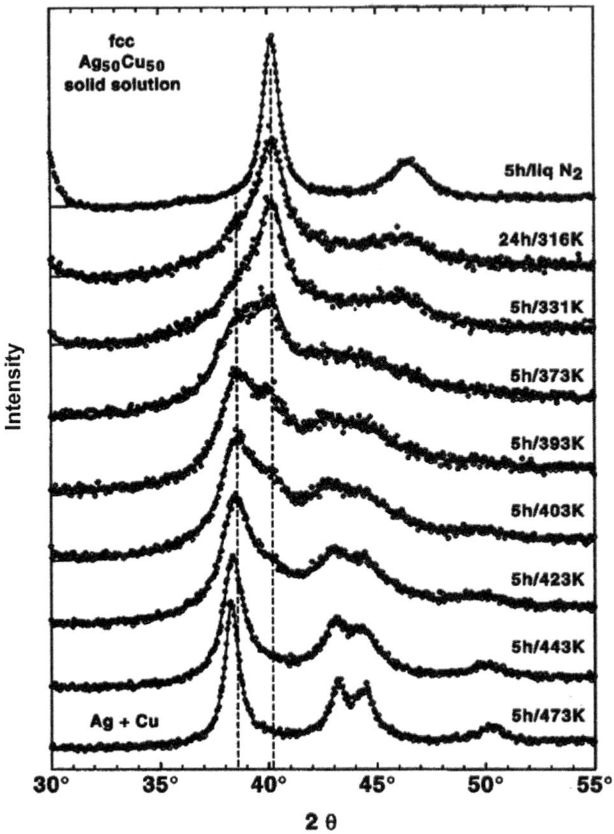

Figure 4.37 The XRD patterns of steady states obtained after milling an $Ag_{50}Cu_{50}$ at different temperatures. The precursor alloy was produced by ball milling the elemental powder blend for 24 h at 316 K.[46]

milling temperature on enhancing the solid-state reaction of the diffusion couples: Cu and Ta.

Another excellent example for studying the effect of the milling temperature on the formation of metastable phase of Ag-Cu system, which has positive heat of mixing, is given by Klassen et al.[46] when they mechanically alloyed $Ag_{50}Cu_{50}$ alloy powders in a ball mill at different temperature, ranged from 85 to 473 K. They investigated that milling at subambient temperatures led to the formation of homogeneous fcc solid solution, while milling at 473 K led to a fully decomposed, two-phase mixture. Partially decomposed systems were obtained during milling at intermediate temperatures with three coexisting phases all present (Fig. 4.37). The milling-induced decomposition process had a very low apparent activation enthalpy, =0.11 eV, which is attributed to a first-order kinetic process involving nonequilibrium vacancies.[46]

References

1. Campbell, S. J., and Kaczmarek, W. A., *Mössbauer Spectroscopy Applied to Materials and Magnetism*, (G. J. Long and F. Grandjean, eds.) Plenum Press, New York, 2:273 (1989)
2. El-Eskandarany, M. S., *Solid State Amorphization of Ta-Based Alloy Powders by Mechanical Alloying*, Ph.D. Thesis, Tohoku University (1992)
3. El-Eskandarany, M. S., *Mechanical Alloying for Fabrication of Advanced Engineering Materials*, 1st Edition, William Andrew Publishing, Inc, New York 13815, the U.S.A, (2001)
4. Martinez-Sanchez, R., Estrada-Guel, I., Jaramillo-Vigueras, D., De la Torre, S. D., Gaona-Tiburcio, C., and Guerrero-Paz, *J. Mater. Sci. For.*, 386–388:135 (2002)
5. Harringa, J. L., Cook, B. A., and Beaudry, B. J. *J. Mater. Sci.* 27:801 (1992)
6. Frazier, W. E., Koczak, M. J. *Scripta Metall.* 21:129 (1987).
7. Suryanarayana, C., Mechanical alloying. In ASM Handbook, Vol. 7. ASM International Publishers, 1998, pp. 80 –90.
8. El-Eskandarany, M. S., Aoki, K., and Suzuki, K., J. Less-Common Met., 167:113 (1990)
9. Smeulders, D.E., et al., J. Materials Science, 40:655 (2005)
10. Cook, T.M., and Courtney, T.H., *Metallurgical Transactions*, 26A:2389 (1995)
11. Vaezi, M.R. et al, *Journal of Theoretical and Applied Physics*, 6:29 (2012)
12. Shi, F., *Minerals Engineering*, 17:1259 (2004)
13. Suryanarayana, C., Mechanical Alloying and Milling, Marcel Dekker, New York, the U.S.A, (2004)
14. C. Suryanarayana, *Prog. Mater. Sci.* 46:1 (2001)
15. Wills, Barry A., and Napier-Munn, Tim, Will's Mineral Processing Technology: An Introduction to the Practical Aspects of Ore Treatment and Mineral, 7th, Edition, Elsevier Science & Technology Books (2006). Available from: https://www.elsevier.com/books/wills-mineral-processing-technology/wills/978-0-7506-4450-1
16. Powell, M.S., and Nurick, G.N., *Minerals Engineering*, 9:343(1996)
17. Abdellaoui, M, and Gaffet, E, *Acta. Metall. Mater.*, 43:1087 (1995)
18. Abdellaoui, M, and Gaffet, E, *J. Alloys Compd.*, 209:351 (1994)
19. Mioa, H., Kanoa, J., Saitoa, F., Kaneko, K., *Int. J. Miner. Process.*, 74S: 85 (2004)
20. Calka, A. and Kaczmarek, W.A., *Scripta METALLURGICA et MATERIALIA*, 26:249 (1992)
21. El-Eskandarany, M. Sherif, Aoki, K., Sumiyama, K., Suzuki, K., *Metallurgical and Materials Transactions A*, 30A:1877 (2002)
22. El-Eskandarany, M. Sherif, Aoki, K., Sumiyama, K. and Suzuki, K., *Appl. Phys. Let.*, 70:1679 (1997)
23. El-Eskandarany, M. Sherif, Aoki, K., Sumiyama, K., and Suzuki, K., *Scripta Metall.*, 36:1001 (1997)
24. El-Eskandarany, M. Sherif, Aoki, K., Sumiyama, K., and Suzuki, K., *Acta Mater.*, 50:1113 (2002)
25. Bab, M. A., Mendoza-Zelis, L., and Damonte, L. C., *Acta Mater.* 49:4205 (2001) 49:4205
26. Rochman, N. T., Kuramoto,S., Fujimoto, R., and Sueyoshi, H., *Journal of Materials Processing Technology*, 138:41 (2003)
27. Ebrahimi-Purkani, A., and Kashani-Bozorg, S.F., *J. Alloys Compd.*, 456:211 (2008)
28. El-Eskandarany, M. Sherif, Matsushita, M., and Inoue, A., *J. of Alloys Comp.* 329:239 (2001)
29. Amirkhanlou, S., Ketabchi, M., and Parvin, N., *Materials Letters*, 86:122 (2012)

30. El-Eskandarany, M. Sherif, Saida, J., and Inoue, A., *Acta mater.* 51:1481 (2003)
31. El-Eskandarany, M. Sherif, Zhang, W, and Inoue, A., *J. Alloys and Compounds*, 350:232 (2003)
32. Koch, C. C., Cavin, O. B., McKamey, C. G., and Scarbrough, J. O., *Appl. Phys. Lett.*, 43:1017 (1983)
33. Yavari, A. R., and Desré, P. J., *Mater. Sci. Forum*, 88–90:43 (1992)
34. Wang, K.Y., Shen, T.D., Wang, J.T., and Quan, M.X., Journal of Materials Science, 28:6474 (1993)
35. El-Eskandarany, M. Sherif, Aoki, K., Sumiyama, K., and Suzuki, K., *Mater. Sci. Forum*, 88:801 (1992)
36. El-Eskandarany, M. Sherif, Ahmed, H.A., Ahmed, Sumiyama, K., and Suzuki, K., *J. Alloys Compounds*, 218:36 (1995)
37. Gilman, P.S., and Nix, W.D., *Met. Trans.*, 12A:813 (1981)
38. Lu, L., and Zhang, Y.F., *J. Alloys Compd.*, 290: 279 (1999)
39. Rocha, C.J., and Leal Neto, R.M., *Mater. Sci. Forum*, 416–418:144 (2003)
40. Huang, H.S., Lin, Y.C., and Hwang, K.S., *Int. J. Refract. Met. Hard Mater.*, 20:175 (2002)
41. Shaw, L., Zawrah, M., Villegas, J., Luo, H. and Miracle, D., *Metall. Mater. Trans. A*, 34A: 159 (2003)
42. El-Eskandarany, M. S., Aoki, K., Itoh, H., and Suzuki, K., *J. Less-Common Met.*, 169:235 (1991)
43. Romankova, S., Kaloshkinb, S.D., Hayasakac, Y., Hayashia, N., Kasaia, E., and Komarov, S.V., *J. Alloys Compd.*, 484:665 (2009)
44. Waje, Samaila Bawa, Hashim, Mansor, Yusoff, Wan Daud Wan, and Abbas, Zulkifly, *Applied Surface Science*, 256:3122 (2010)
45. Lee, C.H., Fukunage, T., and Mizutani, U., *Mat. Sci. and Eng.*, A134:1334 (1991)
46. T. Klassen, U. Herr, R.S. Averback, Acta Mater. 45 (1997) 2921

Ball milling as a superior nanotechnological fabrication's tool

5

5.1 Introduction

5.1.1 Types of nanomaterials

In general, nanomaterials may be classified into three categories, namely (1) one-dimensional (1D), (2) two-dimensional (2D), and (3) three-dimensional (3D) nanomaterials. In the 1D nanomaterials, such as thin films and layer coating, one of their dimensions (usually the thickness) should be in the nanoscale level (1–100 nm). Nanotubes, nanorods, and nanowires are considered as typical examples of the second group of nanomaterials (2D), in which two of their dimensions (usually the length and diameter) should be less than 100 nm. In the last group (3D nanomaterials) including nanoparticles, nanopowders, nanocrystalline powders, quantum dots, Bucky balls, and nanocrystals, the materials have nanoscale in the 3D aspects.

5.1.2 Methods for preparing nanomaterials

Almost all the categories of nanomaterials mentioned earlier can be prepared by two nanofabrication methods, known as (1) top-down, and (2) bottom-up.

A top-down approach refers to the nanofabrication technique that is controlled by external experimental parameters to produce nanoscaled structures/functional devices with the desired shapes and characteristics starting from larger dimensions and reducing them to the required values.[1] Rolling, atomization, electrospinning, evaporation, laser ablation, RF sputtering, and high-energy sonication are some typical examples of nanofabrication methods used in top-down approach. Need not to be mentioned, ball milling has been considered as a powerful cost-effectiveness tool for fabrications of wide varieties of nanoscaled powders via top-down approach.

In contrast to the top-down method, bottom-up approach aims to have molecular or atomic components built up into more complex nanoscale assemblies or directed self-assemblies based on complex mechanisms and technologies.[2] Chemical vapor deposition (CVD), atomic layer deposition (ALD), molecular beam epitaxy (MBE), gas phase condensation thermolysis, and nanolithography: dip-pen methods (DPNL) are few examples of those techniques used to fabricate nanomaterials via bottom-up approach Fig. 5.1.

Mechanical Alloying. http://dx.doi.org/10.1016/B978-0-12-818180-5.00005-4

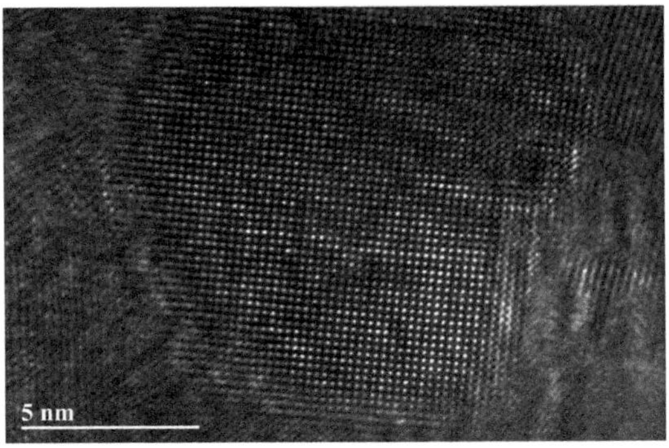

Figure 5.1 A micrograph of atomic resolution transmission electron microscope for a single grain of ß-MgH_2 obtained after reactive ball milling of pure magnesium under hydrogen gas atmosphere for 200h. *(The photo was taken by the author and has not been published yet.)*

5.2 Nanocrystalline materials

The design of nanocrystalline solids with novel properties different from those coarse-grained counterparts is the most fruitful contribution of ball milling method. The term *nanocrystalline materials* (also known as nanostructured or nanophase materials)[3,4] is used to describe those materials that have a majority of grain diameters in the typical range from ~1 to 100 nm.[5] Whether it can be called a revolution or simply evolution, the nanocrystalline materials have received much attention as advanced engineering materials with unique physical and mechanical properties. The mechanical properties of the nanocrystalline materials at room temperature are always superior (i.e., higher strength and toughness) to those of coarse-grained ones. Intensive research on these materials has been performed since the pioneering investigations of Gleiter.[1,6]

Like any fabricated products, the properties of as-prepared nanocrystalline materials via ball milling should be investigated using different characterization techniques. The techniques usually used to characterize the ball-milled nanocrystalline materials can be classified according to the following properties:

- Crystal structure (Table 5.1)
- Thermal stability and phase transformations (Table 5.2)
- Morphology, topological, and surface properties (Table 5.3 and Table 5.4)
- Mechanical properties (Table 5.5)

5.2.1 Influence of nanocrystallinity on mechanical properties: strengthening by grain size reduction

One of the very basic results of the physical chemistry of solids is the insight that most properties of solids depend on the microstructure. The size of the grains (average grain diameter) in a polycrystalline material has pronounced effects on many of its physical and mechanical properties.[7] A fine-grained material, or the so-called nanostructured

Table 5.1 Characterization techniques used to investigate the crystal structure and nanoelemental composition of nanocrystalline powders and bulk materials.

Technique	Analytical value	Example
X-ray diffraction (XRD)	Average crystal structure	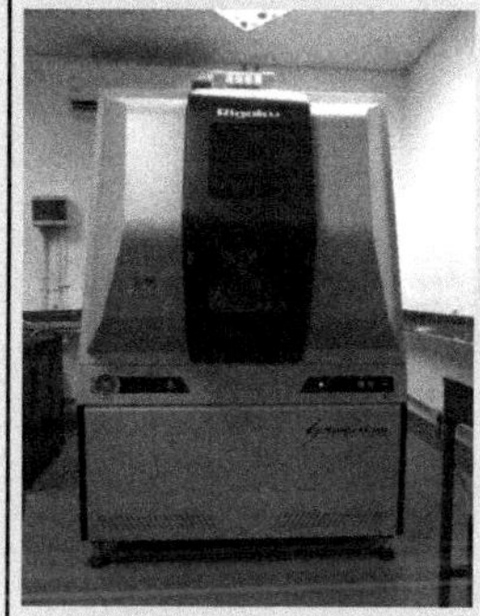 9 kW-SmartLab-XRD/SAXS (Rigaku-Japan) equipment housed in the Nanotechnology Laboratory, Energy and Building Research Center, Kuwait Institute for Scientific Research.
Small angle X-ray scattering (SAXS)	Surface analysis/particle sizing (1–100 nm)	
Field emission high-resolution transmission electron microscope/energy dispersive X-ray spectroscopy (FE-HRTEM/EDS)	Investigate the local structure (~5 nm), particle and grain shapes-sizes and elemental analysis	 200 kV FE-HRTEM/EDS/STEM (JEOL-Japan) equipment housed in the Nanotechnology Laboratory, Energy and Building Research Center, Kuwait Institute for Scientific Research.
Scanning transmission electron microscope (STEM)	Particle shape and size in nano-level. When it is equipped with TEM/EDS system, it can be employed to investigate the local composition of the sample and to design elemental X-ray dot mapping.	

Table 5.2 Characterization techniques used to investigate the thermal stability and the phase transformation temperatures of nanocrystalline powder and bulk materials.

Technique	Analytical value	Example
• Differential scanning calorimeter (DSC) • Thermogravimetry/differential thermal analysis (TG/DTA) • Thermal mechanical analysis (TMA)	• DSC/DTA is used to investigate the thermal stability, glass transition temperature (T_g), crystallization (T_x), melting (T_m), and phase transformation temperatures • TG is used to investigate the mass loss versus temperatures • TMA is employed to investigate the mechanical properties of bulk materials versus temperatures	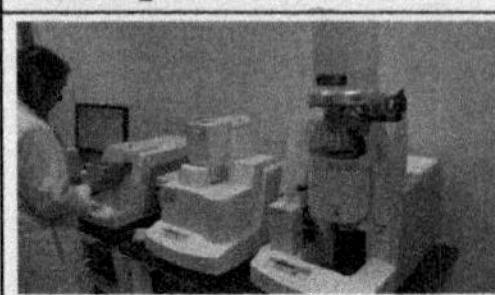 DSC/TG/DTA/TMA unit (Shimadzu Thermal Analysis System) housed in the Nanotechnology Laboratory, Energy and Building Research Center, Kuwait Institute for Scientific Research.

Table 5.3 Techniques used to investigate the morphological, topological, and surface characterizations of nanocrystalline powder and bulk materials.

Technique	Analytical value	Example
Field emission scanning electron microscope/EDS (FE-SEM/EDS) Electron Probe Micro-Analyzer (EPMA)	FESEM/EDS and EMPA are used to investigate the morphology (e.g., grain sizes and shapes) and topology of nanocrystalline powders and bulk materials. They also employed to investigate the elemental compositions.	 FESEM/EDS unit (JSM-7800F, JEOL-Japan) housed in the Nanotechnology Laboratory, Energy and Building Research Center, Kuwait Institute for Scientific Research.
Atomic force microscope (AFM)	To investigate the topology and surface structure of nanopowders and nanocrystalline bulk materials.	 AFM system (Agilent 5600LS, USA) housed in the Nanotechnology Laboratory, Energy and Building Research Center, Kuwait Institute for Scientific Research.

Table 5.3 Techniques used to investigate the morphological, topological, and surface characterizations of nanocrystalline powder and bulk materials. (*Cont.*)

Technique	Analytical value	Example
Fourier transform-infrared spectroscopy (FT-IR)	Surface characterizations of nanomaterials.	 FT-IR (FTIR-8400S, Shimadzu, Japan) equipment housed in the Nanotechnology Laboratory, Energy and Building Research Center, Kuwait Institute for Scientific Research.

Table 5.4 Techniques used to investigate the average particle size, surface area, and pore size distributions of nanocrystalline powders.

Technique	Analytical value	Example
Laser Diffraction Particle Size Analyzer	For check and measure the average particle size distributions of spherical and nonspherical micro- and nanoparticles.	 The Microtrac S3500-Laser Diffraction Particle Size Analyzer (USA) housed in the Nanotechnology Laboratory, Energy and Building Research Center, Kuwait Institute for Scientific Research.
Brunauer, Emmett, and Teller Approach (BET)	BET method is used to investigate the surface area and pore size distribution of nanoparticles and nanopowders.	 BET (BELSORP-max, BEL-Japan) equipment housed in the Nanotechnology Laboratory, Energy and Building Research Center, Kuwait Institute for Scientific Research.

Table 5.5 Techniques used to investigate the mechanical properties of nanocrystalline bulk materials.

Technique	Analytical value	Example
Nanoindentation	Employed to investigate the hardness, nanoscratch, and elastic modulus of the nanocrystalline bulk materials	 The nanoindentation equipment (Bruker, Germany) housed in the Nanotechnology Laboratory, Energy and Building Research Center, Kuwait Institute for Scientific Research.
Pin-on-disk	To investigate the wear resistance of nanocrystalline bulk materials at different forces, sliding speed, and temperature.	 Pin-on-disk machine (Microtest, Spain) housed in the Nanotechnology Laboratory, Energy and Building Research Center, Kuwait Institute for Scientific Research.

material, is harder and stronger than one that is coarse-grained (Fig. 5.2), since the former has a greater total grain boundary area to impede dislocation motion. For many materials the yield strength, σ_y, varies with grain size according to:

$$\sigma_y = \sigma_0 + k_y d^{-1/2} \tag{5.1}$$

In this expression, termed the Hall–Petch equation, d is the average grain diameter, and σ_0 and k_y are constants for a particular material.

The information for the average grain diameters can be primarily obtained from TEM and/or HRTEM observations, and X-ray line broadening.

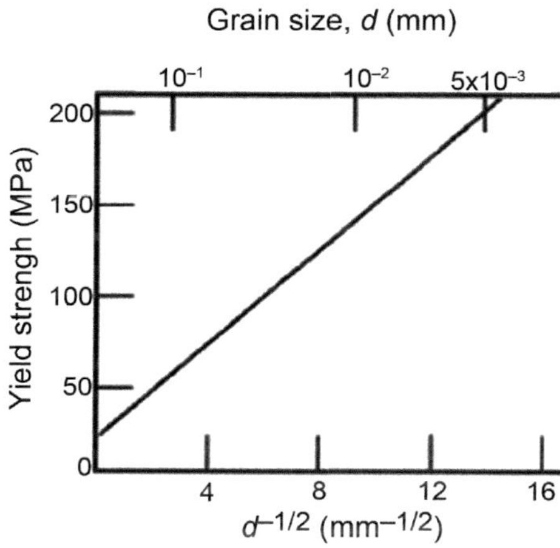

Figure 5.2 The influence of grain size on the yield strength of Cu-30% Zn brass alloy. *(After Suzuki.)*[7]

5.3 Formation of nanocrystalline materials by ball milling technique

Nanocrystalline materials can be successfully synthesized by several techniques, including inert gas condensation,[8] rapid solidification,[9] electrodeposition,[10] sputtering,[11] crystallization of amorphous phases,[12] and chemical processing[13] (Fig. 5.3).

Among the different options for preparations, the mechanical alloying method has been considered the most powerful tool for nanostructured materials[14,15] because of its simplicity, relatively inexpensive equipment, and the possibility of producing large quantities that can be scaled up to several tons.[16] The formation of nanocrystalline materials during MA of ceramics or metallic powders is attributed to the intense cold working on the ball-milled powders. This leads to a dramatic increase in the number of imperfections (e.g., point and lattice defects) that leads to decreasing the thermodynamic stability of the starting materials. Based on the type of defects applied, different kinds of nanocrystalline materials, with different physical and mechanical properties, can be obtained.[14,17]

It is believed that the reduction in grain size during MA takes place similar to that suggested by the model for nanocrystalline materials fabricated by gas condensation.[8] It is worth noting that the enthalpy stored via high-energy ball milling is far above than that for the conventional cold working technique.[18] For example, the enthalpy stored through cold welding of metals and alloys does not exceed 2 kJ/mol and is only a small fraction of the heat of fusion, ΔH_f.[19] In the MA method, however, the enthalpy is larger and can reach higher values of crystallization enthalpies[19] of a metallic glass, as high as 0.4 times the heat of fusion. We should emphasize that such enthalpy storage, in the form of lattice and point defects, cannot be achieved in traditionally processed materials. Hence, the grain boundary energy of milled

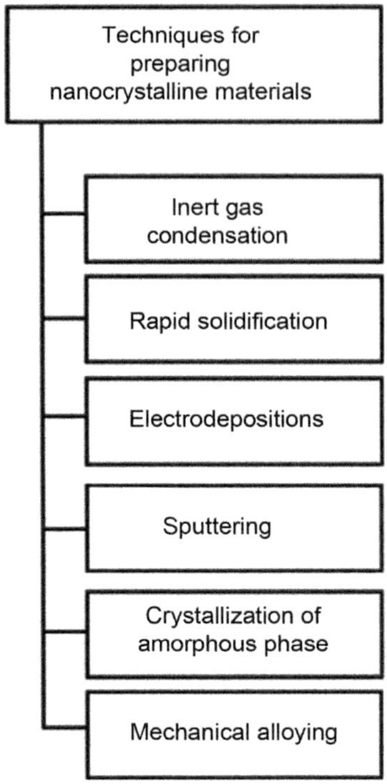

Figure 5.3 Nanocrystalline materials are a unique class of the advanced engineering materials that can be prepared by several methods.

nanocrystalline powders is larger than the grain boundary energy of fully equilibrated grain boundary.[20]

5.3.1 Mechanism(s)

The mechanism for formation of nanocrystalline materials by the ball milling technique has been proposed by Fecht,[21] who summarized the phenomenology of the grain size reduction into three stages.

5.3.1.1 First stage

Plastic deformation is produced in the crystal lattices of the ball-milled powders by slip and twinning. This deformation is localized in shear bands containing a high dense network of dislocation. The local shear instability of a crystal lattice can be triggered by material heterogeneity and enhances instabilities.[18] It is worth noting that these instabilities result from a nonuniform heat transfer during the mechanically induced deformation of the milled powders. During this stage of milling (early stage), the atomic-level strain increases as a result of increasing the dislocation density.

5.3.1.2 Second stage

Due to the successive accumulation of the dislocation density, the crystals are disintegrated into subgrains that are initially separated by low-angle grain boundaries. The formation of these subgrains is attributed to the decrease of the atomic-level strain.

5.3.1.3 Third stage

Further ball milling time leads to further deformation occurring in the shear bands located in the unstrained parts of the powders which lead to subgrain size reduction so that the orientation of final grains becomes random in crystallographic orientations of the numerous grains, and hence the direction of slip varies from one grain to another.

5.4 Selected examples

5.4.1 Formation of nanocrystalline Ni_xMo_{100-x} (x = 60 and 85 at.%)

Trudeau et al.[22] have carried out a high-resolution electron microscope (HRTEM) study of mechanically alloyed Ni_xMo_{100-x} (x = 60 and 85 at.%) powders which were prepared by milling pure Ni and Mo in a high-energy ball mill. The end product for both compositions is nanocrystalline fcc alloys. They have followed the reduction in the grain size for both alloys as a function of the milling time (Fig. 5.4A).

Their results have indicated that the reduction in crystal size is taking place by increasing the ball milling time. By correlating the crystal size (Fig. 5.4A) and the strain values (Fig. 5.4B) we can conclude that a drastic fragmentation of the crystallites is achieved during the beginning of the MA process and the grains refined to be about 10–20 nm with a large increase on the internal strain. Upon further milling, the strain still increases in the lattice and reaches values that provoke, after 15 h of milling (Fig. 5.4B). The strain drops afterward and the crystal size decreases below 10 nm (Fig. 5.4B). They have concluded that a large number of dislocations are present in the crystals when the internal elastic strain reaches its maximum value (10–15 h). The results of their work show that when the size of the crystallites is less than 10 nm, the crystals are dislocation-free. Plastic deformation, in this case, takes place through a glide along the grain boundaries. After a long milling time, high-angle grain boundaries are formed.[22]

5.4.2 Formation of nanocrystalline fcc metals

Eckert et al.[23] carried out a systematic study for the grain reduction of pure metals with fcc structure during ball milling. The milling was carried out in a high-energy ball mill under an argon and hydrogen gas atmosphere, using hardened steel balls with ball-to-powder weight ratio as 4:1 (for Al, Cu, Ni, and Pd) or 6:1 (for Rh and Ir). The structure of the milled powder was characterized by means of XRD and TEM techniques.

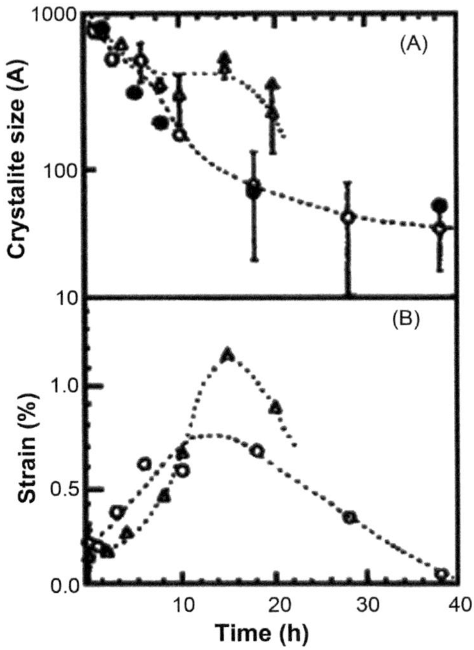

Figure 5.4 (A) Average grain size of the nickel crystallites size, L as a function of milling time measured by X-ray diffraction (open symbols) and direct TEM observations (full symbols) for $Ni_{85}Mo_{15}$ and (open symbol) for $Ni_{85}Mo_{15}$, (B) strain measured by X-ray diffraction for the same alloys. *(After Trudeau et al.)*[22]

The XRD patterns of Pd powders before and after 20 h of milling are shown in Fig. 5.5. The sharp Bragg peaks are remarkably broadened after ball milling, due to the refinement of the crystal size and an increase in internal strain, and they have shown a correlation between the melting points of the metallic milled powders and the final average of the grain size (Fig. 5.6). Whereas the grain size can be reduced to only 22 nm for Al, it decreases to values below 10 nm for the elements with high melting temperatures. This suggests that the recovery rates during the milling process correlate with the melting point of the specific metal, thus preventing very small grain sizes for low melting point elements.

5.5 Effect of ball milling on the structure of carbon nanotubes

In 1991, Iijima discovered a unique allotropes of carbon material, the so-called CNTs.[24] His investigation of multiwall carbon nanotubes (MWCNTs) is carbon soot produced by arc-discharge technique that opens a new horizon for improving the physical and chemical properties of such a unique class of materials that combines wide

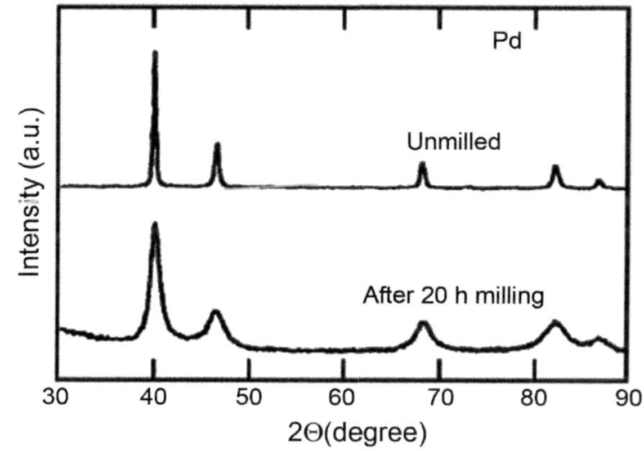

Figure 5.5 The XRD patterns of Pd powders before milling and after 20 h of milling. *(After Eckert et al.)*[23]

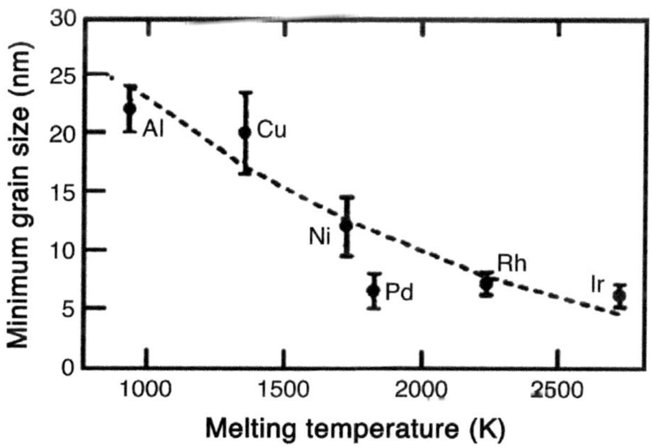

Figure 5.6 Minimum grain sizes for different metallic fcc metals versus melting temperature. *(After Eckert et al.)*[23]

range of interesting characterizations seldom found together in any other known materials. In 1993, Iijima and Ichihashi[25] reported another unique observations of another carbon material, the so-called single wall carbon nanotubes (SWCNTs). Whereas the SWCNTs are generally composed of at least two graphite cylindrical layers with an outer diameter ranging from 3 to 30 nm, the SWCNTs consist of a single graphite layer with a diameter of about 1–2 nm.

High-energy ball milling method has already been used for preparing of carbon nanoparticles,[26] nanoporous carbon,[27] and curving CNTs[28] materials. In 2001, Pierard et al.[29] report the first paper on preparations of short CNTs with open tips by high-energy ball milling of long MWCNTs. In their experiments, an agate vial and a single ball (5 cm in diameter) were used as the milling tools. The vial containing the

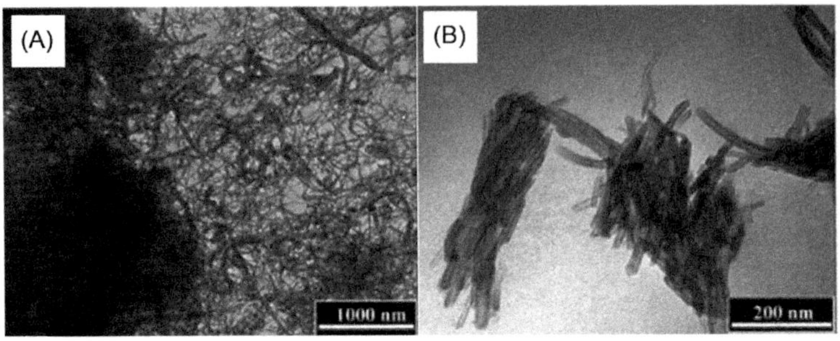

Figure 5.7 Low magnification BFI of MWCNTs as milled for (A) 0 h and (B) 12 of high energy ball milling.[29]

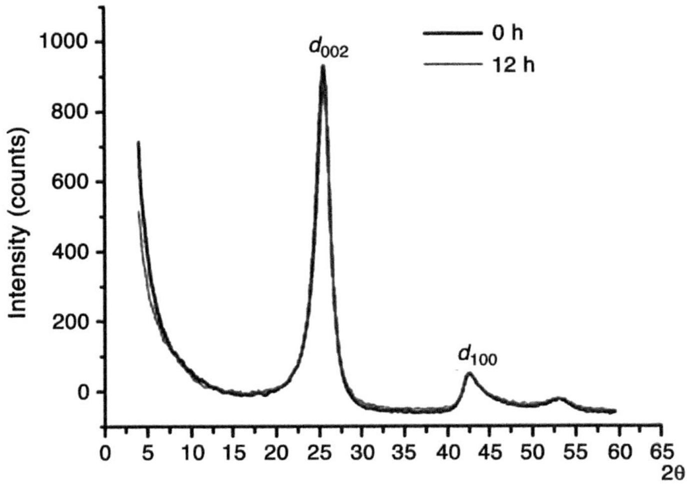

Figure 5.8 XRD patterns of MWCNTs after ball milling for 0 h and 12 h.

CNTs materials (8 or 3 g) and the ball were placed on Pluveriser vibration ball mill (Fritsch) at a vibration speed of 3000 min^{-1}.[29]

The BFI of the MWCNTs obtained after 0 h and 12 of high-energy ball milling are shown in Fig. 5.7A and B, respectively. The length of the nanotubes at the starting stage (0 h) decreased from tens of micrometers (Fig. 5.7A) to less than 1 μm (Fig. 5.7B) upon milling for 12 h.[29] Moreover, the XRD patterns of the samples before and after ball milling for 12 h (Fig. 5.8) showed similarity of the positions of the d_{002} and d_{100} peaks, indicating that the graphitization remained same. This also implies that the fracture taken place upon ball milling was very localized.[29]

Same as MWCNTs, high-energy vibratory ball milling technique can be successfully employed for preparations of several grams of short SWCNTs, as demonstrated by Pierard et al.[30] In contrast to the MWCNTs, SWCNTs do not resist very long milling times. For the first 2 h of ball milling (Fig. 5.9B) there was no significant destruction of nanotubes; however, further milling time for 8 h leads to the formation

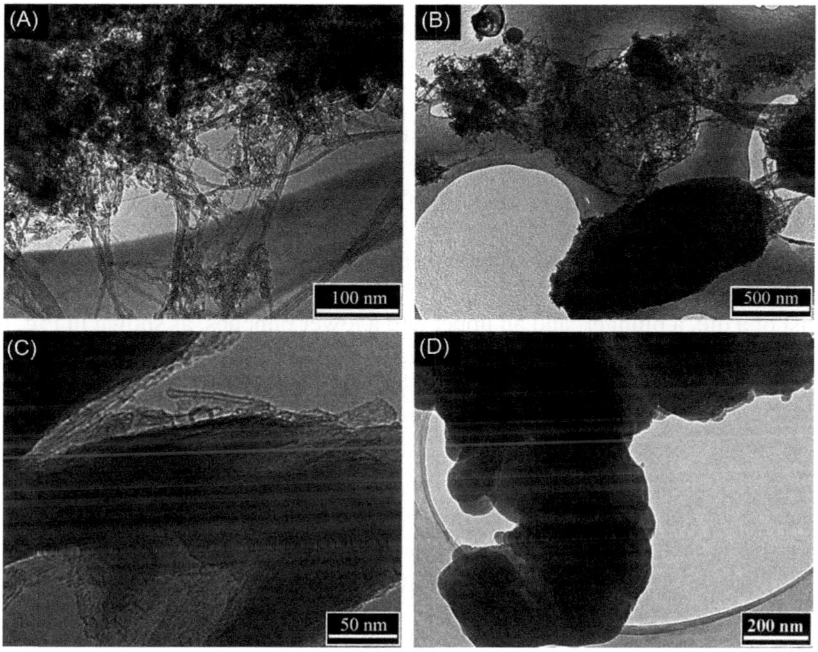

Figure 5.9 BFI images of SWCNTs: before ball milling (A), ball milled for 2 h (B), 8 h (C), and 50 h (D).[30]

of multilayered polyaromatic carbon structures, as shown in Fig. 5.9C. The authors investigated that the SWCNTs can be transformed into an amorphous carbon upon ball milling for 50 h (Fig. 5.9D).[30]

5.6 Pressing and sintering of powders materials

In general, powder metallurgy/technology can be classified into two parts: first, powders production and, second, the production of solid objects from the powders, which is called powder consolidation. Now, there are many optional approaches used for preparing of the feedstock powders with controlled nano- and microstructure and for consolidations of the powders into the desired shapes and sizes. In fact, powder pressing or consolidation process is not new and considered to be a classic procedure basically used either at room (cold pressing) and low temperatures (warm pressing) or at high temperature (hot pressing) to obtain dense compacts. However, although the consolidation of traditional powders with their microstructured grains is well established, the densification of nanostructured-, nanocomposite-, amorphous-, and metallic glassy powders resulted from ball milling process carries out many challenges. In this chapter, we will present a quick overview on the methodology and techniques used to consolidate the ball-milled powder particles via different approaches.

5.6.1 Classic powder metallurgy

In classic powder technology, pressing and sintering are two consequent processes by which the fabricated microscaled powder materials are densified into solid objects at temperatures laid below their melting point (T_m). Dry pressing (either uniaxial die or isostatic pressing) aims to consolidate the powders into green compacts with relative densities ranging from 80% to 95% (depends on the materials of the densified powder). In a typical powder pressing, certain amount of powder (depending on their bulk density), which is either weighed or measured volumetrically, is introduced into a die with the shape of the end product (Fig. 5.10A). For simple parts, the pressing tools consist of a die, a plunger, and an upper and a lower punches, as shown in Fig. 5.10B. In some cases, and based on the desired geometry of the end product, complex-shape pressing tools are used, as presented in Fig. 5.10C. The die containing the powders is mounted on cold press system and subjected to the applications of high pressure (Fig. 5.10D).

Density gradients seen in the materials after consolidation usually result from differences in the compaction pressure inside the compact, due mainly to wall friction effects. Wall friction can be avoided with using suitable lubricant for coating the internal wall of the die. Once the plunger is inserted into the die the powder can be gradually loaded to reach the desired maximum load that is usually in the range between 70 MPa

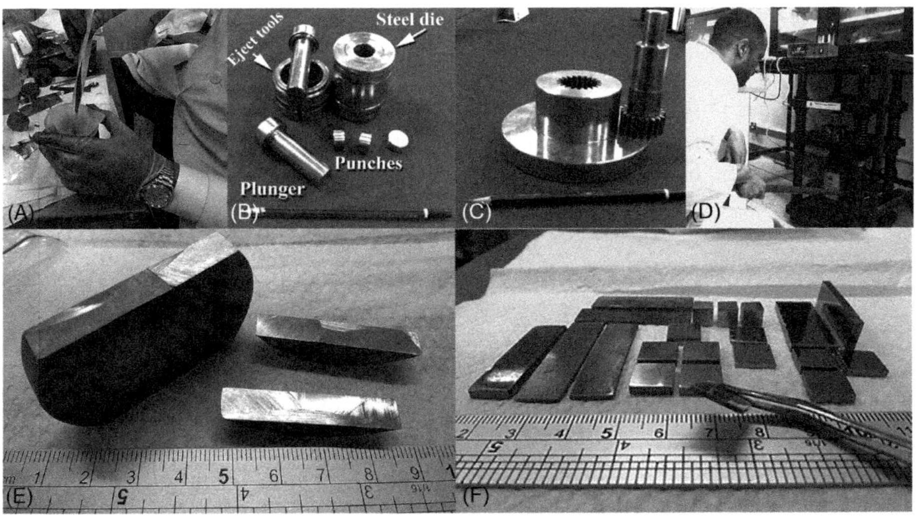

Figure 5.10 Powder cold pressing procedure showing inserting the powders into a cylindrical steel die (A) that consists plunger, and upper and lower punches (B) the die system can be designed and manufactured in a complex shape (C). Then the die together with the powders are mounted on cold press where the system is subjected to high pressure ranging from 70 MPa to 1 GPa (D). The as-consolidated dense bulk objects (E) can be cut and rolling into the desired shapes and preferred thickness (F) (Nanotechnology Laboratory, Energy and Building Research Center (EBRC), Kuwait Institute for Scientific Research (KISR)).

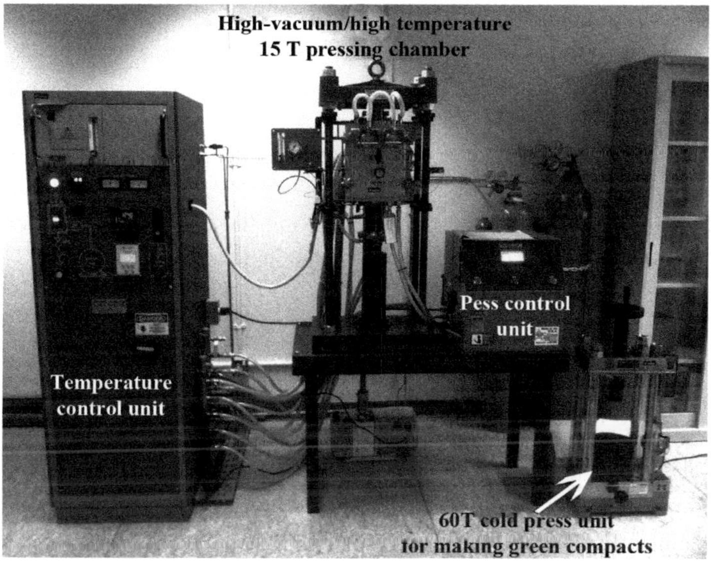

Figure 5.11 High vacuum/temperature hot press (OXY-GON, USA) employed for powder consolidation in the Nanotechnology Laboratory, Energy and Building Research Center (EBRC), Kuwait Institute for Scientific Research (KISR).

and 1 GPa or more. The loading rate used should be minimal as much as possible to overcome the formation of voids filled with trapped if the load is applied rapidly.

After completion of the process, the samples are usually kept at certain load for few minutes up to several hours to improve their density. The density of the as-consolidated bulk objects obtained after the cold processing usually is low so they are called as green compacts. Powder sintering is the second process applied to the green compacts to raise their relative density to more than 99.5% (Fig. 5.10E). The sintering process can be classified into two categories: the first is pressureless and the second is known as pressure-assisted sintering (e.g., hot pressing; Fig. 5.11). Sintering, which consumes surface energy to build bonds between the powder particles, aims to produce sintered parts with reproducible and designed microstructure through control of many sintering variables. Heating the powders during sintering is a combination of several atomic-level events that include diffusion, creep, viscous flow, plastic flow, and evaporation. It is well understood that applying moderate/high temperature (depending on the materials) leads to accelerate the powders sinterability due to increasing the atomic motion of the powder particles. Thus, significant strengthening occurs in the consolidated compacts due to sintering process. Usually, the sintering temperature is selected to be in the range between 0.7 and 0.9 of Tm for the powder materials.

Classic powder sintering for microscaled powders varies from few minutes to several hours (10 h or more). In fact, the sintering process is governed by several parameters, such as applied temperature, time, morphological characterization of the grains of the green compacts, relative density, and the protective atmosphere in the sintering furnace. Then, the solid objects obtained upon sintering can be mechanically cut into the desired shapes and dimensions (Fig. 5.10F) where cold or hot rolling can be employed in order to increase the sample density and to decrease the samples' thickness.

5.7 Consolidation of nanocrystalline powders

Consolidation of nanocrystalline powders, taking the ball-milled powders as an example, is an important step for reliable and reproducible determination of the physical/mechanical properties of the fabricated materials and is required for most industrial and structural applications.[31,32] The end product resulted from a ball milling process is usually consists of fine powder particles (~ 0.5–5 μm in diameter or less). These powders composite nanocrystalline sizes of about 3–8 nm in diameter. Therefore, they are considered to be ideal feedstock materials that can be used for fabrications and manufacturing of complex and high-performance full-dense bulk compacts. The advantage of using the ball-milled powders in sintering comes from the fact that they are small particles and enjoying higher surface energy making them easily compacted. However, consolidation of such category of fine powders into bulk, full-dense compacts while retaining nanoscale grain size is a major challenge. Since the purpose of nanocrystalline and nanoparticles powder consolidation is to achieve the densification process with a minimal particle coarsening and grain growth, the traditional way of powder consolidation (e.g., cold pressing + sintering) is not suitable for the ball-milled powders. Number of traditional sintering techniques, for example, hot pressing,[33,34] hot extrusion,[35] sintering forging,[36] and HIPing,[37] have been employed to consolidate the ball-milled powders into full-dense compacts with nanostructured characteristics. Most of these classic consolidations regime lead to fabricate either green compacts (50%–60% of the theoretical density) with nanograin structured, or full-dense compacts with microstructured grains. In general, the difficulties of consolidation of mechanically alloyed and ball-milled powders are attributed to several problems such as:

- Powder agglomeration,
- Irregular shape,
- Wide range of particle size distribution,
- High reactivity, and
- Surface contamination.

Unfortunately, and due to the large surface area of nanocrystalline materials, powders are highly unstable and undergo to drastic grain growth when subjected to high temperature. Sintering, which is a thermal treatment for bonding particles into a coherent, predominantly solid structure via mass transport events that often occur on the atomic scale[4] of such metastable nanocrystalline powders, presents a significant challenge. Likewise traditional powders (microscaled powders), densification of nanocrystalline powder particles is taking place through solid particle bonding or the so-called neck formation. At this starting stage of sintering the surface energy is consumed through particle bonding and this leads to an increase in the compact strength and dimensional change. Increasing the sintering duration of time performs larger neck and continuously closing the pores. Among the many factors that govern the rate of particle bonding during sintering process, the major parameters are temperature, materials, and particle size.[6] The consolidation procedure of nanopowders takes place through three stages: initial, intermediate, and final.[38] These major stages combine different mechanisms that are involved during the sintering promotion such as evaporation condensation, surface diffusion, grain boundary diffusion, bulk diffusion, viscous flow, and plastic deformation.

Among the proposed sintering mechanisms, the grain size of the nanoparticles is very sensitive and highly affected by surface and grain boundary diffusion.[39] It is understood that densification of large grained powders is not achieved by surface diffusion, which just leads to grains growth.[40] In contrast, surface diffusion plays a major role on the neck formation of nanocrystalline powders, which are characterized by high surface area and curved surfaces.[41] It was suggested by Thuenissen et al.[42] that the neck area increases to surface diffusion in nanoscale Y-TZP powders. Elimination of small pores in TiO_2 nanoparticles was achieved by rapid surface diffusion (425 and 775 K), as investigated by Hahn et al.[43] In addition, it was concluded that surface diffusion was the main mechanism for neck growing in nanoscaled Al_2O_3[44] and ZrO_2.[45] From these studies one can say that the fast neck formation during the sintering process is promoted and highly affected by surface diffusion that is considered to be the main mechanism for sintering of nanoscaled powder particles.

5.7.1 Approaches used for consolidation of the ball-milled powders

Powder consolidation, which is considered as one of the most important factors controlling the mechanical properties of the end product, can be achieved by different compaction and sintering approaches. However, although there are many ways to produce full-dense compacts with controlled grain size, we can roughly classified the consolidation process into two major categories known as (1) cold densification, (2) powder densification and sintering, and (3) pressure-assisted sintering. Some typical examples of the corresponding techniques used for each category can be given as follows:

- Cold densification
  - Cold uniaxial pressing[46]
  - Cold isostatic pressing[47]
- Powder densification and sintering
  - Densification
    - Magnetic pulse compaction (MPC)[48]
    - Equal channel angular press (ECAP)[49]
    - Rapid omnicompaction (ROC)[50,51]
  - Sintering
    - Vacuum sintering[52]
    - Microwave sintering[53]
    - High frequency induction heated sintering[54]

- Pressure-assisted sintering (simultaneous compaction and sintering)
  - Hot isostatic pressing (HIP)[55]
  - Pulse plasma sintering (PPS)[56]
  - Spark plasma sintering (SPS)[32,57–59] (known also as field-assisted sintering technique [FAST])

In the present chapter, we will focus on the applications of SPS method for fabrications of bulk nanocrystalline materials using the ball-milled powders as feedstock powders.

5.8 Spark plasma sintering for consolidation of ball-milled nanocrystalline powders

Due to the capability of SPS method to maintain the consolidated nanopowders in the nanostructured level with a minimal grain growth, SPS technique has attracted many researchers from the worldwide society of nanotechnology, materials science, and powder metallurgy. Accordingly, SPS system has mainly been dedicated for consolidation of metastable materials, which are very sensitive to temperature such as nanocrystalline,[60] amorphous, and metallic glassy materials.[61,62] With SPS, sintering and sinter bonding of mechanically alloyed/milled ultrafine powders can be achieved at low temperatures within a short period of time. The sintering process takes place by charging the intervals between powder particles with electrical energy and effectively applying a high-temperature spark plasma generated at an initial stage of energizing momentarily, and an electromagnetic field and/or joule heating by continuous ON-OFF DC pulsed high electric current with a low voltage.[63]

5.8.1 Components and system configurations of SPS system

A photo of a typical SPS system and its configurations are presented together in Fig. 5.12A and B, respectively. Obviously presented in Fig. 5.12B the system consists of a sintering press unit with a vertical single-axis pressurization, special designed punch electrodes incorporating a water cooler, a water-cooled vacuum chamber, a vacuum/air/argon-gas atmosphere control mechanism, a special (direct current) DC pulse sintering power generator, a cooling-water control unit, *Z*-axis position measuring and control unit, temperature measuring and control units, an applied pressure display unit, and various safety interlock devices.

5.8.2 Powder specimen filling procedure

For achieving a successful SPS consolidation process, preparations, the powders are inserted into a graphite die and stacked between upper and lower punches (Fig. 5.13A). To ensure an easy ejection of the sample after sintering and in order to avoid any reactions between the internal surfaces of the die tools (die and punches) with the powders (sample), graphite sheets are used as spacers (Fig. 5.13A). For reducing the amount

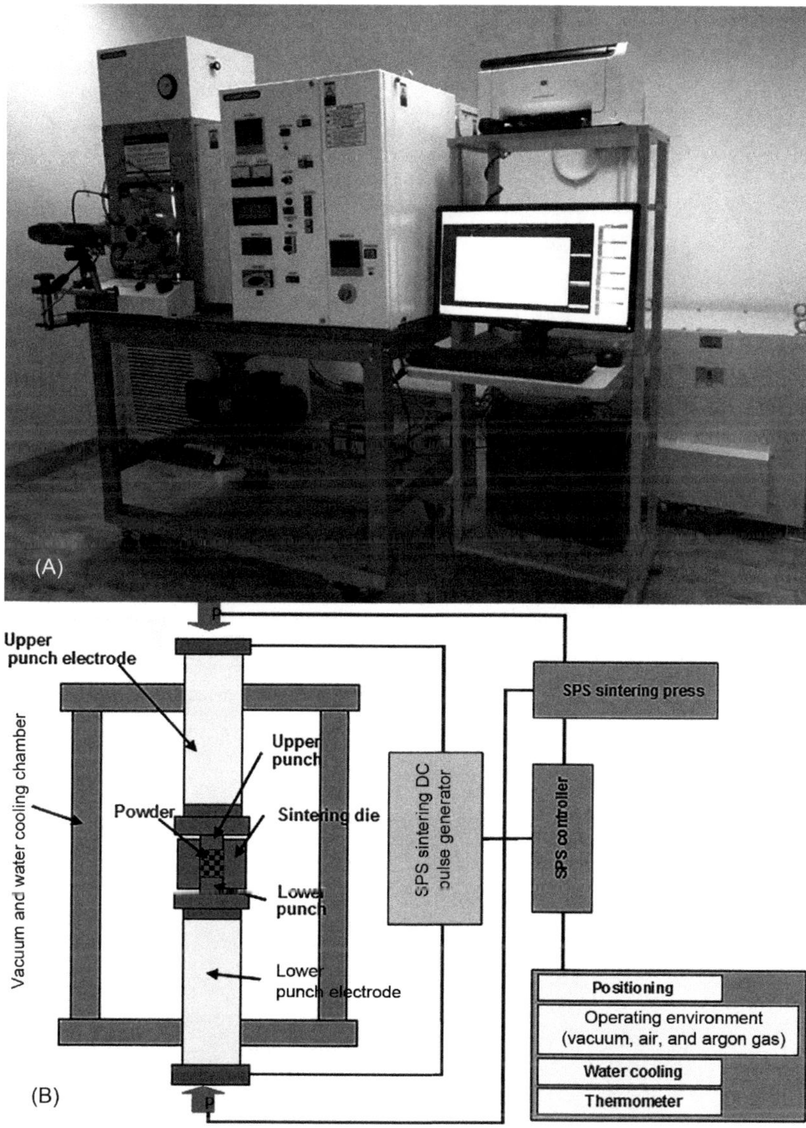

Figure 5.12 (A) A complete SPS system produced by Dr. Sinter Lab. Instrument, Japan, housed in the Nanotechnology Laboratory, EBRC- KISR, (B) schematic illustration of the system configuration. The system consists of an SPS sintering machine with a vertical single-axis pressurization mechanism, specially designed punch electrodes incorporating a water cooler, a water-cooled vacuum chamber, a vacuum/air/argon-gas atmosphere control mechanism, a special DC-pulse sintering powder generator, a cooling water control unit a position measuring unit, a temperature measuring unit, an applied pressure display unit various interlock safety units.

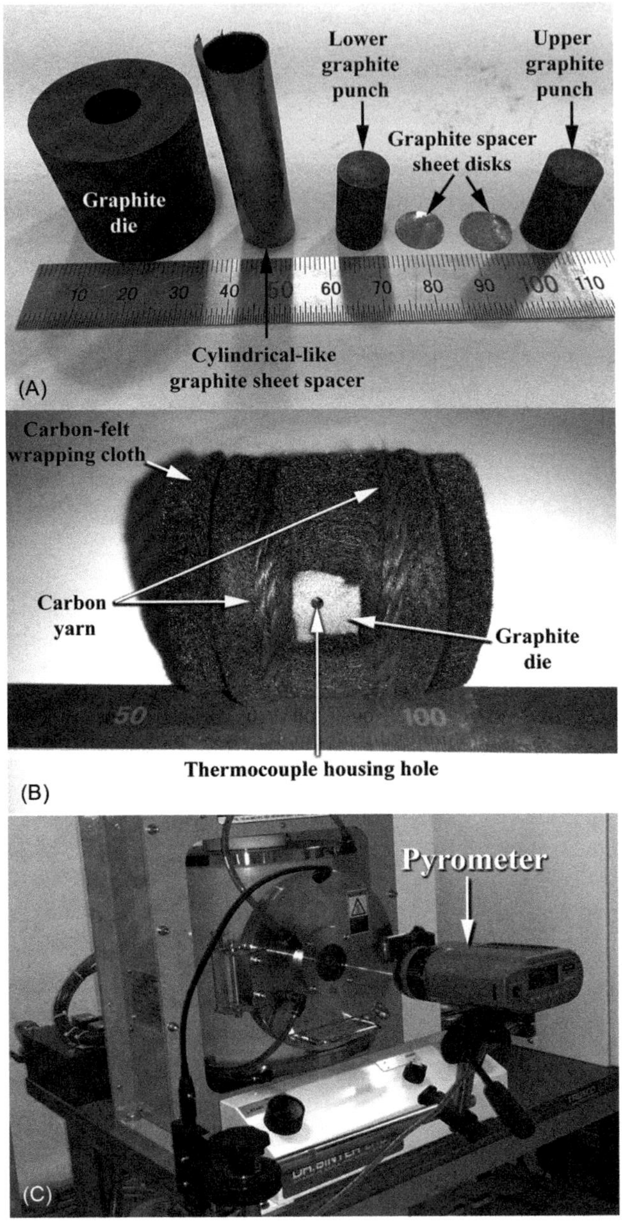

Figure 5.13 (A) Powder loading kits used in SPS consisting of graphite die and punches, cylindrical graphite spacer, and graphite disk spacer, (B) the die containing the powder should be wrapped by carbon-felt and tight with using carbon yarn, (C) the surface temperature of the die can be monitored during the SPS process with using a pyrometer (Dr. Sinter Lab. Instrument, Japan housed in Nanotechnology Laboratory, EBRC-KISR).

of radiant heat transfer to the machine, the entire die and punch assembly should be wrapped with carbon felt, which is held closed using carbon yarn as shown in Fig. 5.13B. The die can be then mounted on the sintering stage in the chamber and held between the upper punch and lower punch electrodes (Fig. 5.12B). During the SPS process, the temperature can be measured at the die surface with using a pyrometer, as shown in Fig. 5.13C. It is worth to be mentioned here that the actual processing temperature in SPS can be quite different from the temperatures measured by the pyrometer for many reasons[64]:

- The pyrometer measures the temperature at a niche in the die which is neither exactly on the sample surface nor in the surface interior,
- The measured temperature presents an average value and gives no indication of the local temperature that can actually exist between the particles, and
- The overall electrical resistance—including the internal resistance of the voltage source and resistance of the bulk of the apparatus—controls the current flow and, consequently, the joule heating generated in the sample.

5.8.3 Procedure

The powder sintering principals in SPS process are based on the presence of an electric field that is generally called FAST. In contrast with the conventional sintering techniques in which the sample is heated from the outside, the sintering procedure in SPS takes place upon heating the sample internally by the passage of an electric current. When SPS is compared with the conventional hot pressing technique, one can realize that the SPS process is taking place with extremely high heating rates that can reach up to 2000 Kmin^{-1}. Moreover, external pressures, which can also be applied during the sintering process in the range between few MPa and 1000 MPa, make lead to a rapid and effective consolidation process. SPS is an extraordinary rapid sintering process in which the whole processes (including temperature rise and holding times) are completed in a short period ranging from 5 to 20 min. This direct way of heating allows the application of very high heating and cooling rates, enhancing densification over grain growth promoting diffusion mechanisms, allowing maintaining the intrinsic properties of nanopowders in their fully dense products.

5.8.4 Mechanism

However, although there are intensive research efforts that have been carried out in consolidation of metal, ceramics, and their composites by SPS, the mechanism of this sintering procedure is still not well established. As mentioned in the earlier section, the main difference between SPS and other sintering approaches is that both die and powder are directly heated by the joule effect of the DC current (Fig. 5.14A). The ON-OFF DC pulse energizing method generates spark plasma, spark impact pressure, joule heating, and an electrical field diffusion effect.[63] In this process, the surfaces of the powders are activated and purified when compared with the other traditional sintering methods. Fresh-powder surfaces lead to easy sintering process, indexed by a lower sintering temperature and a shorter processing time.

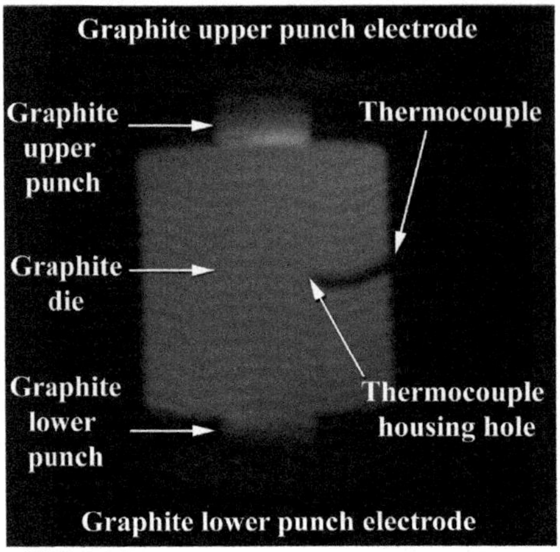

(A)

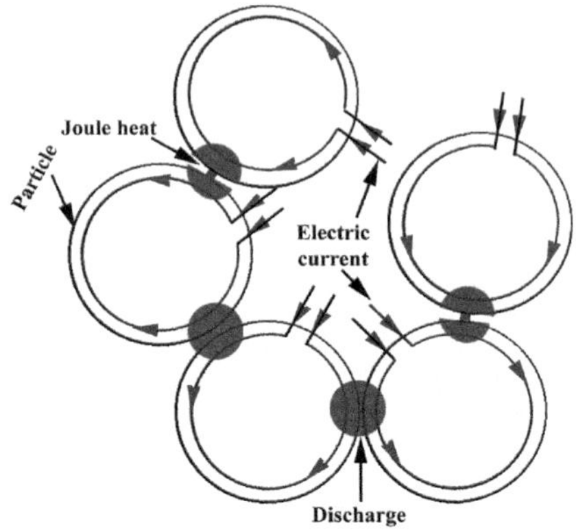

(B)

Figure 5.14 (A) Sintering stage and graphite mold in SPS water-cooling vacuum chamber at 1250°C, (B) schematic presentation of ON-OFF DC pulsed current path and pulsed current flow through powder particles taking place in the graphite die during SPS process.

As presented in Fig. 5.14B, upon starting ON-OFF DC current path the pulse current flows through the powder particles that are loaded into conductive graphite sintering tools (die and punches). Such pulsed current has cleaning effect on the surfaces of the powder particles based on the observation of grain boundaries without oxidation formed between particles. Such intense joule heating effect at the surface of the conductive powder particles results in reaching the boiling point and therefore leads to localized vaporization or cleaning of powder surfaces. Moreover, it has been suggested that joule heating resulted upon the passage of electric current through the powder particles assists and enhances the welding of the particles under mechanical pressure.[68,69] In fact, application of mechanical pressure during the SPS sintering process leads to the removal of pores and therefore enhances the diffusion between the powder particles. Moreover, applying mechanical pressure force during the sintering procedure increases the density of the compact sample and this leads to decrease the distance over which mass transfer must occur. It has been shown in some studies that the applications of mechanical pressure during SPS of Al_2O_3 and ZrO_2 led to decrease the applied sintering temperature and this minimizes the grain growth occurred upon sintering.

5.9 Fabrication of nanodiamonds and carbon nanotubes by milling

Graphite, fullerenes, graphene, carbon nanotubes, nanodiamonds, and new diamond are some carbon-based materials prepared by different techniques.[67,69] In 1959, Bovenkerk et al. reported the first study on the preparation of diamond.[70] A few years later, Danilenko reported the preparation of nanodiamonds, using a detonation technique.[71] Nanodiamonds have been synthesized by several routes including the detonation technique,[70] laser ablation,[71] plasma-enhanced chemical vapor deposition (PECVD),[72] autoclave synthesis from supercritical fluids,[73] chlorination of carbides,[74] ion irradiation of graphite and a plasma process.[75]

Research activities on carbon-carbon phase transformations including investigations of nanodiamond stability have received great attention.[68,77–81] At the macroscale level, it was believed that graphite is the most stable member of carbon family under ambient temperatures and pressures, whereas diamond was known as a metastable carbon phase. Now, it is well established that the fine nanoscale particles of nanodiamonds (3–5 nm in diameter) are more stable compared to graphite, as suggested by Barnard et al.[82] These nanoparticles are always described as a crystalline diamond core with a perfect diamond lattice surrounded by an amorphous shell with a combination of sp^2/sp^3 bonds or an onion-like graphite shell.[83]

Apart from the common approaches used for preparing nanodiamond materials, which require the applications of extremely high temperature and/or high pressures,[67,72,84–85] El-Eskandarany reported the first study for obtaining ultrafine nanodiamonds started from commercial graphite powders, using a high-energy ball milling technique operated at ambient temperature and normal pressure.[86]

5.9.1 Method

5.9.1.1 Materials and equipment

Commercial graphite powders 20 μm in diameter and of 99.99 wt.% purity, PubChem Substance ID 24857078, were provided by SIGMA-Aldrich, USA, helium gas (He) of 99.999% purity, provided by Kuwait Industrial Gases Company W.L.L. (KIGC). A stainless steel vial made of Cr-steel alloy with an internal diameter of 10 cm and capacity of 110 mL, supplied by Evico Magnetic, Germany, and 50 Cr-steel alloy balls with a 14 mm diameter, a density of 7.89 g/cm^3 and 10 g in weight, were used as the milling media. A high-energy planetary-type ball mill, models PM100, was provided by Retsch, Germany, and a He gas atmosphere-glove box (UNILAB Pro Glove Box Workstation) was provided by mBRAUN, Germany.

5.9.1.2 Nanodiamonds syntheses

Three grams of elemental graphite powders were placed inside a high purity He gas atmosphere-glove box and then sealed together with 40 Cr-steel balls (14 mm in diameter), a ball-to-powder weight ratio as 167:1. The vial was then mounted on a planetary-type high-energy ball mill operated at ambient temperature with a rotation speed of 800 rpm for different ball milling times, ranging from 1 to 28 h. The as-ball-milled powders were regularly discharged from the vial completely in the glove box after selected milling runs (1, 6, 10, 12, 14, 16, 20, 23, 26, and 28 h). A new batch of graphite powders was charged and sealed into the vial for new milling runs. The samples obtained after different ball milling stages were characterized by different techniques to understand the progress of the graphite-nanodiamonds phase transformations. The powders were obtained after up to 28 h.

5.9.1.3 Results

XRD and HRTEM were employed to monitor the phase transformations occurring upon high-energy ball milling of graphite powders at different states of milling. Fig. 5.15 shows the XRD patterns of graphite powders obtained during the early stage of ball milling (0–16 h). The starting graphite (G) powders consisted of large-scale crystallites, indicated by the sharp Bragg peaks corresponding to hexagonal close

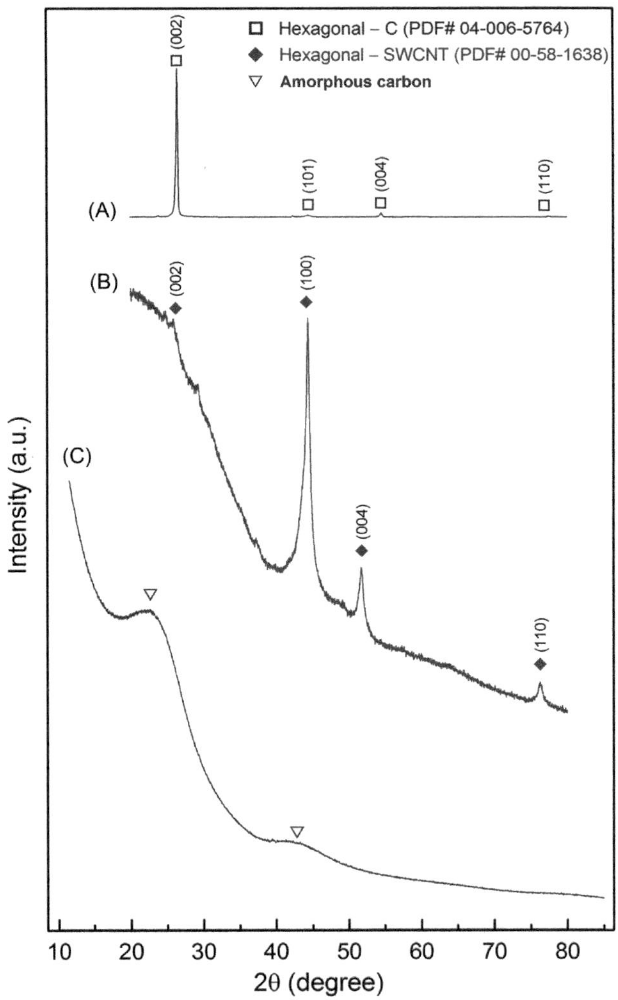

Figure 5.15 XRD diffraction patterns of graphite powders obtained after high-energy ball milling for (A) 0, (B) 10, and (C) 16 h. (*After El-Eskandarany.*[89])

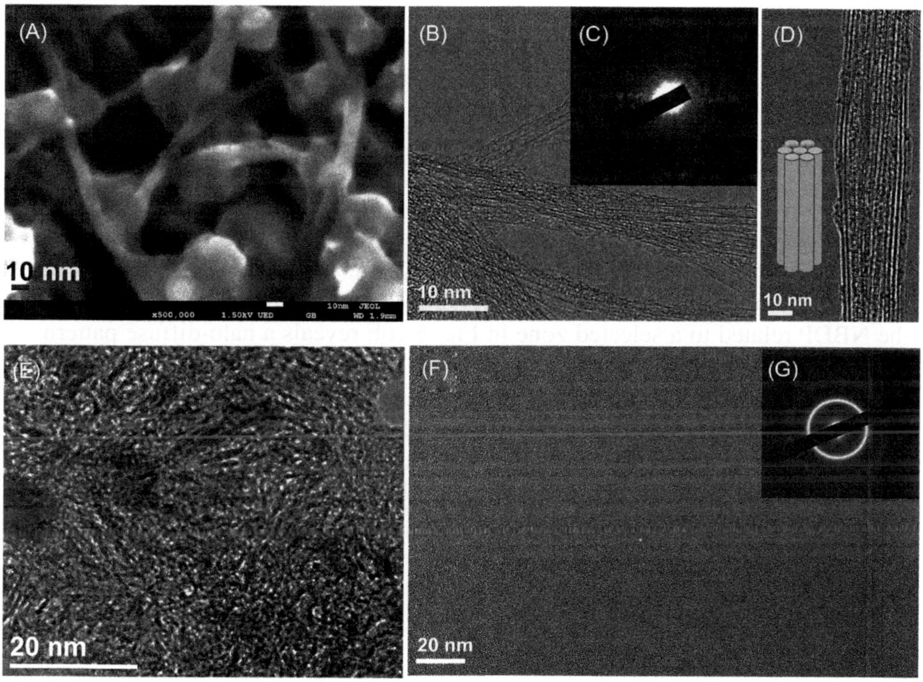

Figure 5.16 (A) FE-SEM micrograph of graphite powders obtained after 12 h of ball milling time. The HRTEM image of graphite powders obtained after 10 h of milling time is shown with different scale in (B) and (D), where the corresponding fast Fourier transform (FFT) images of a selected zone of (B) is presented in (C). (E) This BFI elucidates the defects taken place in the SWCNTs upon further milling for 14 h, where the HRTEM image and the corresponding NBDP of graphite powders obtained after ball milling for 16 h are shown in (F) and (G), respectively. (*After El-Eskandarany.*[89])

packing, hcp-(002), -(101), -(004), and -(110), as elucidated in Fig. 5.15A. After 10 h of ball milling, the XRD pattern revealed Bragg peaks corresponding to SWCNTs (PDF# 00-58-1638), as presented in Fig. 5.15B.

The high-magnification FE-SEM micrograph of the powders obtained after this stage of milling (12 h) showed the existence of untransformed G-nanoparticles (10–30 nm in diameter) attached to the surface of the micron-size aggregates SW-CNTs (Fig. 5.16A). Formation of metastable SWCNTs was confirmed by HRTEM (Fig. 5.16B) and corresponding nanobeam diffraction pattern (NBDP) (Fig. 5.16C). The SWCNTs possessed raft-like bundle (Fig. 5.16D), where the reflection intensities form continuous arcs related to hcp-SWCNTs that can be notified (Fig. 5.16D).

After milling for 12 h, the SWCNTs formed during the early stage of milling were subjected to severe lattice imperfections generated by the milling media (balls), resulting in a heavy deformation occurred of the SWCNTs, as shown in Fig. 5.16B. These defected carbon nanotubes were dramatically disintegrated into short-random fibers with irregular shapes and sizes after increasing the ball milling time (14 h), as shown in Fig. 5.16E.

After 16 of ball milling, those crystalline peaks related to hcp-SWCNTs were completely disappeared and replaced by homogenous halo-diffuse peaks, with first and second maxima detected at 22.56 and 42.94 degrees, respectively (Fig. 5.15C). This implies SWCNTs into amorphous carbon (a-C) phase transformation. The formation of a-C phase was confirmed by HRTEM image, showing a dense random-packing structure with maze-like morphology, as elucidated in Fig. 5.16F. In this figure, the existence of untransformed G-nanoparticle, SWCNTs, and/or nanocarbon fibers can be hardly seen, suggesting complete SWCNTs to a-C phase transformation. Moreover, the NBDP related to a selected zone in Fig. 5.16F reveals a halo-diffuse pattern of a typical amorphous phase, as displayed inset of Fig. 5.16G.

In order to investigate the effect of the electron beam irradiation time on the stability of the formed a-C, the sample obtained after 16 h of milling was continuously exposed to 160 kV-TEM beam irradiation for different period of times, ranging from 0.5 to 30 min. The HRTEM images of as-irradiated samples (Fig. 5.17) revealed outstanding resistance against the irradiation effects in the range between 0.5 and 5 min (Fig. 5.17A–D). No remarkable change in the local structure of the a-C could be notified when the exposure time applied for 10–20 min, as shown in Fig. 5.17E–H.

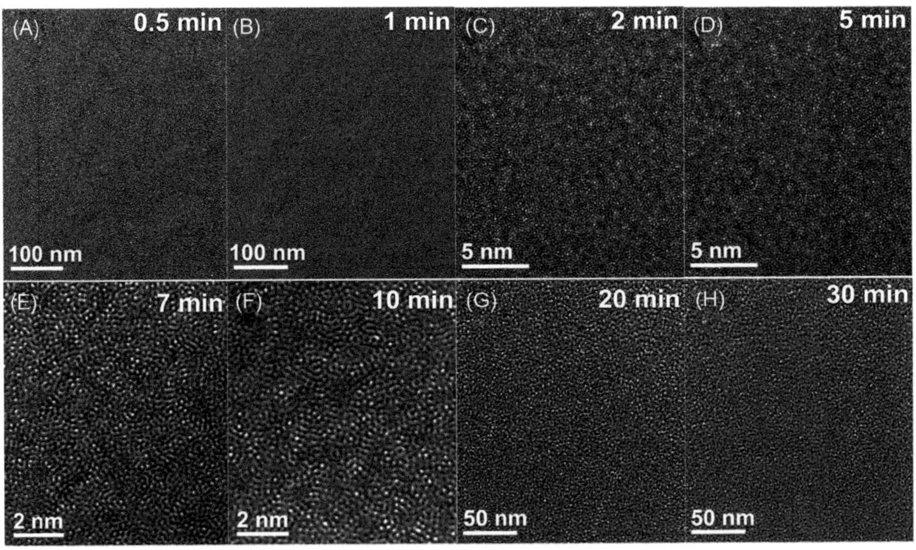

Figure 5.17 Effect of exposure time of HRTEM electron beam irradiation operated at 160 kV on the local structure and morphology of amorphous-carbon (a-C) powders obtained after 20 h of ball milling. Sequence of HRTEM micrographs of a-C powders obtained after 20 h of milling and then subjected to a continuous electron beam irradiation generated by HRTEM/160 kV for (A) 0.5, (B) 1, (C) 2, (D) 5, (E) 7, (F) 10, (G) 20, (H) 30 min. (*After El-Eskandarany.*[89])

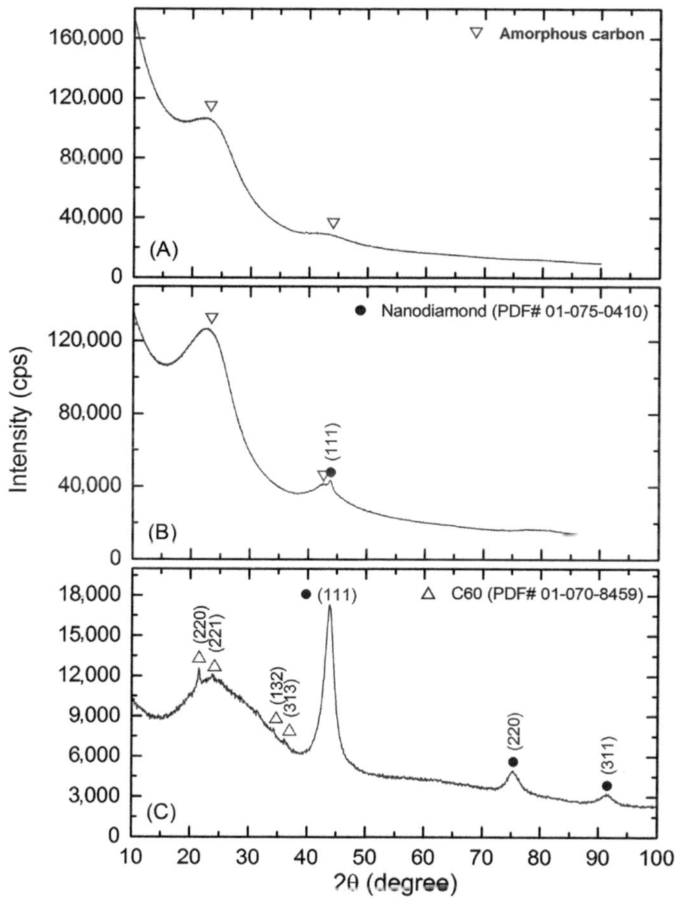

Figure 5.18 XRD patterns of the powders obtained after 20, 23, and 28 h of high-energy ball milling are displayed in (A), (B), and (C), respectively. (*After El-Eskandarany.*[89])

Further ball milling (20 h) of the a-C powders did not lead to any remarkable phase transformation, where the sample maintained its structure as a short-range order metastable phase, as displayed in Fig. 5.18A. However, after 23 h of ball milling time, a crystalline peak related to nanodiamonds, NDs (PDF# 01-075-0410) appeared, indicating a partial a-C to NDs phase transformations, as shown in Fig. 5.18B. The HRTEM and corresponding fast Fourier transform (FFT) for the powders obtained after 23 h are shown together in Fig. 5.19A and B, respectively. The powders consisted of fine structure related to a-C phase, indicated by the maze-like morphology (Fig. 5.19A) where nanodimensional particles (~10 nm in diameter) are embedded into the a-C matrix, as indicated by the arrows shown in Fig. 5.19A. The FFT (Fig. 5.19B) taken from the circular zone in Fig. 5.19A implying the existence of halo-diffraction peak related to a-C coexisted with sharp-spots corresponding to NDs with zone axis orientation to [111], as elucidated in Fig. 5.18B.

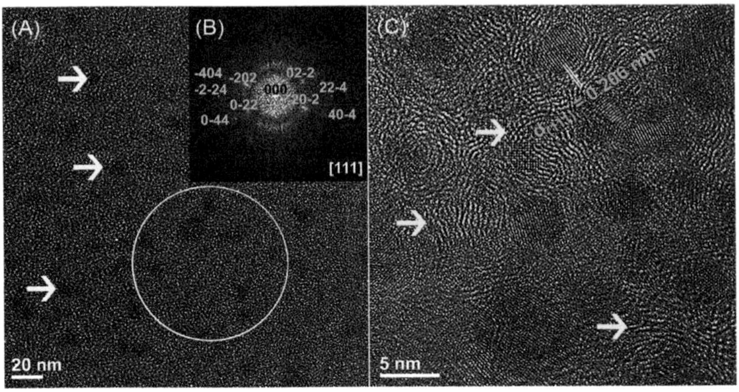

Figure 5.19 The HRTEM and fast Fourier transform (FFT) images of the powders milled for 23 h are shown together in (A) and (B), respectively. The *white* arrows shown in (A) refer to the nanodiamonds (NDs) embedded into the amorphous-carbon matrix, where the *circular* symbol indicates the selected zone taken for the FFT. (C) Partial formation of carbon nano-onions (indexed by the *yellow* symbols). (*After El-Eskandarany.*[89])

Increasing the ball milling time to 25 h led to the formation of concentric graphite-like shells (carbon onions), resulting a dramatic decrease in the volume fraction of a-C, as shown in Fig. 5.19C. This implies a partial a-C–carbon onion phase transformations. Moreover, an increase in the volume fraction of NDs obtained after 25 h of milling can be notified, as shown in Fig. 5.18C.

Toward the end of the ball milling time (28 h), the halo-diffraction peaks related to a-C decreased in the intensity, indicating a decrease in the amorphous phase in the powders of the end product, as shown in Fig. 5.18C. In contrast, the Bragg-diffraction lines related to NDs phase became pronounced, suggesting a significant increase in the NDs volume fractions against the untransformed a-C phase, as shown in Fig. 5.18C. The broadening shown in the Bragg peaks of (111), (220), and (311) implies the formation of nanocrystalline NDs after this stage of ball milling. The NDs/a-C nanocomposite powders contained a very small volume fraction of metastable C60 (PDF# 01-070-8459), as shown in Fig. 5.18C. This may suggest that the a-C consisted of nanoclusters of C60 that did not transformed into NDs phase upon further milling to 28 h. Intensive TEM experimental work has been done to investigate the structure and morphology of C60 formed during ball milling; we could not get clear images to support the XRD analysis. Further efforts and detailed TEM study should be achieved to investigate the formation of this small volume fraction of metastable phase and to realize whether they can be grown in volume and sizes upon further ball milling time or not.

The HRTEM image of the powders obtained after 26 h of milling consisted of NDs particles with an average particle distribution of 5 nm in diameter, as shown in Fig. 5.20A. The dark filed image (DFI) of the powders obtained after 28 h of ball milling is shown in Fig. 5.20B. Obviously, the particles possessed spherical-like crystallinity with rather wide particle size distribution, ranging between 1.8 and 4.8 nm in

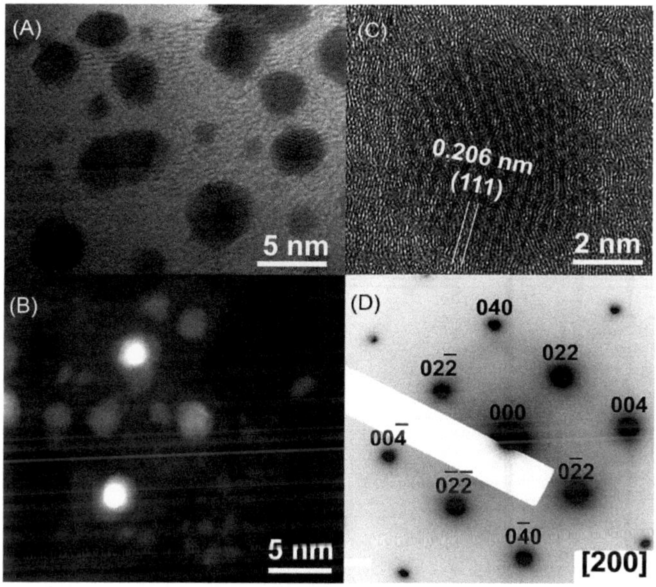

Figure 5.20 Morphological and structural characterizations of the graphite powders obtained after the final stage of milling (26–28 h). (A) HRTEM image of graphite powders obtained after high-energy ball milling for 26 h. The DFI and HRTEM micrographs for the powders obtained after 28 h of milling (end product) are displayed in (B) and (C), respectively. (D) NBDP elucidates the crystal structure of a single nanodiamond particle shown in (C) and oriented to zone axis (200). (*After El-Eskandarany.*[89])

diameter (Fig. 5.20B). These nanoparticles are embedded into a fine matrix of a-C, as shown in Fig. 5.20A and B. The interplanar spacing (*d*; 0.206 nm) of a selected single nanoparticle well matches with NDs (111), as shown in Fig. 5.20C. Moreover, the NBDP of this fine particle is corresponding to NDs, as indexed by the zone axis [200], shown in Fig. 5.20D.

Raman scattering approach was employed in order to monitor the structural changes of G-powders upon high-energy ball milling for different time more clearly. Figure 5.21A shows the Raman spectrum for the sample obtained after ball milling for 10 h. Obviously, G and G' (2D) bands were observed at frequency of 1584 cm^{-1} and 2659, respectively, as elucidated in Fig. 5.21A. In addition, the D band (disordered band), which is produced from hybridized vibrational mode associated with graphite edges, can clearly be seen at a frequency related to 1340 cm^{-1} (Fig. 5.21A) corresponding to the responses of sp^2 and disorder carbon fine particles. Thus, the presence of this band in the Raman spectrum of the sample obtained after 10 h of milling indicates the existing of disordered graphite particles. The volume fraction of these defected nanocrystalline carbon particles is very small, suggested by the very low intensity of D band presented in Fig. 5.21A. The low-intensity broadband can be seen a frequency of about 501 cm^{-1} is related to the existence of a small volume fraction of a-C phase. We should emphasize that the clear band laid at a frequency

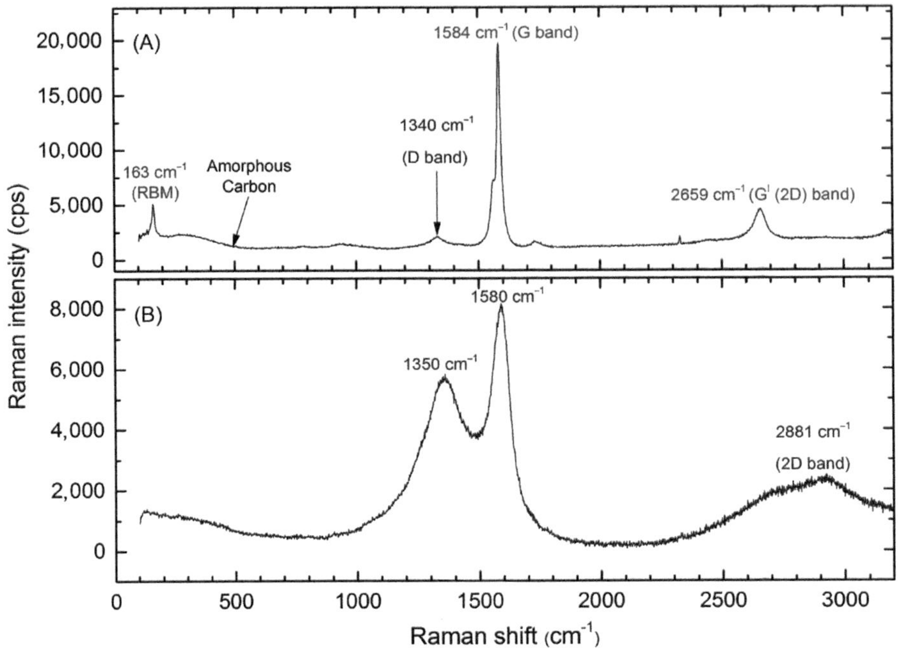

Figure 5.21 Raman spectra of graphite powders obtained after (A) 10 h, and (B) 16 h of the ball milling time. (*After El-Eskandarany.*[89])

of 163 cm^{-1} in Fig. 5.21A is corresponding to the Radial Breathing Mode (RBM), which is unique for SWCNTs, and resulting due to the expansion and contraction of the nanotubes corresponding to the coherent vibration of the carbon atoms in radial direction. Based on the Raman spectrum analysis of the powders obtained after this stage of stage (10 h) one can say that SWCNTs can be obtained upon high-energy ball milling of G-powders. This metastable carbon phase is coexisted with a small volume fraction of a-C phase.

The Raman spectrum of the G-powders after ball milling for 16 h is displayed in Fig. 5.21B. In contrast to the sample obtained after 10 h of ball milling, which is mainly SWCNTs, the Raman spectrum of 16-h sample does not show the RBM peak, implying the disappeared of SWCNTs upon increasing the ball milling time from 10 to 16 h. Moreover, those rather sharp Spectrum lines shown in Fig. 5.21B became broad upon increasing the milling time, as indicated by the two broad lines related to G (1580 cm^{-1}) and D (1350 cm^{-1}) bands, as elucidated in Fig. 5.21B. This broadening is resulted by structural disorder with an unsymmetrical Raman band in the wavenumber region between 900 and 1800 cm^{-1}. We should emphasize that the D band located at 1350 cm^{-1} resembles structural disorder due to the presence of sp^3 carbons, whereas the G band laid at 1580 cm^{-1} corresponds to the E_{2g} mode of sp^2-hybridized carbon frameworks.

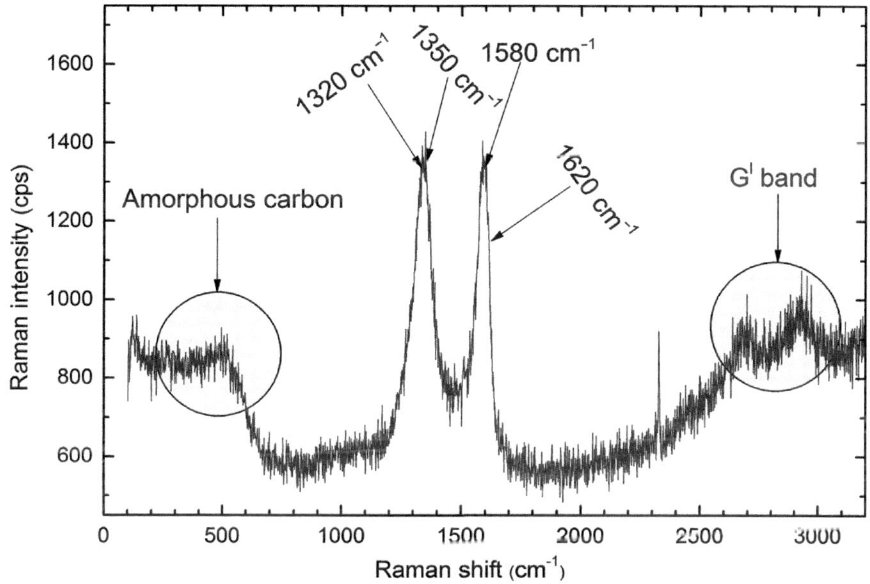

Figure 5.22 Raman spectra of the final product obtained after 28 h of high-energy ball milling time. (*After El-Eskandarany*[89])

Fig. 5.22 presents a full-range Raman spectrum (0–3200 cm^{-1}) of the graphite powders obtained after 28 h of high-energy ball milling time. The spectrum revealed different bands related to the phases existed in the sample. The very broadband laid at 506 cm^{-1} is related to the untransformed amorphous sp^3 bonded carbon phase existed in the final product. At the high wavelength side, the overlapped broad peaks laid at 2669 and 2927 cm^{-1} are corresponding to G^{l} band. The two major peaks shown in Fig. 5.22 with Raman bands of 1320 and 1620 cm^{-1} are related to NDs as they are simply referred in analogy to sp^2. The slight downshifted tetrahedral sp^3 band seen in the Raman spectrum of this sample, indicating the distortion and lattice imperfection introduced to the NDs particles during a long term of high-energy ball milling (28 h). These bands are overlapped with D band located at 1350 cm^{-1} and G band laid at 1580 cm^{-1} related to the untransformed a-C powders.

5.9.1.4 Discussion

Fig. 5.23 presents a schematic illustration elucidating the possible mechanism of G-to-NDs phase transformations taken place in a high-energy ball mill at room temperature. The forces generated by the balls used as milling media affecting the stability of G-powders are shown in Fig. 5.23A and B that presents the possible phase transformation of G-powders achieved upon changing the free energy (ΔG) of G (point 1), SWCNTs (point 2), a-C (point 3), and NDs (point 4). The mechanically induced disordering (MID) shown in Fig. 5.23B refers to the energy generated by the ball milling process, which is required to elevate the ΔG of the starting G-powders from point 1

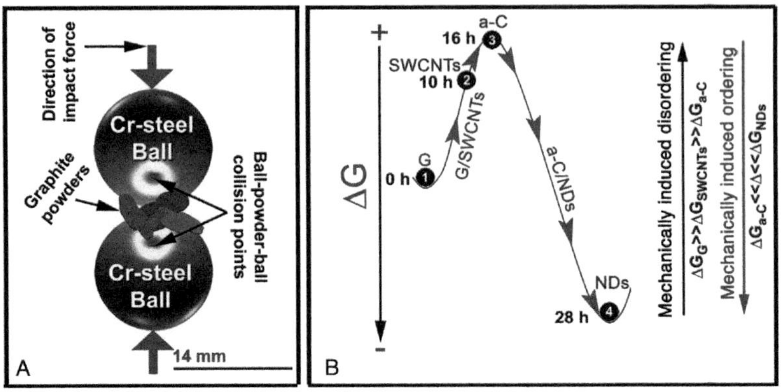

Figure 5.23 Schematic illustration proposing the graphite–nanodiamonds phase transformations conducted in a high-energy ball mill. (A) Ball-powder-ball collisions achieved during high-energy ball milling of elemental graphite powders, and (B) changes on free energy of formation (ΔG) leading to graphite–nanodiamonds phase transformations at different milling times. The possible phase transformations of G-powders achieved upon changing the free energy (ΔG) of G (point 1), SWCNTs (point 2), a-C (point 3), and NDs (point 4) are shown in (B). The mechanically induced disordering (MID) shown in (B) refers to the energy generated by the ball milling process, which is required to elevate the ΔG of the starting G-powders from point 1 (stable G-phase, ΔG_G) to less stable phases of SWCNTs (point 2, ΔG_{SWCNTs}) and a-C (point 3, $\Delta G_{a\text{-}c}$), respectively. On the opposite direction, the mechanically induced ordering (MIO) denotes to the energy generated by the ball milling process leading to stabilize the amorphous-phase (point 3) into a stable phase NDs (point 4).

(stable G-phase, ΔG_G) to less stable phases of SWCNTs (point 2, ΔG_{SWCNTs}), and a-C (point 3, $\Delta G_{a\text{–}c}$), respectively, as shown in Fig. 5.23B. On the opposite direction, and in contrary to the MID process, the mechanically induced ordering (MIO) denotes to the energy generated by the ball milling process leading to stabilize the amorphous phase (point 3) into a stable phase NDs (point 4).

In order to attain the energy of the ball milling process required to achieve the present reported phase transformations, 50 Cr-steel balls with diameter of 14 mm, density of 7.89 g/cm^3 and about 10 g in weigh, were used as milling media. The ball-to-powders weight was maintained at constant ratio of 167:1, using 3 g of graphite powders for each milling run. Moreover, the ball milling process was achieved, using a high-energy ball mill rotated at 800 rpm. In order to minimize the Cr-steel contamination content introduced to the powders during the milling process, the balls (milling tools) were sprayed by graphite and then annealed at 250°C overnight. However, although the coating process was performed on each milling run, the a-C powders obtained after 16 h of milling time contained about 1.45 wt.% Cr-steel contamination content. When the powders were furtherly ball milled to 28 h, this metallic contamination content was significantly increased to 2.1 wt.%. The existence of such undesired contamination content may play a catalytic-assisted role for the formation of nanodiamond phase.

During the early stage of the MID (0–10h), the starting G-powders were subjected to severe continuous stresses generated by the ball-powders-ball collusions (Fig. 5.23A).

These applied stresses were translated into lattice and points defects affecting the stability of G-powders and led to raise the ΔG from point 1 (G) to point 2 (SWCNTs), as shown in Fig. 5.23B. During the second stage (10–16 h) the as-obtained SWCNTs were continuously subjected to severe lattice imperfections, resulting a serious disordering on their surfaces. Such continuous disintegrations of the nanotubes led to the formation of distorted short curved fiber-like morphology (Fig. 5.16E). After 16 h of ball milling, the ΔG_{SWCNTs} raised to a less stable energy state ($\Delta G_{a\text{-}C}$), as illustrated by point 3 shown in Fig. 5.20B. It can be concluded that high-energy ball milling is capable to produce an amorphous fine powders in which the fraction of sp^3 bonded carbon is enhanced due to the damage of sp^2 network on the SWCNTs. Again, this metastable amorphous phase was suffered from the applications of sever mechanical forces generated by the milling media and tended to transform into a less stable phase (nanodiamonds) upon increasing the ball milling time to 28 h (Fig. 5.23B). Thus, applying such drastic mechanical deformation on the unstable sp^2 phase (amorphous) led to stabilize the sp^3 phase (NDs) to obtain nanocomposite SP^2/SP^3 bonds.

References

1. Gates, B. D., Qiaobing, X., Stewart, M., Ryan, D., Willson, C. G., & Whitesides, G. M. (2005). *Chem. Rev.*, *105*, 1171.
2. Ariga, K., Hill, J. P., Lee, M. V., Vinu, A., Charvet, R., & Acharya, S. (2008). *Sci. Technol. Adv. Mater.*, *9*, 014109.
3. Gleiter, H. (1989). *Prog. Mat. Sci.*, *33*, 223.
4. Hahn, H. (1997). *Nanostruct. Mater.*, *9*, 3.
5. McHenry, M. E., & Laughlin, D. E. (2000). *Acta. Mater.*, *48*, 223.
6. Gleiter, H., Deformation of polycrystals. In: Proceeding of the Second Riso International Symposium on Metallurgy and Materials Science (N. Hansen, A. Horsewell, T. Leffers, and H. Lilholt, eds.), Roskilde, Denmark, p. 15 (1981).
7. Suzuki, H., The Relation Between the Structure and Mechanical Properties of Metals, National Physical Laboratory, vol. II, Symposium No. 15:524 (1963).
8. Birringer, R., Gleiter, H., Klein, H. P., & Marquardt, P. (1984). *Phys. Lett. A*, *102*, 3423.
9. Inoue, A. (1994). *Mater. Sci. Eng. A*, *179/180*, 57.
10. Hughes, R. O., Smith, S. D., Pande, C. S., Johnston, H. R., & Armstrong, R. W. (1986). *Scripta Metall.*, *20*, 93.
11. Li, Z. G., & Smith, D. J. (1989). *Appl. Phys. Lett.*, *55*, 919.
12. Lu, K., Wei, W. D., & Wang, J. T. (1995). *Scr. Metall. Mater.*, *24*, 2319.
13. Kear, B. H., & Strutt, P. R. (1995). *Nanostruct. Mater.*, *6*, 227.
14. Gleiter, H. (1995). *Nanostruct. Mater.*, *6*, 3.
15. Koch, C. C. (1993). *Nanostruct. Mater.*, *2*, 109.
16. Koch, C. C. (1997). *Nanostruct. Mater.*, *9*, 13.
17. Froes, F. H., & Suryanarayana, C. (1989). *J. Metals*, *6*, 12.
18. Campbell, S.J., and Kaczmarek, W.A., Mössbauer Spectroscopy Applied to Materials and Magnetism, vol. 2 (G.J. Long and F. Grandjean, eds.), Plenum Press, New York, p. 273 (2000).
19. Johnson, W. L. (1986). *Prog. Mater. Sci.*, *30*, 81.
20. Johnson, A., New Materials by Mechanical Alloying Techniques (E. Arzt and L. Schults, eds.), Informationsgesellschaft Verlag, Calw-Hirsau, p. 354 (1988).

21. Fecht, H. J. (1995). *Nanostruct. Mater.*, *6*, 33.
22. Trudeau, M. L., & Schulz, R. (1991). *Sci. Eng. A*, *134*, 1361.
23. Eckert, J., Holzer, J. C., Krill, C. E., III, & Johnson, W. L. (1992). *Mater. Sci. Forum*, *88–90*, 505.
24. Iijima, S. (1991). *Nature*, *354*, 56.
25. Iijima, S., & Ichihashi, T. (1993). *Nature*, *363*, 603.
26. Chen, Y., Gerald, J. F., Chadderton, L. T., & Chaffron, L. (1999). *Appl. Phys. Lett.*, *74*, 2782.
27. Huang, J. Y., Yasuda, H., & Mori, H. (1999). *Chem. Phys. Lett.*, *303*, 130.
28. Harringa, J. L., Cook, B. A., & Beaudry, B. J. (1992). *J. Mater. Sci.*, *27*, 801.
29. Pierard, N., Fonseca, A., Konya, Z., Willems, I., Van, T. G., & Nagy, B. (2001). *J. Chem. Phys. Lett.*, *335*, 1.
30. Pierard, N., Fonseca, A., et al. (2004). *Carbon*, *42*, 1691.
31. He, L., & Ma, E. (1996). *J. Mater. Res.*, *11*, 73.
32. El-Eskandarany, M. S., Omori, M., Konno, T. J., Sumiyama, K., Hirai, T., & Suzuki, K. (1998). *Metall. Trans. A*, *29*, 1973.
33. Nash, P., Kim, H., Choo, H., Ardy, H., Hwang, S. J., & Nash, A. S. (1992). *Mater. Sci. Forum*, *88–99*, 603.
34. El-Mahdy, A. A., El-Eskandarany, M. S., Ahmed, H. A., & Amer, A. A. (2000). *J. Alloys Comp.*, *299*, 244.
35. Hwang, S. J., Nash, P., Dollar, M., & Dymek, D. (1992). *Mater. Sci. Forum*, *88–99*, 611.
36. Jain, M., & Christman, T. (1994). *Acta. Metall. Mater.*, *42*, 1901.
37. Oehring, M., Appel, F., Pfullmann, T., & Bormann, R. (1995). *Appl. Phys. Lett.*, *66*, 941.
38. German, R. M. (1996). *Sintering Theory and Practice*. New York, NY: John Wiley and Sons.
39. Lee, P.W., Trudel, Y., Iacocca, R., German, R.M.B., Ferguson, L., Eisen, W.B., Moyer, K., Madan, D., and Sanderow, H. (eds.), Powder Metallurgy Technologies and Applications, vol. 7, ASM Handbook, ASM International, Materials Park, OH (1998).
40. German, R.M., Sintering of Advanced Materials: Fundamentals and Processes (Z.Z. Fang, ed.), Woodhead Publishing Limited, Philadelphia, PA (2010).
41. Groza, J.R., Non-Equilibrium Processing of Materials (C. Suryanarayana, ed.), Pergamon, Elsevier Science, Oxford (1999).
42. Thuenissen, G. S. A. M., Winnubst, A. J. A., & Burggraaf, A. J. (1993). *J. Eur. Ceram. Soc.*, *11*, 319.
43. Hahn, H., Logas, J., & Averbach, R. S. (1990). *J. Mater. Res.*, *5*, 609.
44. Porter, D. A., & Easterling, K. E. (1992). *Phase Transformations in Metals and Alloys* (2nd Edition). London: Chapman and Hall 113.
45. Rankin, J., & Sheldon, B. W. (1995). *Mater. Sci. Eng. A*, *204*, 48.
46. Bourell, D. L., Parimal, & Kaysser, W. A. (1993). *J. Am. Ceram. Soc.*, *76*, 705.
47. Ng, L. S., Loh, N. L., & Boey, F. Y. C. (1997). *J. Mater. Process. Technol.*, *67*, 143.
48. Imine, S., Schoenstein, F., Mercone, S., Zaghrioui, M., Bettahar, N., & Jouini, N. (2011). *J. Eur. Ceram. Soc.*, *31*, 2943.
49. Prokofiev, E. A., Jorge, A. M., Jr., Botta, W. J., & Valiev, R. Z. (2014). *Compr. Mater. Process.*, *3*, 187.
50. Pyzik, A.J., and Hchenik, A., Rapid omnidirectional compaction of ceramic-metal composites. In: Proceedings of the 12th Annual Conference on Composites and Advanced Ceramic Materials, Part 1 of 2: Ceramic Engineering and Science Proceedings, vol. 9, Issue 7/8 (J.B. Wachtman, ed.), John Wiley & Sons, Inc., Hoboken, NJ, USA (1988), doi: 10.1002/9780470310496.ch49.

51. Fang, Z.Z., 2010. Sintering of Advanced Materials, first ed. Woodhead Publishing, Elsevier, pp. 356–388.
52. Haizhou, Y., Wenjun, L., Ping, F., & Yong, Z. (2014). *Int. J. Refrac. Metals Hard Mater.*, *45*, 48.
53. Cheng, Y., Sun, S., & Hu, H. (2014). *Ceram. Int.*, *40*, 16761.
54. Shon, I. -J., Jeong, I. -K., Ko, I. -Y., Doh, J. -M., & Woo, K. -D. (2009). *Ceram. Int.*, *35*, 339.
55. Dadbakhsh, S., & Hao, L. (2012). *J. Mater. Process. Technol.*, *212*, 2474.
56. Rosinski, M., Fortuna, E., Michalski, A., Pakiela, Z., & Kurzydlowski, K. J. (2007). *Fusion Eng. Des.*, *82*, 2621.
57. Ishiyama, M., Proceedings of the Powder Metallurgy World Congress, Japan Society of Powder and Powder Metallurgy (JSPM), Kyoto, p. 931 (1993).
58. Tokita, M., Ceram. Trans., 194:51, American Ceramics Society, Wiley-Interscience (2006).
59. Cheng, L., Xie, Z., Liu, G., Liu, W., & Xue, W. (2012). *J. Eur. Ceram. Soc.*, *32*, 3399.
60. El-Eskandarany, M. S., & Al-Hazza, A. (2014). *Mater. Charact.*, *97*, 92.
61. El-Eskandarany, M. S., & Inoue, A. (2006). *J. Mater. Res.*, *21*, 976.
62. El-Eskandarany, M. S., Ishihara, S., Zhang, W., & Inoue, A. (2005). *Metall. Mater. Trans. A*, *36A*, 141.
63. Tokita, M., Spark plasma sintering (SPS) method, systems, and applications. In: Handbook of Advanced Ceramics, Elsevier Inc., Amsterdam (2013).
64. Bernard-Granger, G., & Guizard, C. (2007). *Acta Mater.*, *55*, 3493.
65. Shen, Z. J., Johnsson, M., Zhao, Z., & Nygren, M. (2002). *J. Am. Ceram. Soc.*, *85*, 1921.
66. Anselmi-Tamburini, U., Garay, J. E., & Munir, Z. A. (2006). *Scr. Mater.*, *54*, 823.
67. Kumar, A., et al. (2013). *Nat. Commun.*, *4*, 2618. doi: 10.1038/ncomms3618.
68. Xiao, J., Li, J. L., Liu, P., & Yang, G. W. (2014). *Nanoscale*, *6*, 15098. doi: 10.1039/c4nr05246c.
69. Bovenkerk, H. P., Bundy, F. P., Hall, H. T., Strong, H. M., & Wentorf, R. H. (1959). *Nature*, *184*, 1094.
70. Danilenko, V. V. (2004). *Phys. Solid State*, *46*, 595.
71. Yang, G. W., Wang, J. B., & Liu, Q. X. (1998). *J. Phys. Condens. Matter*, *10*, 7923.
72. Frenklach, M., et al. (1991). *Appl. Phys. Lett.*, *59*, 546.
73. Gogotsi, Y. G., et al. (1996). *J. Mater. Chem.*, *6*, 595.
74. Welz, S., Gogotsi, Y., & McNallan, M. J. (2003). *J. Appl. Phys.*, *93*, 4207.
75. Daulton, T. L., Kirk, M. A., Lewis, R. S., & Rehn, L. E. (2001). *Nucl. Instrum. Methods B*, *175*, 12.
76. Sun, L.T., et al., Appl. Phys. Lett., 84:2901 (2004).
77. Galli, G., Structure, Computer-Based Modeling of Novel Carbon Systems and Their Properties Beyond Nanotubes, Chapter 2, vol. 3 (L. Colombo and A. Fasolino, eds.), Springer, Berlin, p. 37 (2010).
78. Caudillo, R., et al., Radiat. Phys. Chem., 73:334 (2005).
79. Xiao, J., Ouyang, G., Liu, P., Wang, C. X., & Yang, G. W. (2014). *Nano Lett.*, *14*, 3645.
80. Du, G., Song, C., Zhao, J., Feng, S., & Zhu, Z. (2008). *Carbon*, *46*, 92.
81. Barnard, S., Russo, S. P., & Snook, I. K. (2003). *Phys. Rev. B*, *68*, 073406-0.
82. Tyler, T., Zhirnov, V. V., Kvit, A. V., Kang, D., & Hren, J. J. (2003). *Appl. Phys. Lett.*, *82*, 2904.
83. Vandenbulcke, L., Gries, T., & Rouzaud, J. N. (2009). *Appl. Phys. Lett.*, *94*, 044106.
84. Mochalin, V. N., Shenderova, O., Ho, D., & Gogotsi, Y. (2012). *Nat. Nanotechnol.*, *7*, 11.
85 El-Eskandarany, M.S., Method for synthesizing nanodiamonds, US Patent: 9,540,245 B1 (2017).

Mechanochimical process for fabrication of 3D nanomaterials

6

$$WO_3 + 3Mg + C \xrightarrow{\text{Room temperature ball milling}} WC + 3MgO$$

(After El-Eskandarany et al.[16])

6.1 Introduction

As was shown in the previous chapters, most solid-state chemical reactions involve the formation of one (or more) product phases between the reactants, which, therefore, become specially separated. All the solid-state reaction (SSR) rates are controlled by diffusion of the reacting species through the product material. Therefore, the activities of solids are dependent on initial contact areas, hence, particle size, on factors which influence diffusion rates, such as defect densities and local temperatures, and on product morphology.[1–3] In conventional thermally activated systems, the reactants remain substantially unchanged and specially separated during the course of the reaction. A significant increase in reaction rate could result if a means were found to agitate the reaction mass and continually bring unreacted material into contact, preventing the products from dividing the reactants. In fact, mechanical alloy processing is able to provide such a potential (Fig. 6.1).

The displacement reactions of the type

$$A_xO + vB \rightarrow xA + B_vO$$

have been observed during MA processing of a simple metal oxide and a strong metallic reducing agent (RA), such as elemental Ca, Mg, Ti, Fe, Ni, and Al powders.[4–7] This process is known as mechanical solid-state reduction (MSSR).[8] In addition, the MSSR has been employed to fabricate nanocomposite WC/MgO materials by milling elemental Mg with WO_3 powders in a high-energy ball mill.[9]

6.2 Reduction of Cu_2O with Ti by room temperature rod milling

In 1995, the room temperature rod-milling technique was employed for reducing Cu_2O with elemental Ti powders.[8] In this experiment, Cu_2O (99.9%, 50 μm) and Ti (99.5%, 100 μm) powders were mixed in a glove box under a purified argon atmosphere to give the desired starting average composition of $(Cu_2O)_xTi_{1-x}$. The mixed

Mechanical Alloying. http://dx.doi.org/10.1016/B978-0-12-818180-5.00006-6

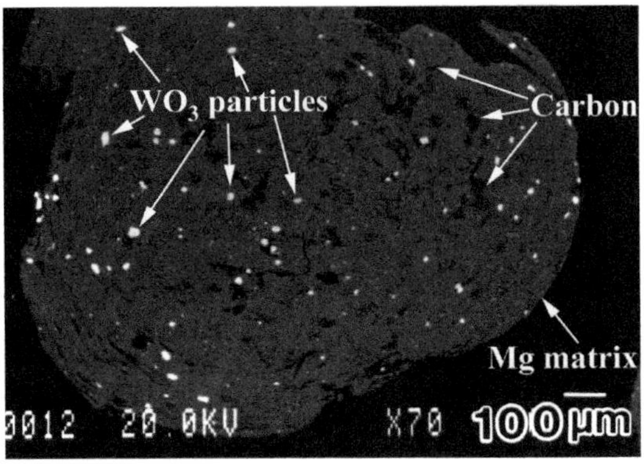

Figure 6.1 SEM micrograph of ball-milled WO3/3Mg/C powders.[16]

powders were then charged and sealed in a cylindrical stainless steel shell (SUS 304, 120 mm in diameter) together with stainless steel rods (SUS 304, 10 mm in diameter). The rod-to-powder weight ratio was controlled to about 30:1. The inlet of the shell was connected with a rotary pump and evacuated for about 4 ks. After evacuation, a flow of argon gas was passed into the rod mill through a plastic pipe. Once the rod mill was filled with 1 atm of argon the inlet of the vial was closed and the MSSR process was carried out at room temperature by mounting the rod mill on a rotator at the rate of 1.4 s^{-1}. The MSSR was interrupted at selected intervals and a small amount of the rod-milled powder was taken out of the vial in the glove bag. The powders were characterized by X-ray diffraction (XRD) with CuKα, transmission electron microscopy (TEM) using a 200 kV microscope and differential thermal analysis in an argon gas atmosphere at a heating rate of 0.33 K s^{-1}. The metallographic characterizations of the milled powders were studied by an optical microscope. Moreover, the induction coupled plasma (ICP) emission method was used to detect the degree of Fe contamination in the milled powders. After 360 ks of milling, the iron contamination content in the alloy powders was determined to be 0.25 at. %.

6.3 Properties

6.3.1 Structural changes with the milling time

The XRD patterns of Cu_2O milled with Ti are presented in Fig. 6.2 after selected MSSR times. At the initial stage of MSSR process (0 ks), the powders consist of unreacted polycrystalline fcc-Cu_2O and hcp-Ti (Fig. 6.2A). The intensities of the peaks corresponding to Cu_2O and Ti crystals decrease remarkably after 43 ks, as displayed in Fig. 6.2B. Increasing the milling time to 173 ks enhances the SSR of the reactants (Cu_2O and Ti) which translates into an MSSR between Cu_2O and Ti, as shown in Fig. 6.2C. At this intermediate stage of milling, new phases corresponding to bcc-Cu and tetragonal-TiO_2 appear, suggesting an increase in the chemical kinetics of MSSR

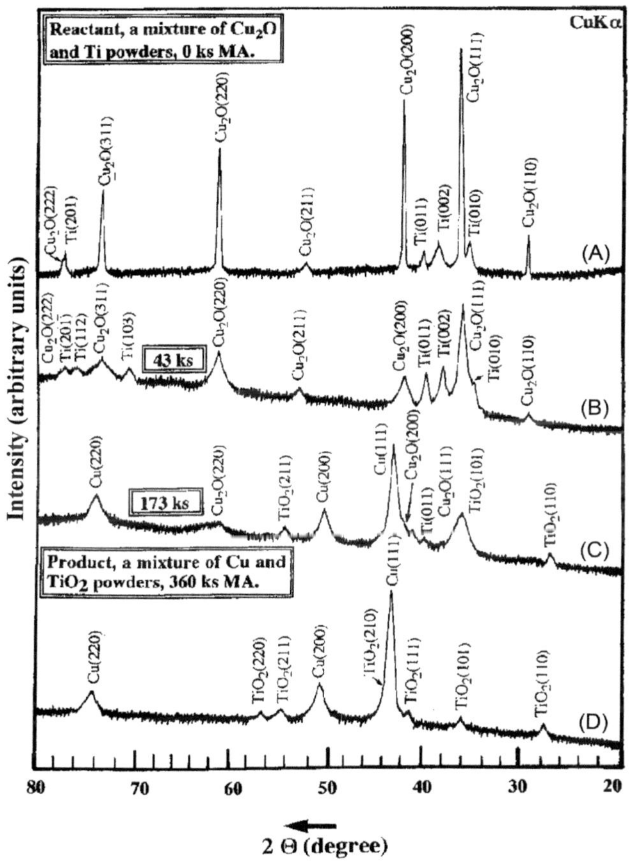

Figure 6.2 XRD patterns of mechanically alloyed Cu_2O and Ti powders after (A) 0 ks, (B) 43 ks, (C) 173 ks, and (D) 360 ks of the rod-milled times. *(After El-Eskandarany.)*[8]

reaction. The yield product at this stage of milling coexisting with unreacted Cu_2O and Ti powder particles is displayed in Fig. 6.2C. Fig. 6.2D shows the XRD patterns of the end product of the MSSR process after 360 ks of rod milling time. Obviously, the Bragg peaks that correspond to bcc-Cu become more pronounced, suggesting an increase in the mole fraction of metallic Cu in the yield product. Remarkably, these peaks of bcc-Cu are broad, indicating the formation of nanocrystalline Cu with an average crystallite size of 7 nm. Moreover, the starting materials (reactants) of Cu_2O and Ti powders no longer appear, suggesting the completion of MSSR process.

Fig. 6.3 displays the change in the concentration of a reactant (Ti) and a product (Cu) with the rod milling time. The concentrations of Ti and Cu were calculated by measuring the areas under the diffraction peaks of Ti (011) and Cu (111) reflections. During the first few kiloseconds (0–43 ks), no remarkable change in the concentrations of Ti and Cu can be observed. Increasing the milling time (43–256 ks) increases the rate of the reaction. Thus, the concentration of Cu increases monotonically with a

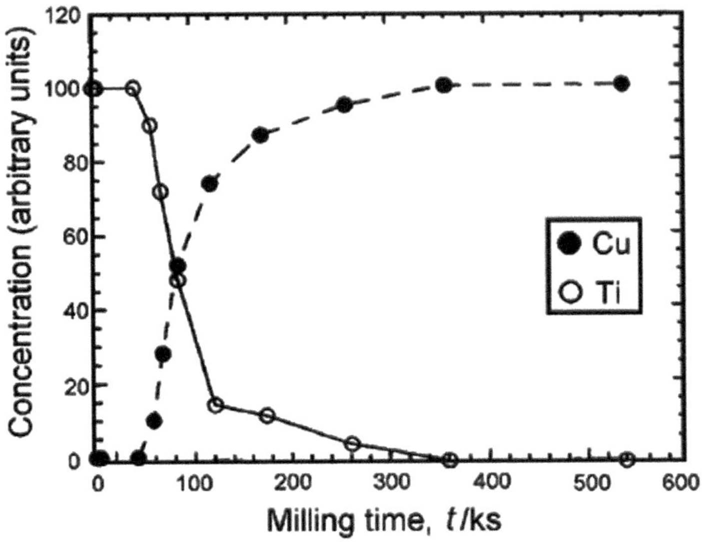

Figure 6.3 Effect of the milling time on the concentration of Cu and Ti in the yield product during rod milling of a mixture of Cu_2O and Ti powders. The concentrations of Ti and Cu were calculated by measuring the areas under the diffraction peaks of the Ti (011) and Cu (111) reflections.

dramatic decrease of the concentration of Ti, as shown in Fig. 6.3. Toward the end of the MSSR process (360 ks) the reduction of Cu_2O to Cu is completed.

Detailed TEM analyses were performed in order to follow the progress of the MSSR process with variation in the rod milling time. The bright-field images (BFIs) and the selected area diffraction patterns (SADPs) of rod-milled Cu_2O-Ti powders after selected milling times are shown in Figs. 6.4–6.6.

Fig. 6.4 shows the (A) BFI and (B) corresponding SADP of a mixture of Cu_2O-Ti powders (reactant) at the starting stage of milling. Obviously, the powders contain

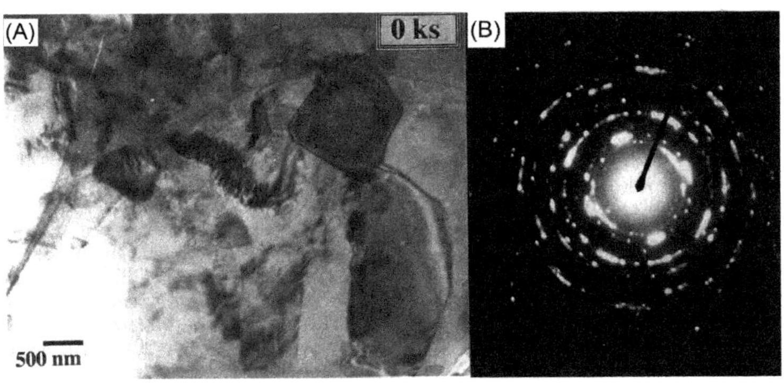

Figure 6.4 (A) BFI and (B) SADP of the reactant materials of Cu_2O and Ti powders at the starting stage (0 ks) of rod milling time. *(After El-Eskandarany.)*[8]

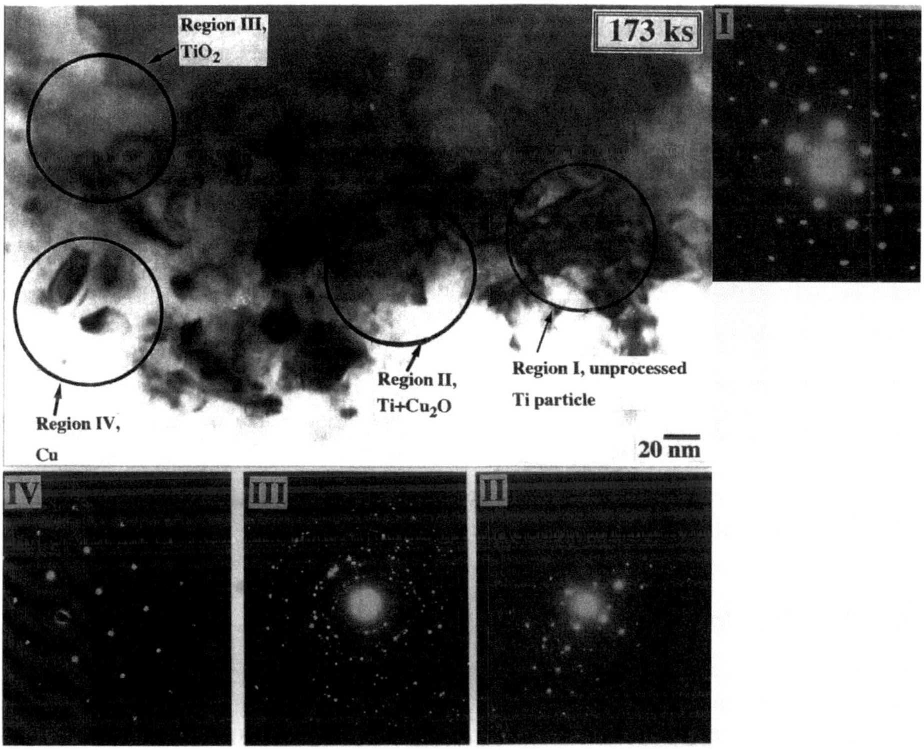

Figure 6.5 BFI and the corresponding SADPs of several regions for mechanically alloyed Cu_2O and Ti powders after 173 ks of the rod milling time. *(After El-Eskandarany.)*[8]

large grains of about 500 nm in diameter. The SADP shows a ring spot pattern that is characteristic of the polycrystalline fcc-Cu_2O and hcp-Ti.

The BFI and the corresponding SADPs of the powders milled for 173 ks of rod milling are shown in Fig. 6.5. At this intermediate stage of the MSSR process, the structures of the particles are extremely heterogeneous, shown by the different features of the powders in Fig. 6.5, suggested by SADPs taken at regions I (unprocessed Ti), II (diffracting polycrystalline Ti and Cu_2O), III (TiO_2), and IV (Cu). It is worth noting that the powder particles consist of somewhat fine grains, ranging from 10 to 50 nm in diameter. Increasing the rod milling time leads to an increase in the shear force generated by the rods and this causes several mechanical defects in the powder particles which translate into a continuous decrease in the unprocessed Cu_2O and Ti powders. Thus, atomically clean surfaces are created and brought into intimate contact so that MSSR takes place and the mole fraction of Cu and TiO_2 increase against the concentration of the reactant material of Cu_2O and Ti.

The BFI and corresponding SADP for the end product of the MSSR process are shown in Fig. 6.6A and B, respectively. The powders are a mixture of polycrystalline Cu (agglomerated dark grains) and TiO_2 (the light fine spherical grains). The reactant powders no longer appear, suggested by the ring diffraction patterns of fcc-Cu coexistent with tetragonal-TiO_2, as shown in Fig. 6.6B.

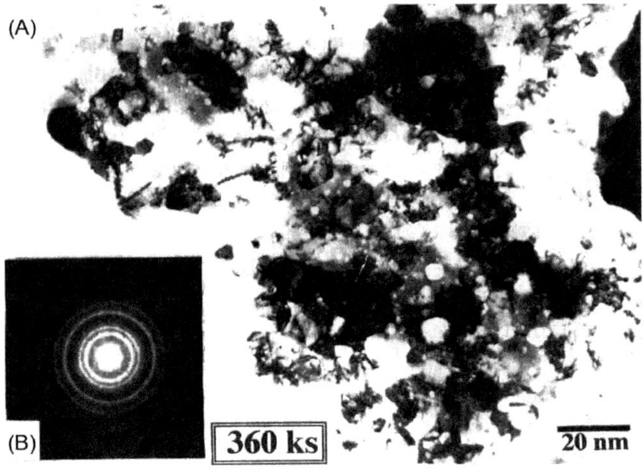

Figure 6.6 (A) BFI and (B) SADP of the powder particles at the end stage of the MSSR process (360 ks of the rod milling time). *(After El-Eskandarany.)*[8]

6.3.2 Metallography

Fig. 6.7 shows a cross-sectional view of polished and etched powder particles of Cu_2O and Ti, the reactant materials that were milled together for (A) 43 ks and (B) 360 ks of the MSSR time. Obviously, after 43 ks of MSSR processing time, the powder is a mixture of Cu_2O grains (dark gray submicron grains) and flaky Ti particles, as presented in Fig. 6.7A. At the final stage of MSSR (360 ks), Cu_2O was reduced completely by Ti and the end product of this stage contained fine powders of Cu (gray particles) together with a relatively fine equiaxed cross-section of TiO_2 grains (dark gray), as presented in Fig. 6.7B.

6.3.3 DTA measurements

The DTA curves for Cu_2O milled with Ti are presented in Fig. 6.8 after (A) 0 ks and (B) 360 ks of rod milling times. All the samples were heated to about 1370 K (first heating run) and cooled to about 310 K. Then, second heating runs (dashed lines) were performed, in order to get a base line. At the starting stage of the MSSR process (0 ks), two reactions between the reactant materials (Cu_2O and Ti powders) took place, characterized by two exothermic peaks which appear at about 773 K and 973 K, as displayed in Fig. 6.8A. These reactions occur due to the reduction of Cu_2O with Ti at such high temperatures. The products of these reactions, as indicated by XRD, contain TiO_2 and metallic Cu that melt at about 1356 K, characterized by a sharp endothermic peak (Fig. 6.8B). At the second heating run, a single sharp endothermic peak appears, due to the melting of pure Cu powders in the yield product. Moreover, no exothermic reactions could be detected, suggesting the completion of the reduction process via

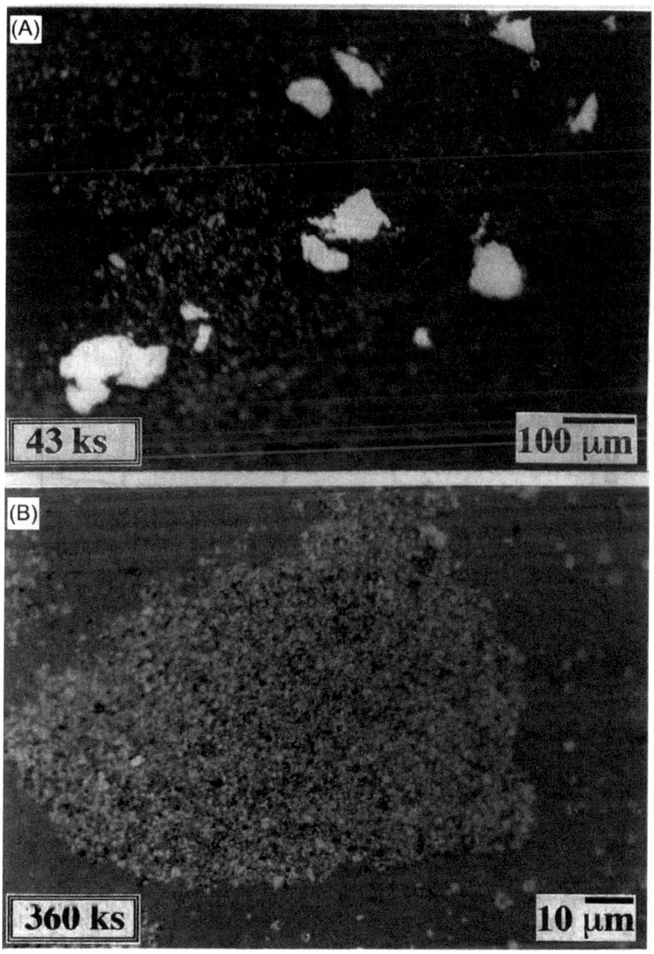

Figure 6.7 Optical micrograph of a cross-sectional view of polished and etched powder particles of Cu_2O and Ti, being the reactant materials that were milled together for (A) 43 ks and (B) 360 ks of the MSSR time. *(After El-Eskandarany.)*[8]

DTA. We conclude that Cu_2O can be reduced to Cu using Ti as an RA at temperatures higher than 973 K.

Fig. 6.8B displays the DTA traces for the end product of milled Cu_2O with Ti powders after 360 ks of rod milling time. Obviously, during heating of the sample, an endothermic reaction takes place, characterized by a single sharp endothermic peak centered at 1356 K. This is attributed to the melting of Cu powders in the end product. Moreover, the exothermic reaction peaks that relate to the reaction of unprocessed Cu_2O with Ti powders (Fig. 6.8A) have already disappeared. This indicates

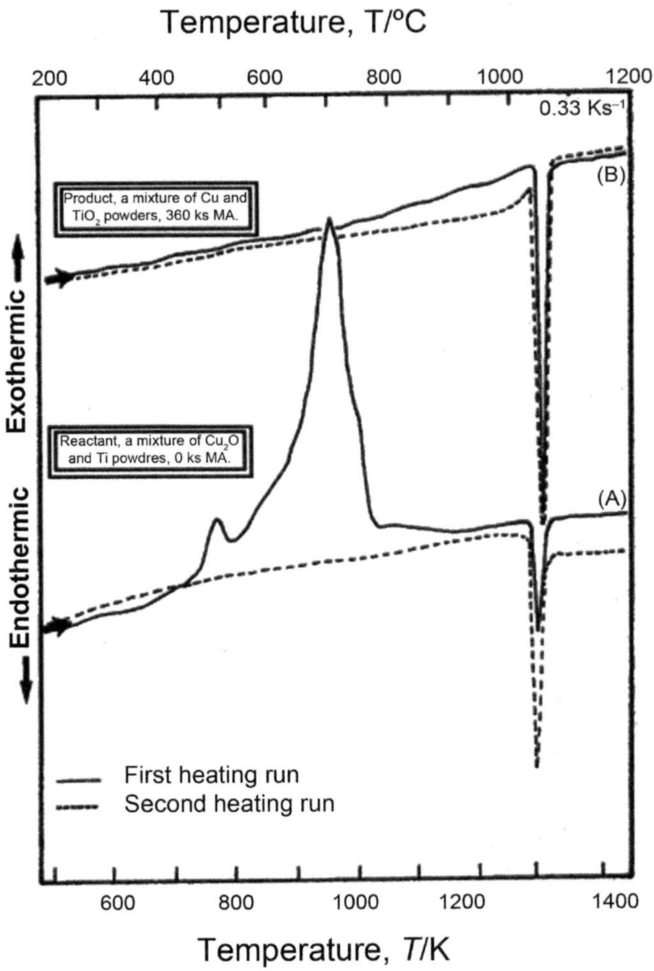

Figure 6.8 DTA curves for Cu_2O milled with Ti after (A) 0 ks and (B) 360 ks of rod milling times. The *solid* and *dashed lines* represent the first and the second heating runs, respectively. *(After El-Eskandarany.)*[8]

the formation of new phases (Cu and TiO_2) without the existence of any other unprocessed reactant materials of Cu_2O or Ti.

6.4 Mechanism of MSSR

The reduction process reported in the previous section is considered a new application of the MA method for reducing metal oxide using a metallic RA via rod-milling technique. This application was reported earlier by Schaffer et al.[4] when they reduced cupric oxide (CuO) to pure copper metal by several metallic RAs, for example, Ca and Mg. Our results[8] have shown that Cu_2O could completely be reduced to metallic Cu by milling a mixture of Cu_2O and elemental Ti powders (due to the stoichiometry of reaction)[10] in a rod mill filled with argon atmosphere at room temperature. The completion of the so-called MSSR process takes 360 ks of milling time. For MSSR, the basic change that occurs during milling the reactants (Cu_2O and Ti powders) can be described in terms of the transfer of electrons, as presented in the following reaction:

$$2\,Cu_2O(s) + Ti(s) \rightarrow TiO_2(s) + 4\,Cu(s)$$

Principles of physics and physical chemistry always lie at the root of every individual application of the MA technique. Obviously, the previous reaction is a typical oxidation-reduction reaction that involves the oxidation of Ti to TiO_2. The oxygen atoms required for this oxidation come from Cu_2O which is reduced to Cu. The formation enthalpy, ΔH^{for}, for TiO_2 (−315 kJ/mol)[11] is about 5 times larger than that for Cu_2O (−57 kJ/mol).[11] In other words, TiO_2 is more stable than Cu_2O so that the aforementioned reaction is a forward reaction and occurs during the rod milling of Cu_2O and Ti. It is shown that the chemical kinetics of the reaction indexed by the rate of MSSR increases monotonically by increasing the milling time (the rate at which the concentration of Cu and TiO_2 increases). However, this rate depends on several known factors such as the nature and concentration of the reacting species, and it can be controlled by the type of mill used. A high-energy ball mill (attritor, vibrating, or planetary mills) may increase the reduction rate due to the generated temperature caused by the collisions between the balls. One drawback of using these high-energy mills comes from the fact that iron contamination could be brought from the milling media during the milling process, especially when using hard reactant materials.[12] Thus, the MSSR will face a serious problem presented by the existence of such foreign material in the end product. Moreover, this foreign material can be alloyed with the end product and a complex phase may be formed. In addition, the gradual oxidation-reduction reaction can be followed easily by low-energy mills which allow us to study the characteristics of the yield product during the different stages of the MSSR process. For these reasons, the rod-milling technique has been used in the present investigation.

For a successful MSSR via rod-milling technique, the elemental particles of RA powders should disintegrate during milling into smaller particles in which fresh surfaces are continuously created. These new particles, with their fresh surfaces, are very active and react immediately with Cu_2O. We should emphasize that these new surfaces play a very important role in the MSSR process; no reduction can occur without the presence of these surfaces. For example, magnesium (Mg) is a well-known reactive RA with two electrons in the outermost shell [(Ne)$3s^2$]; however, during rod milling of a mixture of Cu_2O and Mg under the same mentioned milling conditions, no reaction between Cu_2O and Mg could be detected.[13] This is attributed to the fact that Mg is a very soft metal with a low melting temperature (compared with Ti) and is stacked on the milling tools (due to the cold welding). In contrast to Schaffer et al.[4] during rod milling, Mg powders are being isolated from Cu_2O powders. We should emphasize that the milling or grinding processes are necessary steps in extractive metallurgy. The possibility of performing prereduction stages by the MSSR method may be applied.

6.5 Fabrication of nanocrystalline WC and nanocomposite WC-MgO refractory materials by MSSR method

Among hard alloys, WC alloys find wide industrial applications as tips for cutting tools and wear-resistant parts. The intrinsic resistance to oxidation and corrosion at high temperatures also makes them desirable as a protective coating for devices at elevated temperatures. In the industrial scale of production, the hexagonal phase of WC is usually obtained by a direct alloying of elemental W and graphite at temperatures ranging from 1673 to 1873 K.[14,15] Accordingly, the high cost of preparation is a disadvantage of this process. It has been shown by El-Eskandarany[9,16] that nanocrystalline WC can be obtained by reducing WO_3 through ball milling this material in the presence of Mg and graphite. The Mg reduces the oxide, and the resulting metallic W reacts with graphite to form WC. The oxide phase of MgO is simply removed by selective leaching. Mi et al.[17] have reported the formation of fine-grained WC by ball milling the elemental powders of W and C.[17] They have also shown the possibility of synthesizing composite WC-Co by milling their previously reacted WC particles with elemental Co powder.

However, because the hexagonal phase of WC alloy possesses high hardness (23 GPa),[16] it is rarely used in industrial applications in its native form. This is due to its relatively poor shock resistance, high values of elastic moduli and bulk density, and low fracture toughness. Thus, a binding material (usually Co) is often added to WC during sintering in order to improve the mechanical properties. Co is desirable because it will wet the surface of WC while maintaining negligible solubility in the carbide phase. The present section will present the formation of a new nanocomposite WC/MgO material for the use in industrial applications. In this method, a WC/MgO end product forms during ball milling of WO_3, Mg, and C, and is subsequently consolidated into a nanocomposite bulk compact, using a plasma-activated sintering (PAS) method.

One purpose of the present section is to describe a unique process for forming fully dense, nanocomposite materials. The significant technological potential of the present work derives from a relatively inexpensive and simple process for synthesizing ceramic/ceramic composite materials for applications in cutting tools, tips for drilling tools, and wear-resistant parts used in wire drawing, extrusion, and pressing dies. Finally, this work demonstrates a viable process for preparing bulk, fully dense, nanocomposite refractory material using MSSR as a source of nanophase powder.

In the experiments, elemental (99.5%) powders of WO_3 (5 μm), Mg (750 μm), and graphite (5 μm) with an atomic ratio of 1:3:1, respectively, were mixed in a glove box under argon gas atmosphere and sealed in a cylindrical sapphire vial together with 50 sapphire balls (10 mm in diameter). The ball-to-powder weight ratio was maintained at the level of 10:1. The ball-milling experiments were carried out at room temperature, using a Fritsch P6 high-energy ball mill at a rotation speed of 4.2 s^{-1}. The milling experiments were interrupted at regular intervals and small amounts of the milled powders were taken out of the vial in the glove box. The powders were characterized by means of XRD with CuKα radiation, scanning electron microscope (SEM), TEM using 200 kV microscope, and/or high-resolution transmission electron microscope (HRTEM). The powders were chemically analyzed to determine their concentrations of W and Mg, using the ICP emission method. The sample of the end product contains about 18 at. % MgO. The oxygen and carbon concentrations were investigated by the helium carrier fusion-thermal conductivity method.

Since WC is chemically stable at room temperature and cannot be attacked, even by strong acid solutions, the MgO powder is simply removed from selected samples, using a mixture of HCl (7 mol) and H_3PO_4 (3 mol). The as-leached product, which is pure WC powder, is consolidated into a bulk sample using a PAS method. The same sintering technique has been employed to consolidate the as-milled powder of composite WC-18% MgO material (before leaching). All the consolidation procedures of the present work took place in a vacuum at 1963 K with a pressure of 19.6–38.2 MPa. In order to avoid undesired grain growth, the sintering process was applied for only 0.18 ks without adding any binding materials. The densities of consolidated WC and WC-MgO materials were determined by Archimedes' principle, using water immersion, and found to be 15.60 and 13.10 g/cm^3, respectively. Comparing these values with those for pure WC (theoretical density, 15.77 g/cm^3)[14] and the calculated one for WC-18% MgO (using the rule of mixtures principal, 13.21 g/cm^3) indicates the formation of fully dense bulk materials. Vickers indenter with a load of 50 kg was employed to determine the hardness of the compacted samples. The size of the indentation cracks was used to determine the fracture toughness (K_c) of the sample.[18] The hardness and K_c values reported later are averaged from at least 10 indentations. The elastic properties of the bulk samples were determined by a nondestructive test using the pulse-echo overlap ultrasonic technique using an ultrasonic detector.

6.5.1 Properties of ball-milled powders

6.5.1.1 Structural changes with the milling time

Fig. 6.9 presents the XRD patterns of the (A) starting reactant materials, (B)–(D) as-milled, (E) as-leached powders (pure WC powder) and (F) consolidated sample (bulk WC material). At the starting stage of milling (0 ks) the powders are a mixture of elemental WO_3, Mg, and C materials with sharp Bragg peaks (Fig. 6.9A). After 22 ks of milling, Bragg peaks of metallic W and MgO crystals appear, indicating the starting of an MSSR, which takes place between Mg (RA) and WO_3 powders (Fig. 6.9B). In parallel to the MSSR process, an SSR occurred between W and C powders to yield a refractory phase of hcp-WC. The Bragg peaks of these reacted phases have become pronounced and their intensities increased after 43 ks (Fig. 6.9C), suggesting an increase in the product materials. No free atoms of the starting reactant materials and/or metallic W powders could be detected after 173 ks of milling, suggesting the completion of both MSSR and SSR processes (Fig. 6.9D). When the end product of this stage

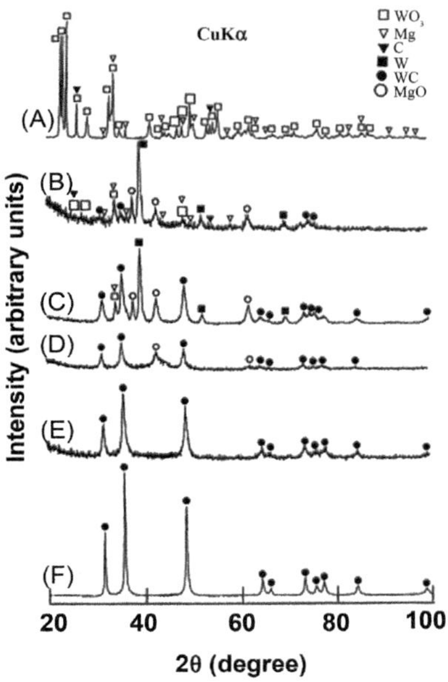

Figure 6.9 Typical XRD patterns of ball-milled WO_3, Mg, and C powders at (A) 0 ks, (B) 22 ks, (C) 43 ks, (D) 173 ks of the milling time, (E) the end product after leaching, and (F) the end product after leaching and consolidation. *(After El-Eskandarany.)*[9]

(nanocomposite WC/MgO material) is leached with an acidic agent (Section 6.5), the Bragg peaks of MgO completely vanish and the residual product is nanocrystalline WC with an average grain size of 7 nm (Fig. 6.9E). The XRD pattern of the as-consolidated WC powder is shown in Fig. 6.9F. Comparing this pattern with that one shown in Fig. 6.9E, one can conclude that this consolidation step does not lead to a dramatic change in the grain size and the as-sintered sample still maintains its unique nanocrystalline properties. Moreover, this sintering procedure does not lead to any phase changes and/or transformations and the sample is a single phase of hcp-WC.

Fig. 6.10 displays the XRD pattern of the as-consolidated WC-18% MgO powder (before leaching). Obviously, the compact sample contains two phases of WC and MgO without the presence of any reactive products. Again, one can conclude that the compacted sample is nanocomposite, characterized by a broadening in the Bragg peaks for both WC and MgO phases.

The BFI and the dark-field image (DFI) of the powder that milled for 1.8 ks are shown, together with the SADP, in Fig. 6.11. Obviously, the powders still contain the reactant elemental phases of WO_3, Mg, and graphite, with sharp spot diffraction patterns, as displayed in Fig. 6.11C.

Fig. 6.12 shows the (A) BFI of the milled powders after 3.6 ks, (B) and (C) the corresponding SADPs of two regions. The hard oxide phase of WO_3 and cubic graphite powders (Fig. 6.12B) are embedded into the soft metallic matrix of the agglomerated Mg (Fig. 6.12C) particles to form a composite-reactant material with ellipsoid-like morphology (Fig. 6.12A).

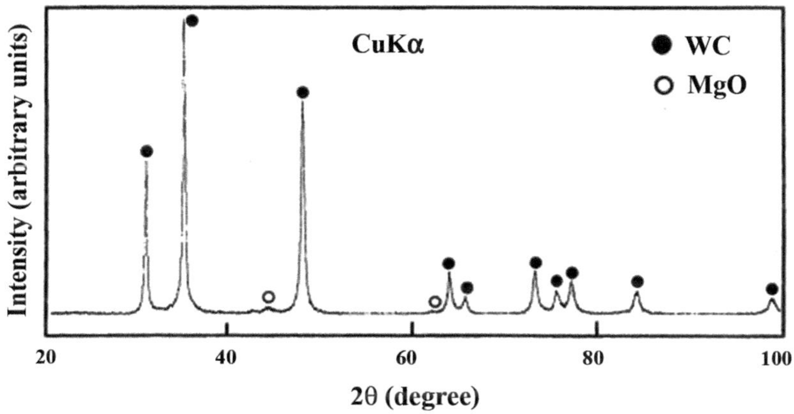

Figure 6.10 XRD pattern of as-consolidated WC-18% MgO (end product, 173 ks of milling) before leaching. *(After El-Eskandarany.)*[9]

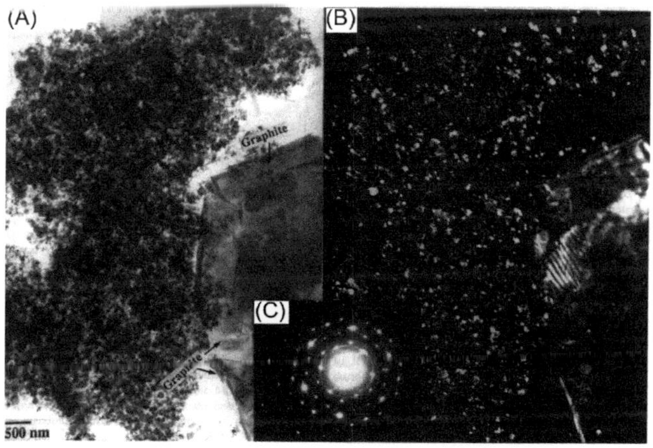

Figure 6.11 (A) BFI, (B) DFI, and (C) the corresponding SADP of the as-milled powders after 1.8 ks of the ball-milling time. *(After El-Eskandarany.)*[9]

Fig. 6.13 presents the BFI of the final product (173 ks) of the milled powders (before leaching). The micrograph shows a typical nanocomposite material consisting of the host ceramic matrix of WC and uniform spherical nanoscale grains of MgO (reinforcement materials) that are embedded into the matrix.

The BFI and SADPs of the as-leached powder are shown together in Fig. 6.14. The powder has the fine structure (Fig. 6.14A) of a single phase of WC, as indicated by SADPs which were taken from several zones (Fig. 6.14B–D).

Fig. 6.15 shows the BFI of the cross-sectional view of the as-consolidated WC (A) and composite WC/MgO (B) powders. Obviously, the consolidation procedure leads to the formation of nanocrystalline WC with an average grain size of about 25 nm in diameter. It can be noted that the consolidated sample of composite WC/MgO

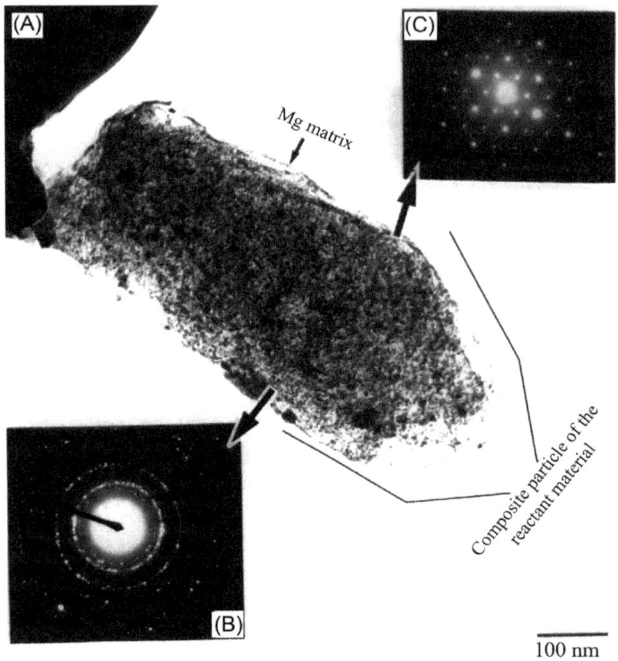

Figure 6.12 (A) BFI, (B) and (C) the corresponding SADPs of the as-milled powders after 3.6 ks of the ball-milling time. The micrograph presents a typical individual agglomerated reactant particle. *(After El-Eskandarany.)*[9]

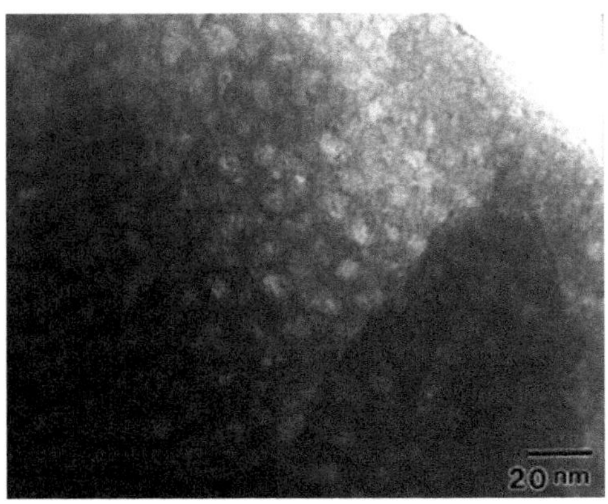

Figure 6.13 BFI of as-milled powder after 173 ks (end product). The spherical grains of MgO, with about 7 nm in diameter, are embedded in the WC matrix to form nanocomposite WC-MgO powders. *(After El-Eskandarany.)*[9]

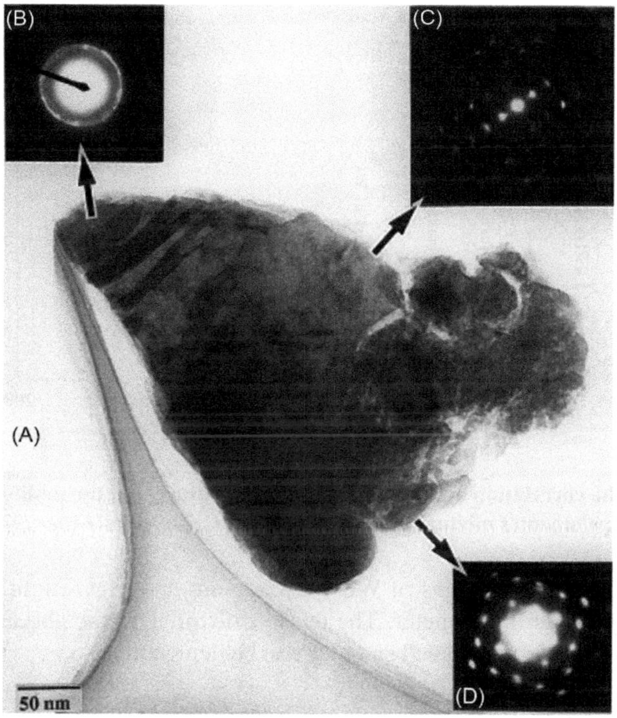

Figure 6.14 (A) BFI and (B), (C), and (D) the corresponding SADPs of the end product after leaching. *(After El-Eskandarany.)*[9]

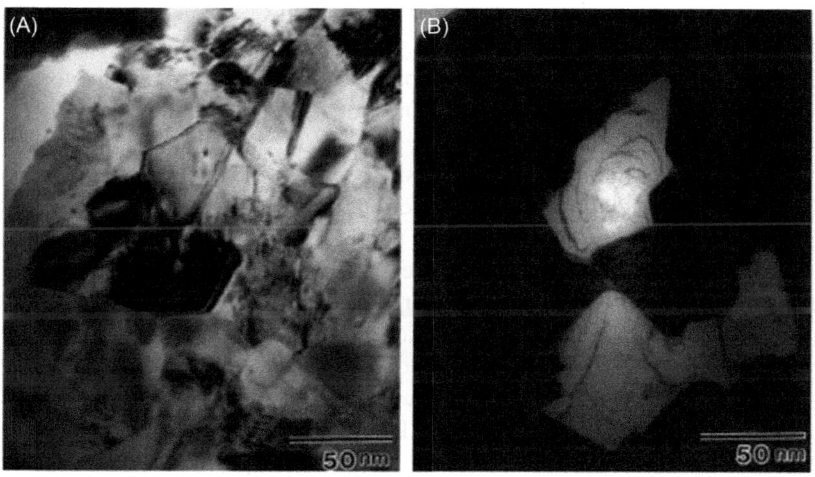

Figure 6.15 BFI planner view of the as-consolidated end product (A) after leaching and (B) before leaching. *(After El-Eskandarany.)*[9]

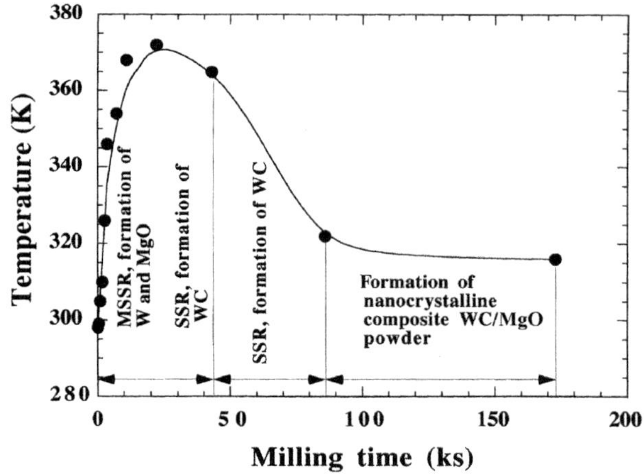

Figure 6.16 The correlation between the vial's temperature and the milling time during the ball milling elemental mixture of (WO_3 + 3 Mg + C) powders. *(After El-Eskandarany.)*[9]

contains nanocrystalline grains of WC (dark region) and MgO (light gray region) of not more than 50 nm in diameter. The dense structure and the absence of any voids and/or cracks in the consolidated samples can be depicted.

6.5.1.2 Temperature change with the milling time

Fig. 6.16 shows the change of the vial's temperature during the milling time. It has been shown in a previous work[19] that this temperature did not increase more than 314 K, when the same mill was operated without charging the reactant powders. During the first few kiloseconds (0–3.6 ks), the temperature of the vial increased slightly, and the starting reactant materials of WO_3, Mg, and C, which have irregular shape and wide particle size distributions (Fig. 6.17A), agglomerated to form larger particles. The cross-sectional view for these particles (Fig. 6.17B) shows the formation of composite powders consisting of WO_3 (fine white grains about 20 μm in diameter), C (black grains about 100 μm in diameter), and Mg (the gray matrix about 1800 μm in diameter). The formation of such composite powder is attributed to the cold welding performed at this stage of milling in which the ductile Mg particles are deformed to form soft matrices containing rather hard particles of WO_3 and C powder. Increasing the milling time (3.6–43 ks) enhances the MSSR and SSR between the reactant materials, hence the heat liberated by these reactions is transferred to the surface of the vial and leads to an increase in its temperature to about 370 K (Fig. 6.16). Toward the end of this stage (43 ks), a small mole fraction of WC is formed as a result of an SSR between the W and C powders. A complete SSR takes place between W and C powders at the second stage of milling (43–86 ks) and the vial's temperature drops to about 325 K. At the end of the MA time (173 ks), the vial's temperature remains at a constant value of about 320 K, suggesting the completion of both reactions and the formation of the final product, which is a composite WC/MgO powder which has spherical-like morphology, with an average particle size of about 0.5 μm (Fig. 6.17C).

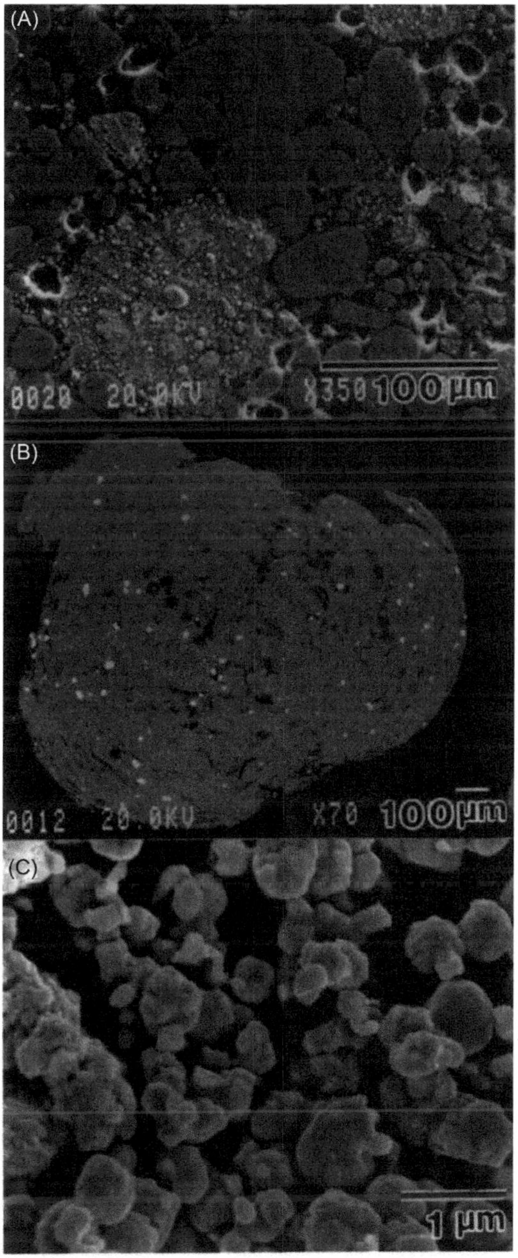

Figure 6.17 SEM micrographs of the as-milled powders after (A) starting stage, (B) 3.6 ks, and (C) 173 ks of the ball-milling time. *(After El-Eskandarany.)*[9]

6.5.1.3 Hardness, toughness, and elastic moduli of consolidated WC and WC/MgO

A SEM micrograph of Vickers hardness indentation of the consolidated sample appears in Fig. 6.18. The hardness of this compact sample was measured to be 15 GPa, taking the average of at least 10 indentations. It is worth noting that this value is higher than that for bonded WC-14% Co (13.2 GPa).[20] The cracks displayed in the micrograph, which developed in the product during indentation and extended 250–350 μm away from it, were used as indicators for estimating the fracture toughness, K_C, via the model that was suggested by Anstis et al.[18] This value is found to be 14 MPa·m$^{1/2}$, well above that for commercial WC/Co composites (12 MPa·m$^{1/2}$). This indicates the possibility of producing high fracture toughness WC/MgO nanocomposite materials.

Table 6.1 summarizes the results of some mechanical properties for the consolidated samples. However, pure nanocrystalline WC possesses extraordinarily high values of hardness (23 GPa), Young's and shear moduli (813.5 and 405.1, respectively), and has a low value of K_C (4 MPa·m$^{1/2}$), making it difficult for use as a cutting tool. Contrary to the WC material, composite WC/MgO combines unique properties of high hardness value (15 GPa) and an extremely high value of K_C (14 MPa·m$^{1/2}$) with reasonable values of Young's (413 GPa) and shear (171 GPa) moduli. In fact, composite WC/MgO material is a unique material in that it can combine two opposite properties, high hardness, and high fracture toughness. This may be understood when recognizing that the fabricated WC/MgO material consists of fine grains (less than 50 nm in diameter) that are able to improve the mechanical properties of the fabricated materials and introduce materials with unique properties.

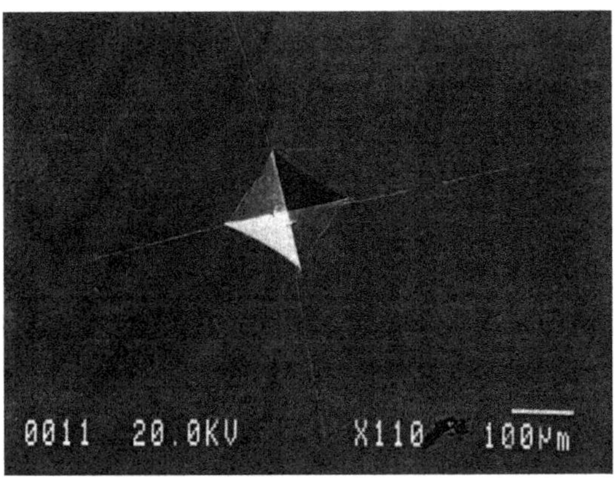

Figure 6.18 SEM micrograph of Vickers hardness indentation developed by applying a load of 50 kg on the consolidated sample of the end product (173 ks). The presented radial cracks, induced during the indentation procedure with extensions of 250–350 μm away from the indentation, were used to estimate the fracture toughness, K_C. *(After El-Eskandarany.)*[9]

Table 6.1 Some measured physical and elastic moduli of the fabricated nanocrystalline WC and nanocomposite WC/MgO materials. *(After El-Eskandarany.)*[9]

	Consolidated sample of as-mechanically reacted and then leached WC	**Consolidated sample of as-mechanically reacted WC-18 at. % MgO**
Density (g/cm^3)	15.6	13.1
Poisson's ratio	0.004	0.209
Young's modulus (GPa)	813.5	413
Shear modulus (GPa)	405.1	171
Bulk modulus (GPa)	273.4	237
Hardness (GPa)	23	15
Fracture toughness (MPa·m$^{1/2}$)	4	14

6.6 c-BN

Cubic phase (c) of Born nitride (BN) has been receiving great attention as an advanced technological materials that find wide range of useful industrial applications. Due to its high hardness value (the second hardest material after diamond), it is mainly used as grinding and cutting tools, especially for ferrous materials. In contrast to diamond, BN does not react with iron. In addition, BN with possesses high thermal conductivity and chemically very stable at very high temperature. Accordingly, BN is used as crucibles for single crystal growth.

6.6.1 Synthesis of BN-nanotubes by RBM

In 1999, Chen et al. reported the formation of BN nanotubes using a reactive ball milling (RBM) method coupled with isothermal annealing processes.[21] In their experiments, the ball-milling procedure was conducted using a steel rotating ball mill. They loaded several grams of elemental B powders together with four steel balls with diameter of 25.4 mm. Thus, the vial was then filled with reactive ammonia gas (NH_3) at a positive pressure of 300 kPa (3 bar) prior to milling. They used pressure gauge to monitor the pressure inside the vial during the ball-milling process.[21] The pressure changes recorded during over the whole ball-milling process are illustrated in Fig. 6.19. Obviously, a rapid reduction on the pressure value from 300 to 190 kPa occurred during the first 36 h of milling, which was caused by absorption of ammonia gas onto fresh surfaces of B particles newly created by high-energy impacts.[21] The pressure inside the vial started to rise during further milling and exceeded the starting pressure at the end of 150 h (315 kPa), as shown in Fig. 6.19. The increase in the pressure values implies the occurrence of nitriding reactions between boron powder and the absorbed ammonia gas. They attributed such pressure increase due to the hydrogen realizing during the gas-solid reaction ($2B + NH_3 = 2BN + 3H_2$) that took place inside the vial. The nitrogen gas content in the ball-milled B powders was analyzed as a function of the milling time, as shown in Fig. 6.20. The gradual increase of the N content over the whole milling process is consistent with ammonia absorption and a nitriding

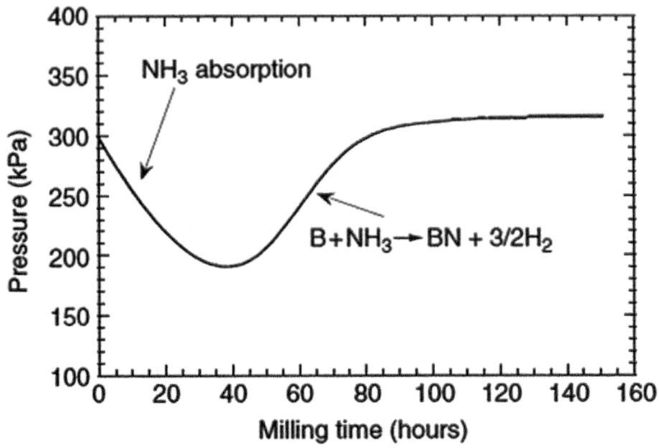

Figure 6.19 Pressure changes inside the milling chamber of a steel rotating mill during milling of B powder in NH_3 at room temperature.[21]

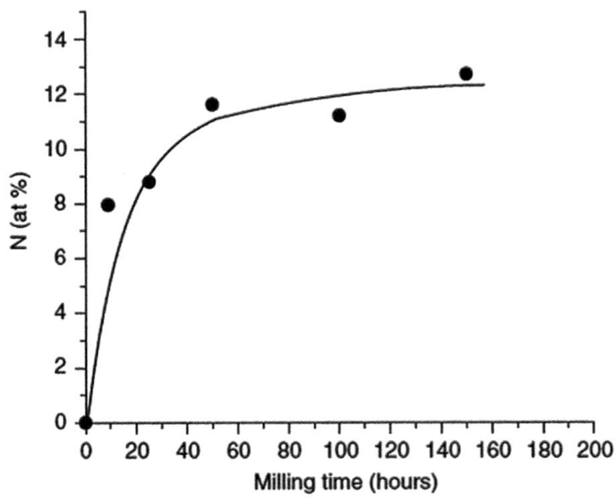

Figure 6.20 Nitrogen content of milled B samples in the rotating mill as a function of milling time.[21]

reaction that results in the formation of BN. The high N content (10.1 wt%) and low H content (0.9 wt%) in the 150 h milled sample confirm the presence of nitriding reactions during further milling stages.

The XRD patterns of the end product of BN, which was obtained upon RBM of B for 150 h under NH_3 gas atmosphere and the corresponding milled sample subsequently annealed at 1300°C for 6 h in N_2, are presented together in Fig. 6.21. After 150 h of RBM time, the XRD pattern (Fig. 6.21A) shows broadened peaks from unreacted B and a high background (a possible very large peak at around 25 degree) associated with amorphous B and newly formed nanostructural BN phases.[21] The

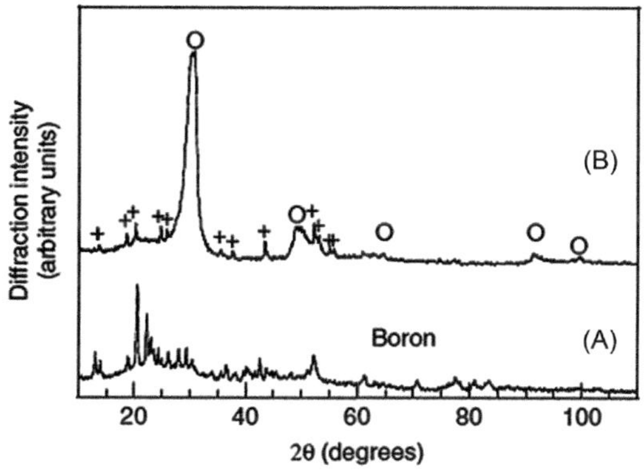

Figure 6.21 XRD patterns taken from (A) the 150-h milled B sample, (B) the milled sample subsequently annealed at 1300°C for 6 h in N_2. O = hexagonal BN; + = Fe_xB_y.[21]

new diffraction peak at around 51 degree was found to be associated with Fe, which came from the steel milling balls and the cell as milling contamination. It is worth to be mentioned here that when the as-ball-milled sample was examined by TEM, the authors could detect a large number of Fe and Cr particles dispersed in the disordered B matrix with concentration reached up to 7 wt.% and 1.4 wt.%, respectively.[21] They, however, believed that such milling contaminations may have positive role to play in nanotube formation since they can act as catalysts to help nanotube nucleation.[21]

The as-ball-milled BN powers with then subjected to isothermal annealing at 1300°C for 6 h in a tube furnace under nitrogen gas flow (0.5 l/min). This annealing step led to a full nitridation with the formation of BN nanotubes. Fig. 6.22 shows a typical SEM image taken from a sample milled for 150 h and heated at 1200°C for 16 h in N_2. Filaments with a diameter of 20–200 nm are observed. TEM revealed that these filaments have tubular structure. Typical TEM images in Fig. 6.23 show that thin (diameter less than 20 nm) nanotubes have typical parallel-walled, cylindrical structures. The parallel fringes on the wall areas of the nanotube suggest eight atomic layers. The authors have investigated the chemical nature of these nanotubes using

Figure 6.22 SEM image of a B sample first milled for 150 h in NH3 and then heated at 1200°C for 16 h in N_2.[21]

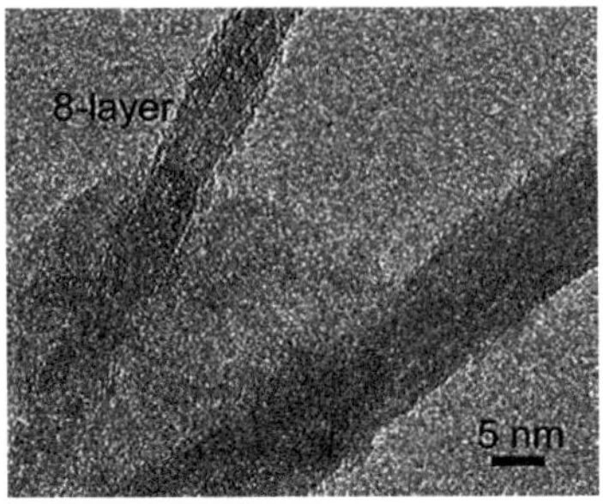

Figure 6.23 TEM image of two thin BN nanotubes with diameter less than 20 nm. Parallel fringes suggest a multiwalled, cylindrical structure.[21]

electron energy loss spectroscopy (EELS) attached to a field-emission TEM. They reported that two characteristic K-shell ionization edges (188 and 401 eV) of B and N are observed and the B/N ratio is 0.98.[21]

More recently, Li et al. prepared BN nanosheets, using ball-milling technique.[22] In their process, they employed wet ball-milling process to prepare a large quantity of BN nanosheets. In their work, they used benzyl benzoate as the milling agent. They pointed out that using small ball-milling media (0.1–0.2 mm in diameter) was much more effective in exfoliating BN particles to BN nanosheets. Under the optimum condition, the production yield can be as high as 13.8% and the BN nanosheets are 0.5–1.5 mm in diameter and a few nanometers thick and of relative high crystallinity and chemical purity.[22]

6.7 NbN

A metastable phase of fcc-NbN alloy powders has been fabricated by milling pure Nb powders in a high-energy ball mill under a nitrogen gas flow.[23] The nonequilibrium phase that shows a superconducting behavior with high T_c is considered as a promising industrial material that can be used in the field of electronics.

The XRD patterns of elemental bcc-Nb powders and fcc-NbN alloy powders after 720 ks of the RBM time are shown in Fig. 6.24A and B, respectively. Toward the end of the RBM processing time (720 ks), the Bragg peaks of pure bcc-Nb powders have disappeared completely and a new phase that related to the fcc-NbN alloy appeared clearly. The lattice parameter, a_0, of the unit cell for this new phase is calculated to be 0.4330 nm. The grain size of this alloy powder that was calculated from TEM and XRD shows nanocrystalline characteristics (less than 5 nm in diameter) of the fabricated alloy. It is worth noting that no remarkable change in the crystal structure of this metastable phase can be observed, even after heating to 950 K, as illustrated in Fig. 6.24C. At 1150 K, however, the metastable fcc-NbN phase transforms to mixed phases of hcp-Nb_4N_3 and hcp-NbN, as presented in Fig. 6.24D.

The DTA technique has been used to get more information about the thermal stability and the behavior of the NbN alloy powders at several temperatures. Fig. 6.25

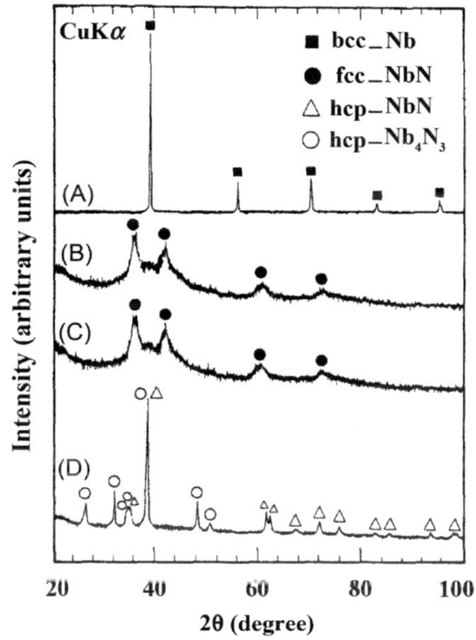

Figure 6.24 The typical XRD patterns of (A) pure Nb, (B) milled NbN alloy powders after 720 ks, (C) milled NbN alloy powders after 720 ks, then heated up to 950 K, and (D) milled NbN alloy powders after 540 ks, then heated up to 1150 K. *(After El-Eskandarany et al.)*[23]

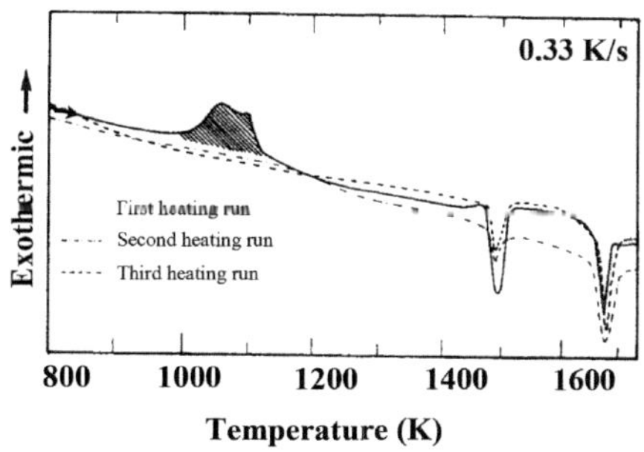

Figure 6.25 DTA curves for Nb powders milled in a nitrogen gas flow for 720 ks of the RBM time. *(After El-Eskandarany et al.)*[23]

illustrates the DTA curve of the metastable-fcc-NbN after 720 ks of RBM time. During the first heating run (solid line), four reactions are detected, at about 1025, 1100, 1490, and 1680 K. The first two reactions are exothermic and attributed to the phase transformation of metastable-fcc-NbN to hcp-Nb_4N_3 and hcp-NbN, as was shown in Fig. 6.25D. These phases melt at 1490 and 1680 K, and are characterized by clear sharp endothermic peaks, as shown in Fig. 6.25. During the second and third heating runs (dashed lines), the exothermic reactions have already disappeared and two endothermic reactions are permanent and appear clearly, as shown in Fig. 6.25. This suggested a complete disappearance of the metastable-fcc-NbN alloy powders and formation of a mixture of hcp-Nb_4N_3 and hcp-NbN.

References

1. Gomes, W. P., and Dekeyser, W., in: *Treatise on Solid State Chemistry*, (N. B. Hamny, ed.) Plenum Press, New York, 4:61 (1976)
2. Carter, R. E., *J. Chem. Phys.*, 34:2010 (1961)
3. Rapp, R. A., Yurek, A., and Yurek, G. J., *Metall. Trans.*, 4:1283 (1973)
4. Schaffer, G. B., and McCormick, P. G., *Appl. Phys. Lett.*, 55:45 (1989)
5. Schaffer, G. B., and McCormick, P. G., *Scripta Metall.*, 23:835 (1989)
6. McCormick, P. G., Whartom, V. N., and Schaffer, G. B., *Proc. Int. Symp. Phys. Chem. Powder Metall. Production and Processing*, TMS, (1990)
7. Schaffer, G. B., and McCormick, P. G., *Metall. Trans. A*, 21:2789 (1990)
8. El-Eskandarany, M. S., *Materials Transactions*, JIM, 36:182 (1995)
9. El-Eskandarany, M. S., *J. Alloys Comp.*, 296:175 (2000)
10. Benjamin, J. S., *Metall. Trans. A*, 1:2943 (1970)
11. de Boer, F. R., Boom, R., Mattens, W. C. M., Miedema, A. R., and Niessen, A. K., *Cohesion in Metals-Transition Metal Alloys*, North-Holland, Amsterdam (1988)
12. El-Eskandarany, M. S., Aoki, K., and Suzuki, K., *J. Less-Common Met.*, 167:113 (1990)
13. El-Eskandarany, M. S., unpublished work.
14. Yih, S. W. H., and Wang, C. T., *Tungsten*, Plenum Press, NY, p. 392 (1979)
15. Toth, L. E., *Transition Metal Carbides and Nitrides*, Academic Press, NY and London (1971)
16. El-Eskandarany, M. S., Omori, M., Ishikuro, M., Konno, T. J., Takada, K., Sumiyama, K., Hirai, T., and Suzuki, K., *Metall. Trans. A*, 27:4210 (1996)
17. Mi, S., and Courtney, T. H., *Scripta Metall.*, 38:171 (1997)
18. Anstis, G. R., Chantikul, P., Lawn, B. R., and Marshall, D. B., *J. Am. Ceram. Soc.*, 64:533 (1981)
19. El-Eskandarany, M. S., *J. Alloys Comp.*, 279:263 (1998)
20. *Fracture Mechanics Applied to Brittle Materials*, (S. W. Freiman, ed.) ASTM Spec. Tech. Publ., No. 678, Philadelphia (1979)
21. Chen, Y., and Williams, *J. Mater. Res.*, 6:1500 (1996).
22. Li, Deepika, Glushenkov, Alexey, Hait, Samik, Hodgson, Peter, & Chen Ying, *Scientific Reports*, 4: 7288 (2014)
23. El-Eskandarany, M. S., Sumiyama, K., Aoki, K., Masumoto, T., and Suzuki, K., *J. Mater. Res.*, 9:2891 (1994)

Fabrication of nanocrystalline refractory materials

7

Carbides have a unique set of properties that are necessary for cutting tools. This is not only because they have great hardness and wear resistance, but also because they have good thermal shock resistance and thermal conductivity. In addition, they have good oxidation resistance. All of these properties are necessary for considering materials to serve as cutting tools (the photo presented a typical tricone drilling toolwhere tungsten carbide materials used as drilling bits. source; http://www.acewel.com/product75.html) (Fig. 7.1).

7.1 Introduction

Carbides, especially those of the transition-metal carbides, possess unusual properties that make them desirable and useful engineering materials for many industrial applications.[1,2] They have great hardness values, ranging between 20 and 30 GPa, laid between alumina and diamond. Most of the carbides have extremely high melting points (2000–3000°C). For example, TaC has the highest melting point known for any material (about 3983°C), within a few degrees less than the sublimation point of graphite (4000°C).[3]

One important property of the metal carbides is their chemical stability at room temperature; they are attacked slowly when subjected only to very concentrated acidic media. The metal carbides are generally very strong at high temperatures with extremely high values of Young's modulus; however, they are brittle at room temperature. They undergo a brittle-to-ductile transformation at around 1000°C. In fact, metal carbides are unique in that they combine the characteristic properties of metals and ceramics.[4]

From the aforementioned characteristics of metal carbide, one can describe them as hard-stable materials. Since the World War II, this class of materials has been placed on top of the refractory materials, high-temperature and high stress-resistant materials, surface protective coating, rotors in gas turbines, and coating for fusion reactors,[5] which can find a wide range of industrial applications as structure materials used for cutting tools. In fact, carbides have a unique set of properties that are necessary for cutting tools. This is not only because they have great hardness and wear resistance, but also because they have good thermal shock resistance and thermal conductivity. In addition, they have a good oxidation resistance. All of these properties are necessary for considering the materials to be serviced as cutting tools.[6,7]

Mechanical Alloying. http://dx.doi.org/10.1016/B978-0-12-818180-5.00007-8

Figure 7.1 Typical drilling bits used for drilling in oil and water sectors.

The melting point (T_m) in °C, density (ρ) in g/cm^3, and hardness (H) in GPa for each metal carbide are presented as:

$$= \frac{\frac{T_m}{\rho}}{H} \tag{7.1}$$

As well as their applications in cutting tools, metal carbides are employed as wear-resistant parts in wire drawing and extrusion and pressing dies. They are also used as drilling tools and bits for mining and petroleum field applications.

7.2 Preparation challenges and difficulties

In the industrial scale of production, the technique of powder metallurgy is usually used for preparing polycrystalline metal carbides. In this process, a direct reaction between the powder form of metallic elements and their hydrides with graphite takes place at extremely high temperatures. Tables 7.1 and 7.2 summarize the different ways of obtaining the metal carbides, and the reaction temperature of selected systems.

Recently, a significant amount of attention has been focused on another technique for preparing metal carbides, using a method called self-propagating high-temperature synthesis (SHS).[8,9] In this process, a successful reaction between the reactant materials that are a mixture of elemental metals and graphite takes place only at a temperature well above the melting points of the metallic elements.[9]

Table 7.1 The possible methods for fabrication of transition-metal (TM) carbides via the technique of powder metallurgy.[a]

Method	Reaction
Direct reaction between metallic elements or metallic hydrides and graphite, under vacuum or inert gas.	$TM + C \rightarrow TMC$ $TMH + C \rightarrow TMC + H_2$
Reduction of the metal oxide by graphite, under vacuum or inert gas.	$TMO + C \rightarrow TMC + CO$
Reaction of the metal with carbonizing gas.	$TM + C_n\ H_{1-n} \rightarrow TMC + H_2$ $TMC + CO \rightarrow TMC + CO_2$

[a] (After Schwarzkopf et al.)[7]

Table 7.2 The reaction temperatures for the formation of selected transition-metal carbides from elements via powder metallurgy way of fabrication[a]

Reaction	Temperature (°C)
Ti + C → TiC	1700–2100
Zr + C → zrC	1800–2200
Hf + C → HfC	1900–2300
v + c → vc	1100–1200
Nb + C → NbC	1300–1400
Ta + C → TaC	1300–1500
W + C → WC	1400–1600

[a] (After Schwarzkopf et al.)[7]

From the earlier, it can be concluded that the preparation of homogeneous metal carbides with high purity via the traditional ways of fabrications is a difficult task. This is attributed to the application of high temperatures that may restrict the fabrication process, especially in developing countries where the facilities of high-temperature furnaces are limited. Accordingly, the high cost of this process is considered a major disadvantage for these processes and this increases the price of the end product.

7.3 Synthesizing and properties of mechanically solid-state reacted tic powders

The worldwide interest in TiC, particularly TiC-nanophase, is attributed to the unusual properties of this material, including a set of unique characteristics such as low density, high melting point, high hardness, excellent resistance to wear and abrasion, and infusibility. These desirable properties always support the use of this material for useful applications such as cutting tools and surface protective coating against severe erosion and corrosion. However, the consolidation of pure TiC nanograined powders into nanocrystalline bulk materials with grain sizes less than 100 nm involves many challenges.

Since all metal carbides are prepared at high temperatures (Table 7.2), the possibility for preparing large amounts of homogeneous nanocrystalline TiC powders at room temperature has been proposed by El-Eskandarany.[4,6] Unlike all other techniques that are used for fabrication of metal carbides (Table 7.1), MA leads to the production of metal carbides with nanoscale dimensions. These nanostructured refractory materials must receive tremendous attention due to their expected unique physical and mechanical properties that can add more advanced properties to this class of important material.

In this method, the starting materials of pure elemental Ti metal (<45 μm, 99.98%) and carbon powders (<10 μm, 99.99%) supplied by Sigma Aldrich were balanced and well mixed in a glove box (UNILAB Pro Glove Box Workstation, mBRAUN, Germany) under a helium gas atmosphere (99.99%) to give the desired nominal composition of $Ti_{55}C_{45}$. The mixed powders were then sealed together with 25 balls made of hardened Cr-steel into a hardened Cr-steel vial (220 mL in volume) under a helium gas atmosphere, using gas-temperature-monitoring system (GST) supplied by Evico Magnetic, Germany. The ball-to-powder weight ratio was 40:1. The milling process was conducted at room temperature using a high-energy ball mill (Planetary Mono Mill PULVERISETTE 6, Fritsch, Germany). The ball-milling experiments were interrupted after selected milling times and a small amount of the powder was taken from the vial in the glove box. An individual ball-milling run was conducted to determine the milling temperature and pressure inside the vial during the solid-state reaction between the diffusion couples of Ti and C. The average crystal structure of all samples was investigated by X-ray diffraction (XRD) with CuKα radiation, using a 9 kW Intelligent XRD system, SmartLab-Rigaku, Japan. The local structure of the synthesized material powders at the nanoscale was studied by 200 kV-field emission high-resolution transmission electron microscopy/scanning transmission electron microscopy (HR-TEM/STEM) supplied by JEOL-2100F, Japan. The morphological properties of the powders after the selected ball milling times were determined by 15 kV-field emission scanning electron microscope (SEM), supplied by JSM-7500F, Japan.

The ball-milled powders were binderless consolidated in vacuum at 1500°C under a pressure of 34.5 MPa, using spark plasma sintering (SPS; Fuji SPS-615, Japan) or uniaxial hot press (HP-Oxy-gon, United States). The times needed for the consolidation process to obtain full-density samples with a relative density above 99.5% (5.185 g/cm^3) using the SPS and HP techniques were 10 and 800 min, respectively. The heating rates used for SPS and HP processes were 500 and 20°C/min, respectively. The as-consolidated samples were also characterized by means of XRD, HRTEM, and FE-SEM. The density of the consolidated TiC was determined by Archimedes' principle using water immersion. A Vickers indenter (Wilson Vickers Hardness Tester, Germany) with a load of 20 kg was employed to determine the hardness of the consolidated samples. The hardness values reported later were averaged from at least 60 indentation results. In addition, Young's and shear moduli as well as Poisson's ratio of the consolidated samples were investigated at room temperature by resonant frequency and damping analysis techniques, using RFDA HTVP1600 (IMCE, Genk, Belgium) equipment.

Fig. 7.2 presents the dependence of the vial's temperature and pressure on the milling time upon the high-energy ball milling of elemental $Ti_{55}C_{45}$ (atomic %) mixed powders. Both temperature and pressure were obviously and drastically increased during the first few minutes of the milling time (0–150 min). After 172 min of milling, pronounced sharp pressure and temperature peaks were observed at about 225 kPa and

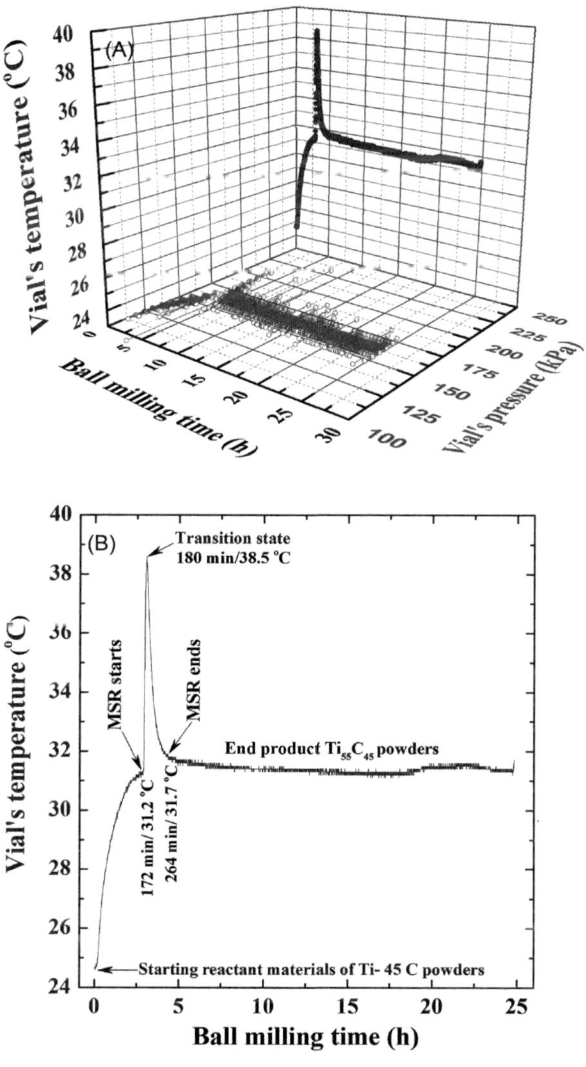

Figure 7.2 (A) 3D plot presenting the dependence of the vial temperature and pressure on the ball milling time, and (B) the reaction stages performed upon high-energy ball milling of $Ti_{55}C_{45}$. The data were obtained by using a pressure-temperature-measuring system (GTM).[10]

38.5°C, respectively, as displayed in Fig. 7.2A. The sudden increase in temperature during high-energy ball milling is attributed to the mechanically induced self-propagating reactions (MSR)[10] performed between the highly exothermic $Ti_{55}C_{45}$ system, which has a large negative heat of formation (−77 kJ/mol).[11,12] We should emphasize that the recorded temperature here resulted from the heat liberated during the reaction of Ti with C, but it cannot be used to investigate the actual temperature occurring at the ball-powder-ball collision zone. This is because the actual temperature decreased when transferred from the reacted zones (between the two collisional balls) in the inner portion of the vial to the cap of the vial where the thermocouple is installed. Thus, the measured temperature here can only be used to indicate the trend of the reaction kinetics during the high-energy ball milling of Ti and C powders.

During the early stage of ball milling time (0–172 min) the vial's temperature increased monotonically from 24.6°C to reach a value of 31.2°C, as shown in Fig. 7.2B. The XRD pattern of the sample after 172 min (2.8 h) of ball milling indicates that the powder after this stage of milling was polycrystalline mixture of Ti and C, as shown in Fig. 7.3A. The absence of a TiC phase from the powder mixture, which was milled for 172 min, indicates that the temperature increases during the early stage of milling (0–172 min) can be attributed to the ball-to-ball and ball-to-vial collisions. At a critical time (172 min), namely the ignition time, the vial's temperature surprisingly increases in a drastic fashion to reach a maximum value of 38.5°C within a very short ball milling time (8 min), as displayed in Fig. 7.3B. This drastic increase in the temperature is attributed to the heat liberated during the strong exothermic reaction between the reactant material powders, and this leads to a self-sustaining process in a similar fashion to the reported process of thermally ignited SHS.[8] After the reaction was completed, the vial's temperature tended to decrease dramatically to approximately 31.7°C after 264 min (4.4 h) of ball milling, suggesting the completion of the MSR process, as shown in Fig. 7.2B.

In order to confirm the structure of the product yielded from the MSR reaction, the crystal structure of a sample obtained after 5 h of milling was investigated by XRD analysis (Fig. 7.3B). The analysis indicated that all the Bragg peaks corresponded to the starting reactant materials of Ti and C (Fig. 7.3A), which completely disappeared, whereas a single phase of TiC was present, characterized by sharp Bragg peaks of TiC (111), TiC (200), TiC (220), TiC (311), and TiC (222) reflections, as shown in Fig. 7.3B. The lattice parameter (a_o) of the formed phase for TiC was calculated and found to be 0.433 nm, in good agreement with the reported value (0.4327 nm).[13] Moreover, the intensity ratio of these Bragg peaks was almost in a good agreement with those of the TiC powder,[13] suggesting that the crystal structure of the obtained powders was NaCl type. No remarkable increase in the vial's temperature, which was saturated at 31.4°C, could be detected upon ball milling for 25 h or even for the samples that were ball-milled for 200 h (not presented here), implying the absence of any reactions, as presented in Fig. 7.2B.

To investigate the effect of further ball milling on the stability of $Ti_{55}C_{44}$ system, two samples were ball-milled for longer times (50 and 200 h). Remarkable broadening of the Bragg peaks of TiC was observed, suggesting a decrease in the grain size of the powders obtained after 50 h of ball milling (Fig. 7.3C). Increasing the ball milling

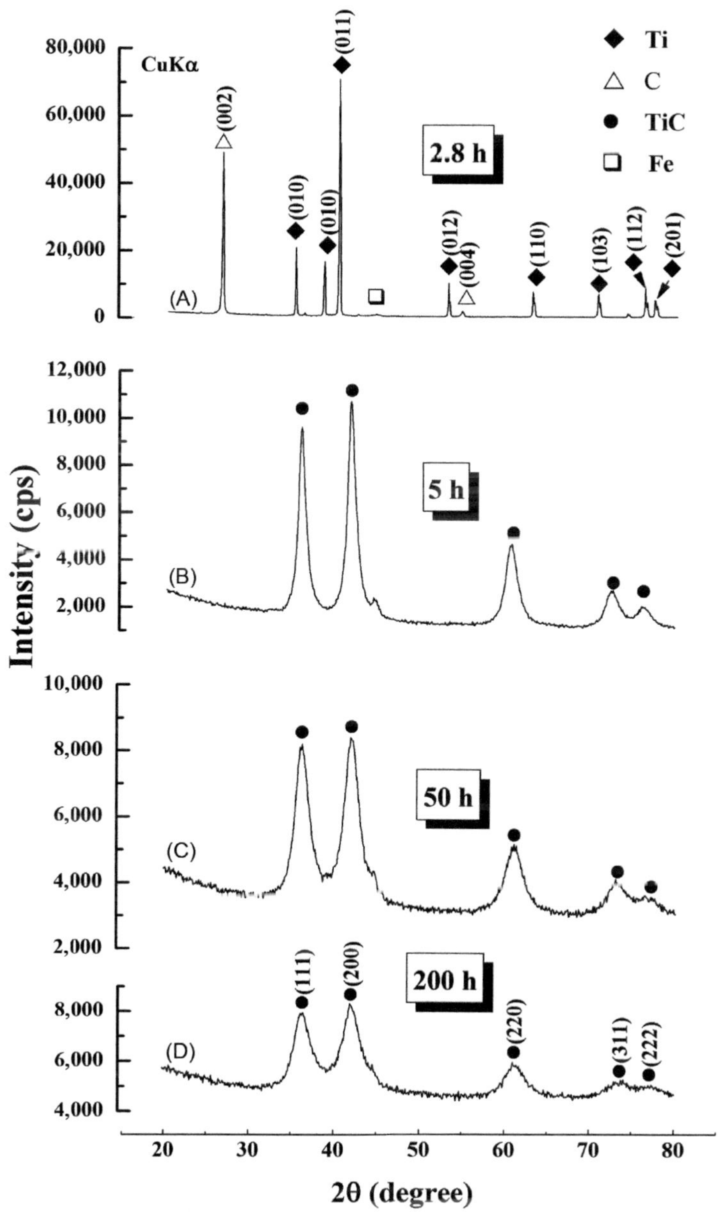

Figure 7.3 XRD patterns of $Ti_{55}C_{45}$ powders after (A) 2.8 h, (B) 5 h, (C) 50 h, and (D) 200 h of high-energy ball milling.[10]

time to 200 h (Fig. 7.3D) leads to further increases in the mechanical deformation that is generated by the milling tools, resulting in a remarkable broadening of the Bragg peaks for TiC, associated with an increasing in the full width at half maximum (FWHN) values. This implies the formation of nanoscaled-TiC powders.

The bright field image (BFI) and the corresponding selected area diffraction pattern (SADP) of the powders obtained after 5 h of ball milling are displayed in Fig. 7.4A and B, respectively. The powders have large grains size distributions, ranging from 40 to 200 nm in diameter with of irregular shapes (Fig. 7.4A). No free Ti and/or C crystals could be observed after this stage of milling (Fig. 7.4B), as indicated by the Debye–Scherrer rings corresponding to NaCl-like TiC. The STEM images for the powders milled for 150 h (Fig. 7.4C) indicate a dramatic decrease in their grain sizes upon milling for 150 h, with a grain size distribution in the range between 2 and 18 nm in diameter. As shown, the powders obtained after this stage of milling present

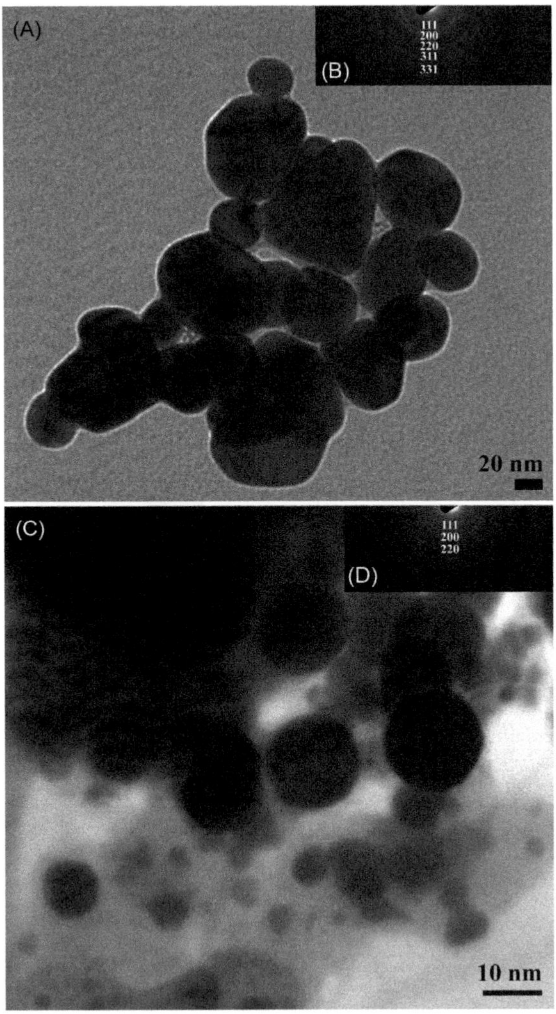

Figure 7.4 The BFI and the corresponding SADP of the powders obtained after 5 h of ball milling are shown in (A) and (B), respectively. The STEM and SADP of the powders that were ball-milled for 200 h are presented in (C) and (D), respectively.[10]

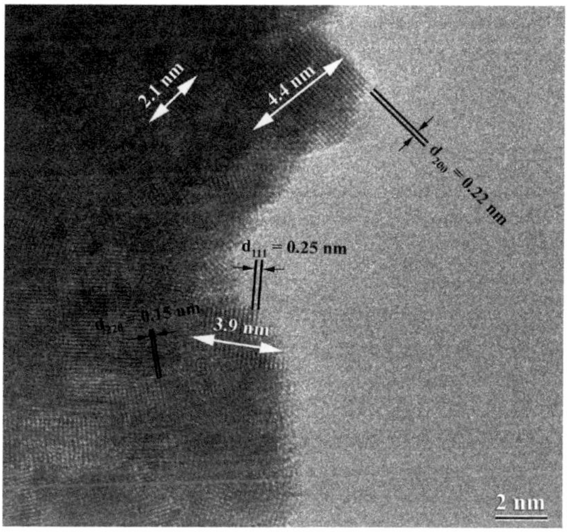

Figure 7.5 HRTEM image of the sample ball-milled for 200 h. The micrograph shows the lattice fringe spacing of 0.15, 0.22, and 0.25 nm, which matches well with the interplanar spacing of TiC (220), (200), and (111), respectively.[10] The grain size of some selected grains is indicated in the figure.[10]

homogeneous shapes with a spherical-like morphology (Fig. 7.4C). Moreover, the SADP of the powders obtained after 150 h still present clear fcc rings, as displayed in Fig. 7.4D.

The HRTEM image of the powders obtained after 200 h of ball milling is shown in Fig. 7.5. The lattice fringe spacing of 0.15, 0.22, and 0.25 nm indexed in the image match well with the interplanar spacing of TiC (220), (200), and (111), respectively.[13] The average grain size calculated from the size of 47 grains (some of these grains are indexed in the figure) was 4.2 nm.

The high-angle annular dark field (HAADF)/STEM for the sample obtained after 200 h of ball milling and the corresponding energy-dispersive X-ray spectroscopy (EDS) dot mapping of elemental Ti and C are shown in Fig. 7.6A–C, respectively.

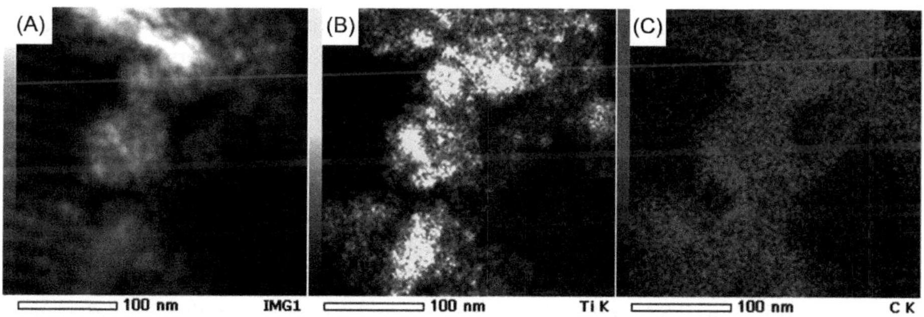

Figure 7.6 (A) Image of the HAADF/STEM for the sample obtained after 200 h of ball milling. The corresponding EDS dot mapping of elemental Ti and C is shown in (B) and (C), respectively.[10]

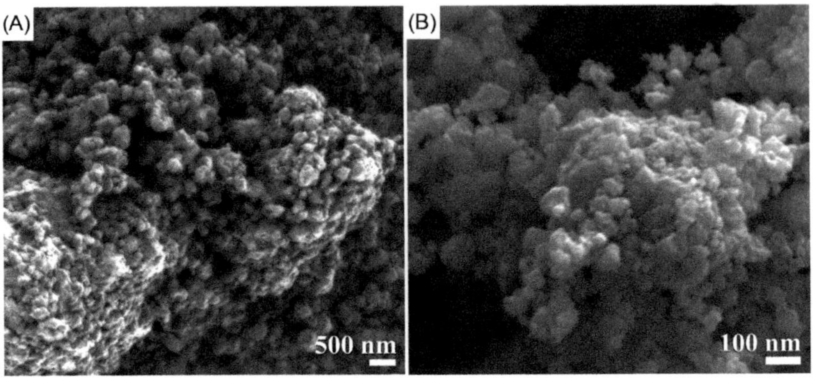

Figure 7.7 SEM micrographs of $Ti_{55}C_{45}$ powder particles obtained after ball milling for (A) 50 h and (B) 200 h.[10]

Obviously, the intensity of the dots corresponding to particular Ti and C is proportional to their concentration at the particular locations. No segregations on the nanoscale of the elements can be detected, and the elements are distributed homogeneously, indicating the formation of a homogeneous TiC material and the absence of any unprocessed Ti and/or C powders. The spot elemental analysis results of the powders are close, indicating nominal composition of $Ti_{55}C_{45}$.

Fig. 7.7 shows the SEM micrographs of the powders obtained after 50 h (Fig. 7.7A) and 200 h of ball milling (Fig. 7.7B). After 50 h, the powders consist of agglomerated spherical clusters with an average particle size of approximately 500 nm in diameter (Fig. 7.7A). Increasing the ball milling time to 200 h leads to a significant particle refinement to obtain fine powder particles with an average size of 45 nm, as shown in Fig. 7.7B. We should emphasize that the observed agglomeration of the nanopowder particles here is dominated by van der Waals, Coulomb forces, and electrostatic forces.[14]

7.3.1 Consolidation ball-milled $Ti_{55}C_{45}$ nanopowder particles

To investigate the grain size effect and the influence of consolidation techniques on some of the mechanical and physical properties of $Ti_{55}C_{45}$ powders, the ball-milled products obtained after selected milling time (5, 50, 100, and 200 h) were consolidated in vacuum at 1500°C/34.5 MPa into bulk samples, using SPS and HP techniques. Figure 7.8 shows the BFIs of the cross-sectional view and the corresponding SADPs for the powders obtained after ball milling for 200 h and then consolidated by SPS (Fig. 7.8A and B) and HP (Fig. 7.8C and D), respectively. Both consolidation approaches led to the obvious production of full-dense bulk materials, suggested by the existence of high-dense structure of equiaxed grains and the absence of any pores beyond the nanoscale, as displayed in Fig. 7.8A and C. Moreover, it can be said that both consolidation approaches did not lead to any structural changes, as implied by the Debye–Scherrer rings corresponding to NaCl-type TiC, which are shown in Fig. 7.8B and D.

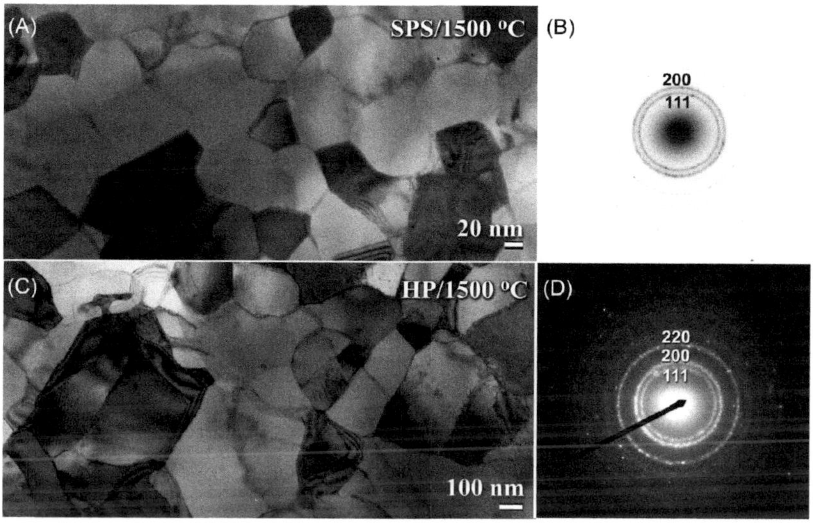

Figure 7.8 BFI images (A and C) and the corresponding SADPs (B and D) for the consolidated 200-h samples obtained by spark plasma sintering and HP techniques, starting from the same feedstock powders presented in (B).[10]

It is worth mentioning here that the grains of as-SPSed $Ti_{55}C_{45}$ sample still maintained their nanocrystalline structure with an average grain size of 28 nm in diameter, as displayed in Fig. 7.8A. In contrast, the grains of the sample produced by HP under the same sintering conditions of same material suffered from severe grain growth that led to the disappearance of their original nanocrystalline characteristics and production of sub-microstructured grains with an average size of 436 nm, as shown in Fig. 7.8C.

7.3.2 Mechanical properties of consolidated $Ti_{55}C_{45}$

7.3.2.1 Microhardness

Fig. 7.9 elucidates the dependence of the Vickers microhardness of as-SPSed and HPed $Ti_{55}C_{45}$ samples ($Y1$-axis) on the ball milling time (X-axis) and the grain sizes ($Y2$-axis). Based on the present results, increasing the ball milling time, which leads to decreases in the powder and grain sizes of ball-milled powders, improves the microhardness characteristics of the consolidated samples, suggested by the monotonic increase in the values of Vickers microhardness, as shown in Fig. 7.9. One can conclude that increasing the ball milling time leads to a decrease in the grain size of the powders, and this leads to a significant decrease in the grain size of the sintered samples. Thus, the grain sizes of the consolidated samples are dependent on the original grain size of the as-milled powders used as feedstock materials for the consolidation procedure. The powders obtained after 100 and 200 h of ball milling and then consolidated by SPS presented obviously higher microhardness values (31.7 and 32.2 GPa, respectively) when compared with the same samples obtained upon powder hot pressing

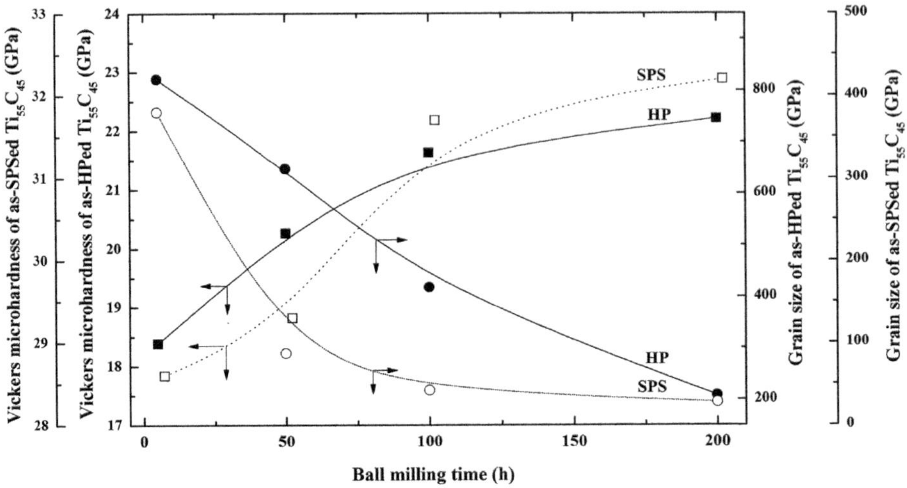

Figure 7.9 Dependence of the Vickers microhardness and grain size of the consolidated samples, which were obtained by spark plasma sintering of the milled powders, on the ball milling time.[10]

samples (21.6 and 22.2 GPa, respectively). The average grain sizes of $Ti_{55}C_{45}$ material obtained by SPS of 100 and 200 h samples were 42 and 28 nm, respectively. In contrast to the SPSed samples, the grain sizes for the same materials obtained upon HP consolidation of 100 and 200 h samples were 418 and 208 nm, respectively, as shown in Fig. 7.8. The dependence of the microhardness on the grain size is attributed to the grain boundary hardening effect due to Hall–Petch assumption.[15,16] Moreover, it can also be concluded from the figure that optimizing of the ball milling time for preparing nanopowders particles should be combined with a prober consolidation process to prevent the dramatic grain growth of the consolidated materials that always led to the minimization of their harness values.

Fig. 7.10A shows the outer shape of three different $Ti_{55}C_{45}$ samples obtained by consolidation of the 200-h powders by SPS. The surfaces of the samples on each face were classified into grids to facilitate the selections of the points examined by the Vickers tester. Some of the micro-indentations resulted from these measurements are shown in the corresponding optical micrograph, which is presented in Fig. 7.10B. The indents are homogeneous in shape and size, suggesting closed microhardness values. Moreover, the clear subsurface lateral cracking and secondary radial cracking shown in Fig. 7.10B were used to estimate the fracture toughness (K_{Ic}),[17] which was calculated to be 6.4 MPa $m^{1/2}$.

7.3.2.2 *Elastic moduli*

The elastic moduli of Young's modulus (E) and the shear modulus (G) of as-HPed (E^{HPed}, G^{HPed}) and as-SPSed (E^{SPSed}, G^{SPSed}) samples are displayed as a function of the ball milling time and grain size in Fig. 7.11A and B, respectively. An effect of grain

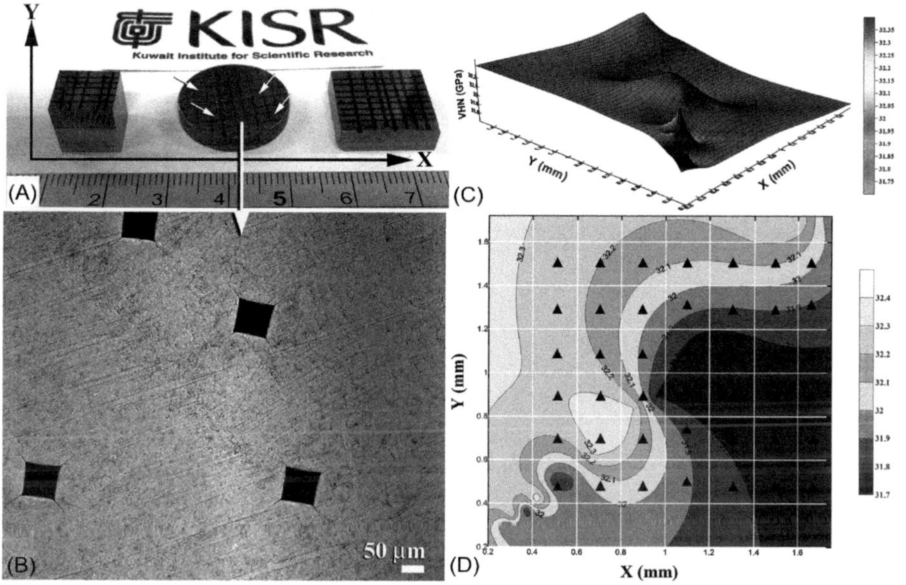

Figure 7.10 (A) Outer shape of three selected consolidated samples obtained by spark plasma sintering of the powders, which were ball-milled for 200 h. The average value of the microhardness for the cubic sample (left-hand side in (A)) was calculated from 60 indents (10 for each face). The upper and lower faces of the cylindrical-shaped (middle in (A)) and tetragonal-shaped (right-hand side in (A)) samples were subjected to independent hardness tests to 25 zones, as indicated in (A). The optical micrograph of the cross-sectional view for selected zones in the middle sample (A) that shows typical developed Vickers indents is presented in (B). The variation on the microhardness values is presented by three-dimensional plot (C) and isomicrohardness contour mapping (D). The *triangle-closed symbols* in (D) refer to the indented locations of the middle sample in (A).[10]

size on the elastic moduli is scarcely observed for those samples obtained using HP approach (Fig. 7.11A). The decrease of the grain size from 822 nm (5 h of ball milling) to 208 nm (200 h of ball milling) had no pronounced influence on the corresponding E^{HPed}, which maintained very close values of 452 and 449 GPa, respectively. In addition, G^{HPed} had close values of 189 and 184 GPa for the 822 and 208 nm samples, respectively, as shown in Fig. 7.11A.

7.4 Other carbides produced by mechanical alloying

In this section, selected examples are given of carbide synthesis by mechanical alloying of elemental component powders which produce the phase predicted by the equilibrium phase diagram.

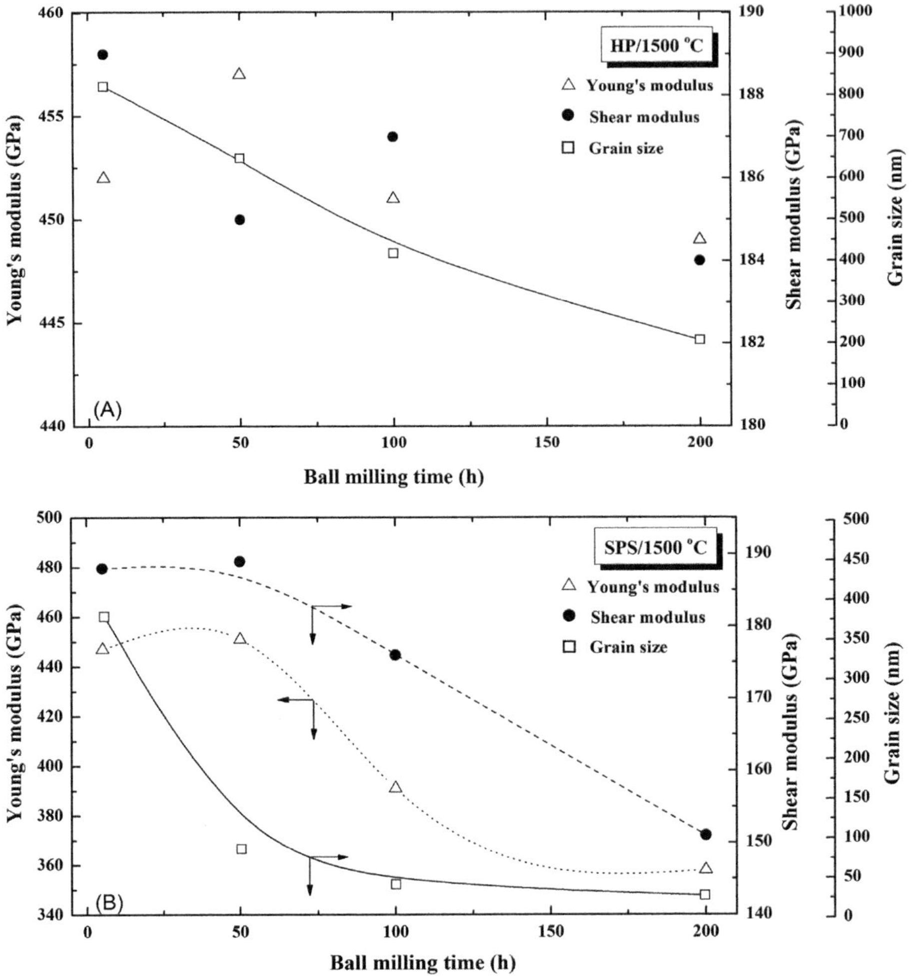

Figure 7.11 Dependence of the Young's modulus (*open triangles symbols*) and shear modulus (*closed circles symbols*) of the consolidated samples, which were obtained by spark plasma sintering of the milled powders, on the grain size (*closed diamond symbols*) and ball milling time.[10]

7.4.1 Fabrication of β-SiC powders

Among ceramics composites, the cubic form of SiC (β-SiC) has received much attention as a promising semiconductor for high-temperature, high-frequency, and high-power electronic devices, due to its wide band gap, high-saturated electron velocity, excellent thermal stability, and high-breakdown field.[18] Moreover, its intrinsic resistance to oxidation, corrosion, and creep at high temperatures also makes it a desirable protective coating for devices operating at elevated temperatures.[19] Single crystal

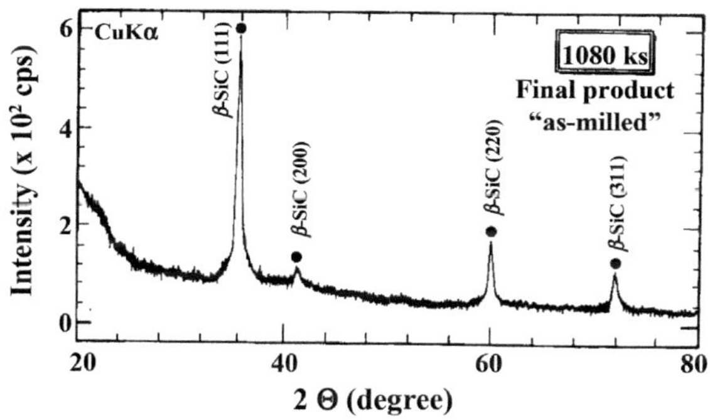

Figure 7.12 The XRD pattern of as-ball-milled $Si_{50}C_{50}$ powders after the final stage of mechanical alloying (1080 ks).[20]

β-SiC thin films are conventionally grown on Si(100) substrate by the chemical vapor deposition (CVD) process using cold-wall or hot-wall techniques.

In 1995, El-Eskandarany et al.[20] could prepare stoichiometric β SiC powders by solid-state reaction of elemental silicon and carbon powders via the room temperature mechanical alloying process. The MA process was performed in a high-energy ball mill similar to the one used for preparing TiC powders. Complete fcc-SiC alloy powders were obtained after 1080 ks of milling (300 h). The XRD patterns of the end product of as-ball-milled $Si_{50}C_{50}$ powders is shown in Fig. 7.12, obtained after 1080 ks of MA time. The Bragg peaks for this end product are broad, suggesting the formation of fine grain alloy powders. The lattice parameter (a_o) of the formed β-SiC was calculated to be 0.4357 nm,[21] in excellent agreement with the reported value (0.4358).[22]

Fig. 7.13 illustrates the infrared transmission of $Si_{50}C_{50}$ powders at (A) starting stage (0 ks), (B) final stage (as-milled for 1080 ks), and (C) as-milled, for 1080 ks then annealed under vacuum at 1773 K for 86 ks. At the starting stage of milling (Fig. 7.13A), there is no evidence of any oxide phase(s), such as SiO_2. Contrary to this,

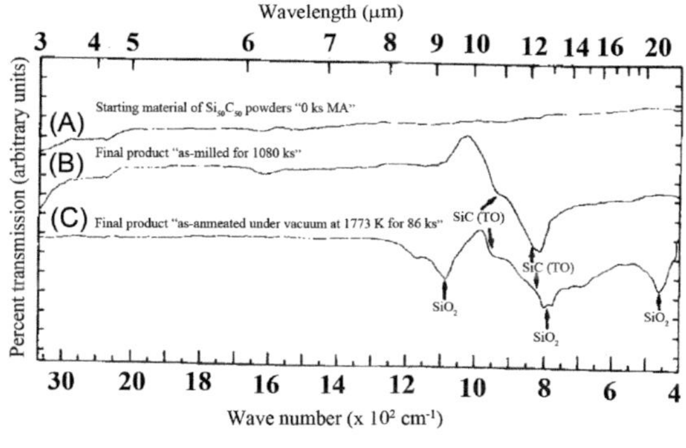

Figure 7.13 Infrared transmission spectra of (A) unmilled $Si_{50}C_{50}$ powders, (B) final product of β-SiC powders which were obtained after 1080 ks of milling, and (C) final product of β-SiC powders after 1080 ks of milling, then annealed under vacuum at 1773 K for 86 ks.[20]

a narrow absorption band is observed at about $9.5 \times 10^2\ cm^{-1}$, suggesting the existence of longitudinal optical (LO)-like phonon mode. In addition, the existence of transverse optical (TO)-like phonon mode is observed at lower energy (about $8.4 \times 10^2\ cm^{-1}$), as shown in Fig. 7.13B. It can be concluded that a complete β-SiC phase is formed after 1080 ks. This formed phase has a nanocrystalline structure with an average grain size of about 7 nm in diameter (Fig. 7.14). We should emphasize that almost the same TO- and LO-like phonon modes corresponding to the β-SiC phase are observed in the alloy powders after annealing the final product in vacuum at 1773 K for 86 ks. The other absorption bands centered at 10.9×10^2, 7.8×10^2, and $4.75 \times 10^2\ cm^{-1}$ are due to the existence of the SiO_2 phase in the as-annealed powders (Fig. 7.13C).

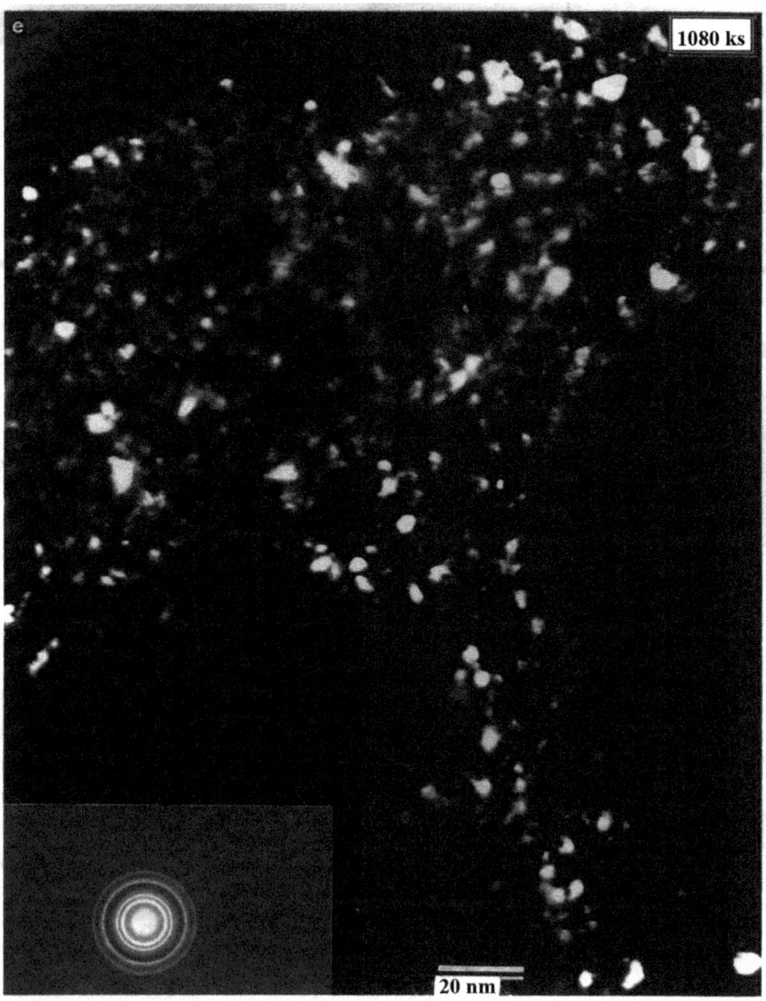

Figure 7.14 The DFI and corresponding SADP of the end product of β-SiC phase after 1080 ks of ball milling time.[20]

7.4.2 Fabrication of nanocrystalline WC powders

Among the hard materials WC, which is an interstitial compound of C atoms filling into W crystal, has been receiving a great attention due to its unique chemical and mechanical properties.[24] This technological material enjoys a set of interesting properties that cannot be found together in a different hard material.[25] It has high hardness, good wear-resistance, good fracture resistance, and high-temperature strength.[26,27] Based on these properties, WC can find wide spectrum of applications as desirable materials used intensively in chemical plants and reactors, national defense, electronic applications, manufacturing of hard cutting tools in oil and water sectors, as well as in surface protective coatings.[28–30]

Synthesizing of nanocrystalline WC powders has attracted different materials science schools in the world due to the expected improvement on their mechanical behavior, such as hardness, elastic modulus, and fracture toughness. The production of polycrystalline superfine WC powders with an average grain size ranging between 300 and 40 nm can be achieved by different ways, such as chemical route,[23] mechanically induced solid-state reduction,[28] plasma-chemical reaction,[29] and CVD.[30] However, mechanical milling (MM) approach, using high-energy ball mill is considered to be the most powerful process employed to fabricate an industrial scale of nanocrystalline WC ultrafine powders.[27,31–33]

During 1990 and 2000[34] single hcp phase of WC can be fabricated by high-energy ball milling of elemental W and carbon powders. The end product (432 ks) of as-mechanically alloyed WC powders possesses nanocrystalline characteristics, with an average grain size of less than 5 nm.[31] Hot pressing of the mechanically alloyed WC powders at 1773 K leads to slight grain growth and the compact sample still retains its nanocrystalline characteristics, with an average grain size of about 45 nm. Contrary to this, consolidation of the powders at a higher temperature (1973 K) leads to a dramatic grain growth and the compact sample loses its unique nanocrystalline characteristics.

7.4.2.1 Top-down approach combined with spark plasma sintering for fabrication of superhard bulk WC nanocrystalline materials

Consolidation of nanocrystalline WC powders is an important step for reliable and reproducible determination of the physical/mechanical properties of the fabricated powders and is required for most industrial and structural applications. However, consolidation of such category of ultrafine powders into bulk, full-dense compacts while retaining nanoscale grain size is a major challenge. Classically, hot pressing (HP) and hot isostatic pressing (HIP) techniques were mainly used to consolidate the WC powders at temperature between 1700 and 2000°C.[31–33] Such traditional consolidation regimes led to produce either green compacts (~75% of the theoretical density) with nanograin structured, or full-dense compacts with microstructured grains. Due to the large surface area of nanocrystalline material powders they are highly unstable and undergo to drastic grain growth when subjected to high-temperature such that one applied in HP and HIP. In the conventional methods, pure form of elemental metallic powders such as Co, Cr, and Ni is added to the WC in order to promote sintering process and to ensure the formation of full-dense compacts. Unluckily, the existence of such low melting point metallic matrix always led to several problems such as lowering the hardness and corrosion/oxidation resistance and enhanced grain growth.[35]

Due to the capability of SPS method to maintain the consolidated nanopowders in their nanostructured level with a minimal grain growth, SPS technique has attracted many researchers from the worldwide society of nanotechnology, materials science, and powder metallurgy.[24] Accordingly, SPS system has mainly been dedicated for consolidation of those metastable materials (*e.g.*, nanocrystalline powders), which are very sensitive to temperature. With SPS, sintering and sinter bonding of ultrafine powders can be achieved at lower temperatures when compared with the HP/HIP in a short period of time. The sintering process takes place by charging the intervals between powder particles with electrical energy and effectively applying a high-temperature spark plasma generated at an initial stage of energizing momentarily, and an electromagnetic field and/or joule heating by continuous ON-OFF DC-pulsed high electric current with a low voltage.[36] The most effective parameters of the SPS consolidation process are the current effects and the high heating rates.[35–37]

In the experiment,[38] pure WC (100 μm, 99.5% purity) powders provided by Alfa Aesar, United States, were used as starting materials. The powders were balanced inside a helium (He) gas atmosphere (99.99%) glove box (UNILAB Pro Glove Box Workstation, mBRAUN, Germany) to give the starting charge (50 g). The powders were then sealed together with 15-WC balls (10 mm in diameter) into a WC steel (250 mL in volume, Fritsch, Germany), using a ball-to-powder weight ratio as 50:1. The milling process was carried out, using high-energy planetary ball mill (Fritsch, P5, Germany) at a speed of 400 rpm for 20 h. However, in another experimental set the milling process was interrupted periodically after 3, 6, 12, 15, and 18 h, and the powders were completely discharged from the vial inside the glove box for several analysis. Then, new powder charge was introduced to the vial.

The XRD patterns of mechanically milled (MMed) pure hcp-WC powders are shown in Fig. 7.15 after selected ball milling time. The as-received powders (0 h) had sharp Bragg peaks corresponding to polycrystalline hcp-WC (PDF# 00-025-1047), as shown in Fig. 7.15A. This starting material was bulky in shape and composed of large particles with irregular-like morphology, having particle sizes ranging between 50 and 600 nm in diameter (Fig. 7.16A). The Bragg peaks of the powders milled for 3–6 h of MM time possessed significant broadening due to the grain refining process resulted from the milling effect (Fig. 7.15B and C). The powders obtained after 3 h of MM were experienced from lattice imperfections of nano-twins and stuck faults appeared in their internal structure, as indexed in Fig. 7.16B. Moreover, the sizes of the powder particles obtained after 3 h of MM were drastically decreased to be about 28 nm, as displayed in Fig. 7.16B. The powders after this stage of MM contain several grains varied in sizes (8–10 nm in diameter). During the intermediate stage of MM (6–12 h), further broadening was observed in the Bragg peaks with considerable decreasing in their intensities (Fig. 7.15D and E), implying a continuous grain refining and the formation of finer particles. Moreover, the WC powders were suffered from great amount of lattice imperfections, where clear stacking faults and dislocations were existed inside the dark-contrast grains (Fig. 7.17A). The lattice resolution TEM micrograph of the bright-contrast region shown in Fig. 17.17A shows the existence of

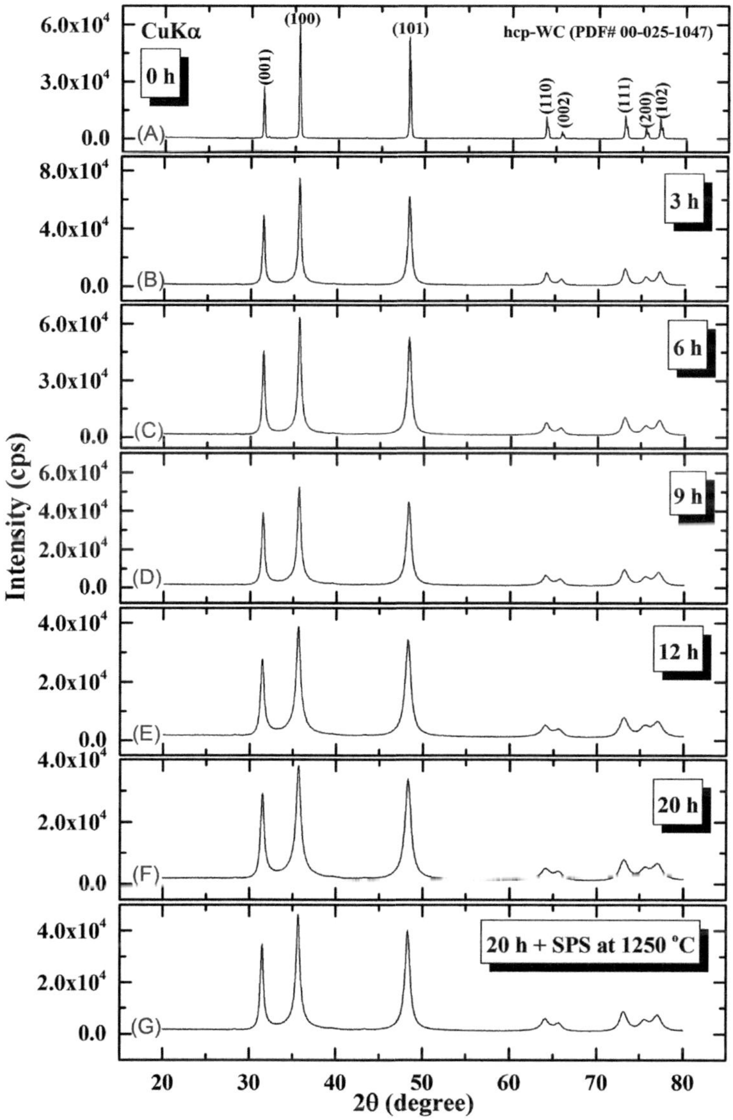

Figure 7.15 XRD patterns of mechanically ball-milled hcp-WC powders obtained after (A) 0 h, (B) 3 h, (C), 6 h, (D) 9 h, (E) 12 h, and (F) 20 h of high-energy ball milling time. The XRD of the powders obtained after 20 h of milling and then consolidated into bulk button 1250°C, using SPS technique is shown in (G).[38]

severe dislocations, indexed by the symbols shown in Fig. 7.17B. These imperfections were continuously propagated with increasing the MM time to 12 h (Fig. 17.17C), implied by the clear staking faults shown in Fig. 17.17D. In fact, the plastic deformation introduced to the WC crystal lattice upon during the early stage of MM resulted

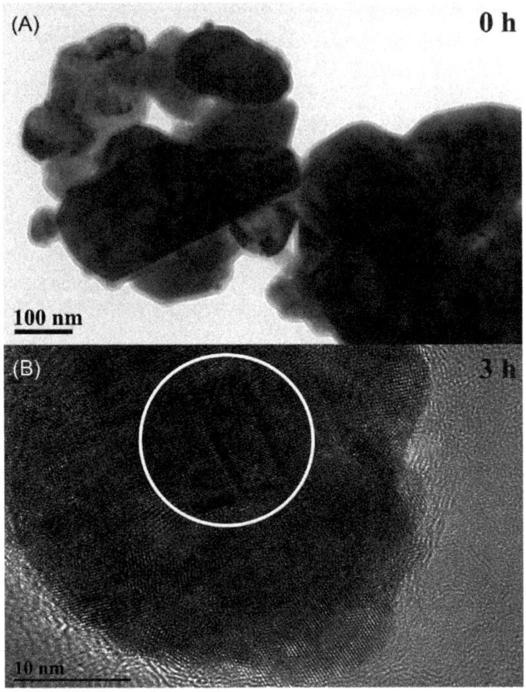

Figure 7.16 BFIs of mechanically ball-milled hcp-WC powders obtained after high-energy ball milling for (A) 1 h and (B) 3 h.[38]

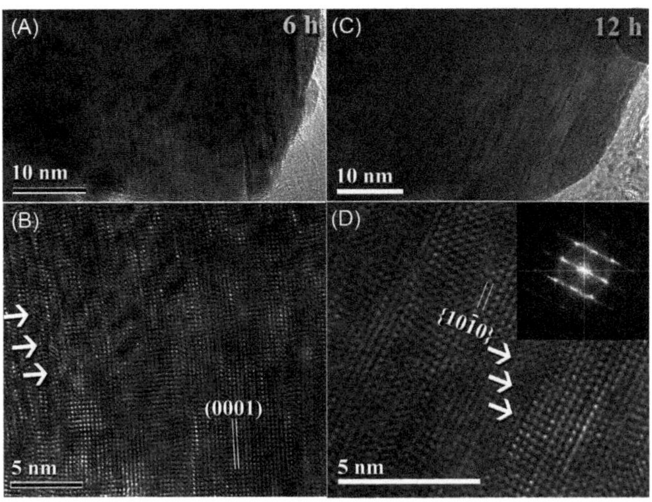

Figure 7.17 HRTEM micrograph of as-mechanically milled WC powders obtained after (A) 6 h and (C) 12 h of ball milling time. The lattice resolution TEM image for a selected zone in (A) oriented to (0001) is shown in (B). The lattice resolution image and NBDP of a selected zone in (C) are displayed together in (D).[38]

from the slip and twinning, as exemplified by Fig. 7.17A. This mechanical deformation preferred to be localized into the shear bands containing a high-dense network of dislocations,[39] as presented in Fig. 7.17B. At this stage of MM, the atomic level strain increases as a result of increasing the dislocation density.

The FE-SEM micrograph explicates that the powders obtained after 18 h of MM time were widely varied in sizes (50–550 nm) having different morphological characterizations (bulky-, rugby ball-, and spherical-like morphology), as presented in Fig. 7.18A. After this stage of MM, the WC powders shown in Fig. 7.18A consisted of nano-sized grains (~5–10 nm) of polycrystalline hcp-WC with different lattice planar orientations ((110), (002), (001), and (100)) as shown in Fig. 7.19A and B. Based on the successive accumulation of the dislocation density stored in the milled powders, the WC crystals were fragmented along their grain boundaries into nano-dimensional subgrains (Fig. 7.19A) due to a decrease of the atomic level strain.

Toward the end of MM processing time, the powders reaped the benefits of being milled for 20 h, revealing broad Bragg peaks (Fig. 7.15F). This implies the completion of MM process for obtaining WC nanopowders. Moreover, the intensive ball-powders-ball collisions led the powders to continue their disintegration tendency for forming ultrafine nanospherical powders with an apparent size ranging between 20 and 400 nm in diameter, as elucidated in Fig. 7.18B. The HRTEM micrograph of the WC powders obtained after 20 h of milling (Fig. 7.19C) confirmed the formation of polycrystalline (Fig. 7.19D) hcp-WC nanocrystalline cells. The obtained WC-nanocrystalline grains

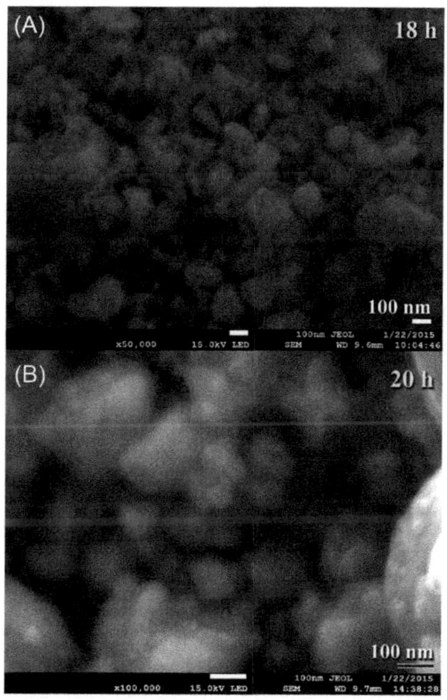

Figure 7.18 FE-SEM micrographs of as-mechanically ball-milled WC powders after high-energy ball milling for (A) 18 h and (B) 20 h.[38]

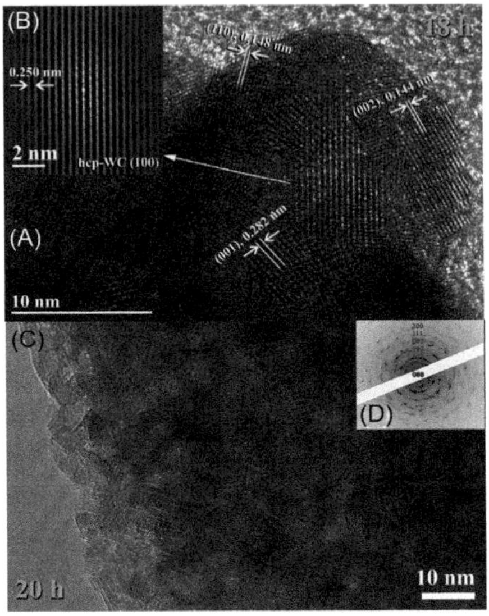

Figure 7.19 HRTEM images of mechanically ball-milled WC powders obtained after (A) 18 h and (B) 20 h ball milling time.[38]

after 20 h of MM had numerous gain boundaries with random crystallographic orientations, as shown in Fig. 7.19C. We should emphasize that increasing the MM time to 20 h was a necessary step to obtain homogeneous WC cells with narrow grain size distribution (5–9 nm), as presented in Fig. 7.19C.

Based on the purpose of the present study focused on consolidation and characterizations of nanocrystalline WC powders with a minimal particle coarsening and grain growth, SPS densification approach was selected as the most reliable consolidation technology used for sintering of metastable materials. This technique was used in the present work to consolidate the WC powders obtained after 20 h of MM at 1250°C, into several buttons having diameter of 20 mm with different thicknesses varied from 5 to 20 mm (Fig. 7.20A). The XRD pattern of as-consolidated WC-bulk button is shown in Fig. 7.15G. Obviously, the SPS-consolidation step did not lead to any phase transformation, as indicated by the Bragg lines related to hcp-WC (Fig. 7.15G). In general, the diffraction lines displayed in Fig. 7.15G revealed considerable broadening, suggesting the existence of nanocrystalline WC grins.

In order to ensure the formation of nanocrystalline bulk materials, a small piece of the as-consolidated WC button was prepared by an ion-slicer for TEM investigations. The STEM-BFI and the corresponding STEM-dark field image (DFI) for the as-consolidated bulk WC are shown together in Fig. 7.20B and C, respectively. The as-consolidated powders composed of dense structure of fine equiaxed grains, as shown in Fig. 7.20B. However, the SPS process was conducted in a very short time and at temperature (1250°C), which is far below than the one required for consolidation of WC by hot pressing (above 1650°C), with a very high heating rate (~300°C/min), the WC grains suffered from moderate grain growth (Fig. 7.20A). The grain sizes of those

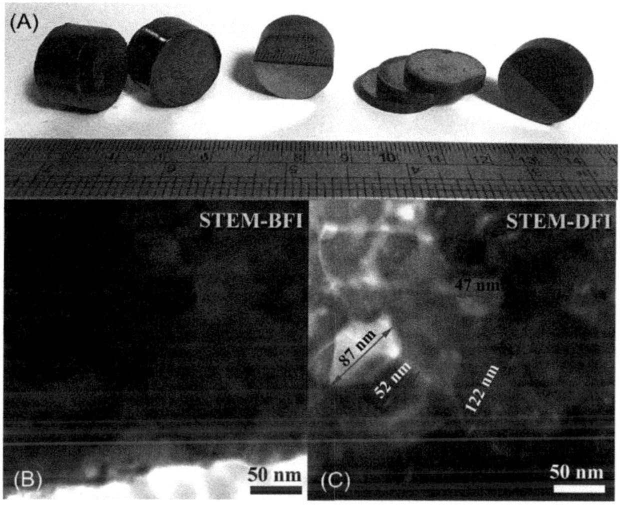

Figure 7.20 (A) A photograph showing the shapes and sizes of different WC consolidated buttons obtained upon consolidation of the powders of the final product (20 h) at 1250°C, using SPS technique. The STEM-BFI and the corresponding STEM-DFI of a selected zone of one button are shown in (B) and (C), respectively.[38]

full-dense WC buttons showed rather wide grain size distribution (~47–122 nm) with an average value of 77 nm in diameter. Comparing this value with that for as-milled WC powders (7 nm) it can be concluded that SPS process led to a moderate grain growth of about 10 times when used to consolidate WC nanocrystalline powders under the present experimental conditions used for powders preparations and consolidation. This grain growth may be attributed to the programmed cooling procedure taken place in ~5 min to reach the temperature of the SPS system to room temperature and to eject the WC button from the graphite mold. Efforts have been done to consolidate the powders at lower temperatures (800–1100°C). Unluckily, those buttons consolidated at such low temperature showed poor densities, ranging from 12.38 to 14.26 g/cm^3. In contrast, the buttons consolidated at 1250°C possessed high-density values ranging between 15.56 and 15.58 g/cm^3. Moreover, the FE-SEM micrographs of the fracture surfaces for selected buttons consolidated at 1250°C showing a very dense nanostructure morphology indicating the formation of high-dense fully developed WC prism-shaped fine grains, as presented in Fig. 7.21A–D. The figures indicated that the four samples obeyed to anisotropic grain growth with apparent grain sizes laid in the range between 84 and 358 nm in diameter.

Nanoindentation, which has become a common tool for studying of the mechanical properties on small scales, was used in the present work to investigate the nanohardness (*NH*) and Young's modulus (*E*) of bulk WC buttons. The investigated *NH* and *E* values for 20 selected grains of as-received WC powders (0 h) and then consolidated by SPS is shown in Fig. 7.22A. The *NH* values varied from 20.7 to 25.3 GPa (~21.73), where the *E* showed very high values in the range between 510 and 547

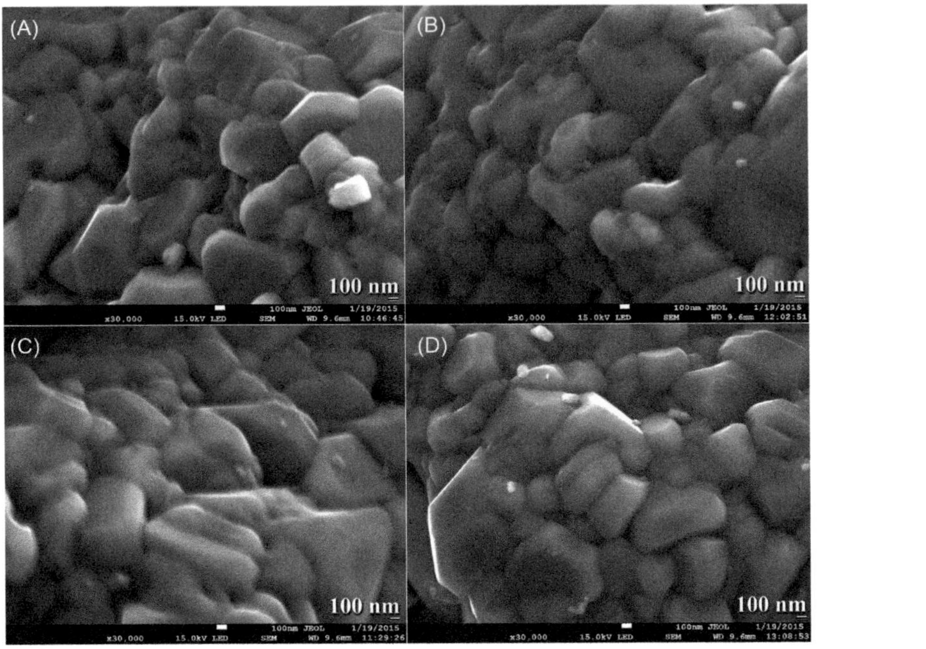

Figure 7.21 FE-SEM micrographs showing the fracture surface for four individual consolidated WC bulk samples.[38]

GPa (~566.02 GPa), as shown in Fig. 7.22A. Increasing the MM time to 12 h led to decrease the grain size of powders used as feedstock materials in the SPS process. The *NH* of the 12-h-bulk sample showed significant increasing (~23.08 GPa) with sharp decreasing in the *E* value (~482.06 GPa), as shown in Fig. 7.22B. The as-consolidated 20-h-WC sample possessed high *NH* (~24.41 GPa) and low *E* (332.02 GPa) values, as shown in Fig. 7.22C. More information related to the nanoindentation examination of the 20-h-consolidated sample was obtained upon intensive test. The optical micrograph of the sample after conducting more than 80 nanoindentation tests is shown in Fig. 7.22D. The arrows shown in that figure refer to some of those nanoindents developed after conducting the test. The indexed circular symbols shown in Fig. 7.22D were the zones taken as references. The AFM images of selected nanoindents of this sample are shown in Fig. 7.23. The indents had almost the same dimensions, indicating the homogeneity in the hardness and elastic properties of bulk WC beyond the sub-nanoscale. Moreover, the nanoindents taken at higher magnification (Fig. 7.23B) do not show sink-in behavior. The short dimension of the nanocracks shown around the indents indicates a high toughness value of the as-consolidated WC bulk material. The data collected from the nanoindentation tests were employed to design iso-nanohardness and iso-elastic modulus contour mapping, as elucidated in Fig. 7.23E and F, respectively. The *NH* showed narrow distribution values along the examined area (~300 × 300 μm), varied from 21 to 25 GPa without sever fluctuation, as shown in Fig. 7.22E. Likewise the *NH*, *E* values show very closed values in the range between

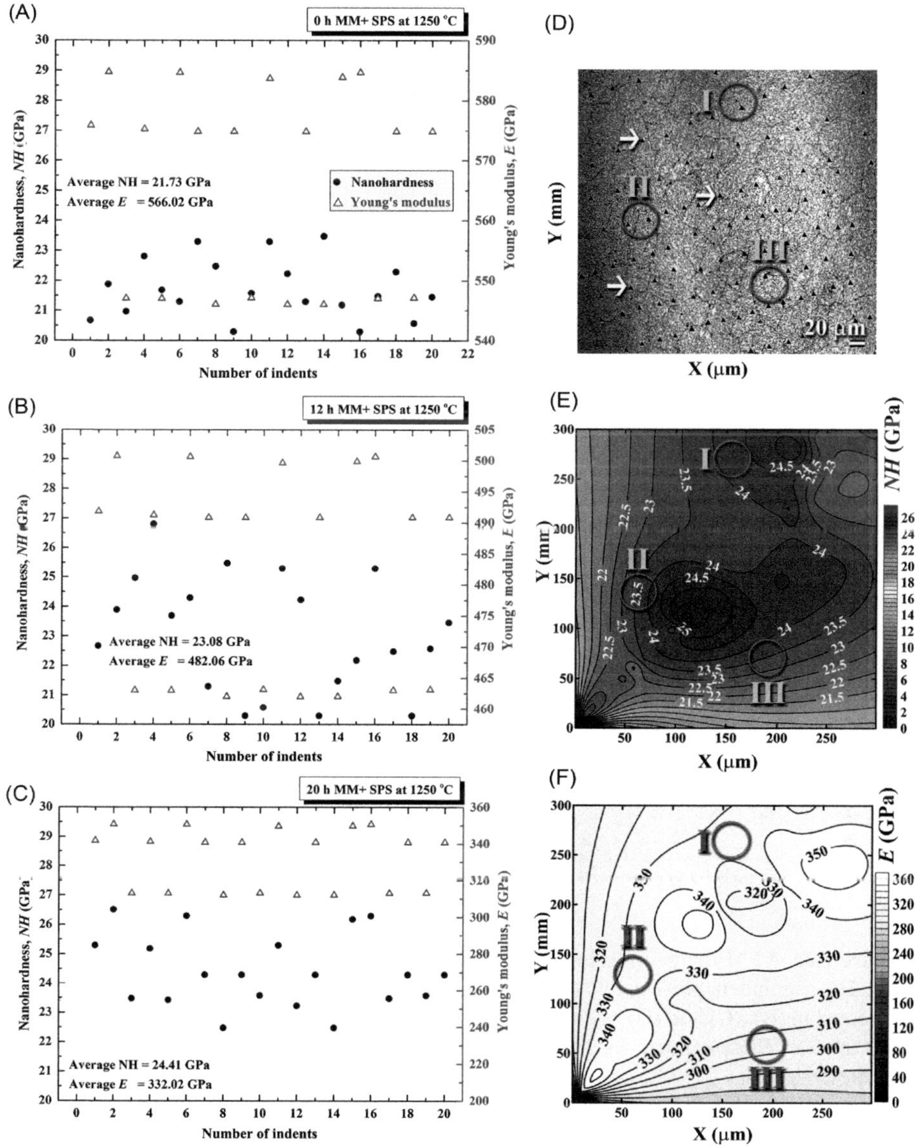

Figure 7.22 Dependence of the nanohardness (NH), Young's modulus (E) on the selected nanoindents tested points for WC bulk samples fabricated upon consolidation of the powders mechanically milled for (A) 0 h, (B) 12 h, and (C) 20 h, using SPS approach. The optical micrograph of the 20 h bulk sample after developing the nanoindentation test is presented in (D). The corresponding iso-nanohardness and iso-elastic modulus contour mapping are shown in (E) and (F), respectively.[38]

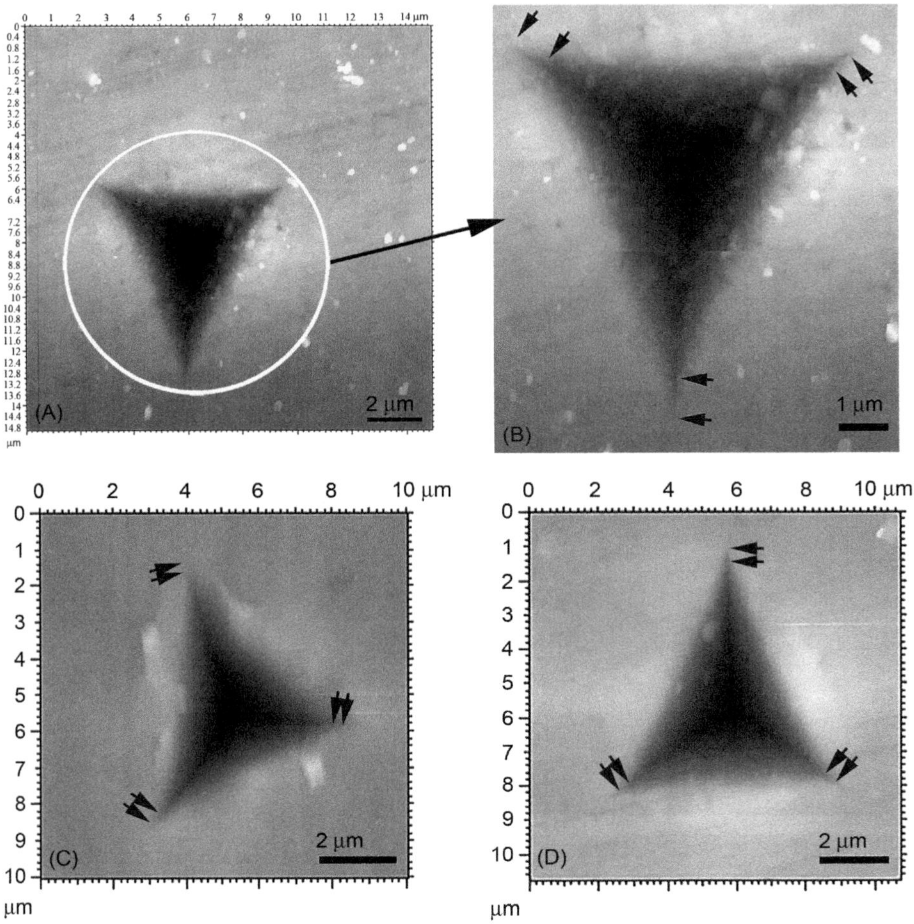

Figure 7.23 AFM images of selected WC bulk samples after developing nanoindentation test. The nanoindent presented in (A) is displayed in (B) with different scale. The *arrows* presented in (B), (C) and (D) refer to the nanoradial cracks generated around the nanoindents after the test. These nanocracks were used to estimate the fracture toughness of the samples.[38]

310 and 350 GPa, without sharp degradations or fluctuations in the counter intervals, as presented in Fig. 7.22F. It can be concluded from the nanoindentation test that the bulk WC sample is enjoying high degree of homogeneity beyond the sub-microscale level.

Fig. 7.24 summarizes the nanomechanical properties of WC consolidated after different stages of MM. The fracture toughness (K_{c1}) was calculated from the radial cracks existed around the nanoindents. During the early and intermediate stages of MM, both NH and K_{c1} were monotonically increased upon increasing the MM time. They then approached to saturated values of about 24 GPa and 10 MPa $m^{1/2}$, respectively, as

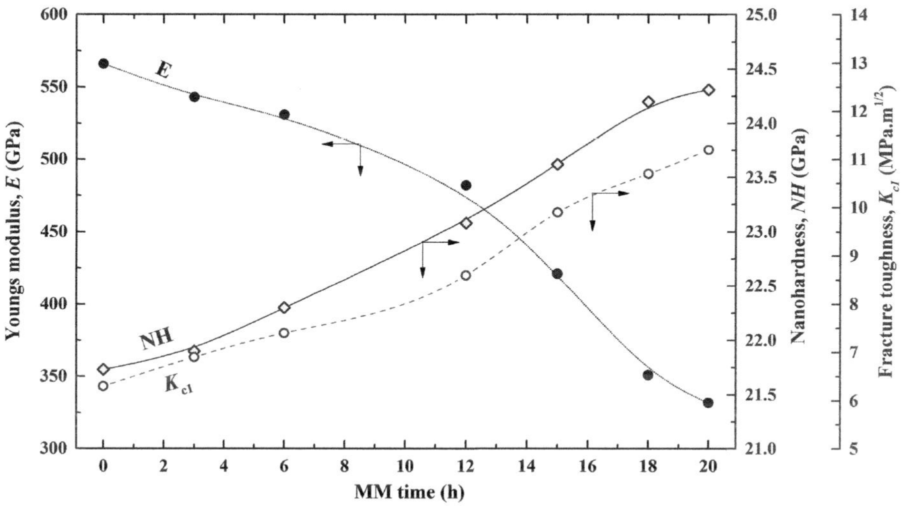

Figure 7.24 Correlationship between the mechanical milling time, Young's modulus (E), nanohardness (NH), and fracture toughness (K_{c1}). The figure elucidates the effect of mechanical milling time on the mechanical properties indexed by E, NH, and K_{c1} for bulk nanocrystalline WC prepared for different ball milling time and then consolidated at 1250°C, using SPS technique.[38]

shown in Fig. 7.24. The outstanding increase in *NH* for those consolidated buttons obtained upon SPS of the sample of the final product (18–20 h) is attributed to the grain refinement achieved after this stage of milling. Accordingly, these as-SPSed WC samples of the end product were harder when compared with the coarse-grained WC of the starting material (Fig. 7.24).

Apart from the dependence of *NH* and K_{c1} on the WC grain size, *E* showed a different correlation fashion. It tended to decrease slightly during the early stage of MM from 565 GPa (0 h) to 532 GPa (6 h). During the intermediate stage of MM, where a drastic decreasing in the grain size of WC, the *E* values were dramatically decreased from 442 GPa (12 h) to 351 GPa (18 h), as shown in Fig. 7.24. Toward the end of MM processing time (20 h), *E* reached to the level of 332 GPa, with a decrease amplitude of about 59%. Such drastic decreasing in *E* comes in contrast with the classical Hall–Petch (H–P) relationship, in which the strength of polycrystalline materials is expected to increase with decreasing grain size. It has been pointed out[40,41] that *E* of nanomaterial is far below than those of micro-grained materials, with amplitude reached up to 70%.[42] It was demonstrated by Qing and Xingming[43] that in interface regions, there are profuse defects and the atomic spacing deviates from the equilibrium state of the regular lattice. We have attributed the severe decreasing of *E* shown for WC bulk samples to the defects existed in the nanocrystalline WC lattice that were propagated upon increasing the MM time. Moreover, and based on the STEM observations for as-consolidated nanocrystalline WC (Fig. 7.20B and C), the sample contains large number of grain boundaries when compared with the microscaled WC material. The density of the grain boundaries in this sample should be smaller than those for bulk WC. Moreover, the existed grain boundaries in nanocrystalline WC reveal more open structure, leading to a significant decrease in E values with increasing the MM time, as indicated in Fig. 7.24.

7.4.3 Fabrication of nanocrystalline ZrC powders

Mahdy et al.[44] successfully synthesized nanocrystalline powders of $Zr_{56}C_{44}$ by high-energy ball milling of elemental Zr and C powders at room temperature. They have reported that a complete single phase of fcc-ZrC was obtained after 173 ks of milling time. The fabricated ZrC powders possess excellent morphological characteristics, such as a homogeneous shape (spherical-like morphology) with fine and smooth surface relief, and uniform size (less than 0.5 μm in diameter). The fabricated refractory ZrC powders have fine cell-like structure with nanoscale dimensions of about 5 nm in diameter.[44] Cold- and hot-pressing techniques have been used to consolidate the end product of the milled powders (259 ks) into fully dense (>99.5%) compacts with grain size of 95 nm. In the consolidation procedure, the powders were pressed at 1573 K with a pressure of 1.0–1.5 GPa for 43 ks.

7.4.4 Fabrication of nanocrystalline TiN powders

Owing to its high hardness, stability at high temperatures, chemical inertness, corrosion resistance, excellent thermal conductivity, and electrical and optical properties, TiN has become a desirable material for tremendous advanced applications in different areas.[45–47] The nanocrystalline powder form of TiN can be used as wear-resistance surface protective coating for cutting and machinery tools and for solar control films, and it is also a potential candidate for fusion reactors.[48]

In fact, different routes can be used to prepare TiN such as a sol-gel approach,[49] direct reaction of $TiCl_4$ with ammonia,[50] activated reactive evaporation,[51] and a self-propagating combustion method under high nitrogen pressure and high temperature.[52] CVD can also be used to prepare TiN by reacting the $TiCl_4$ with ammonia at temperatures above 1000°C.[46] In addition, the low-temperature physical vapor deposition (PVD) approach allows the deposition of TiN on metallized wafers.[46] The high costs of these preparations, utilization of toxic precursor gases (e.g., $TiCl_4$, HN_3), the toxic exhaust gases (e.g., HCl, Cl), and the excess of contamination in the end product are considered to be the disadvantages of these traditional methods. In contrast to the traditional chemical and physical routes used for the preparation of TiN, Calka et al.[53] and El-Eskandarany et al.[54] prepared TiN powders at ambient temperature under a flow of nitrogen gas atmosphere using a reactive ball milling (RBM) technique.[55] Since then, a wide variety of metal nitrides and hydrides have been successfully prepared by the RBM technique.[24]

Where most of the reported work described the preparation of a few grams (~5 or less) of nanocrystalline TiN powders, El-Eskandarany has recently proposed a roller-milling technique for the preparation of a large amount (500 g) of nanocrystalline TiN powder upon RBM of polycrystalline of Ti powders under 10 bar of nitrogen gas.[56]

The short RBM time required to prepare single-phase fcc-TiN powder with uniform nanocrystalline grains together with the consolidation of the bulk TiN nanophase to full density to achieve high mechanical properties are the advantages of his proposed method.

7.4.4.1 Powder preparation

In his method, commercial Ti powders with particles 45 μm in diameter and 99.95 wt.% purity, provided by Sigma-Aldrich-GF14536537, United States, as well as helium (He) and nitrogen (N) gases with purity of 99.999 wt.%, were used as starting materials. The reactive ball mill vial supplied by ZOZ GmbH, Germany, was made of Cr steel alloy with an internal diameter of 12 cm and capacity of 1200 mL (Fig. 7.25A and B).

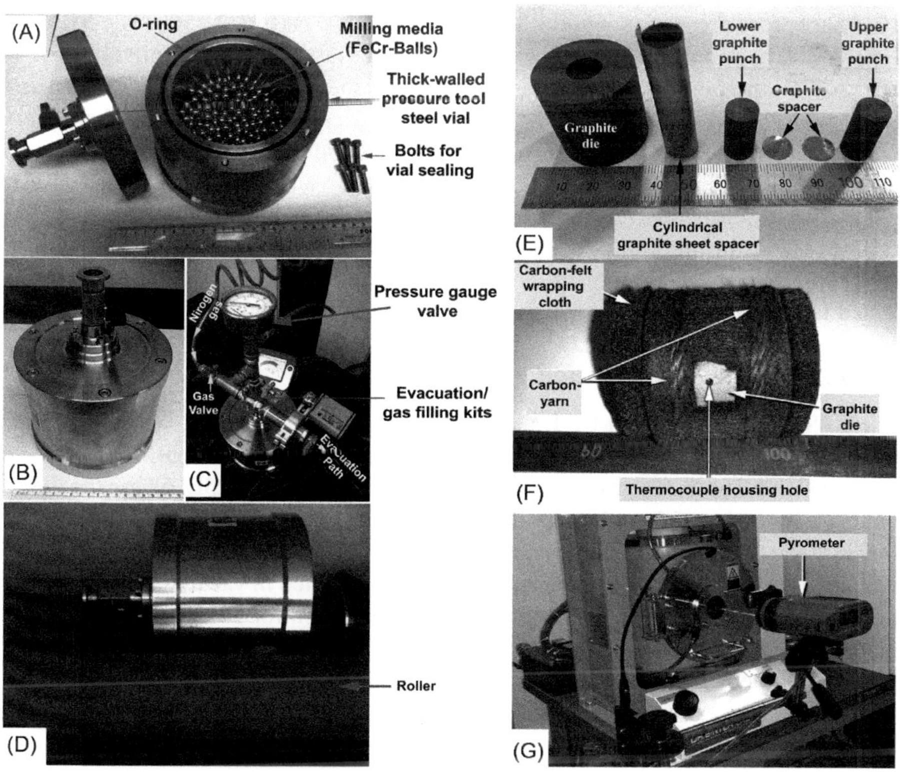

Figure 7.25 Setup of a laboratory-scale roller mill (1200 mL in volume) showing (A) the milling tools including the balls (milling media) and vial, (B) evacuation setup showing gas valve for (C) evacuation and inert gas filling, and (D) ball-milling process, using the roller mill. The powder charging kits used in SPS consisted of (E) graphite die and punches, cylindrical graphite spacer, and graphite disk spacer. The die charged with the powders was wrapped by carbon felt and tied with carbon yarn (F). A pyrometer used to measure the die's surface temperature during the SPS process is shown in (G).[56]

The milling media were 4 kg of Cr steel balls, 14 mm in diameter, and a density of 10 g/cm^3 (Fig. 7.25A). High-pressure gauges and valves were used as evacuation/gas charging kits (Fig. 7.25C). A horizontal-type roller mill, models RM200-ZOZ GmbH, Germany, was used to carry out the RBM process at room temperature (Fig. 7.25D). Weighting and handling of the powders were done inside a He gas atmosphere-glove box (UNILAB Pro Glove Box Workstation), provided by mBRAUN, Germany.

To minimize the Fe contamination introduced to the powders upon ball milling using the FeCr milling tools, 5 g of Ti was first milled with the FeCr balls under an Ar gas atmosphere for 50 h. After performing this milling run, the milling tools (the vial and the balls) were completely coated by thin Ti layers due to the cold welding process that occurred during the milling process. 50 g was massed inside the glove box and then sealed together with 400 Cr steel balls (Fig. 7.25A), using a ball-to-powder weight ratio of 80:1.

The sealed vial was then removed from the glove box (Fig. 7.25B) and evacuated to 10^{-3} bar before pressurizing under 10 bar of nitrogen gas (Fig. 7.25C). The RBM process started by mounting the nitrogen-pressurized vial onto a roller mill (Fig. 7.25D) operated at ambient temperature with a rotation speed of 350 rpm. The processing RBM time varied from 1 to 20 h. The RBM process was periodically halted after selected milling times (1, 3, 6, 12, and 20 h) where 2 g of the ball-milled powder was discharged from the vial inside the glove box. The vial was then pressurized again with nitrogen gas before resuming the RBM process. This RBM process was repeated 6 times to ensure the reproducibility of the end product for each stage of milling. However, the powder of the end product obtained after 20 h of RBM time was collected from five individual runs, giving a total amount of approximately 223 g.

7.4.4.2 Powder consolidation

The SPS system, provided by Dr. Sinter Lab. Instrument, Japan, was employed to consolidate the powder samples obtained after each stage of the RBM runs (1, 3, 6, 12, and 20 h). The TiN powders (5 g) were charged into Fig. 7.25E, where the die was wrapped by carbon felt and tied with carbon yarn (Fig. 7.25F). A pyrometer used to measure the die's surface temperature during the SPS process was used (Fig. 7.25G). All powder samples without exceptions were consolidated in vacuum at 1600°C under a vertical pressures ranging between 19.6 and 38.2 MPa for 5 min. The total amount of the powder samples obtained after each RBM time was consolidated into six bulk samples with equal dimension of diameter (15 mm) and aspect ratio (1:1).

7.4.4.3 Results

XRD pattern of the starting elemental hcp-Ti powder samples milled under 10 bar of nitrogen gas atmosphere for different RBM time is presented in Fig. 7.26A–D. The powder obtained after 1 h consisted of large crystallites, indicated by the sharply narrowed Bragg peaks corresponding to hcp-Ti, as identified by the Miller indices of the (100), (002), (101), and -(110) shown in Fig. 7.26A. After this early stage of RBM time, a small mole fraction of the TiN phase was detected, as indicated by the very low intensity Bragg peaks related to fcc-TiN (#PDF file 01-071-9845), as presented in

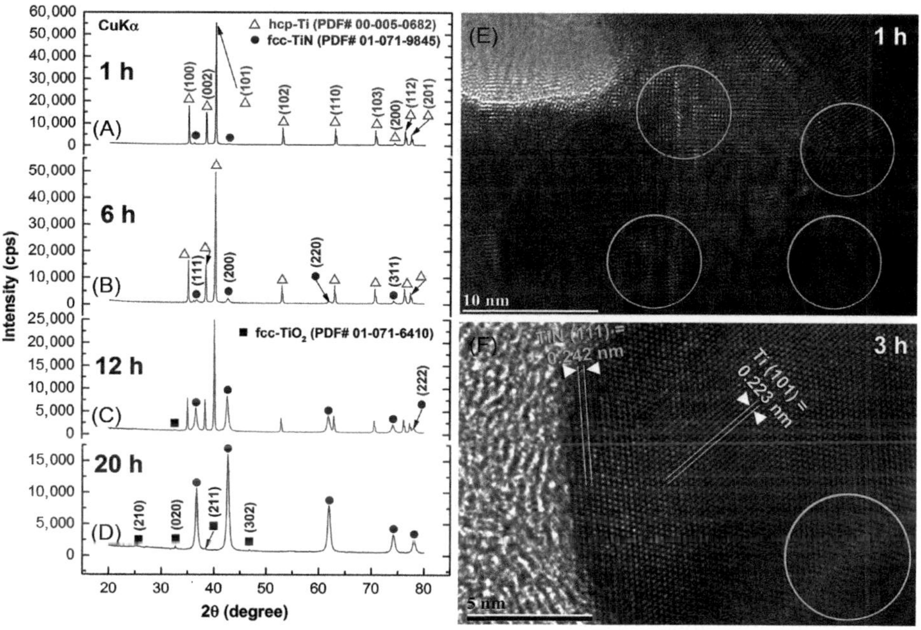

Figure 7.26 XRD diffraction patterns of Ti powder samples obtained after RBM for (A) 1 h, (B) 6 h, (C) 12 h, and (D) 20 h. The HRTEM images taken for the samples milled for 1 and 3 h are shown in (E) and (F), respectively.[56]

Fig. 7.26A. The powder samples at this early stage of milling experienced lattice imperfections, exemplified by the nanotwins, and stuck faults in their internal structure, as indexed by the circular symbols shown in Fig. 7.26E. The HRTEM micrograph taken at the near edge of an aggregated powder particle obtained after 3 h of RBM revealed the existence of reacted TiN phase embedded into a Ti matrix, as elucidated in Fig. 7.26F.

During the intermediate stage of RBM (6–12 h), the Bragg peaks corresponding to the TiN phase became pronounced (Fig. 7.26B and C), indicating an increase in the mole fraction of the reacted TiN phase. The HRTEM micrograph of the powder obtained after 6 h of RBM time shows two TiN particles, with diameter of approximately 20 nm on the unprocessed Ti matrix zone (Fig. 7.27A). Both the particles and matrix suffered from significant amounts of lattice imperfections implied by the presence of stacking faults and dislocations, as denoted by the circular symbols shown in Fig. 7.27A. As the RBM time increased (12 h), the intensity of the Bragg peaks of unreacted hcp-Ti decreased, implying a decrease in the mole fraction of Ti metal, as presented in Fig. 7.26C. After 20 h of RBM, all the Bragg peaks corresponding to unprocessed metallic Ti powders disappeared, a single phase of nanocrystalline TiN appeared, characterized by broad Bragg peaks displayed in Fig. 7.26D. We should emphasize that those Bragg peaks corresponding to the fcc-TiO_2 phase came from the local oxidation of TiN powder surfaces during the preparations of XRD samples. The lattice parameter a_o of the TiN phase obtained after 20 h (end product) was calculated from (111) and (200) reflections and found to be 0.423 nm, matching well with the reported value of pure TiN (0.4235 nm, PDF# 01-071-9845).

The HRTEM image of TiN powder samples obtained after 20 h of RBM is presented together with fast Fourier transformations (FFT) in Fig. 7.27B. Obviously, the powder consisted of fine grains of about 8 nm in diameter (or less) with different crystallographic orientation, as shown in Fig. 7.27B. Moreover, the corresponding FFT of the selected zone denoted by the ring symbol (Fig. 7.27B) implied the existence of TiN grains oriented to the zone axis [200]. Under this resolution of FE-HRTEM, no hcp-Ti crystals could be detected in the powder of the end product, suggesting the completion of the gas-solid reaction via the RBM process.

The homogeneity of the TiN powders obtained after 20 h of RBM time was intensively examined by EDS, using the HRTEM/STEM technique. The STEM-dark field (DF) image of the near edge of an aggregated TiN particles is shown in Fig. 7.27C. Based on the corresponding EDS elemental mapping of Ti (Fig. 7.27D) and N (Fig. 7.27E), it can be concluded that both elements were homogeneously distributed in the examined area ($\sim$100 $\times$ 150 nm) without elemental segregation beyond the nanoscale level, as presented in Fig. 7.27D (Ti-Kα1) and 7.26(e) (N-Kα1). Moreover, the average composition (in at.%) of this end product calculated from EDS spot analysis of 36 different zones was $Ti_{51.68}N_{48.32}$.

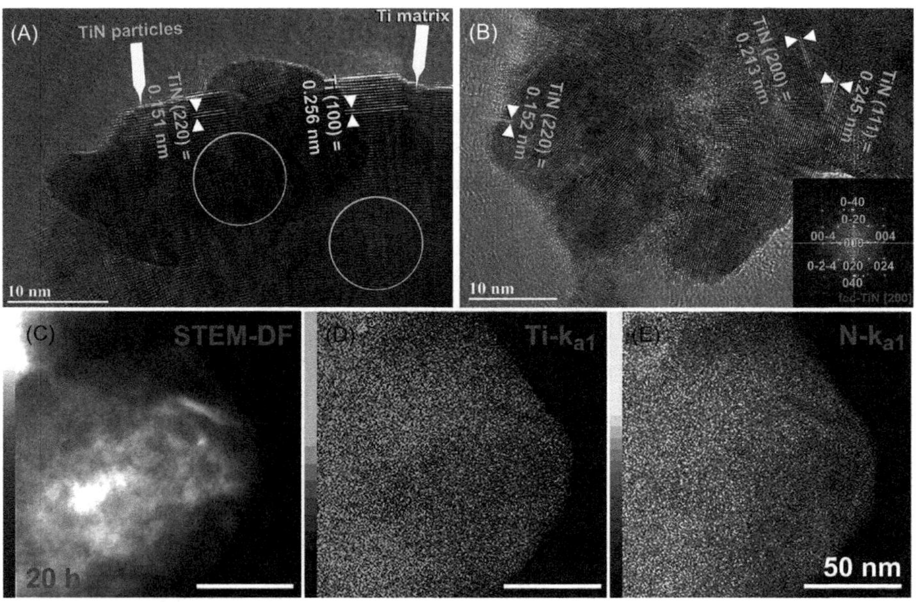

Figure 7.27 HRTEM images of the samples obtained after (A) 6 h and (B) 20 h. The FFT of the selected *circle* shown in (B) is displayed as an inset of (B). The STEM-DFI of the powders obtained after 20 h of RBM is displayed in (C) where the corresponding EDS map of elemental Ti and N is shown in (D) and (E), respectively.[56]

Figure 7.28 FE-SEM micrograph of the powder samples obtained after (A) 1 h, (B) 6 h, (C) 12 h, and (D) 20 h of RBM.[56]

Selected FE-SEM micrographs of the powder samples obtained after the early, intermediate, and final stages of RBM are shown together in Fig. 7.28. The sample obtained after 1 h of milling was bulky in shape and had wide particle size distribution, ranging between 20 and 110 μm, as displayed in Fig. 7.28A. Increasing the RBM time to 6 h led to a considerable decrease in the particle size and the formation of significant large volume fractions of particles with spherical morphology, as presented in Fig. 7.28B. The local-spot EDS analyses of those large particles in Fig. 7.28B show negligible values of N content; however, the spherical fine particles (under 10 μm) possessed nearly equiatomic compositions of TiN. Increasing the RBM time to 12 h led to increasing the fine TiN particles (less than 1 μm in diameter) against the unprocessed Ti (1–1.7 μm in diameter) powders, as shown in Fig. 7.28C. Toward the completion of RBM processing time (20 h), the end-product TiN powders possessed very smooth surfaces, and spherical morphology with narrow particle size distribution in the range between 190 and 400 nm in diameter, as shown in Fig. 7.28D.

The BFI of ion-sliced SPS-consolidated powder samples obtained after 1, 12, and 20 h of RBM is shown in Fig. 7.29. The bulk sample of the powder milled for 1 h (Fig. 7.29A) consisted of large grains (300 nm in diameter) of polycrystalline hcp-Ti, as confirmed by the corresponding SADP shown in Fig. 7.29B. As previously shown (Fig. 7.26C), the powders obtained after 12 h of RBM were composed of unreacted Ti and fully reacted TiN fine powders. The bulk-consolidated sample corresponding to these powders revealed a wide grain size distribution (50–280 nm in diameter), as shown in Fig. 7.29C. The SADP taken from the center of Fig. 7.29B confirms the presence of Ti and TiN phases, as shown in Fig. 7.29D. Consolidation of the end-product TiN powders obtained after 20 h of RBM into a bulk sample did not lead to severe

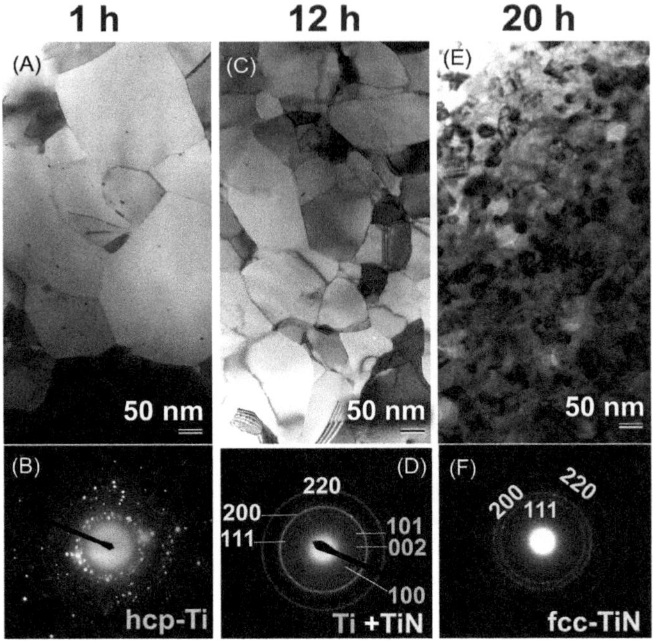

Figure 7.29 BFI and corresponding SADP of the planar view for as-consolidated and then ion-sliced powder samples obtained after (A and B) 1 h, (C and D) 12 h, and (E and F) 20 h of RBM.[56]

grain growth, as characterized by the fine nanocrystalline grains ranging between 42 and 86 nm in diameter shown in Fig. 7.29F. Comparing the grain size of the consolidated bulk TiN compact with the original size obtained after 20 h of RBM (8 nm (Fig. 7.27B)), we can conclude that the SPS can only lead to a moderate growth of TiN grains by approximately 5 times. Moreover, the SPS consolidation step did not lead to the formation of any other reactive phase rather than the original fcc-TiN phase, as confirmed by the SADP displayed in Fig. 7.29G. In fact, the powder sintering principles in the SPS process are based on the presence of an electric field called the field-assisted sintering technique (FAST). In contrast with the conventional sintering techniques in which the sample is heated from the outside, the sintering procedure in SPS takes place upon heating the sample internally by the passage of an electric current. When SPS is compared with the conventional hot pressing technique, one can realize that the SPS process is taking place with extremely high heating rates. Moreover, external pressures, which can also be applied during the sintering process in the range between few MPa up to 1000 MPa, lead to a rapid and effective consolidation process. SPS is an extraordinary rapid sintering process in which the whole processes (including temperature rise and holding times) are completed within a short period ranging from 5 to 20 min. This direct way of heating allows the application of very high heating and cooling rates, enhancing densification over grain growth promoting diffusion mechanisms, maintaining the intrinsic properties of nanopowders in their fully dense products.

Although, intensive research efforts have been carried out regarding the consolidation of metal, ceramics, and their composites by SPS, the mechanism of this sintering procedure is still not well established. As mentioned in the earlier section, the main difference between SPS and other sintering approaches is that both die and powder are directly heated by the joule effect of the DC current. The ON-OFF DC-pulse energizing method generates spark plasma, spark impact pressure, joule heating, and an electrical field diffusion effect.[36] In this process, the surfaces of the powders are activated and purified when compared with the other traditional sintering methods. Fresh powder surfaces lead to an easy sintering process, characterized by a lower sintering temperature and a shorter processing time. Upon starting the ON-OFF DC current path, the pulse current flows through the powder particles that are loaded into conductive graphite sintering tools (die and punches). The pulsed current has a cleaning effect on the surfaces of the powder particles based on the observation of grain boundaries without oxidation formed between particles. Such an intense joule heating effect at the surface of the conductive powder particles results in reaching the boiling point and therefore leads to the localized vaporization or cleaning of powder surfaces. Moreover, it has been suggested by Tiwari et al.[57] that joule heating resulting from the passage of electric current through the powder particles assists and enhances the welding of the particles under mechanical pressure.

Fig. 7.30A presents the bulk densities taken from six consolidated independent samples obtained after different milling times. The density of the samples obtained after 1–3 h of RBM varied from 4.1 to 4.8 g/cm^3, as shown in Fig. 7.30A. This wide variation was attributed to the composition variations of the powders after this early stage of milling containing large volume fractions of pure Ti against a small fraction of the processed TiN reacted phase. As the RBM time increased (6–12 h), the density of the consolidated sample tended to have larger value (4.5–4.9 g/cm^3) suggested an increasing in the volume fraction of the dense TiN phase (5.4 g/cm^3)

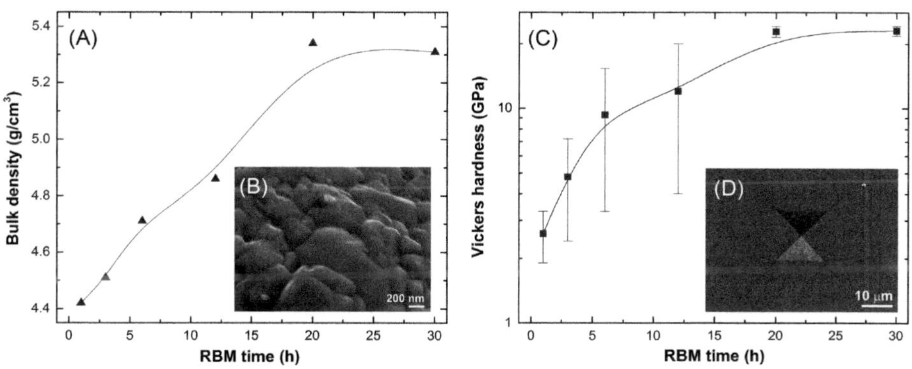

Figure 7.30 Effect of RBM time on (A) the bulk density and (C) Vickers hardness for the powders obtained after different stages of milling. The FE-SEM micrograph of the fracture surface and Vickers indentation for the consolidated samples obtained after 20 h of RBM are shown in (B) and (D), respectively.[56]

against the light Ti phase (4.5 g/cm^3), as shown in Fig. 7.30A. After 20 h of RBM, the density ranged between 5.29 and 5.37 g/cm^3 with an average value of 5.34 g/cm^3. Comparing this value with the theoretical reported value of pure TiN (5.38 g/cm^3, PDF# 01-071-9845), we can conclude that SPS of RBMed powders can lead to the formation of a nearly full-dense TiN bulk material, with a relative density of 99.26%. The FE-SEM micrograph of the fracture surface for the powders consolidated after 20 h of RBM shows a very dense structure morphology, implying the formation of a high-density fully developed TiN bulk material, as displayed in Fig. 7.30B. Increasing the RBM time to 30 h did not show a significant variation on the density value that remained in the range of 5.27–5.35 g/cm^3, as shown in Fig. 7.30A.

Vickers hardness measurements taken from six consolidated independent samples obtained after different milling time are displayed in Fig. 7.30C. The hardness of those samples obtained after completion of the early stage of RBM (1–3 h) varied from 1.9 to 7.2 GPa, as shown in Fig. 7.30C. Increasing the RBM time lead to increasing the volume fraction of the hard TiN phase against the unreacted Ti metal. This can be understood by looking at the hardness values, which were varied from 4 to 20.2 GPa, as shown in Fig. 7.30C. Toward the end of the RBM time (20–30 h), the hardness values of the consolidate samples were almost saturated at an average value of 22.9 GPa ±1.25, as shown in Fig. 7.30C. This implies the formation of single TiN phase and the disappearance of Ti metal. The FE-SEM micrograph of the Vickers indentation for the SPSed sample obtained after 20 h of RBM is elucidated in Fig. 7.30D. The micrograph reveals obvious cracks emanating from the indentation corners, implying brittle behavior of the obtained TiN phase.

The elastic modulus values, indexed by the Young's and shear moduli of the consolidated samples obtained after different RBM times, were estimated from the measured sample densities and the constant parameters of the nondestructive testing apparatus. Fig. 7.31 shows the effect of RBM time on the Young's and shear moduli. These values increased significantly during the early and intermediate stages of RBM time (1–12 h), suggesting an increase of TiN phase in the milled powder. Toward the end of the RBM time (20–30 h), both Young's and shear values are almost saturated at 384 and 192 GPa, respectively, indicating the formation of a single homogeneous TiN phase.

The relation between the coefficient of friction (l) and the sliding distance of as-consolidated powder samples obtained after different RBM times is presented in Fig. 7.32. The wear test was achieved using pin-on-disk technique, using a WC-Ni composite ball (pin) under the application of 200 N, and a rotation speed of 150 rpm. All samples had similar behaviors of reaching steady-state stages after approximately 20 m of sliding distance, as shown in Fig. 7.32. During the second stage (20–100 m), the μ value for 1 h sample was dramatically increased to 0.65 after 100 h, as displayed in Fig. 7.32. The samples obtained after 6 and 12 h of RBM that contain larger TiN volume fractions showed slightly increased values after 100 m of approximately 0.49 and 0.41, respectively (Fig. 7.32). The sample obtained after 20 h of RBM (end product) showed a lower μ value of approximately 0.35, as displayed in Fig. 7.32. It can be then concluded that, in addition to the attractive high hardness value demonstrated for the

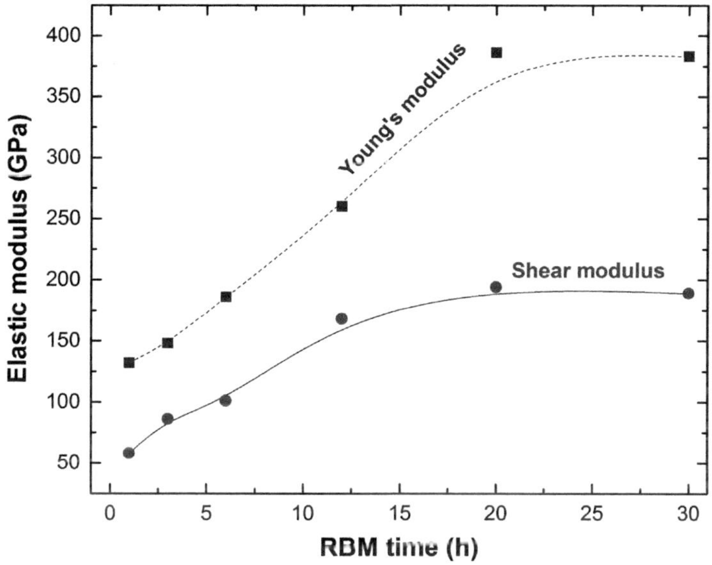

Figure 7.31 Influence of RBM time of the Young's and shear modulus values of the powder samples milled for different stages of milling.[56]

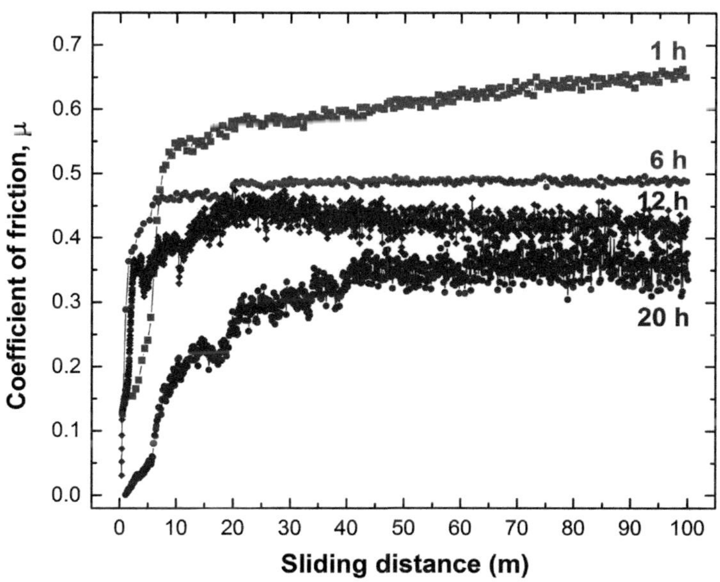

Figure 7.32 Dependence of coefficient of friction (l) on the RBM time and sliding distance.[56]

Table 7.3 Chemical analyses of as-milled and as-consolidated mechanically reacted TiC after 720 ks of ball-milling time.[6]

Sample	Ti content (at.%)	c content (at.%)	Al content (at.%)	0 content (at.%)
As-milled	44.80	54.93	0.03	0.24
As-consolidated	44.45	55.17	0.03	0.35

nanocrystalline TiN bulk material, this system also possesses superior wear-resistance behavior, making it a desirable material for a surface protective coating employed for cutting tool applications in oil and underground water sectors (Table 7.3).

References

1. Storms, E. K., *The Refractory Carbides*, Academic Press, New York (1967)
2. Pierson, Hugh O., *Handbook of Refractory Carbides and Nitrides*, ELSEVIER, Oxford (1996)
3. Segal, D. L., *The Physics and Chemistry of Carbides, Nitrides and Borides*, Chemical Routes for the Preparation of Powders. In: Freer R. (eds). NATO ASI Series (Series E: Applied Sciences), vol 185. Springer, Dordrecht (1990)
4. El-Eskandarany, M. S., *Metall. Trans. A*, 27:2374 (1996)
5. Niu, T., *Nano Today*, 18: 12 (2018).
6. El-Eskandarany, M. S., *J. Alloys Comp.*, 305:219 (2000)
7. Schwarzkopf, P., and Kieffer, R., *Refractory Hard Metals*, Macmillan, New York (1953)
8. Munir, Z. A., *Ceramic. Bull.*, 67:342 (1988)
9. Munir, Z. A., and Anselmi-Tamburini, U., *Mater. Sci. Rep.*, 3:277 (1989)
10. El-Eskandarany, M. S., and Al-Hazza, Abdulsalam, *Materials Characterization*, 97:92 (2014)
11. Laszlo, Takacs, *Progress in Materials Science*, 47:355 (2002)
12. de Boer, F. R., Boom, R., Mattens, W. C. M., Miedema, A. R., and Niessen, A. K., *Cohesion in Metals-Transition Metal Alloys*, 1:127, 1st Ed. North-Holland, Amsterdam (1988) p.127.
13. ASTM Card No. 32-1383, *ASTM*, Philadelphia, PA
14. Sabine Leroch and Martin Wendland, *Langmuir*, 29:12410 (2013)
15. E.O. Hall, *Proceedings of the Physical Society of London*, Vol. B64:747 (1951)
16. Schiotz, F.D. Di Tolla, K.W. Jacobsen, *Nature*, 391:561 (1998)
17. K. Niihara, R. Morena, D.P. H. Hasselman, *J. Mater. Sci. Lett.*, 1:13 (1982)
18. Chiu, C. C., Desu, S. B., and Tsai, C. Y., *J. Mater. Res.*, 8:2617 (1993)
19. Sasaki, K., Samkuma, E., Misawa, S., Yoshida, S., and Gonda, S., *Appl. Phys. Lett.*, 45:72 (1984)
20. El-Eskandarany, M. S., Sumiyama, K., and Suzuki, K., *J. Mater. Res.*, 10:659 (1995)
21. El-Eskandarany, M. S., Konno, T. J., Sumiyama, K., and Suzuki, K., *Mater. Sci. Eng. A*, 217/218:265 (1996)
22. *ASTM* Card No. 1-1119
23. Yan ZX, Cai M, Shen PK., *Sci. Rep.* 2013; 3:1646; DOI:10.1038/srep01646.

24. El-Eskandarany, M. S. Mechanical Alloying: Nanotechnology, Materials Science and Powder Metallurgy, 2nd Ed. Chapter 6 (Elsevier Inc. Oxford, 2015).
25. El-Eskandarany MS., *J. of Alloys Comp.*, 296: 175 (2000)
26. Lee, Jeong-Han, Oh, Ik-Hyun, Jang, Jun-Ho, Hong, Sung-Kil, Park, Hyun-Kuk, *Journal of Alloys and Compounds*, 786: 1 (2019)
27. Grasso, S. et al., *J. Am. Ceram. Soc.*, 1; DOI: 10.1111/jace.12365 (2013).
28. El-Eskandarany MS., *Met. Trans. A.*, 27: 4210 (1996)
29. Kurlov AS. et al., *Rev. Adv. Mater. Sci.*, 27: 165 (2011)
30. Fang ZZ, Wang X, Ryu T, Hwang K Sup, Sohn HY., *Int. Journal of Refractory Metals & Hard Materials*, 27:288 (2009)
31. El-Eskandarany, M. S., Mahdy, A. A., Ahmed, H. A., and Amer, A. A., *J. Alloys Compounds*, 312:315 (2000)
32. Enneti, R. K., *International Journal of Refractory Metals and Hard Materials* 53: 111 (2015)
33. Xun, Y., Mohamed, F. A. *Materials Science and Engineering* A., 527: 5446 (2011)
34. Wang, G. M., Millet, P., Calka, A., and Campbell, S. J., *Mater. Sci. Forum*, 179/181:183 (1995)
35. Maizza, G., Grasso, S., Sakka, Y., Noda, T.,Ohashi, O., *Science and Technology of Advanced Materials* 8: 644 (2007).
36. Tokita, M., Spark plasma sintering (SPS) method, systems, and applications, Handbook of Advanced Ceramics, 1st Ed (Elsevier Inc. Oxford, 2015).
37. El-Eskandarany, M. S., *J. Alloys and Compounds* 391: 228 (2005)
38. El-Eskandarany, M. Sherif, Al-Hazza, Abdulsalam, Al-Hajji, L. A., *Journal of Materials Engineering and Performance*, 26: 124 (2017)
39. Fecht, H. J., *Nanostruct Mater.*, 6: 33 (1995)
40. Gleiter, H., *Acta Mater.*, 48: 1 (2000).
41. Zhou, Y., Erb, U., Aust, K.T., Palumbo, G., *Scripta Mater.*, 48: 825 (2003)
42. Wang, F., Feng, Xi-Qiao, Yu, Shou-Wen, Nan, Ce-Wen. *Mater. Sci. Eng. A*, 363: 1 (2003)
43. Qing, X., Xingming, G., *International Journal of Solids and Structures*, 43: 7793 (2006)
44. El-Mahdy, Amir A., El-Eskandarany, M. Sherif, Ahmed, H.A., Amer, A.A., *J. of Alloys Comp.*, 299: 244 (2000)
45. Oyama, S.T. (ed.), The Chemistry of Transition Metal Carbides and Nitrides, Chapter 11, Chapman & Hall, London, 1996, pp. 215-232.
46. Musher, J.N., Gordon, R.G., *J. Mater. Res.*, 11: 989 (1996)
47. El-Eskandarany, M.S., Omori, M., Hirai, T., Konno, T.J., Sumiyama, K., Suzuki, K. *Metall. Mater. Trans. A.*, 29: 1973 (1998)
48. Anttila, A., Raisanen, J., Keinonen, J., *J. Appl. Phys. Letters.*, 57: 1423 (1983)
49. Kima, I.-S., Kumta, P.N., *J. Mater. Chem.*, 13: 2028 (2003)
50. Kawano, S., Takahashi, J., Shimada, S., *J. Am. Ceram. Soc.*, 86: 1609 (2003)
51. Chavda, M.R., Dave, D.P., Chauhan, K.V., Rawal, S.K., *Proced. Technol.*, 23: 36 (2016)
52. Deevi, S., Munir, Z.A., *J. Mater. Res.*, 5: 2177 (1990)
53. Calka, A., *Appl. Phys. Lett.*, 59: 1568 (1991)
54. El-Eskandarany, M.S., Sumiyama, K., Aoki, K., Suzuki, K., *Mater. Sci. Forum.*, 88: 801 (1992)
55. El-Eskandarany, M.S., Sumiyama, K., Aoki, K., Suzuki, K., *Mater. Res.*, 7: 888 (1992)
56. El-Eskandarany, M.S., *Journal of Materials Engineering and Performance*. 26 (2017) 2954-2962, doi:10.1007/s11665-017-2709-4.
57. Tiwari, D., Basu, B., Biswas, K., *Ceram. Int.*, 35: 699 (2009)

Fabrication of and consolidation of hard nanocomposite materials

8

The idea of combining two or more different materials resulting in a new material with improved properties dates back to 3000 years, when the Pharaohs of Egypt knew that plant fibers helped to strengthen and prevent bricks and pottery from cracking. The figure shows a typical composite material, which consists of brick (dark-gray block matrix) where the plant fiber materials (light-gray needle-like material) are embedded (Fig. 8.1). (*After El-Eskandarany*, unpublished work.)

8.1 Introduction and background

Synthesizing new composite materials with improved physical, chemical, and mechanical properties is one of the basic interests of both fundamental science and applications. The possibility of combining different types of material (metal–ceramic–nonmetal) to produce composite materials offers an excellent opportunity for synthesizing of new categories of materials in which tailored properties are achieved by bringing the combined advantages of both reinforcement and matrix into full play, which gives us a rather high degree of freedom in material design.[1] For instance, combining compliant isotropic material, which has low bulk and shear moduli, with a stiff isotropic material that enjoys high bulk and shear moduli, we can expect the generation of an elastically isotropic composite that enjoys the bulk modulus of the compliant phase and the shear modulus of the stiff phase. The combined two categories of materials existing in any composites are known as matrix and reinforcement(s). The matrix of any composites is usually more ductile and less hard phase, whereas the reinforcements, which are embedded into the matrix, are hard and strong.

Based on the approach employed for introducing the reinforcements to the matrix, composites may be classified into two groups known as (1) ex situ and (2) in situ. Whereas the reinforcement particles in the ex situ composites are introduced to the melt or solid-state matrix materials, the reinforcement phase(s) of the second composite group (in situ) are yielded as a result of a chemical reaction between different components in the mixed phases.

8.1.1 Nanocomposites

Nanocomposites or nanocomposite materials[2] are rather new terminology referring to those advanced classes of composite materials consisting of at least two phases with one dispersed in another to form a three-dimensional network. In contrast to the traditional composites, the dimension of one (at least) of the constituent phases (either

Mechanical Alloying. http://dx.doi.org/10.1016/B978-0-12-818180-5.00008-X

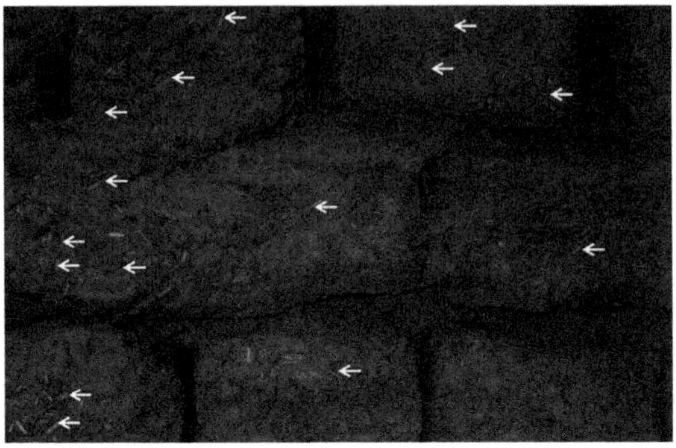

Figure 8.1 A typical composite material dates back to 3000 year, when the Pharaohs of Egypt reinforced brick with plant fibers.

the matrix or the reinforcements) in the nanocomposite materials must be within a nanoscopic scale.[3] In practice, the reinforcements are selected to as the nanoscaled materials that promote the densification of nanocomposites due to their large surface area and short diffusion distance.[4]

Need not to be mentioned, nanocomposite materials have been receiving great attention since 1990s. It has been shown in numerous publications[3] that using nanosized building blocks as matrix and/or reinforcements makes it possible to tailor advanced composites with unprecedented flexibility and improvements in their physical, chemical, and mechanical properties. In comparison with their counterparts of traditional coarse-grained composite materials, the nanocomposites possess pioneering mechanical properties such as high strength, high fracture toughness, stiffness, and wear resistance.

Likewise the traditional types of composites materials, nanocomposites can be also classified into two ways as follows:

- Based on the type of matrix materials,
  - *Metal matrix composite (MMC)* in which the host matrix material can be pure metals or metal alloys and the reinforcements are one phase or more of nanoscaled materials of either metal oxides, carbides, nitrides, borides. Reinforcing phase(s) can be also hard metals and intermetallics compounds of W-, Mo-based alloys.
  - *Ceramics matrix composites (CMC)* where the matrix is one of ceramics materials (oxides, carbides, nitrides) reinforced by other types of ceramics in the form of nanotubes, nanofilaments, nanofibers.
  - *Polymer matrix composites (PMC)* are types of composites where matrix materials are selected to be polymer materials (e.g., epoxy, polycarbonate, polystyrene). Based on the applications, the reinforcement phase(s) can be ceramics materials and graphite or carbon nanoparticles, carbon nanotubes (CNTs), clay nanotubes and nanofilament, or nano-dispersoids metallic phase.
- Based on the structure and arrangement of the reinforcement materials,
  - *Particulate composites*

    It refers to those composites where their matrices are reinforced by nano-dispersoids and nanoparticles materials. Particulate nanocomposites can be classified into two subgroups as follows:

- Composites with random orientation of nanoparticles
- Composites with preferred orientation of nanoparticles
- *Carbon nanotubes reinforced ceramics nanocomposites*
  CNTs are enjoying excellent properties with high aspect ratio. For example, CNTs own unusual mechanical characteristics upon their high ability of absorbing energy through their highly flexible elastic behavior during deformation when they are subjected to the applications of external loads. Based on their advanced properties CNTs have become at the top list of the desirable nanofiller materials used to improve many ceramics matrix composite materials. By CNTs-ceramics nanocomposite we can obtain high-performance ceramic composites with wide spectrum of superior mechanical properties (high mechanical strength and stiffness, superplastic deformability, and high fracture toughness) thermal and electrical conductivity. All of these unique properties promote CNTs-ceramics composites as multifunctional materials that are proposed for applications as structural materials in automotive, aerospace, naval, microelectronic and microwave devices, solid fuel cells, chemical sensors, and many others.
- *Laminate nanocomposites*
  Laminate nanocomposites are multilayered thin films consisting of multi-separated phases, in which the dimensions of one phase (or more) are in the nanoscaled level.

8.1.2 Metal-matrix nanocomposites (MMNCs)

As mentioned earlier, MMNCs are designed to have contrast properties such as ductility and toughness plus hardness and stiffness. This can be attained upon introducing rigid ceramic reinforcements to the metal ductile matrix. The combination between such two types of materials will lead to get new materials that enjoy wide spectrum of characterizations that cannot be seen in single-phase materials. Improving the mechanical, physical, and chemical properties of MMNCs has been considered as hot-research area and discussed in many research publications.[5] The key factors that carry the responsibilities of enhancing the properties of MMNCs can be concluded as: (1) the particle and grain sizes of both the reinforcements and the matrix, (2) morphological properties, and (3) volume fractions of the reinforcement additives and the distributions of the reinforcement in the host metallic matrix.

Niihara[6] has classified the nanocomposite materials based on the location of the reinforcements and their correlation in the matrix. This calcification was also driven according to the dimensions of both reinforcements and the matrix. Accordingly, nanocomposites can be classified into four groups, namely:

- Intergranular-type,
- Intragranular-type,
- Inter/intragranular-type (hybrid-type), and
- Nano/nano-type.

8.2 Fabrications methods of particulate MMNCs

Industry-based nanocomposites have multiplied in recent years. Nanocomposites can find wide area of applications in every industry. As examples, these new nanotechnology-based materials can be employed in medicine, electronics, sports equipment, aerospace, and automobile industries. There are several approaches used to fabricate particulate MMNCs such as solidifications, vertex process, squeeze casting, powder metallurgy (PM), hot pressing, and spray and spray deposition. However, solidification approach is considered to be a cheap one when compared with the other fabrications techniques, some technical problems during the process are usually occurred according to the tendency of the nano-reinforcement particles, which are added to the matrix, to agglomerate accordingly due to Van der Waals forces, electrostatic forces, and surface tension. Moreover, serious technical problems can come as a result of oscillating probe in the ultrasonic casting method.[4] Accordingly, the probe, which is in direct contact with liquid metal, may be dissolved in the liquid metal. Melting and casting methods may lead to obtain inhomogeneous particulate MMNCs where the reinforcement agglomerated nanoparticles are embedded into the metal matrix with a poor distribution fashion.

8.2.1 SiC/Al MMNCs

As mentioned before, the traditional technique of liquid metallurgy, which is the least expensive method for composite fabrications, cannot be successfully applied for the synthesizing of many nanocomposite systems such as due to the extreme gap difference in the thermal expansion coefficients between the two constituents and also due to the poor wettability between molten Al metal (or Al alloys) and SiC. According to various theoretical and experimental studies carried out thus far, SiC reacts with the molten Al to form Al_4C_3 and Si according to the following reaction[7]:

$$4\,Al + 3\,SiC \rightarrow Al_4C_3 + 3\,Si \tag{8.1}$$

This reaction is known to have several undesirable effects on the overall composite properties:

- Mechanical properties of SiC will be degraded due to the formation of Al_4C_3
- Since the reaction product Al_4C_3 is unstable in some environments, such as water, methanol, HCl, etc.,[8,9] the composite can be susceptible to corrosive environments
- In addition, Si, formed as an interfacial reaction product, will produce Al-Si eutectic at the interface and the grain boundary regions, resulting in unintended mechanical properties of the composite

As a result, fabrication of SiC/Al composite devoid of Al_4C_3 has long been one of the major concerns.

Among the traditional ways used to fabricate Al/SiC nanocomposites, two methods have been proven to be effective in achieving the goal for fabricating the MMNCs.

- Si addition into the Al matrix[10–13]
- Artificial oxidation of SiC to produce an SiO_2 layer on the surface of SiC[14–16]

The basic principle behind both methods is to enhance the Si activity and thereby reduce the Al activity, by dissolving Si into the Al matrix. However, considering the forward reaction which is presented in Eq. (8.1), one can say that such interfacial reaction is also dependent on the temperature and holding time, and can be the alternative solution to support the interfacial reaction.

8.2.2 Fabrication of SiC_p/Al MMNCs by mechanical solid-state mixing

The Al-based nanocomposites have recently attracted considerable attention as researchers strive to enhance composite properties and extend their utility by using nanoscale reinforcements instead of the more conventional particulate-filled composites.[17] While smaller reinforcements have a better reinforcing effect than larger ones, applying the ball-milling technique for composite fabrications must have the following merits:

- Since ball milling is processed at room temperature, the disadvantages of the liquid metallurgy method for producing undesirable materials can be avoided
- The ball-milling process can produce homogeneous nanocomposite powders instead of the more conventional coarse particulate-filled composites

In 1998, El-Eskandarany[2] employed a high-energy ball mill to fabricate metal matrix, a composite of Al reinforced with SiC_p with distinct nanocrystalline characteristics. In his experiments, pure elemental powders of Al (99.99%, 10 mm) and SiC ($\beta + \alpha$ phases) (99.9%, 100 mm) were weighed to give the nominal composition of $SiC_xAl_{100\text{-}x}$ ($x = 2$, 5, 7, and 10 vol %) and mixed in a glove box under a purified argon atmosphere, using sapphire mortar and pestle. The initial mixed powders were then charged and sealed in a stainless steel vial (SUS 316, 250 mL in volume) together with 50 stainless steel balls (SUS 316, 10 mm in diameter). The ball-to-powder weight ratio was maintained as 20:1. The ball-milling experiments were performed in a high-energy planetary ball mill at a rotation speed of 3.3 s^{-1}. To avoid the formation of any undesired contaminated phase(s), no lubricant solutions that are mainly hydrocarbon compounds were added to the charge. The ball-milling experiments were stopped periodically (every 1.8 ks) and then resumed when the temperature of the vial decreased to about 300 K. The final product (86 ks of milling) of the mixed powders was then consolidated into compacts in vacuum at 823 K, which is far below the melting point of Al (934 K), with a pressure of 19.6–38.2 MPa for 0.3 ks, using a plasma activated sintering (PAS) method. The as-milled and as-consolidated samples were characterized by means of X-ray diffraction (XRD) with CuKα radiation, scanning electron microscopy (SEM) using a 20 kV microscope, transmission electron microscopy (TEM) using a 200 kV microscope, and chemical analyses, using the induction coupled plasma emission and helium carrier fusion-thermal conductivity methods. The gas (oxygen, nitrogen, and hydrogen) and the iron, which was introduced to the samples from the milling tools, contamination contents in the end product of composite SiC_p/Al were determined to be less than 0.07 and 0.2 (at. %), respectively. The density of the consolidated samples was determined by Archimedes' principle, using water immersion. The hardness of the compacted sample was determined using a Vickers indenter with a load of 10 kg. Moreover, the elastic properties of the bulk samples were investigated by a pulse-echo overlap ultrasonic technique, using an ultrasonic detector.

8.2.2.1 Properties of mechanically solid-state fabricated SiC_p/Al nanocomposites

Fig. 8.2 presents the typical XRD patterns of mechanically mixed SiC_{10}/Al_{90} powder after (A) 0 ks of ball-milling time (starting material mixture of elemental Al and SiC powders) and (B) 86 ks of ball-milling time (the end product). Obviously, the starting materials contain coarse polycrystalline grains of SiC and Al, suggested by the sharp diffraction peaks of the powder mixture. On the contrary, the Bragg peaks of both the reinforcement (SiC) and metallic matrix (Al) powders become broad toward the end stage of milling (Fig. 8.2B), indicating the formation of nanocomposite SiC_p/Al material. Remarkably, this end product is still an intact mixture of polycrystalline SiC and Al and does not coexist with any reactive products, such as Al_4C_3 and Si, suggesting the absence of any undesirable reactions at the SiC/Al interfaces.

SEM technique was used to follow the characteristics of the powders during the several stages of ball milling. Fig. 8.3 may summarize the morphology, topology, and size of mechanically mixed SiC_{10}/Al_{90} (vol %) powder after selected ball-milling time. The powders at the starting stage of milling (Fig. 8.3A) are bulky with random shape and size. The soft Al particles in the initial starting mixture are affected by the cold working and the impact and shear forces generated by the milling media, and tend to agglomerate to form larger powder particles with an average diameter of nearly 1000 μm, after 3.6 ks of the milling (Fig. 8.3B). At this stage of milling, the hard particles of SiC are randomly embedded into the deformed Al soft matrix (host matrix) to form coarse composite SiC_p/Al powders, as shown in Fig. 8.4. Contrary to the Al powders, the SiC powders have been dramatically disintegrated to form finer particles with spherical-like morphology. During the subsequent disintegration stage (22–43 ks), the agglomerated powder particles are subjected to a continuous disintegration with fragmentation to form finer powders with extremely wide particle size

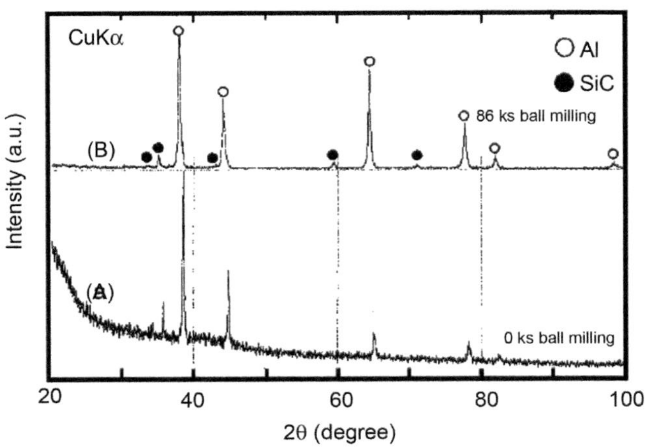

Figure 8.2 XRD patterns of mechanically mixed SiC_{10}/Al_{90} powder after (A) 0 ks and (B) 86 ks of the milling time. *(After El-Eskandarany.)*[2]

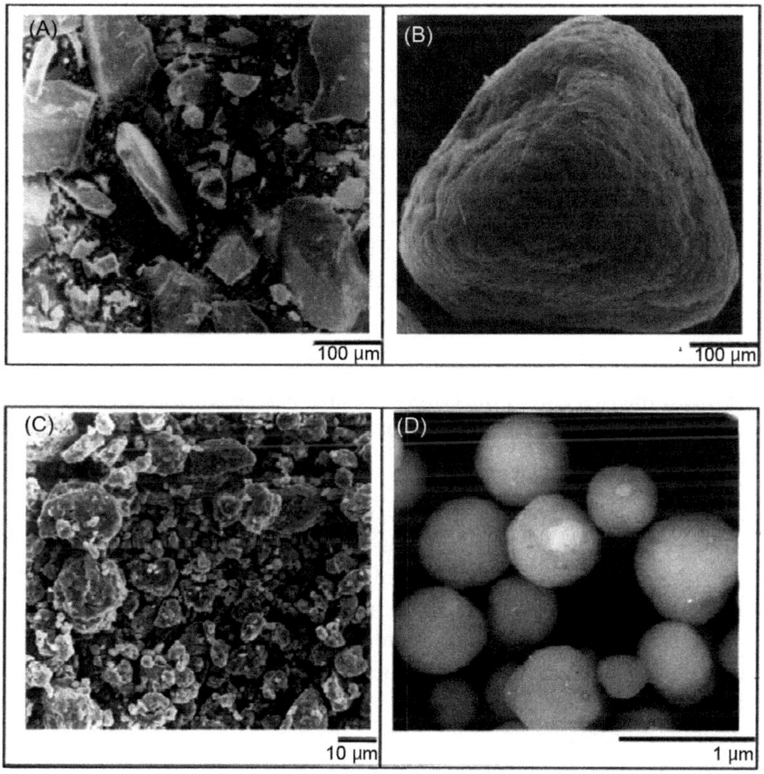

Figure 8.3 SEM micrographs of mechanically solid-state mixed SiC_{10}/Al_{90} powder after (A) 0 ks, (B) 3.6 ks, (C) 22 ks, and (D) 86 ks of the ball-milling time. *(After El-Eskandarany.)*[2]

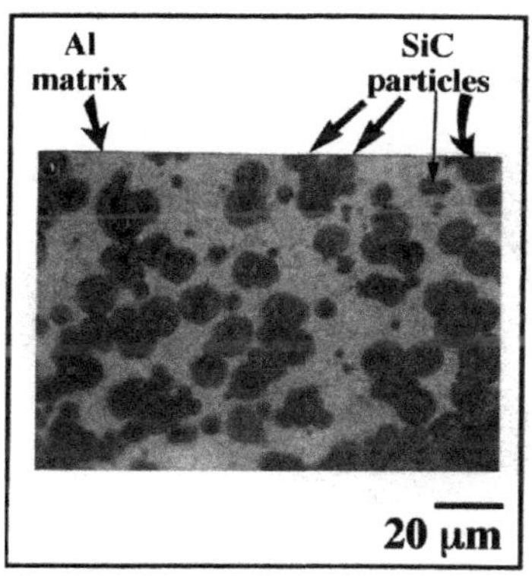

Figure 8.4 Optical micrograph of the polished and etched particle of mechanically solid-state mixed SiC_{10}/Al_{90} after 3.6 ks of the ball-milling time. *(After El-Eskandarany.)*[2]

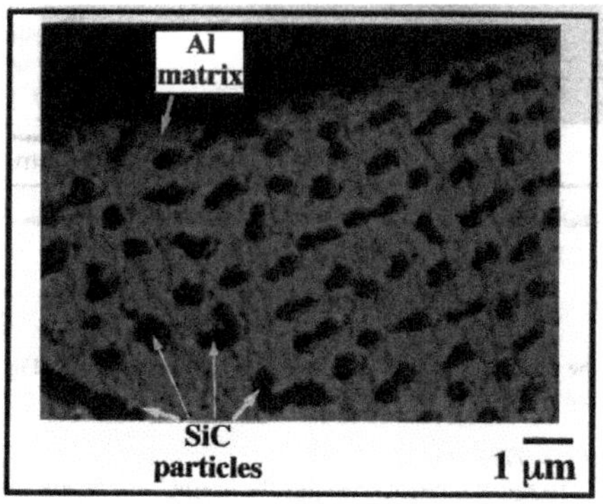

Figure 8.5 SEM micrograph of the cross-sectional view of a selected mechanically solid-state mixed SiC_{10}/Al_{90} particle after ball milling for 43 ks. *(After El-Eskandarany.)*[2]

distribution (Fig. 8.3C). Toward the end of this stage (43 ks), the reinforcement SiC particles become finer (less than 1 μm in diameter) and more uniformly distributed in the Al matrix (Fig. 8.5). The powder of the end product (86 ks of milling time) of SiC_{10}/Al_{90} composite possesses excellent morphological appearance, characterized by the ultrafine (less than 0.5 μm in diameter), smooth (without any detailed structural relief on the powder surface), and the exactly spherical-like morphology, as shown in Fig. 8.3D.

To reveal more information, the internal structure of the mechanically mixed powders has been investigated by the TEM technique. This technique provides information on the structure of the powder particles within small areas (less than 50 nm in diameter) in which SEM is not able to give such information.

Fig. 8.6 shows the bright field images (BFIs) and the selected area diffraction patterns (SADP) of mechanically mixed powders after 3.6 ks of ball-milling time. Obviously, the large grains of SiC are embedded into the Al matrix that are heavily faulted,

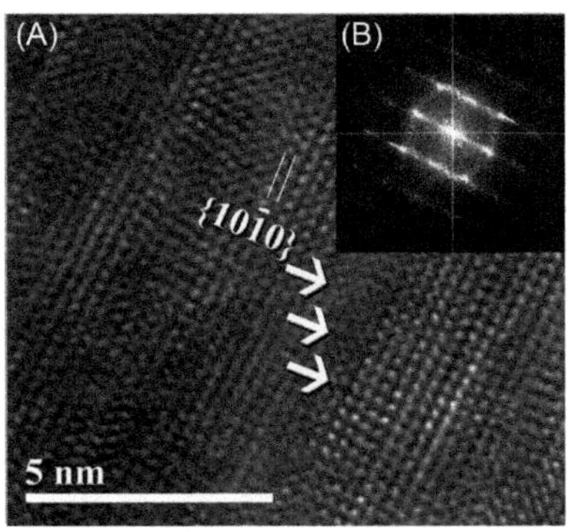

Figure 8.6 (A) BFI and (B) the corresponding SADP of mechanically solid-state mixed SiC_{10}/Al_{90} composite particle after 3.6 ks of the ball-milling time. *(After El-Eskandarany.)*[2]

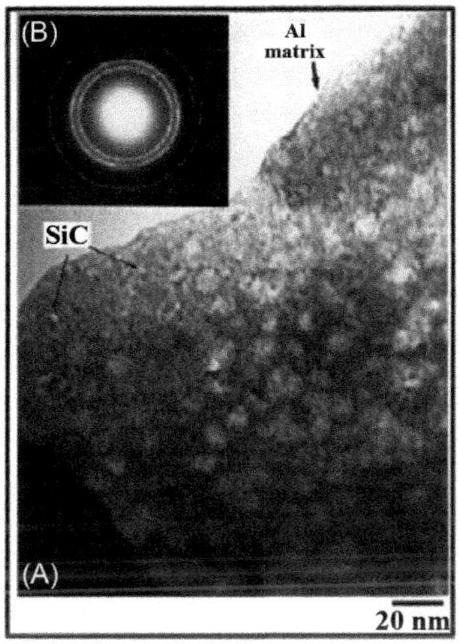

Figure 8.7 (A) BFI and (B) the corresponding SADP of mechanically solid-state mixed SiC_{10}/Al_{90} composite particle after 86 ks of the ball-milling time. *(After El-Eskandarany.)*[2]

characterized by parallel rows of faults which appear as multiple fringes, as shown in Fig. 8.6A. The SADP (Fig. 8.6B) shows a sharp ring-spot pattern that is characteristic of several simultaneity diffracting polycrystalline β-SiC (fcc) and Al (fcc). It is worth mentioning that there is no evidence of the formation of any reactive products such as Al_4C_3 and Si phases.

Fig. 8.7 displays the BFI and the corresponding SADP of mechanically mixed powders after 86 ks of ball-milling time. Obviously, the end product is a typical nanocomposite material, containing only the two starting phases of SiC (granular fine grains) and metallic Al (overall matrix). No more sharp ring-spot patterns appeared (Fig. 8.7B), suggesting the formation of fine powders.

In order to determine some of the physical and mechanical properties of fabricated composite powders, the PAS method has been applied to consolidate the final product of the milled powders into compacts (20 mm in diameter and 20 mm in thickness). The XRD pattern of the end product of as-consolidated mechanically mixed powders, exemplified by SiC_{10}/Al_{90} compact, is shown in Fig. 8.8. Comparing this pattern with that for the milled powders (Fig. 8.2B), one can conclude that this consolidation step does not lead to a remarkable grain growth of the sintered powders, and the materials still maintain their unique nanocrystalline structure.

TEM technique has also been used to investigate both the grain size and the phases that may be formed during the consolidation procedure. The BFI and corresponding SADP of as-consolidated SiC_{10}/Al_{90} are shown together in Fig. 8.9. Comparing this micrograph with the one in Fig. 8.7, we can say that this consolidation step leads to slight or moderate grain growth for both the reinforcements (SiC) material and Al matrix. Since the average grain size of the composite consolidated sample is less than 100 nm (about 20 nm in diameter or less) one can say that the sintered sample

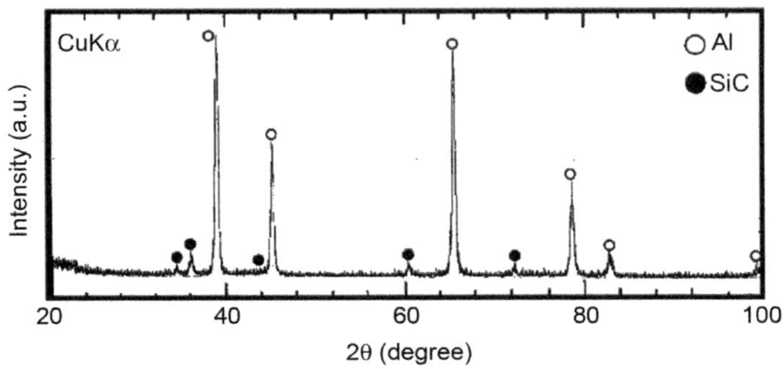

Figure 8.8 The typical XRD pattern of as-consolidated mechanically solid-state mixed SiC_{10}/Al_{90} that ball-milled for 86 ks. *(After El-Eskandarany.)*[2]

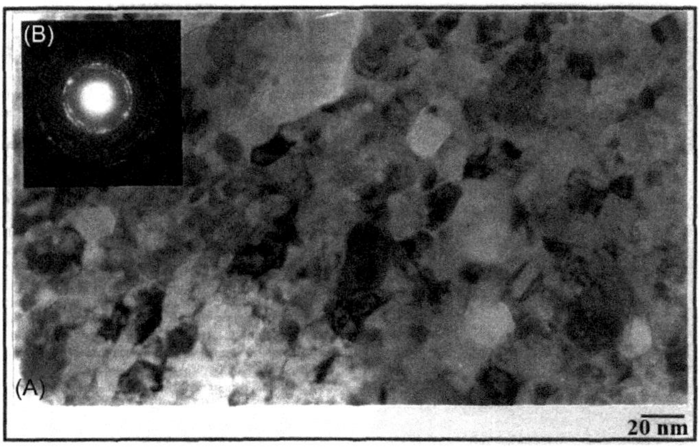

Figure 8.9 The (A) BFI and (B) the corresponding SADP of as-consolidated mechanically solid-state mixed SiC_{10}/Al_{90} that ball-milled for 86 ks. *(After El-Eskandarany.)*[2]

maintains its nanocrystalline character. No reaction layers can be detected at the SiC/Al interfaces, using the shown magnification. It has also emphasized that this consolidation step does not lead to the formation of any reactive product(s), and the fabricated nanocomposite bulk material combines only the two starting materials of SiC and Al, as suggested by the SADP that is shown in Fig. 8.9B.

The SEM (back scattering) micrograph and the elemental dot mapping of Al and Si of as-consolidated SiC_{10}/Al_{90} compact are shown in Fig. 8.10. Remarkably, the SiC particles (sharp-edged light-gray particles) are embedded and distributed along the Al matrix (dark-gray background matrix). Apparently neither voids nor cracks can be detected, indicating an excellent SiC/Al interfacial bonding (Fig. 8.10A). Moreover, no reacted product of amorphous or an intermediate phase can be identified, suggested by segregation of the elements of Al and Si in the composite sample on a micron scale, as shown in Figs. 8.10B and C.

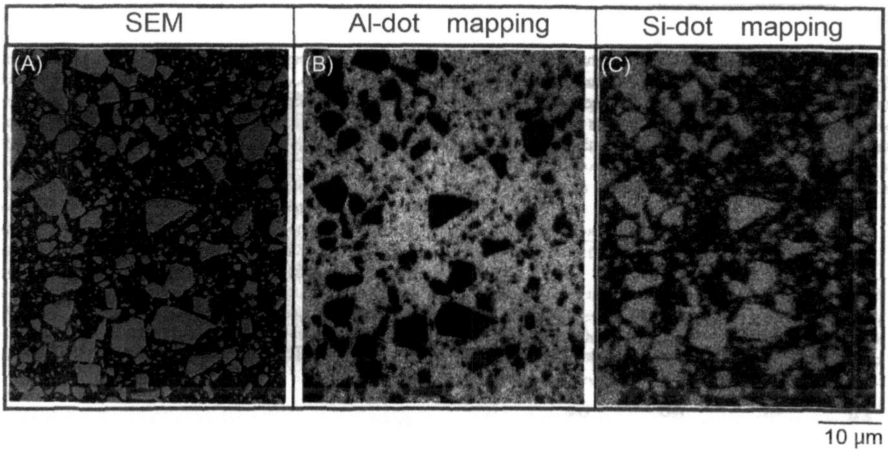

Figure 8.10 (A) The SEM micrograph and dot mapping of (B) Al and (C) Si of as-consolidated mechanically solid-state mixed SiC_{10}/Al_{90} that ball-milled for 86 ks. *(After El-Eskandarany.)* [2]

The bulk density of the as-consolidated SiC_x/Al_{100-x} composite samples is plotted as a function of the SiC content, x, in Fig. 8.11. Obviously, the densities increase monotonically with an increase in the volume fraction of the reinforcement material, to have a maximum value of about 2.77 Mg m^{-3} for the sample which contains 10 vol % SiC, just above the theoretical density of pure Al (2.70 Mg m^{-3}). It is worth noting

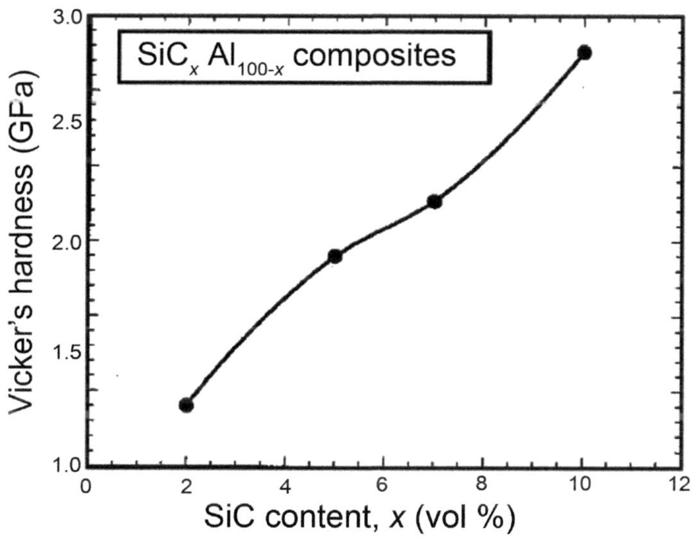

Figure 8.11 Influence of SiC concentration content on the bulk density of as-consolidated mechanically solid-state mixed SiC_x/Al_{100-x} that ball-milled for 86 ks. *(After El-Eskandarany.)*[2]

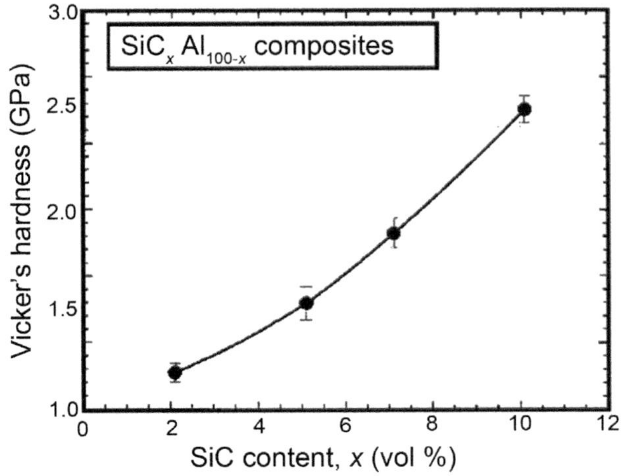

Figure 8.12 Correlation between SiC concentration content, *x*, and Vickers hardness of consolidated SiC_x/Al_{100-x} powders that ball-milled for 86 ks. *(After El-Eskandarany.)*[2]

that the densities of all the consolidated samples presented in Fig. 8.11 are about 99.5% of the theoretical density, suggesting the formation of a fully dense SiC_x/Al_{100-x} composite material.

Fig. 8.12 illustrates the effect of the SiC additives on the Vickers hardness of the consolidated samples. As expected, the hardness increases linearly with an increase in the volume fraction of the harder phase (SiC) in the soft matrix of metallic Al, to have a maximum value of about 2.6 GPa for the sample that contains 10 vol % SiC, which is well above the measured hardness of pure Al (about 0.95 GPa) under the same measurement conditions. Remarkably, the hardness values for all the SiC_x/Al_{100-x} samples have a narrow distribution range indicating a homogeneous distribution of the reinforcement particles in the Al matrix.

The elastic moduli presented by the bulk modulus, Young's modulus, and the shear modulus, of the consolidated mechanically-mixed SiC_x/Al_{100-x} composites, is shown as a function of the SiC content in Fig. 8.13. We should emphasize that the values of the elastic moduli of the consolidated samples have been estimated from the measured samples densities and the constant parameters of the nondestructive testing apparatus, as was explained elsewhere.[18,19] These values increase significantly with increasing the SiC content, suggesting the formation of a composite material that contains a brittle phase (SiC) which embedded in the ductile Al matrix.

8.2.2.2 Mechanism of fabrication

In this section, the mechanism for synthesis of this type of important engineering composite material will be discussed. In addition, the mechanism of powder consolidation into nanocrystalline compacts is explained as well. The challenge in processing SiC/Al composites is to promote wetting and interface formation between

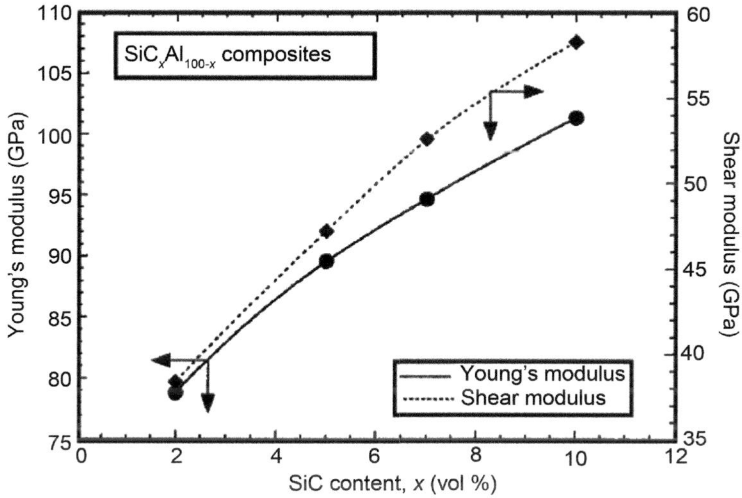

Figure 8.13 Dependence of the elastic moduli on the SiC concentration content, x, for consolidated SiC$_x$/Al$_{100-x}$ powders that ball-milled for 86 ks. *(After El-Eskandarany.)*[2]

SiC and Al. Contrary to the liquid metallurgy technique that may lead to the formation of undesirable brittle phases of Al_4C_3 and Si, the ball-milling technique shows the possibility of fabrication of nanocomposite SiC_xAl_{100-x} without the existence of these reactive products. Synthesizing nanocomposite SiC_p/Al by mechanical mixing via ball-milling technique can receive much attention. This may be due to the simplicity in the technique and the huge amount of homogeneous composite powders that can be produced on an industrial scale at room temperature. Based on the earlier results, well-bonded nanocomposite SiC_xAl_{100-x} powder particles are yielded by milling elemental Al and SiC powders in a high-energy ball mill at room temperature. Neither solid solutions nor reactive products could be detected during the milling process.

First, it may be useful to classify the milling process into four stages according to the structure and morphological characteristics of the powder particles.

8.2.2.2.1 Formation of agglomerates coarse composite SiC_p/Al powder particles

This refers to the first stage of the mechanical mixing (0–22 ks). During the first few kiloseconds of milling, all the Al powder particles are stacked on the surface of the milling tools as a result of cold welding. Meanwhile, most of the hard particles of refractory SiC powders are disintegrated and do not blend with the agglomerated Al powders. As the milling time increased (3.6 ks), those hard particles of SiC that are refined (nearly 20 μm in diameter) and changed in shape to a spherical-like morphology penetrated and embedded into the soft deformed metallic matrix of Al particles to form coarse composite particles of about 1000 μm in diameter.

8.2.2.2.2 Disintegration of the agglomerates composite SiC_p/Al powder particles

The first stage of milling is followed by a second stage (22–43 ks) in which the agglomerated composite particles are shattered and disintegrated into several particles apparently irregular in size (Fig. 8.3C). The disintegration of the powder occurred as a result of continuous shear and impact forces generated by ball-powder-ball collisions. However, the morphology and composition of the individual composite powder at this milling time may differ remarkably from particle to particle and within the particle itself, the SiC particles have not separated from the host Al matrix and no free SiC and/or Al powders could be detected in the mill product at this stage of milling. This is attributed to the well bonding between SiC and Al that was improved by the mechanical mixing.

8.2.2.2.3 Formation of nanocomposite SiC_p/Al powder particles

Toward the end of the mechanical mixing processing (43–86 ks), a complete homogenization of the composite powder particles takes place. At this stage of milling, the powders become ultrafine and the size of the individual particles is less than 0.5 μm in diameter. In addition, the particle size distribution of the powders has become very narrow, indicating the formation of a homogeneous and uniform composite powder. The final product of the powder (86 ks) contains internal extraordinary equiaxed fine sphere grains of SiC about 3 nm in diameter.

8.2.2.2.4 Consolidation of nanocomposite SiC_p/Al powder particles

An attractive technique that has been used for consolidation of several metallic and ceramics powders is the PAS. Nanocrystalline materials with grain size of less than 100 nm are receiving much attention due to their unique mechanical properties. The ball-milling technique has been considered a powerful tool for preparing such a class of materials. The important steps during the PAS process are as follows:

- The application of uniaxial load
- The application of pulsed voltage for plasma activation
- The resistance heating of graphite crucible and powders

As DC pulse voltage is applied to the powder, a micro-discharge takes place among the composite powder particles, which generate plasma. Hence, the atoms on the surface of each particle are activated, and the sintering procedure takes place in a short period of time (consolidation is achieved within a few kiloseconds, 0.3 ks). Accordingly, grain growth during sintering can be minimized and the sintered maintain their nanocrystalline properties.

Another important factor in PAS that governs the grain size of the consolidated samples is the source of the milled materials powders. When the as-milled powders, which are introduced to the PAS, have spherical homogeneous shape with fine nanocrystalline grains, the PAS step leads in parallel to the formation of nanocrystalline fully dense compacts.

It is worth noting that the PAS technique does not result in the formation of any undesirable reactive products, such as Al_4C_3 and Si. These products are usually obtained when the sintering temperature is above the melting point of pure Al.

8.3 WC-based nanocomposites

8.3.1 WC/Al_2O_3 nanocomposite

Among the hard alloys and refractory carbides, WC finds wide range of industrial applications, being used extensively in commercial applications such as tips for cutting and drilling tools, wear-resistant parts in wire drawing, extrusion and pressing dies and wear-resistant surfaces in many types of machines. Since the hexagonal phase of WC is usually obtained by direct union of elemental W and graphite at temperature ranging from 1673 to 1873 K, their preparation costs are very high. In 2005, El-Eskandarany[20] proposed an attractive technique for preparation of large-quality powders of nanocomposite WC/Al2O3 at room temperature, by ball milling a mixture of WO3, Al, and graphite powders were mixed in a glove box under argon gas atmosphere and sealed in a cylindrical sapphire vial together with 50 sapphire balls (10 mm in diameter). The ball-to-powder weight ration was maintained at the level of 10:1. The ball-milling experiments were carried out at room temperature, using Fritsch P6 high-energy ball mill at a rotation speed of 4.2 s^{-1}. The sample of the end product contains about 32 at.% Al_2O_3.

X-ray analysis was employed to monitor the structural changes that occurred upon ball milling of (2Al + WO_3 + C) mixed powders. The XRD patterns of the ball-milled powders after selected ball-milling time are displayed in Fig. 8.14. At the starting stage of milling (0 ks) the reactant powders are mixture of elemental WO_3, Al, and C materials with sharp Bragg peaks (Fig. 8.14A). After 43 ks of milling, however, new Bragg peaks that correlated to metallic W and Al_2O_3 crystals appear, indicating the starting of a mechanical solid-state reduction (MSSR), which takes place between the Al (reducing agent) and WO_3 powders (Fig. 8.14B). In parallel to the MSSR process, a solid-state reaction (SSR) occurred between metallic W and C powders to yield a refractory phase of hcp-WC. The Bragg peaks of these reacted phases have become pronounced and their intensities increased after 173 ks (Fig. 8.14C), suggesting a monotonical increase of the product materials. No free atoms of the starting reactant materials and/or pure metallic W powders could be detected after 360 ks of milling, suggesting the completion of both MSSR and SSR processes (Fig. 8.14D). The XRD pattern of as-consolidated composite WC/Al_2O_3 bulk material, which is characterized by the broadening in the Bragg peaks for both WC and Al_2O_3 phases, is shown in Fig. 8.14E. Comparing this pattern with the one shown in Fig. 11.14D, one can conclude that this consolidation step does not lead to a dramatic change in the grain size and the as-sintered sample still maintains its unique nanocrystalline properties.

Fig. 8.15 shows an SEM micrograph of an individual milled particle that was obtained after 3.6 ks of milling (Fig. 8.15A). At this early stage of milling, all of the ductile Mg particles tend to agglomerate and form larger particles with more than 300 μm in diameter (Fig. 8.15A). The cross-sectional view for these particles (Fig. 8.15B) shows a typical composite particle with ellipsoid-like morphology, which consists of WO_3 (fine white grains about 10 μm in diameter), C (black grains about 30 μm in diameter), and the Mg (the gray matrix about 1800 μm in diameter). The

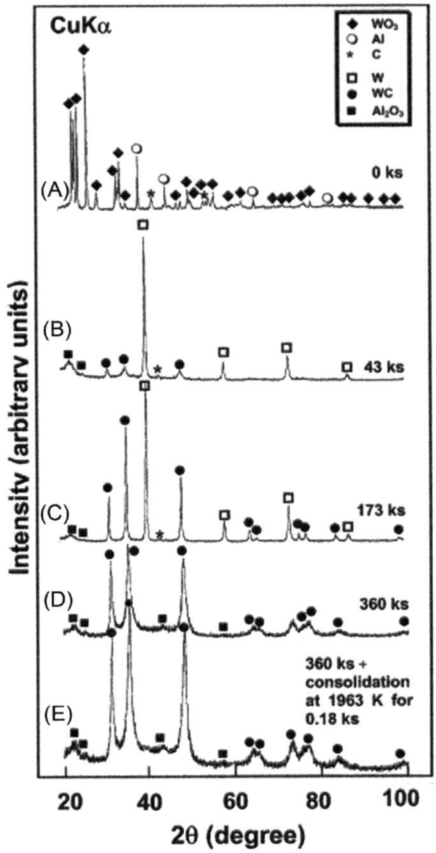

Figure 8.14 XRD patterns of ball-milled WO3, Al, and C powders after (A) 0 ks, (B) 43 ks, (C) 173 ks, and (D) 360 ks of the milling time. The XRD pattern of the end product after consolidation at 1963 K for 0.18 ks is presented in (E). *(After El-Eskandarany.)*[20]

formation of such composite-reactant particles is attributed to the cold welding performed at this early stage of milling in which the ductile Mg particles are deformed to form soft matrices containing the rather hard particles of WO_3 and C powders. These agglomerated powders are disintegrated to form finer particles with grain size of less than 300 nm in diameter, as shown in Fig. 8.16A, which presents the BFI that is taken for the powders after milling for 43 ks. The BFI of the sample after 259 ks of ball-milling time shows the internal structure of the milled powders that consist of WC grains, which are embedded into the matrix of Al_2O_3 (Fig. 8.16B).

The SEM micrograph for the powders end product obtained after 360 ks of ball milling time is shown in Fig. 8.17A. Obviously, the powders have excellent morphological characterizations, indicated by the nearly appearance of having spherical-like morphology with narrow particle size distribution, ranging from 0.5 to 1.2 μm in

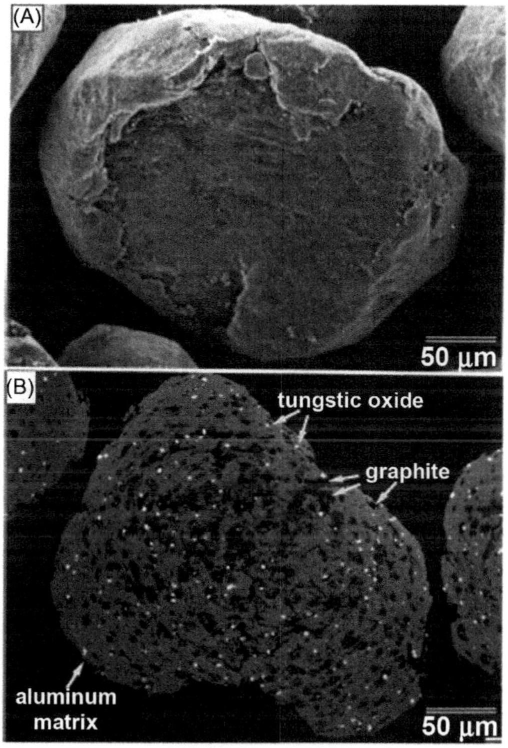

Figure 8.15 (A) SEM micrograph of the ball-milled powders after 3.6 ks of the milling time. The cross-sectional view of the powders at this stage of milling is shown in (B). *(After El-Eskandarany.)*[20]

diameter. The internal structure of this end product has been investigated by HRTEM technique. The HRTEM micrograph and the corresponding SADP of the powders after 360 ks of ball milling are shown together in Fig. 8.17B and C, respectively. The micrograph shows that fine nanograins of WC reinforcement materials (about 5 nm in diameter) are embedded into the host ceramic matrix of Al_2O_3, as indicated by the lattice-fringe image (0.284 nm), which is oriented to the (001) plane (Fig. 8.17B). No evidence of the existence of any foreign materials could be detected at the end product (Fig. 8.17C).

The SEM micrograph and the elemental dot mapping of W and Al for the as-consolidated WC/Al_2O_3 compact are shown together in Fig. 8.18. Remarkably, the WC particles that have sharp edges (Fig. 8.18A) are embedded and homogeneously distributed along the grain boundaries of Al_2O_3 particles. Apparently, neither voids nor cracks can be detected, indicating excellent interfacial bonding (Fig. 8.18A). Moreover, no undesirable reaction product could be identified, as suggested by the segregation of the elemental W and Al in the composite material on a micron scale (Fig. 8.18B and C).

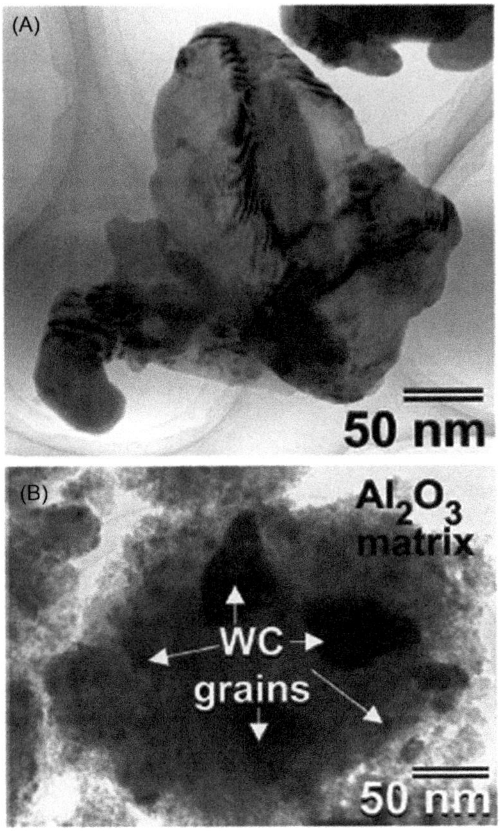

Figure 8.16 BFI of the ball-milled powders after ball milling for (A) 43 ks and (B) 259 ks.

Fig. 8.19 shows a low magnification TEM micrograph of the cross-sectional view for as-consolidated WC/Al_2O_3 powders. Obviously, the consolidation procedure leads to the formation of full-dense nanocomposite materials in which the fine WC grains (dark grains) that have average grain size of less than 50 nm in diameter are perfectly interfacial bonded with the ceramics matrix of Al_2O_3 (light grains).

Fig. 8.20 shows the change of the vial's temperature during the milling time. It has been shown in a previous work[21] that this temperature does not increase more than 314 K, when the same mill was operated without charging the reactant powders. During the first few kiloseconds (0–22 ks), the temperature of the vial increases monotonically, and the starting reactant materials of WO_3, Al, and C tend to agglomerate to form larger particles. Increasing the milling time (22–50 ks) leads to enhance the MSSR and SSR between the reactant materials, hence the heat liberated by these reactions transferred to the surface of the vial and leads to an increasing of its temperature to about 376 K (Fig. 8.20). Toward the end of this stage (50 ks), a small mole fraction of WC is formed as a result of an SSR between the W and C powders. A complete SSR takes place between W and C powder at the second stage of milling (50–324 ks)

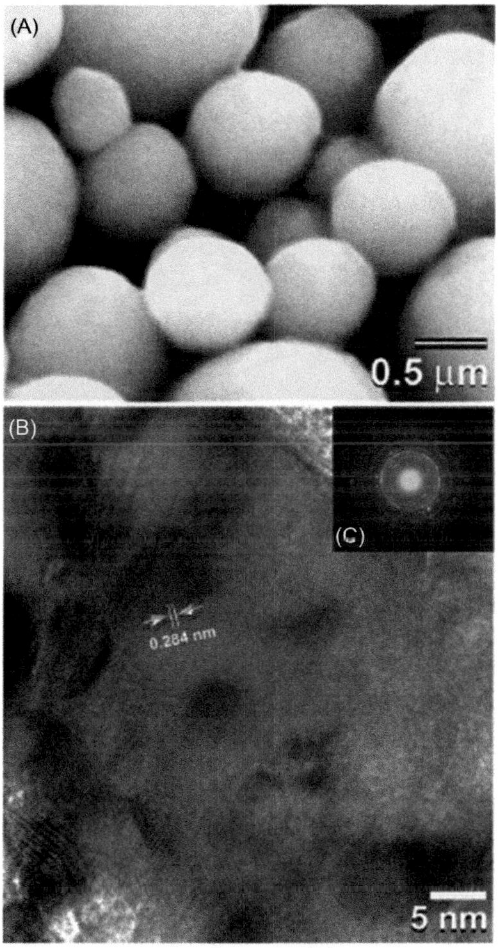

Figure 8.17 (A) SEM micrograph and (B) HRTEM image of the (C) end product of nanocomposite WC/Al_2O_3 obtained after 360 ks of ball-milling time. *(After El-Eskandarany.)*[20]

and the vial's temperature drops to about 319 K. At the end of the ball-milling time (360 ks), the vial's temperature remains at a constant value of about 318 K, suggesting the completion of both reactions and the formation of the final product, which is nanocomposite of WC/Al_2O_3 powder.

An optical micrograph of the Vickers hardness indentation of the consolidated sample appears in Fig. 8.21. The hardness of this compact sample was measured to be 16 GPa, taking the average of at least 10 indentations. It is worth noting that this value is higher than of bonded nanocrystalline WC-14 pct Co (13.2 GPa).[22] The cracks displayed in the micrograph, which developed in the product during indentation and extended to 110 μm away from it, were used as an indicator for estimating

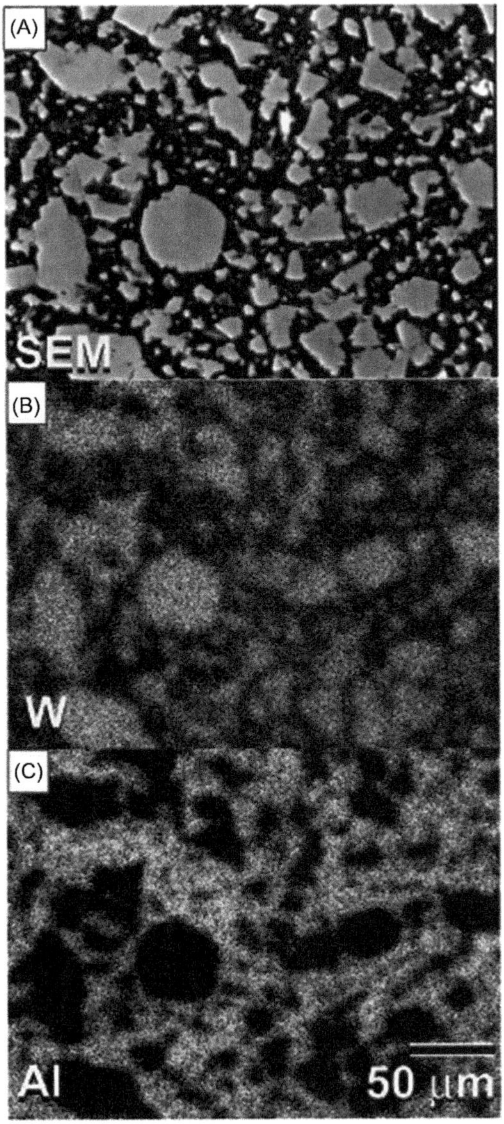

Figure 8.18 (A) SEM micrograph and (B) and (C) dot mapping of elemental W and Al of the end product of nanocomposite WC/Al_2O_3 after consolidation. *(After El-Eskandarany.)*[20]

the fracture toughness (K_{c1}) via the model that was proposed by Anstis et al.[23] This value is found to be 15 MPa√m well above that for commercial WC/Co composites (12 MPa√m). This indicates the possibility of producing high-fracture-toughness WC/ Al_2O_3 nanocomposite materials.

However, pure nanocrystalline WC possesses extraordinary high values of hardness (23 GPa),[24] Young's and shear moduli (813.5 and 405.1, respectively),[24] it has

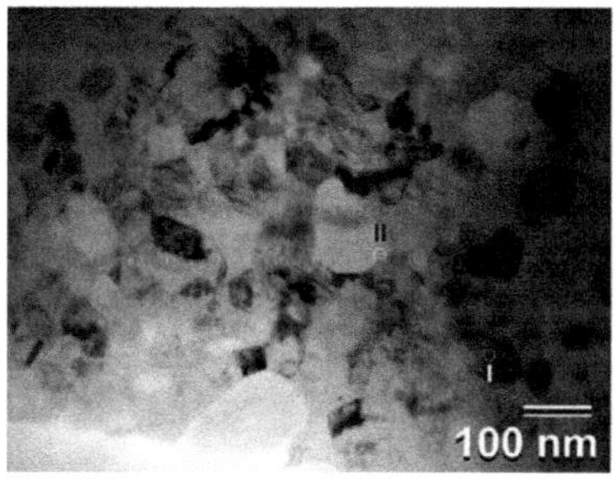

Figure 8.19 BFI planar view of the end product of nanocomposite WC/Al_2O_3 after consolidation. *(After El-Eskandarany.)*[20]

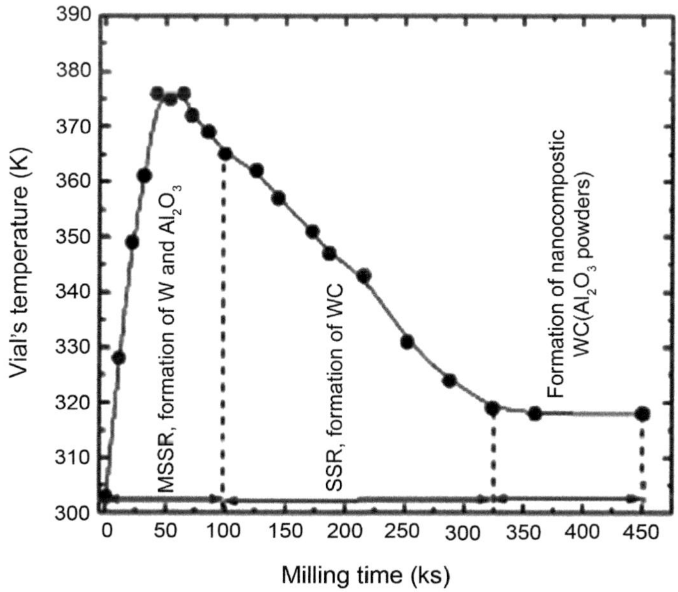

Figure 8.20 The correlation between the vial's temperature and the milling time during the ball-milling elemental mixture of (WO_3 + 2Al + C) powders. *(After El-Eskandarany.)*[20]

very low value of K_{c1} (4 MPa$\sqrt{m}$),[25] making it difficult for use as cutting tools. Contrary to the WC material, composite WC/Al_2O_3 combines unique properties of lower density (11.95 g/cm^3), high hardness value (16 GPa), and extreme high value of K_{c1} (15 MPa$\sqrt{m}$) with reasonable values of Young's (422 GPa) and shear (184 GPa) moduli. In fact, nanocomposite WC/Al_2O_3 material is unique in that it can combine two opposite properties, high hardness, and high-fracture toughness. This may be understood when recognizing that the fabricated WC/Al_2O_3 material consists of fine grains

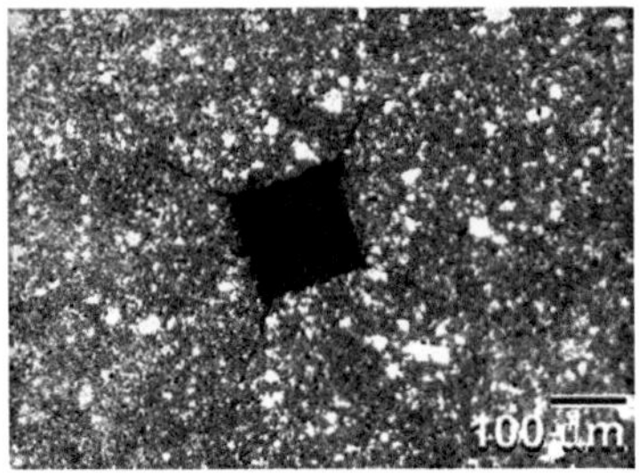

Figure 8.21 Optical micrograph of Vickers hardness indentation developed by applying a load of 50 kg on the consolidated sample of the end product (360 ks). *(After El-Eskandarany.)*[20]

(less than 50 nm in diameter) that are able to improve the mechanical properties of the fabricated materials and introducing materials with unique properties.

8.3.2 WC-5Co-1Cr-3MgO-0.7VC-0.3Cr_3C_2 nanocomposite

More recently, El-Eskandarany et al.[26] have prepared nanocomposite WC-5Co-1Cr-3MgO-0.7VC-0.3Cr_3C_2 powders at room temperature, using a high-energy ball-milling technique. In their experiments, the starting materials of WC, Co, Cr, VC, Cr_3C_2, and MgO were first balanced in a glove box under argon gas atmosphere to give the average nominal composition shown in title of the paper. Then, the mixed powders were charged into WC vials and sealed together with 25 balls made of WC (15 mm in diameter) to give an average ball to powder weight ratio of 20 to 1. The ball-milling process was carried out at room temperature using high-energy ball milling with a rotation speed of 250 rpm for 75 h of milling time.

Fig. 8.22 shows the SEM micrograph of the powders obtained after high-energy ball milling for 75 h. The powders possess excellent morphological characterizations

Figure 8.22 SEM micrograph of the powders obtained after 75 h of ball milling. *(After El-Eskandarany.)*[26]

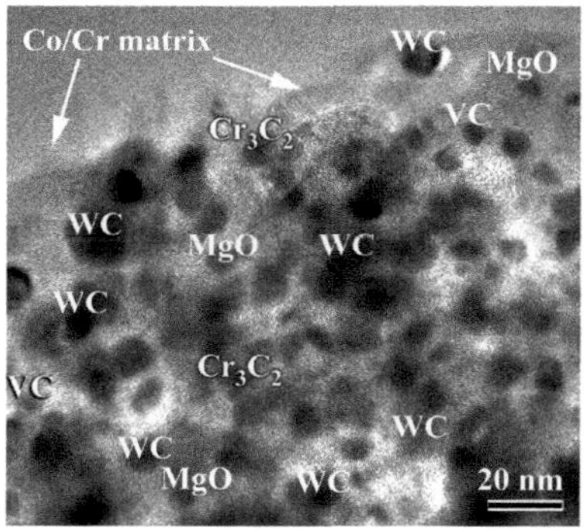

Figure 8.23 BFI of the nanocomposite powders obtained after ball milling for 75 h. *(After El-Eskandarany.)*[26]

indexed by their spherical-like morphology and narrow particle size distributions (0.5–1 μm in diameter).

The BFI of the powder after this stage of ball milling shows the formation of homogeneous nanocomposite powders, indicated by the excellent distribution of the reinforcement materials in the metal matrix (Fig. 8.23). The BFI of the consolidated sample (Fig. 8.24) shows fine grains with nanostructured characterizations of less than 60 nm in diameter. This indicates that the consolidation step did not lead to a serious grain growth and the bulk material still maintains its nanocrystalline characterizations.

The SEM micrograph of the Vickers microhardness indentation of the cubic consolidated sample is presented in Fig. 8.25. The average value of the hardness calculated from 60 measurements (10 measurements of each face of the cube) was found to be 19.32 GPa. The cracks displayed in the micrograph that developed in the product during indentation and extended to 192 μm away from it were used for estimating the fracture toughness, K_c. This value was found to be 17.36 MPa $m^{1/2}$, well above that of commercial WC/Co composites (12–14 MPa $m^{1/2}$).

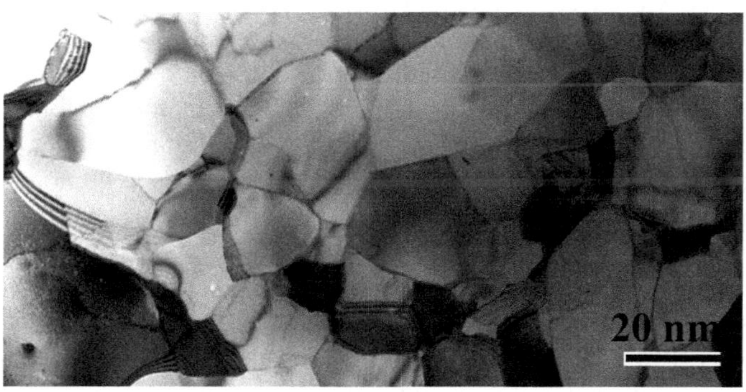

Figure 8.24 BFI of as-consolidated bulk sample. *(After El-Eskandarany.)*[26]

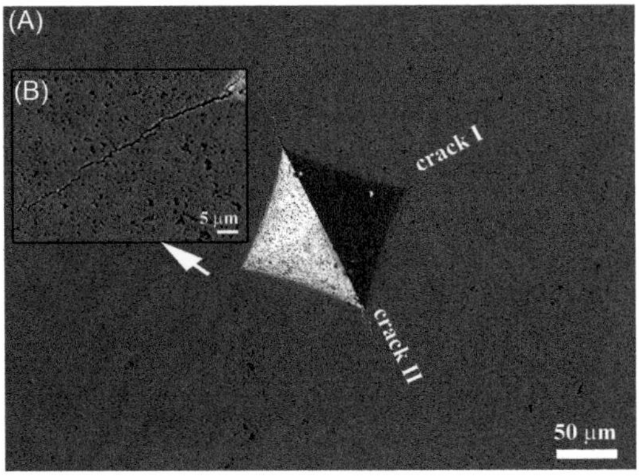

Figure 8.25 (A) SEM of the Vickers hardness indentation developed by applying a load of 20 kg on the consolidated sample. A one-side radial crack emanating from the Vickers indentation is shown in (B) in a different scale. *(After El-Eskandarany.)*[26]

The metallographic examinations of the powders after consolidation (Fig. 8.26A) show fine grain microstructure of WC grains that are embedded into the Co/Cr metal matrix. The average grain size of WC grains was measured and found to be 50 μm in diameter. The triangle indents located in different grains and at the grain boundaries (Fig. 8.26A) refer to the Berkovich indents developed upon examining the sample by nanoindentation. This advanced approach was employed to study the degree of uniformity in the mechanical properties (hardness and Young's modules) at the sub-microscale and nanoscaled. In order to tackle this task over than 300 indents were developed, using single indent continuous multicycle ramp approach. The results have shown that both of the nanohardness and Young's modules are not dramatically varied from grain to grain (Fig. 8.26B) implying the absence of a serious compositional gradient at the microscaled. The average values of nanohardness and Young's modules calculated from at least 300 tests are found to be 18.67 and 406.33 GPa, respectively. These values together with the measured value of the fracture toughness (17.36 MPa $m^{1/2}$) indicate the possibility of fabrication new families of WC-Co-based nanocomposites with superior properties at the micro- and nanoscaled.

Atomic force microscope (AFM) micrographs of the nanocomposite bulk material for selected nanoindents developed during Berkovich probe nanoindentation tests, using a load of 400 mN, are presented in Fig. 8.27. The multi-indents presented in Fig. 8.27A have almost the same dimensions, indicating the homogeneity in composition and nanomechanical properties of the nanocomposite materials. The micrograph of a single nanoindent taken at higher magnification (Fig. 8.27B) does not show sink-in behavior. Sinking-in is expected in materials which exhibit a low value of E/Y (where E is the Young's modulus and Y is the yield strength).

In order to investigate the plastic deformation behavior of as-consolidated WC/Co-based nanocomposite material prepared in the present study, high magnification AFM

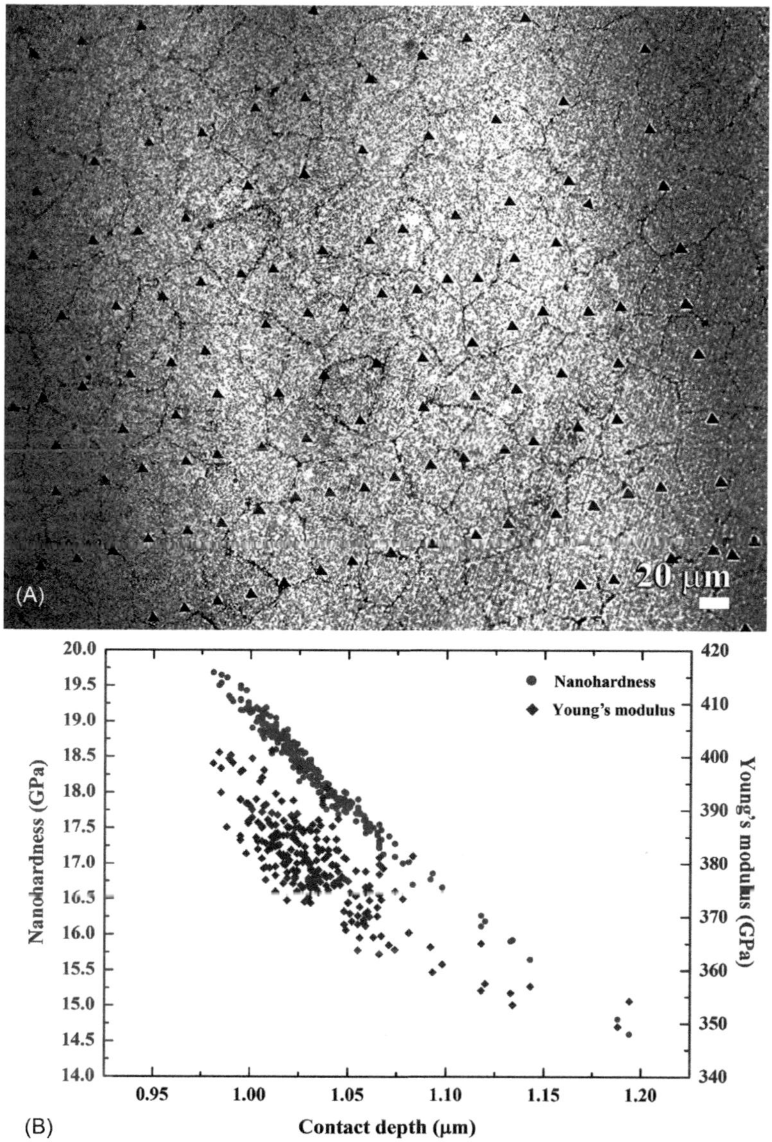

Figure 8.26 (A) SEM micrograph of the cross-sectional view for the consolidated sample, which shows part of Berkovich indents in different examined zones. The nanohardness-Young's modulus-contact depth relationships of the examined sample are presented in (B). *(After El-Eskandarany.)*[26]

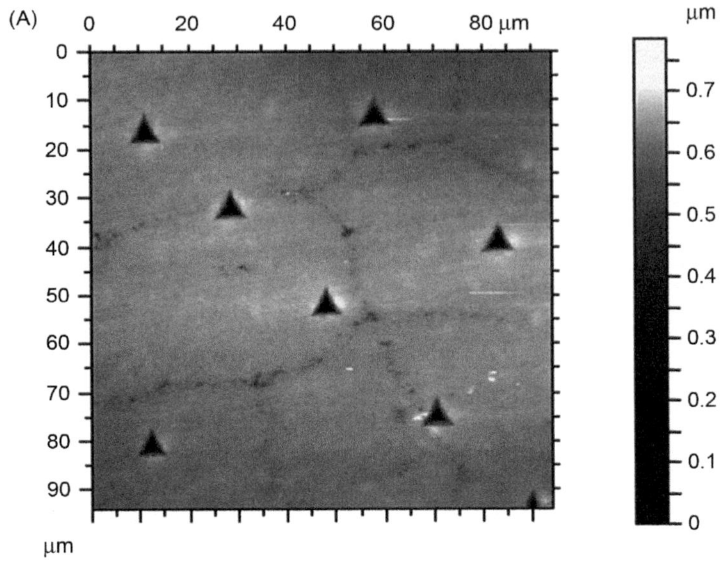

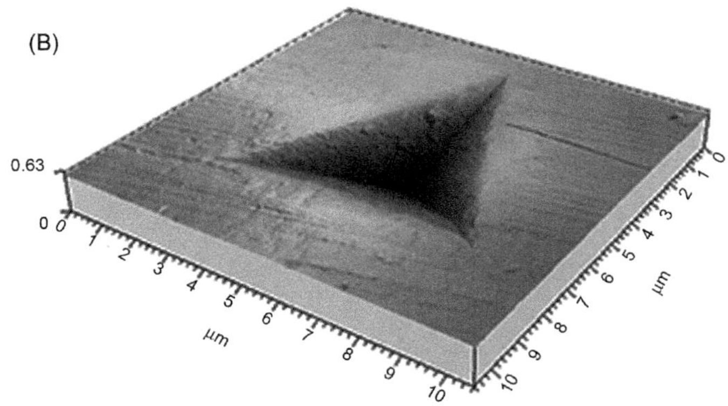

Figure 8.27 AFM of (A) selected nanoindents image and (B) 3D image of a single nanoindent for consolidated WC-Co-based nanocomposite powder. *(After El-Eskandarany.)*[26]

micrographs for the developed nanoindents were performed. The AFM micrographs for selected nanoindents taken at higher magnifications are shown in Fig. 8.28. The nanoindents exhibit pileup behavior as shown in both Figs. 8.28A and B. We should emphasize that such pileup behavior was performed a result of the deflection of WC grains which are pushed upward by the indenter, as suggested by Ndlovu et al.[27]

SEM investigations showed that the hard grains of WC located around the nanoindent contained slip lines (Fig. 8.29), suggesting that the slip was performed according to the plastic deformation. The formations of such slips were also found in WC grains upon Vickers testing.[28]

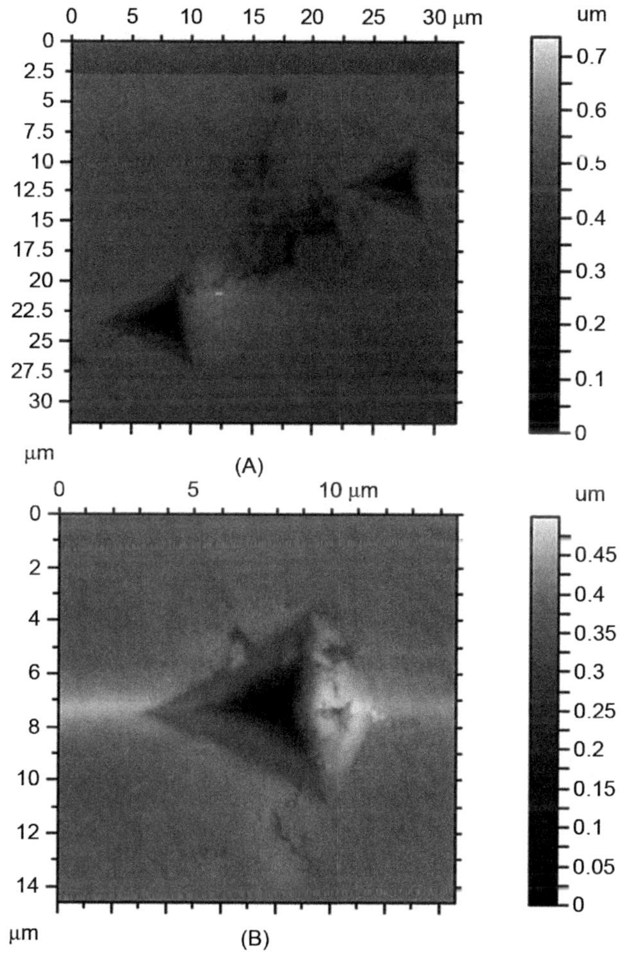

Figure 8.28 AFM micrographs of (A) two selected nanoindents and (B) single nanoindent for consolidated WC-Co-based nanocomposite powder. Both micrographs indicate that the nanoindents exhibit pileup behavior. *(After El-Eskandarany.)*[26]

8.3.3 WC-5Co-1Cr-3MgO-0.7VC-0.3Cr_3C_2 nanocomposite

In 2017, El-Eskandarany et al. prepared nanocomposite powders of WC-5Co-1Cr-3MgO-0.7VC-0.3Cr_3C_2, using high-energy mixer ball mill (Spex 8000-D).[29] The milling process time required to obtain ultrafine nanocomposite powders was 20 h. The powders were then consolidated into bulk compacts with 20 mm in diameter, using SPS technique.

The XRD pattern of as-received WC powders is shown in Fig. 8.30A. The starting material powders consisted of polycrystalline large crystallites, characterized by sharp Bragg peaks corresponding to hcp-WC phase (PDF# 00-025-1047), as elucidated in

Figure 8.29 SEM micrograph of as-consolidated WC/Co-based nanocomposite. The slip line shown in the WC grains, which is indexed by the arrows, refers to the plastic deformation taken place upon nanoindentation testing. *(After El-Eskandarany.)*[26]

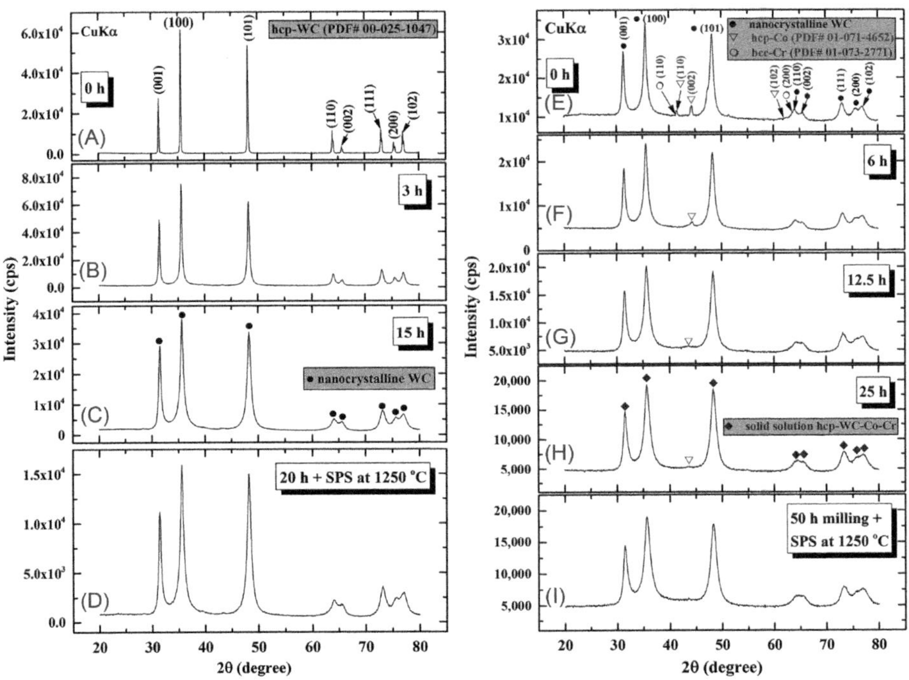

Figure 8.30 XRD patterns of as-received hcp-WC powders after mechanical milling time for (A) 0, (B) 3, and (C) 15 h. The XRD pattern of the powders milled for 20 h and then consolidated at 1250°C, using SPS technique, is presented in (D). The XRD patterns of as-mechanically milled WC powders and then ball-milled with 7 wt.% of (10Co/4Cr) powders for 0, 6, 12.5, and 25 h are shown in (E–H), respectively. The XRD pattern of 93-WC/7 wt.%(10Co/4Cr) powders obtained after 50 h and then consolidated at 1250°C, using SPS technique, is presented in (I).[30]

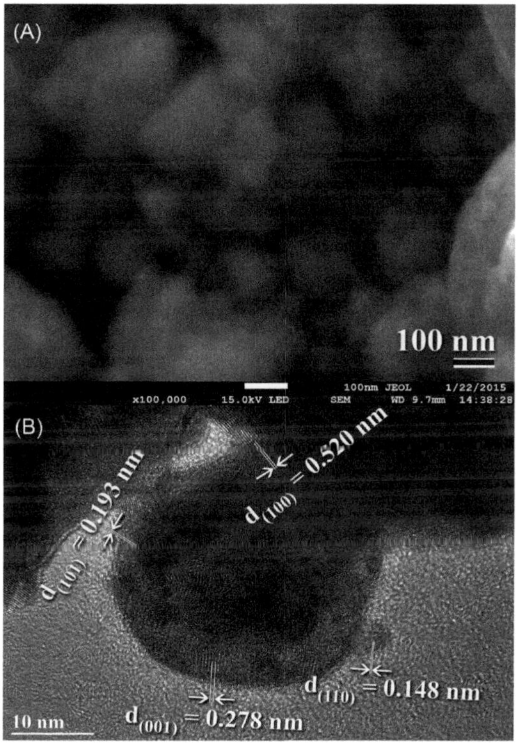

Figure 8.31 (A) FE-SEM micrograph, and (B) HRTEM image of as-mechanically ball-milled hcp-WC for 20 h.[30]

Fig. 8.30A. When the powders subjected to continuous ball milling for 3 h, significant broadening can be notified in the major ((001), (100), (101)) and minor ((110), (002), (111), (200), (102)) Bragg peaks denoting mechanically induced microstructure refinement, as shown in Fig. 8.30B. Increasing the ball-milling time (15 h) assisted the continuity of the grain refining, as indicated by outstand broadening in the Bragg peaks displayed in Fig. 8.30C. We should emphasize that during this stage of milling, WC powders experienced repeated collision, fracture, and cold welding. Such severe plastic deformation induced by high-energy ball milling lead also to accumulation of internal stresses that changes the lattice parameter and formation of cell structure.[30]

The FE-SEM taken at a magnification of 100,000 and FEHRTEM micrographs of the end product of WC powders obtained after 20 h of the mechanical milling time are shown in Fig. 8.31A and B, respectively. At this final stage of ball milling, the powders had smooth featured surfaces and became ultrafine with an average apparent diameter of 185 nm, as shown in Fig. 8.31A. Moreover, the as-ball-milled nanopowders tended to aggregate due to the Van der Waals force to form larger particles with an average diameter of 413 nm in diameter, as presented in Fig. 8.31A. These nanopowder particles consisted of ultrafine grains ranging between 3 and 8 nm in diameter, as shown in Fig. 8.31B. The indexed grains in Fig. 8.31B had different orientations related to interplanar spacing (D) of 0.520, 0.193, 0.278, and 0.148 nm that well matched with WC (100), (101), (001) and (110), respectively (PDF# 00-025-1047).

When the nanocrystalline WC powder particles obtained after 20 h of MM time were consolidated at 1250°C using SPS technique, the materials maintained their nanocrystalline characterizations, suggested by the broadening seen in the hcp-WC Bragg peaks presented in Fig. 8.20D.[30] This indicates the formation of bulk nanocrystalline WC material. These fine WC-nanopowder particles obtained after 20 h of MM time were mixed with 7 wt.% of (10Co/4Cr) powders and then ball-milled for different MM time. The XRD pattern of the starting WC/7 wt.%(10Co/4Cr) powders (before ball milling, 0 h) is shown in Fig. 8.30E. The powders consisted of nanocrystalline hcp-WC (reinforcement materials) coexisted with large crystalline phases of elemental hcp-Co (PDF# 01-071-4652) and bcc-Cr (PDF# 01-073-2771), as indicated in Fig. 8.30E.

After MM time ranged between 6 and 25 h, the elemental metallic powders of Co and Cr were subjected to aggressive grain refining caused by the milling media (WC balls) together with the nano-milling media (WC-nanopowders), as indicated by the significant broadening seen in their Bragg peaks shown in Fig. 8.30F. During the intermediate (12.5–25 h) stage of MM time, the Bragg diffractions related to the elemental powders of Co and Cr can be hardly seen (Fig. 8.30G and H), suggesting the formation of solid-solution WC-Co-Cr nanocrystalline phase. During the ball-milling process, the precursor WC, Co, and Cr powders undergone to different mechanical deformation generated by the milling media such as cold welding, fracturing, rewelding, and lattice imperfections. Such severe defects led to enhance the diffusion of Co and Cr atoms into the WC matrix and the formation of metastable solid-solution phase. The formation of WC-Co-Cr solid solution is indicated by the significant continuous shifts observed in the Bragg peaks of WC to the low-angle side upon increasing milling time from 0 h (Fig. 8.30E) to 25 h (Fig. 8.30H).

The FE-SEM micrograph of nanocomposite WC/7 wt.%(10Co/4Cr) powders obtained after 50 h of MM time is shown in Fig. 8.32A. After this stage of ball milling, the powders had spherical-like morphology with particle size in the range between 0.35 and 1.2 μm in diameter, as displayed in Fig. 8.32A. This sample showed slight grain growth upon consolidation at 1250°C by SPS, as indicated in Fig. 8.32B. The nanocomposite materials obtained upon SPS technique had excellent homogeneous distribution of the sharp angular fine grain (less than 5 μm in diameter) related to the WC (Figs. 8.32B–D) into the metallic matrix of Co (Fig. 8.32E) and Cr (Fig. 8.32F). Neither pores nor reinforcement-free zones were existed in the as-consolidated powders, suggesting the formation of a uniform bulk WC/6 wt.%(10Co/4Cr) nanocomposite material.

The as-mechanically milled 93-WC/7 wt.%(10Co/4Cr) nanocomposite powders obtained after 50 h of ball-milling time were then mechanically mixed with 7 wt.%(ZrO_2/1.5 wt.% Y_2O_3) for different MM time. At the starting stage (0 h), the powders consisted of solid-solution hcp-WC-Co-Cr coexisted with large crystalline grains of ZrO_2 (PDF# 01-083-0810), as shown in Fig. 8.33A. Those Bragg reflections related to ZrO_2 tended to have clear broadening with decrease in their intensities upon

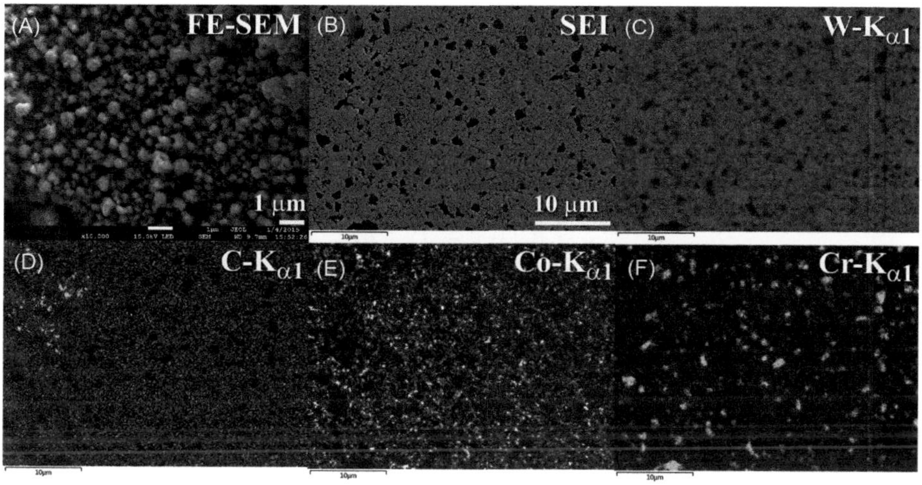

Figure 8.32 (A) FE-SEM of nanocomposite 93-WC/7(10Co/4Cr) powders obtained after 50 of ball-milling time. The cross-sectional view of the consolidated powders achieved at 1250°C, using SPS technique, is presented in (B) together with the corresponding X-ray elemental mapping of (C) W-$K_{\alpha 1}$, (D) C-$K_{\alpha 1}$, (E) Co-$K_{\alpha 1}$, and (F) Cr-$K_{\alpha 1}$.[30]

increasing the MM time between 6 h (Fig. 8.33B) and 12.5 h (Fig. 8.33C). This suggests the diffusion of ZrO_2 phase into the hcp-WC-Co-Cr matrix. The Bragg peaks related to ZrO_2 phase were hardly detected when the MM time increased in the range between 25 h (Fig. 8.33D) and 50 h (Fig. 8.33E), suggesting the formation of metastable hcp-93(93-WC/7 wt.%(10Co/4Cr))/7 wt.% ZrO_2 powders.

The FE-SEM micrograph of the cross-sectional view for 93(93-WC/7 wt.%(10Co/4Cr))/7 wt.% ZrO_2 powders obtained after 50 h of MM time and then consolidated into bulk material at 1250°C, using SPS, is shown in Fig. 8.34A. In general, the composite material had fine structure containing micro-grained WC embedded into the Co/Cr matrix reinforced by ZrO_2 (Fig. 8.34A). The high-magnification FE-SEM micrograph of the indexed region in Fig. 8.34A is shown in Fig. 8.34B. The consolidated sample was dense with no remarkable pores or cavities. In addition, the WC grains (light gray grains) did not suffer from drastic growth upon consolidation, indicated by their fine grain sizes characterizations (ranging between 260 and 887 nm in diameter), as presented in Fig. 8.34B. It can be notified that the black matrix zone (Fig. 8.34B) consisting of Co (Fig. 8.34E) and Cr (Fig. 8.34F) was reinforced by ZrO_2 fine nanopowder particles (Figs. 8.34G and H). The existence of such a hard ZrO_2 particles into the metallic matrix may act as grain growth inhibitors for WC grains (Figs. 8.34C and D) to maintain their nanocrystalline characterizations.

Fig. 8.35 shows the optical micrographs of the cross-sectional view of the consolidated samples of 93WC/7(10Co/4Cr) (A) and 93(93-WC/7 (10Co/4Cr))/7 ZrO_2 (C). The relationships between nanohardness, nano-Young's modulus, and the indentation depth of the examined samples are presented in (B) and (D) for 93WC/7(10Co/4Cr) and

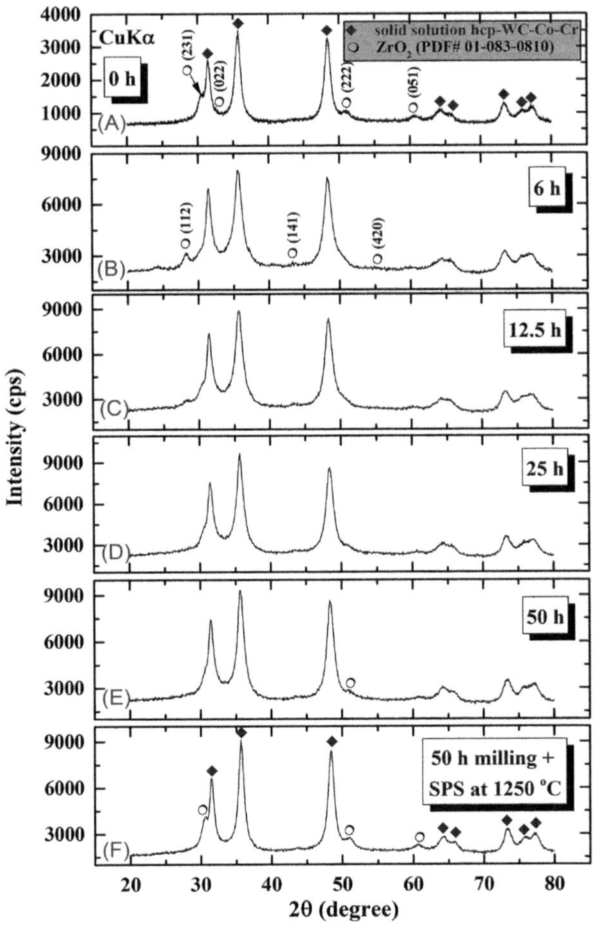

Figure 8.33 XRD patterns of as-mechanically milled 93-WC/7(10Co/4Cr) powders obtained after 50 h of ball-milling time and then mechanically mixed with 7 wt.% (ZrO_2) powders for (A) 0, (B) 6, (C) 12.5, (D) 25, and (E) 50 h. The XRD pattern of nanocomposite 93(93-WC/7 wt.%(10Co/4Cr))/7 wt.% ZrO_2 powders obtained after 50 h and then consolidated at 1250°C, using SPS technique, is presented in (F).[30]

93(93-WC/7(10Co/4Cr))/7 ZrO_2 samples, respectively. The triangle indents located at the grain boundaries and inside the grains refer to the Berkovich indents developed upon examining the sample by nanoindentation. This approach was employed to study the degree of uniformity in the mechanical properties (hardness and Young's modulus) at the sub-microscale level. To tackle this task, more than 100 indents were developed for each composition, using the single-indent continuous multicycle ramp approach. The results showed that neither nanohardness nor Young's modulus varied dramatically from grain to grain (Figs. 8.35B and D), implying the absence of a serious compositional gradient at the microscale. The average values of nanohardness

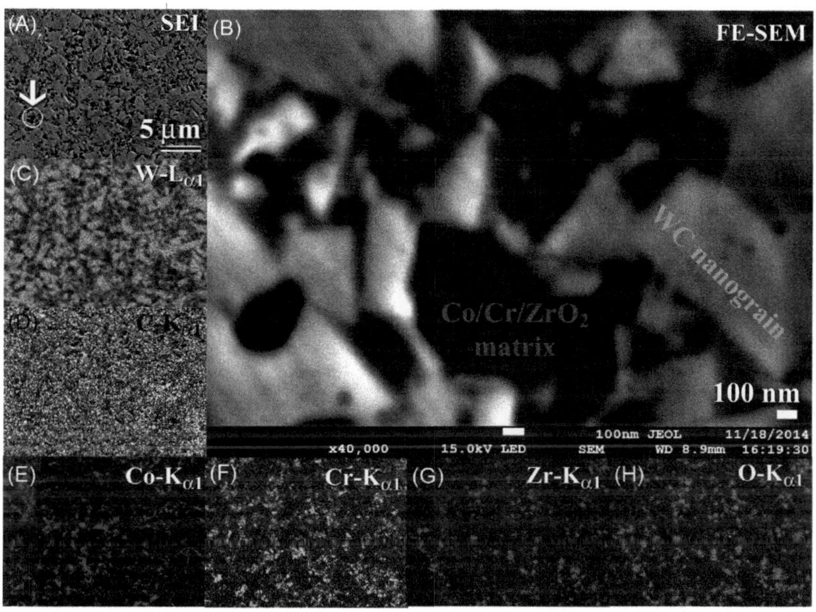

Figure 8.34 (A) FE-SEM micrograph of the cross-sectional view of as-ball-milled nanocomposite 93(93-WC/7 wt.%(10Co/4Cr))/7 wt.% ZrO_2 powders (50 h) and then consolidated at 1250°C, using SPS technique. The high magnification (940,000) for the zone indexed by a circular symbol shown in (B) is presented in (B). The X-ray elemental mapping corresponding to (A) is shown in the figure for (C) W-$L_{\alpha1}$, (D) C-$K_{\alpha1}$, (E) Co-$K_{\alpha1}$, (F) Cr-K_{a1}, (G) Zr-$K_{\alpha1}$, and (H) O-$K_{\alpha1}$.[30]

and Young's modulus obtained from at least 100 tests were found to be 19.67 and 275 GPa for the 93WC/7(10Co/4Cr) sample and 19.26 and 269 GPa for the 93(93-WC/7 (10Co/4Cr))/7 ZrO_2 sample, respectively.[30]

8.4 Fabrication of metal matrix/carbon nanotubes nanocomposites by mechanical alloying

Because of their unusual mechanical properties, with Young's modulus of up to 1 TPa and tensile strength of up to 30 GPa,[30,31] CNTs have received much attention since they were investigated by Iijima in 1991.[32] Due to their excellent mechanical properties, many researchers from different materials science schools proposed CNTs as reinforcement materials for different materials, including polymers, ceramics, and metals.[32–34]

Employing ball-milling technique for fabrication of metal matrix/CNTs nanocomposites has shown promising results. Recently, Phuong et al.[34] used ball-milling approach via planetary mill with rotation speed of 250 rpm for fabrication of CNT/Al

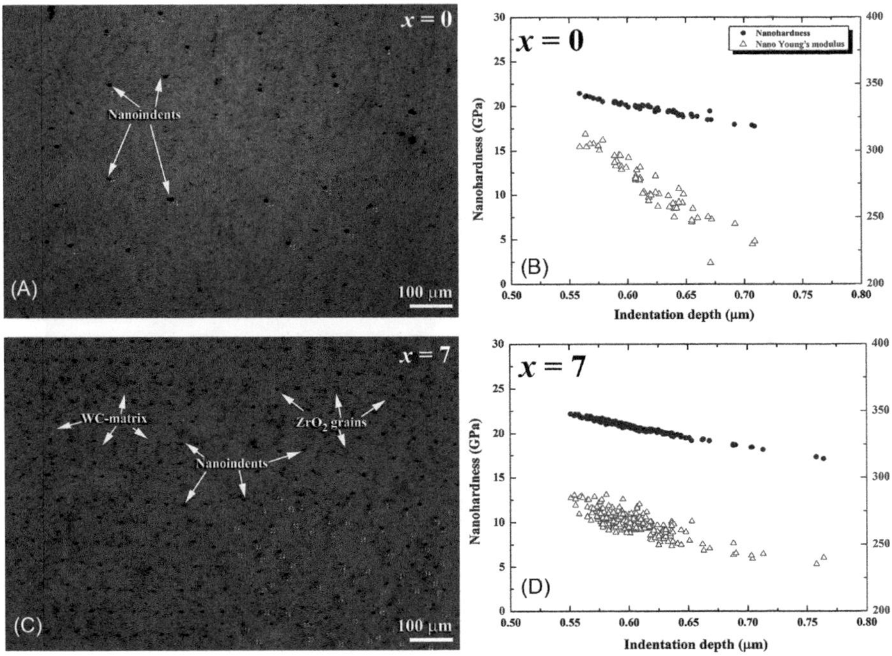

Figure 8.35 Optical micrographs of the planar cross-sectional view of the as-consolidated (A) 93WC/7(10Co/4Cr) and (C) 93(93-WC/7 (10Co/4Cr))/7 ZrO_2 samples, showing Berkovich indents in various examined zones. The corresponding nanoindentations and nano-Young's modulus are shown in (B) and (D), respectively.[30]

with different wt.% of CNTs ranging from 0 to 2 wt%. The end products were then consolidated into dense compacts, using hot isostatic pressing (HTP). Their results showed that annealing the nanocomposite materials at temperatures below 150°C leads to secondary hardening, while annealing at higher temperatures soften the nanocomposites. Moreover, HPT-processed CNT/Al nanocomposites with 1.5 wt.% of CNTs showed the highest hardness (about 1550 MPa) in comparison with other composites containing CNTs from 0 up to 2 wt.%.[34]

More recently, the influence of milling time on the structure, morphology, and thermal stability of ball-milled multi-walled carbon nanotubes (MWCNTs) reinforced EN AW6082 aluminum alloy powders has been demonstrated by Basariya et al.[35] Their results showed the absence of any reactive product even after long milling up to 50 h. Moreover, they have investigated that the nanotubes became shorter in length as they fractured under the impact and shearing action during the milling process. A high hardness of about 436 HV was achieved for the milled powders, due to the addition of MWCNTs, after milling for 50 h. They attributed the increasing of the elastic modulus and nanohardness to the finer grain size evolved during high-energy ball milling and to the uniform distribution of hard CNTs in the Al-alloy matrix.

References

1. Zhang, J. S., Liu, X. J., Cui, H., Duan, X. J., Sun, Z. Q., and Chen, G. L., *Metall. Trans. A*, 28:1261 (1997)
2. El-Eskandarany, M. S., J. Alloys Comp., 279:263 (1998)
3. Suryanarayana C., and Al-Aqeeli, Nasser, *Progress in Materials Science*, 58:383 (2013)
4. CaoSeattle, G., Nanostructures and Nano Materials, Synthesis, Properties, and Applications, *Imperial College Press*, London (2004)
5. Kang, Y.C., and Chan, S.L., *Materials Chemistry and Physics*, 65:436 (2004)
6. Niihara, K., J. Ceram. Soc. Jpn., 99:974 (1991)
7. Lee, J. C., Byun, J. Y., Oh, C. S., Seok, H. K., and Lee, H. I., *Acta. Metall. Mater.*, 45:5303 (1997)
8. *CRC Handbook of Chemistry, Physics*, 74th edition and CRC Press, p. 4–36 (1992)
9. *The MRCK INDEX*, 10th edition, Mrck & Co., p. 50 (1983)
10. Ferro, A. C., and Derby, B., *Acta. Metall. Mater.*, 43:3061 (1995)
11. Viala, J. C., Fortier, P., and Bouix, J., *J. Mater. Sci.*, 25:1842 (1990)
12. Narciso, J., Garacia-Cordovilla, C., and Louis, E., *Mater. Sci. Eng. B*, 15:148 (1992)
13. Ribes, H., Suery, M., L'sperance, G., and Legoux, G., *Metall. Trans. A*, 26:2637 (1995)
14. Lloyd, D. J., Lagace, H., McLeod, A., and Morris, P. L., *Mater. Sci. Eng. A*, 107:73 (1989)
15. Bartos, G., and Xia, K., in: *Proc. 4th Int. Conf. on Semi-Solid Processing of Alloys and Composites*, p. 290, University of Sheffield, England (1996)
16. Witulski, T., Winkelmann, A., and Hirt, G., in: *Proc. 4th Int. Conf. on Semi-Solid Processing of Alloys and Composites*, University of Sheffield, England, p. 242 (1996)
17. Novak, B. M., *Adv. Mater.*, 5:422 (1993)
18. El-Eskandarany, M. S., *J. Jpn Society of Powder and Powder Metallurgy*, 44: 547 (1997)
19. El-Eskandarany, M. S., Omori, M., Sumiyama, K., Hirai, T., and Suzuki, K., *J. Jpn Society of Powder and Powder Metallurgy*, 44:1143 (1997)
20. El-Eskandarany, M. S., *Journal of Alloys and Compounds*, 391:228 (2005)
21. El-Eskandarany, M. Sherif, and Inoue, A., *Met. Trans.A*, 33:135 (2002)
22. El-Eskandarany, M. Sherif, Mahday, A.A., Ahmed, H.A., and Amer, A.H., *J. Alloys Compd.*, 312:315 (2000)
23. Anstis, G.R., Chantikul, P., Lawn, B.R., and Marshall, D.B., *J. Am. Ceram. Soc.*, 64:533 (1981)
24. El-Eskandarany, M. Sherif, Omori, M., Konno, T.J., Sumiyama, K., Hirai, T., and Suzuki, K., *Met. Trans. A* , 29:1973 (1998)
25. El-Eskandarany, M. Sherif, *J. Alloys Compd.*, 296:175 (2000)
26. El-Eskandarany, M. Sherif, Al-Hazza, A., Alduweesh, A., Al-Rowayeh, A., and Shaban, E., Proceeding of NSTI-Nanotech 2014, 1:408 (2014)
27. Ndlovu, Siphilisiwe, Durst, Karsten, and Goken, Mathias, *Wear*, 263:1602 (2007)
28. Engqvist, H., Ederyd, S., Axén, N., and Hogmark, S., *Wear*, 230:165 (1999)
29. El-Eskandarany, M. Sherif, Al-Hazza, A., Al-Hajji, L.A., *Journal of Materials Engineering and Performance*, 26: 1540 (2017)
30. Wong, E.W., Sheehan, P.E., and Lieber,C.M., *Science*, 277:1971 (1997)
31. Salvetat, J.-P., Briggs, G.A.D., Bonard, J.-M., Bacsa, R.R., Kulik, A.J., Stöckli, T., Burnham, N.A., and Forró, L., *Phys. Rev. Lett.*, 82:944 (1999)
32. Harris, P.J.F., *Int. Mater. Rev.*, 49:31 (2004)

33 Cho, J., Boccaccini, A.R., and Shaffer,M.S.P., *J. Mater. Sci.*, 44:1934–1951 (2009)
34 Phuong, Doan Dinh, et al., *Journal of Alloys and Compounds*, 613:68 (2014)
35 Basariya, M. Raviathul, Srivastava, V.C., and Mukhopadhyay, N.K., *Materials and Design*, 64:524 (2014)

Solid-state hydrogen storage nanomaterials for fuel cell applications

9

Owing to the dramatic global environmental changes associated with man-made carbon dioxide emissions and the huge consumption of the limited resources of fossil fuels, developing alternate energy sources is important for a sustainable future. The increase in threats from global warming due to the consumption of fossil fuels requires adopting new strategies to harness the inexhaustible sources of energy. Among the different scenarios proposed for using alternative energies, hydrogen energy holds tremendous promise as a new clean energy option. Hydrogen storage, which cuts across both hydrogen production and hydrogen applications, plays a critical role in initiating a hydrogen economy. Apart from the traditional ways used for hydrogen storage, magnesium (Mg) metal and its based-materials (alloys and composites) have been considered as the most candidate hydrogen storage materials for safe applications. In spite of its availability, low cost, and large hydrogen storage capacity, Mg still holds several technical problems that must be solved first before considering such attractive light metal in real applications. In this chapter, we will explain and discuss the preparation mechanism of reactive ball milling for preparing nanocrystalline MgH_2 powders under high hydrogen gas pressure. Moreover, detailed experimental work dedicated for improving the hydrogenation/dehydrogenation kinetic behavior will be presented and discussed. In addition, some typical examples of MgH_2-based materials employed for fuel cell applications will be represented. The photos are taken in November 2018 from Nanotechnology and Advanced Materials (NAM) Laboratory/Energy and Building Research Center (EBRC) of Kuwait Institute for Scientific Research, presenting a successful application of solid-hydrogen storage MgH_2/5 wt% metallic glassy-$Zr_{70}Ni_{30}Pd_{10}$ powders for charging eight batteries (48 V) of an electric golf cart.

9.1 Introduction

Owing to the rapid growth of human population engaged with escalating development in heavy industrial sector, drastic increase in the mass production of fossil fuel (coal, oil, and natural gas) is required. Based on such magnificent expansion in energy demands, the world production of fossil fuel jumped from 2500 metric tons (Mt) in 1971 to 4321 in 2016,[1] as displayed in Fig. 9.1. Where the world fuel consumption was about 6000 million tons of oil equivalent (Mtoe) in 1995 (Fig. 9.2A and B), it exhibited a dramatic increase when many developed countries started their heavy industries in 1990s to reach in 2015 to 9384 Mtoe.[1] However, the fossil fuel reserves, in particularly oil are sharply decreased; they remain the dominant source of energy and present more than 80% of the world energy supply, as displayed in Fig. 9.2B.

Mechanical Alloying. http://dx.doi.org/10.1016/B978-0-12-818180-5.00009-1

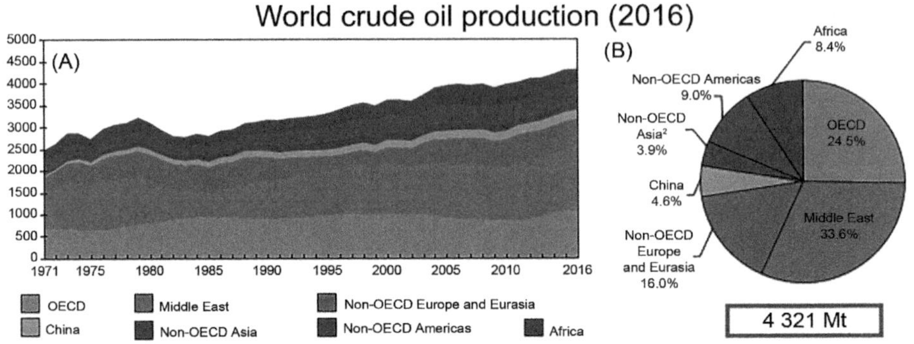

Figure 9.1 (A) World production of crude oil in the period between 1971 and 2016. The percentage of oil production indexed by the regions is presented in (B).[1]

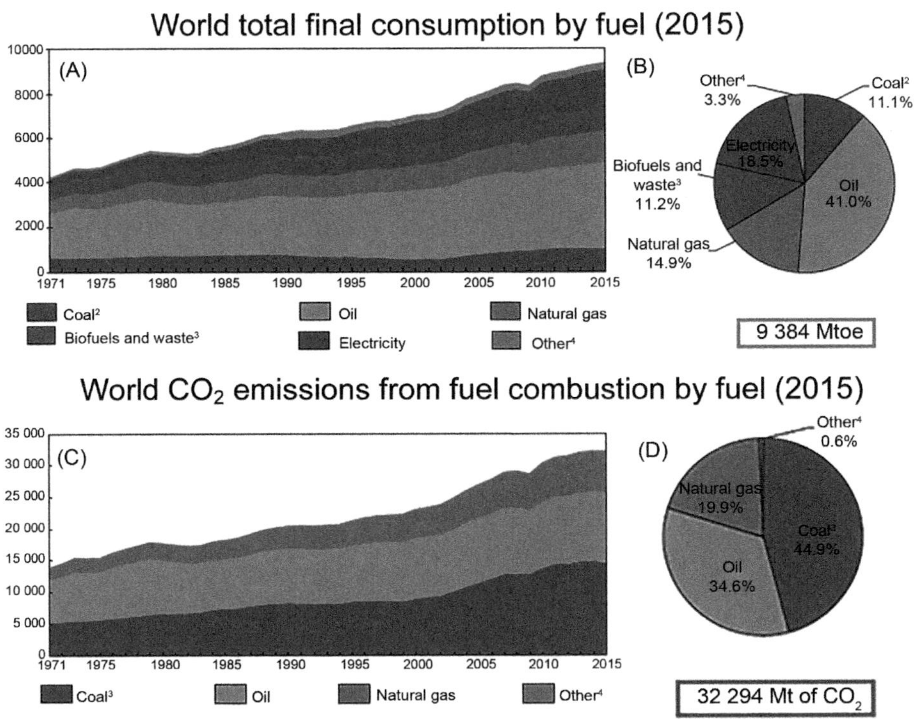

Figure 9.2 World total final consumption of fossil fuel in 2015 by (A) fuel and (B) region. The world CO_2 emissions (Mt of CO_2) from fuel combustion in the period between 1971 and 2015 are presented in (C) by fuel type and (D) percentage of each fuel in CO_2 emissions.[1]

According to environmental and health points of view, fossil fuels emit carbon dioxide and other harmful air pollutants such as SO_2 and NO_x when burned.[2,3] The uncontrolled growing on fossil fuel consumption and their corresponding harmful gas released emissions led to wide varieties of public health and environmental costs at the global levels.[4,5] Fig. 9.2C and D summarizes the world CO_2 emission in the period between 1971 and 2015 on terms of quantity (C) and fuel sources (D).[1] Obviously presented, the amount of CO_2 released upon fossil fuel burning was increased almost linearly from 14,000 Mt in 1971 to 22,500 Mt in 2000. As the fossil fuel consumption dramatically increased in the period between 2000 and 2015 (Fig. 9.2A), the related released CO_2 emission jumped in 2015 to 32,294 Mt, as presented in Fig. 9.2C and D.[1] Although primary energy sources (fossil fuel), which are not infinitely available, and may not be enough to provide the amount of near future energy demanded in the world, their production and consumption are expected to rise with the rapid growth of the world population.

9.2 Hydrogen energy

It is clearly obvious that the energy sector is facing the biggest energy challenges to move toward sustainable and clean energy futures. Maintaining the current linear consumption of fossil fuels (Fig. 9.2A) is unrealistic. The depletions of fossil fuels are the key drivers to find alternative solutions to fuel our energy systems.[6,7] Versatile carbon-free energy carriers with high-energy density such as hydrogen are considered an alternative to fossil fuels. Accordingly, employing renewable and sustainable energy sources such as solar, wind, geothermal energies for developing clean and efficient energy system has become one of the hottest research topics in the world that attracts many scientists and researchers.[8]

Since the last decade, hydrogen has been widely proposed as a promising future alternative for replacing fossil fuels and delivering clean energy required almost in all sectors.[9] The relevance of hydrogen as an energy carrier is attributed to its unique properties[7–16] that can be summarized as follows:

- Hydrogen is the light chemical element that offers the best energy-to-weight ratio of any fuel,
- It is colorless, odorless, nontoxic, carbon-free gas, and its only byproduct is water,
- Hydrogen is the energy vector which could provide the bridge from the fossil energy economy to renewable energy sources,
- It can store the energy until it's needed and can be moved to where it's needed,
- It has the best ratio of valence electrons to nucleons,
- There is a huge range of possible ways to produce hydrogen from different primary energy sources,
- It can readily be produced from renewable energy resources, thus eliminating the net production of greenhouse gases, and
- Recent studies have indicated that hydrogen fuel costs are reasonable and hydrogen is therefore an ideal candidate to replace fossil fueled as an energy carrier.[17–19] In fact, hydrogen has become a favorite energy carrier because it has a very high calorimetric value, with a lower heating value of 120 MJ kg^{-1}, compared to petrol, which is approximately one-third of this at 43 MJ kg^{-1}.[20] However, hydrogen energy shows advanced properties and characteristics when compared with other sources of primary energy (Table 9.1[21]); the key advantage of hydrogen is that when burned, CO_2 is not produced. Thus, hydrogen satisfies all concerns over air pollution and global warming, and the transition to use hydrogen as fuel may occur even well before oil reserves are depleted.

Table 9.1 Selected physical and chemical properties of hydrogen, methane, and petrol.

Properties	Hydrogen (H_2)	Methane (CH_4)	Petrol (-CH_2-)
Lower heating value (kWh kg^{-1})	33.33	13.9	12.4
Self-ignition temperature	585	540	228–501
Flame temperature (°C)	2045	1875	2200
Ignition limits in air (vol. %)	4–75	5.3–15	1.0–7.6
Minimal ignition energy (mW s)	0.02	0.29	0.24
Flame propagation in air (m s^{-1})	2.65	0.4	0.4
Diffusion coefficient in air (cm^3 s^{-1})	0.61	0.16	0.05
Toxicity	Non	Non	High

Source: Compiled from Ref.[21]

9.2.1 Hydrogen economy

Based on its unique behavior as an energy carrier, hydrogen has been holding tremendous promises as a new renewable and clean energy option.[22] It is a convenient, safe, versatile fuel source that can be easily converted to a desired form of energy without releasing harmful emissions. Thus, hydrogen has been considered as the ideal candidate both mobile and stationary applications.[23] Hydrogen fuel cells (FCs) have wide spectrum range of potential applications, operating from a few watts up to gigawatts. Moreover, hydrogen has the potential to run a fuel-cell engine with greater efficiency over an internal combustion engine.[24,25] The same amount of hydrogen will take an FC-car at least twice as far as a car running on gasoline. Hydrogen micro-FCs have great potential for power portable electronics such as laptops and mobile phones.[22,26] The superior energy density of a hydrogen polymer electrolyte membrane (PEM) micro-fuel cell may open a new horizon for manufacturing longer battery lifetimes.[27,28] For mobile applications, fuel cells could provide auxiliary power for electrical units such as air conditioning and refrigerators.[28] Hydrogen-PEM fuel cells have become well-known option for automobile[29,30] and vehicle powertrain applications.[31,32]

9.2.2 Hydrogen storage

The idea of hydrogen economy is based on the assumption that hydrogen gas can be produced by decomposing water using inexpensive primary sources of energy, such as solar energy, wind energy, tidal energy, or biomass decomposition.[33] Recent studies have pointed out that hydrogen fuel costs are reasonable and are therefore can be considered as an ideal candidate to replace fossil fuels as an energy carrier.[34] In fact, hydrogen storage is considered to be one of the most crucial factors of hydrogen economy that governs the future of utilization hydrogen energy for real applications. Once hydrogen is produced, it needs to be stored in order to overcome daily and seasonal discrepancies between energy source availability and demand.[35] Several approaches of storing hydrogen, including gaseous storage, liquid storage, and solid-state storage methods, are available, as discussed briefly in the following. The storage characteristics of each of method indexed by volume density and gravimetric density are presented in Figs. 9.3 [36] and 9.4, [35] where Table 9.2 shows a comparison between these three approaches.

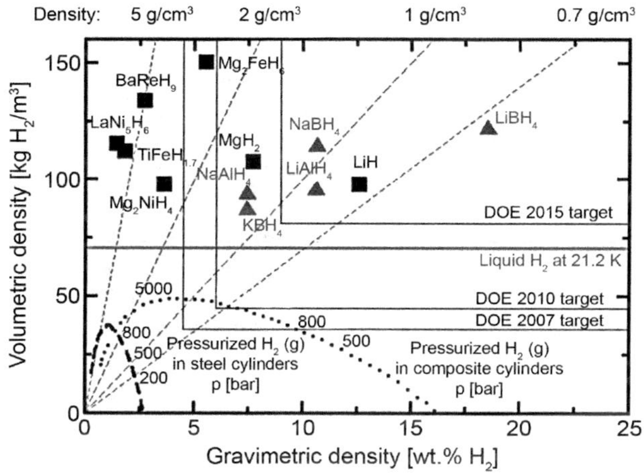

Figure 9.3 Volumetric and gravimetric hydrogen storage densities of different hydrogen storage methods. Metal hydrides are represented with squares and complex hydrides with triangles. $BaReH_9$ has the highest known hydrogen to metal ratio, Mg_2FeH_6 has the highest known volumetric H_2 density, $LiBH_4$ has the highest gravimetric density. Reported values for hydrides are excluding tank weight
Source: Ref.[36]

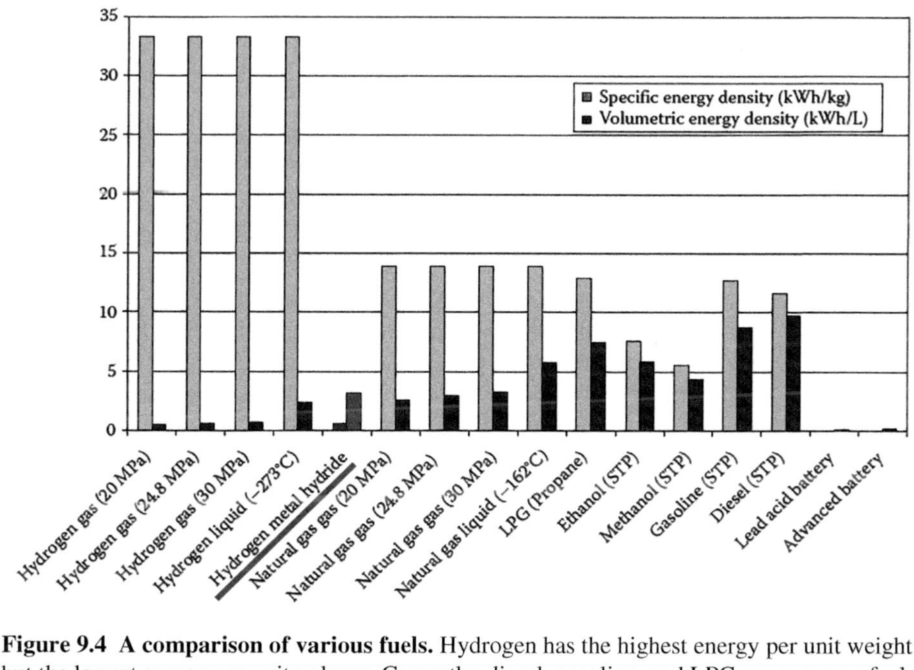

Figure 9.4 A comparison of various fuels. Hydrogen has the highest energy per unit weight but the lowest energy per unit volume. Currently, diesel, gasoline, and LPG are common fuel for transportation application.
Source: Ref.[35]

Table 9.2 Comparison between the three principal approaches used for storing hydrogen.

Storage approach	Volumetric hydrogen capacity (kgH_2m^{-3})	Disadvantages
Gaseous storage method (compressed H_2 gas under 80 MPa (800 bar) pressure	~40	• Safety problem • High cost of pressurization • Large pressure drop during use • Hydrogen embrittlement of storage tanks
Liquid storage method at cryogenic tank at −225°C	~71	• Since it is open, large thermal losses are expected • Safety problem • High cost of liquefaction
Solid-state method via metal hydrides	80–160	None of the above

Source: Compiled from different sources.

9.2.2.1 Gaseous storage method

Physical storing of hydrogen is a mature technology in which hydrogen gas is pressurized under very high pressure (350–700 bar) into thick-wall cylinders made of high strength carbon fiber reinforced polymer (CFRP).[35,37] Obviously, using such pressurized hydrogen gas tanks as source of fuel in vehicles is not safe and still needs many intensive studies dedicated for improving the structural and mechanical properties of the materials used in tank's manufacturing.[38] Nevertheless, the volumetric energy density compressed hydrogen gas systems is low compared to gasoline (7 times lower even at 700 bar).[36] Another drawback of steel cylinders is a relative large contribution from the cylinder material to the overall weight of the storage system reducing the gravimetric energy storage density. Moreover, pressurized hydrogen gas takes a great deal of volume compared with, for example, gasoline with equal energy content—about 30 times bigger volume at 100 atm gas pressure.

It is obvious that while hydrogen gas has many unique and pioneering advantages, there remains a problem with storage and transportation in gaseous state. For instance, commercial modern cars burn around 400 kg of petrol in combustion engines to cover a range of about 400 km.[39,40] In order to cover the same range, an amount of 8 kg of hydrogen is required for combustion engines, whereas 4 kg of hydrogen is required to run an electric car on a fuel cell.[21] If we convert this amount of hydrogen into volume, we realized that it can fill a balloon with a diameter of 5 m or 45 m^3 volume!

9.2.2.2 Liquid storage method

Cryo-compressed hydrogen storage (CcH_2) and liquid hydrogen (LH_2) storage are alternative mature approaches where hydrogen is liquidized at −253°C and compressed into vessels that can be pressurized to 250–350 atm. Accordingly, the size of liquid hydrogen requires larger tanks reaching to about 3 times larger than the current used gasoline tank.[41] Moreover, liquid hydrogen requires well-insulated and expensive cryogenic storage vessels to prevent boil-off and maximize dormancy and to maintain the temperature below to 20 K.[42] Unluckily, the gas-liquid transformation process is extremely expensive, consuming approximately ~25%–30% of the energy content of the stored hydrogen.[43] Thus, employing these traditional storage approaches in real future applications may be difficult due to the high cost and safety issues.[44]

With the beginning of this century, a new approach for hydrogen storage was developed by Aceves and his teamwork.[45] This approach, the so-called cryogenic pressure vessel, tried to combine the existing classic storage technologies (cryogenic and high pressure) to capture the advantages of both methods.[28] The idea behind this Cryo-compressed storage system depends on designing a high-pressure (~350 bar) cryogenic pressure vessel and operates the system at low temperature (<20 K). Based on that, hydrogen could be stored at significantly higher density than compressed hydrogen gas and probably even higher density than liquid hydrogen.[45] This new technology has received much attention because the higher pressure may be attained on the vessel before the vent valve is activated. They expected that this technology may reduce the boil-off from a liquid cryogenic hydrogen.[46]

9.3 Solid-state hydrogen storage

9.3.1 *Nanomaterials for hosting hydrogen*

In contrast to the traditional methods for hydrogen storage, nanomaterials, such as single-walled carbon nanotubes (SWCNTs)[47] and multi-walled carbon nanotubes (MWCNTs),[48] have received great attention as potential synthetic nano-media for solid-state hydrogen storage. It was reported that storage capacity of hydrogen molecules that were physisorption at SWCNTs bundles reached to 8 wt % under 77 K.[49] The superior properties of SWCNTs exemplified by chemical stability, large surface area, hollowness, and light mass were attributed to be responsible of high storage capacity.[48]

In addition to carbon nanotubes, metal organic framework (MOFs) systems have shown beneficial properties of storing hydrogen by either physisorption or chemisorption processes.[50] However, these materials with their extremely large specific surface areas and ultra-porous properties can reach very high gravimetric hydrogen storage capacities; they possess typically very low density and, therefore, poor volumetric hydrogen storage capacity.[50] Likewise MOFs materials, complex hydrides such as amides, alanates, and brohydrides are often not reversible or require high temperatures for hydrogen releasing.

9.3.2 *Metal hydrides*

The possibility of formation metal hydride through chemisorption concept was firstly reported in 1868 by T. Graham, when he discovered that metallic palladium (Pd) wires were readily charged with hydrogen.[51] Since then, it has been demonstrated by many researchers that hydrogen can be chemically stored in wide spectrum of metals, metal alloys, and intermetallic compounds in their solid state through absorption process.

Metal (M) hydrides, including pure metal (e.g., Mg, Ti, Zr, V, Nb) and their intermetallic compounds and alloys, are broad spectrum of materials synthesized by direct exothermic reaction (absorption) between M and hydrogen (H_2) in a gas-solid fashion absorption reaction, achieved at certain pressure and temperature to form metal hydrides (MH_x) according to the following chemical reaction:

$$M+\left(\frac{x}{2}\right)H_2=\mathrm{MH}_x-\Delta Q \tag{9.1}$$

where $-\Delta Q$ denotes to the heat released yielded from the exothermic reaction. This reaction is usually known as "hydrogenation reaction." Metal hydrides decompose (hydrogen desorption) at certain temperature and pressure into M and H_x, as presented in the following equation:

$$\mathrm{MH}_x=M+\left(\frac{x}{2}\right)H_2+\Delta Q \tag{9.2}$$

where $+\Delta Q$ denotes to the heat absorbed to the system in order to release hydrogen (decomposition) from MH_x. This process is known as "dehydrogenation reaction." Combing the two reaction in one equation gives us the so-called reversible reaction, as presented in Eq. (9.3):

$$M+\left(\frac{x}{2}\right)H_2 \rightleftarrows MH_x \tag{9.3}$$

Many of solid-state metal hydrides possess excellent hydrogenation properties, such as high volumetric capacities, capability to absorb hydrogen at low pressure, and temperature, reversibility, and hydrogen releasing through endothermic process at moderate temperature. Moreover, another group of metal hydrides have higher hydrogen storage density (6.5 H atoms/cm^3 for MgH_2) than hydrogen gas (0.99 H atoms/cm^3) or liquid hydrogen (4.2 H atoms/cm^3).[52] When compared with compressed gas or liquid hydrogen storage, metal hydride storage is a safe, volume-efficient storage approach that can be employed for onboard vehicle applications. Many scientific and engineering studies have been carried out in order to develop the absorption/desorption of hydrogen in metals for fuel cell applications.[53] Table 9.3 presents some typical intermetallic solid-state hydrogen storage materials.

In general, metal hydrides can be grouped based on their hydrogenation/dehydrogenation temperatures into two major groups:

- *Low-temperature metal hydrides* that release hydrogen under normal atmospheric pressure around room temperature but with hydrogen storage capacity restricted to be less than 2.15 wt.% H_2 (e.g., $FeTiH_{1.8}$ and $LaNi_5H_6$),
- *High-temperature metal hydrides* that possess high hydrogen storage capacity by weight but they require high temperatures for hydrogen absorption/desorption. For example, commercial Mg can reversibly store about 7.6 wt.%. The problems are the high operating temperature (300°C) and the slow hydrogenation/dehydrogenation kinetics.

Table 9.3 Comparison in hydrogen capacity per weight and cost of some metal hydrides.

Alloy	Capacity (wt. %)	Cost of alloy ($/kg)	Cost of hydrogen ($/g)
$LiNi_5$	1.49	9.89	0.77
$ZrMn_2$	1.77	11.2	1.25
$TiMn_{1.5}$	1.86	4.99	0.44
$TiMn_{1.4}V_{0.6}$	2.15	29.40	2.67
hcp-Ti, bcc-V	3.70	10.63	0.59
TiFe	1.86	4.68	0.31
Mg_2Ni	3.60	6.26	0.19

Source: Collected data from different sources.

9.4 Magnesium hydride as an example of solid-state hydrogen storage material

Metallic Mg and its based-materials have been considering as the most candidate hydrogen storage materials for real applications[20,23,29,33–41,52] The worldwide interest on Mg is attributed to its natural abundance, low cost, light-weight, high gravimetric (7.60 wt.%), and volumetric (110 g/L) hydrogen storage capacities. Recently, it has been reported that waste Mg-based alloys such as Mg-Al[53] and Mg-Gd[54] were successfully used as feedstock materials to produce a good quality of MgH_2. More recently, the formation of Mg_2FeH_6 powders, starting from MgH_2 and plain carbon steel powders, used as precursors has been demonstrated by Polanski et al.[55] They claimed that magnesium iron hydride can be industrially manufactured at high yields using steel scrap, which can lower the cost of the manufacturing process.

9.4.1 *Traditional approach for synthesizing commercial MgH_2*

Theoretically, MgH_2 can be prepared by a direct reaction between Mg metal and hydrogen gas at normal pressure (~1 bar) and low temperature (~50°C).[56] Unluckily, in practice, this reaction cannot be achieved at such modest conditions due to the poor hydrogenation kinetics.[56] In order to achieve a successful forward reaction with fast uptake kinetics, both of pressure and temperature must be increased to 70 bar and 350°C, respectively.[55] These conditions are essential in order to supply the system with the pressure drive and thermal energy required to enhance its hydrogenation kinetics. Moreover, heat treatment of Mg is very important to break down the oxide layer of MgO coating Mg and hence allowing hydrogen atoms to diffuse into Mg fresh/clean surface.

In the classic methods used for preparing bulk MgH_2, absorption process is achieved through two consequence stages of physisorption followed by dissociation of hydrogen molecules that are diffused throughout hydride layer and nucleation of the hydride. Contrary to this, the dehydrogenation process started with nucleation of Mg metallic phase, and then diffusion hydrogen atoms throughout Mg and hydride phase.

Thus, the liberated hydrogen atoms are recombined at Mg surface to form hydrogen molecules.[56] The rate-limiting process for desorption is diffusion of the hydrogen through the hydride material.[57]

9.4.2 Synthesizing of nanocrystalline MgH_2 powders by reactive ball milling

In spite of the traditional approaches used for gas-solid reaction at relatively high temperature, Calka et al.[58] and El-Eskandarany et al.[59] proposed a solid-state approach, the so-called reactive ball milling (RBM), used for preparations different families of meal nitrides and hydrides at ambient temperature. This mechanically induced gas-solid reaction can be successfully achieved, using either high- or low-energy ball-milling methods, as shown in Fig. 9.5. However, high-energy ball mill is an efficient process for synthesizing nanocrystalline MgH_2 powders using RBM technique, it may be difficult to scale up for matching the mass production required by industrial sector. Therefore, from a practical point of view, high-capacity low-energy milling, which can be easily scaled-up to produce large amount of MgH_2 fine powders, may be more suitable for industrial mass production.

In both approaches but with different scale of time and milling efficiency, the starting Mg metal powders milled under hydrogen gas atmosphere are practicing to dramatic lattice imperfections such as twinning and dislocations. These defects are caused by plastics deformation coupled with shear and impact forces generated by the ball-milling media.[60] The powders are, therefore, disintegrated into smaller particles with large surface area, where very clean or fresh oxygen-free active surfaces of the powders are created. Moreover, these defects, which are intensively located at the grain boundaries, lead to separate micro-scaled Mg grains into finer grains capable to getter hydrogen by the first atomically clean surfaces to form MgH_2 nanopowders.

Fig. 9.5 illustrates common lab scale procedure for preparing MgH_2 powders, starting from pure Mg powders, using RBM via (1) high-energy and (2) low-energy ball milling. The starting material can be Mg-rods, in which they are processed via sever plastic deformation,[61] using for example cold-rolling approach,[62] as illustrated in Fig. 9.5. The heavily deformed Mg-rods obtained after certain cold rolling passes can be snipped into small chips and then ball-milled under hydrogen gas to produce MgH_2 powders.[8]

9.4.2.1 High-energy reactive ball milling

Planetary ball mills are the most popular mills used in scientific research for synthesizing MgH_2 nanopowders. In this type of mill, the ball-milling media have considerably high energy, because milling stock and balls come off the inner wall of the vial and the effective centrifugal force reaches up to 20 times gravitational acceleration. The centrifugal forces caused by the rotation of the supporting disc and autonomous turning of the vial act on the milling charge (balls and powders). Since the turning directions of the supporting disc and the vial are opposite, the centrifugal forces alter-

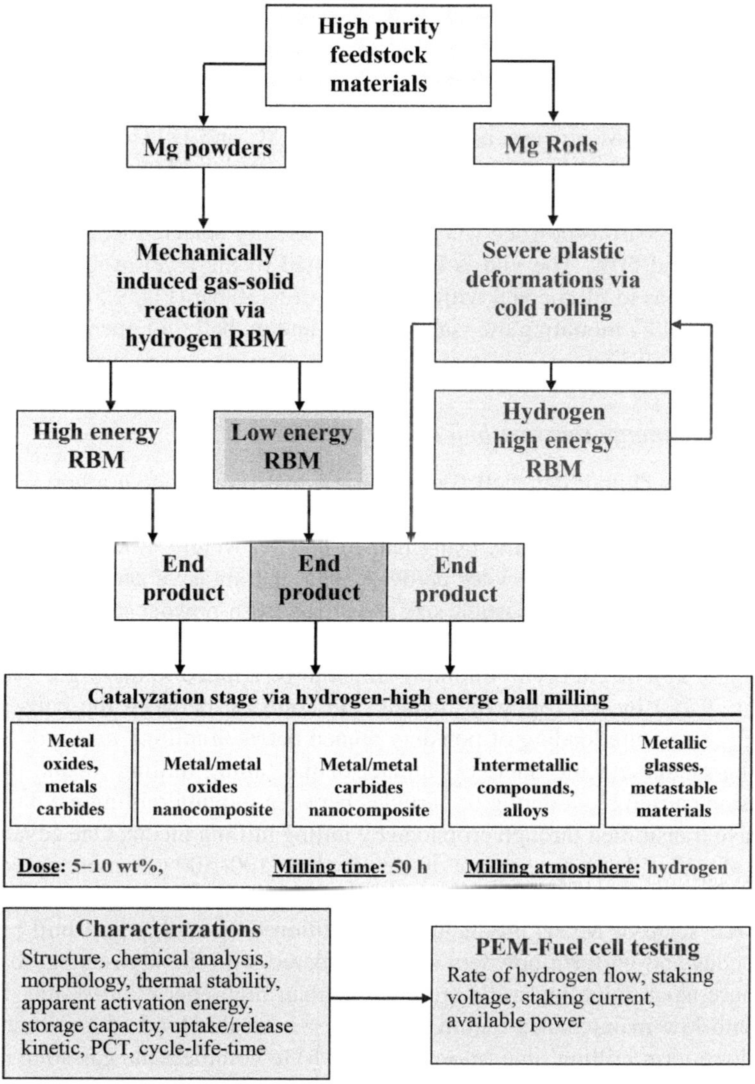

Figure 9.5 Flowchart diagram presents two approaches used in KISR for preparing nanocrystalline MgH_2 powders, using high- and low-energy RBM technique under hydrogen gas pressure. The chart elucidates the two approaches (cold rolling and catalyzation) used to improve the hydrogen storage behavior of MgH_2.

nately are synchronized and opposite. Therefore, the milling media and the charged powders alternatively roll on the inner wall of the vial, and are lifted and thrown off across the bowl at high speed.

In the typical experimental procedure, a certain amount of the Mg (usually in the range between 3 and 10 g based on the vial's volume) is balanced inside an inert gas atmosphere (argon or helium) in a glove box and sealed together with certain number of balls (e.g., 20–50 hardened steel balls) into a hardened steel vial (Fig. 9.5A and B), using, for example, a gas-temperature-monitoring system (GST). With the GST system, it becomes possible to monitor the progress of the gas-solid reaction taking place during the RBM process, as shown in Fig. 9.5C and D. The temperature and pressure changes in the system during milling can be also used to realize the completion of the reaction and the expected end product during the different stages of milling (Fig. 9.5D). The ball-to-powder weight ratio is usually selected to be in the range between 10:1 and 50:1. The vial is then evacuated to the level of 10^{-3} bar before introducing H_2 gas to fill the vial with a pressure of 5–50 bar (Fig. 9.5B). The milling process is started by mounting the vial on a high-energy ball mill operated at ambient temperature (Fig. 9.5C).

9.4.2.2 Low-energy reactive ball milling

Tumbling mill is cylindrical shell (Fig. 9.6A–C) that rotates about a horizontal axis (Fig. 9.6D). Hydrogen gas is pressurized into the vial (Fig. 9.6C) together with Mg powders and ball-milling media, using ball-to-powder weight ratio in the range between 30:1 and 100:1. Mg powder particles meet the abrasive and impacting force (Fig. 9.6E), which reduce the particle size and create fresh-powder surfaces (Fig. 9.6F) ready to react with hydrogen milling atmosphere.

The useful kinetic energy in tumbling mill can be applied to the Mg powder particles (Fig. 9.7E) by the following means: (1) collision between the balls and the powders; (2) pressure loading of powders pinned between milling media or between the milling media and the liner; (3) impact of the falling milling media; (4) shear and abrasion caused by dragging of particles between moving milling media; and (5) shock-wave transmitted through crop load by falling milling media. One advantage of this type of mill is that large amount of the powders (100–500 g or more based on the mill capacity) can be fabricated for each milling run. Thus, it is suitable for pilot and/or industrial scale of MgH_2 production. In addition, low-energy ball mill produces homogeneous and uniform powders when compared with the high-energy ball mill. Furthermore, such tumbling mills are cheaper than high-energy mills and operated simply with low-maintenance requirements. However, this kind of low-energy mill requires long-term milling time (more than 300 h) to complete the gas-solid reaction and to obtain nanocrystalline MgH_2 powders.

9.4.3 Characterization of reacted ball-milled MgH_2 powders

The as-prepared MgH_2 powders are usually characterized by different common techniques in order to investigate general/local crystal structure, morphology, thermal stability, apparent activation energy, hydrogen storage capacity, hydrogenation and dehydrogenation kinetics, pressure–composition–temperature (PCT), and cycle-life-time. Moreover, the hydrogen released from powders can be used to feed a fuel cell (FC) to investigate some useful and important parameters such as rate of hydrogen flow, staking voltage, staking current, available powder.

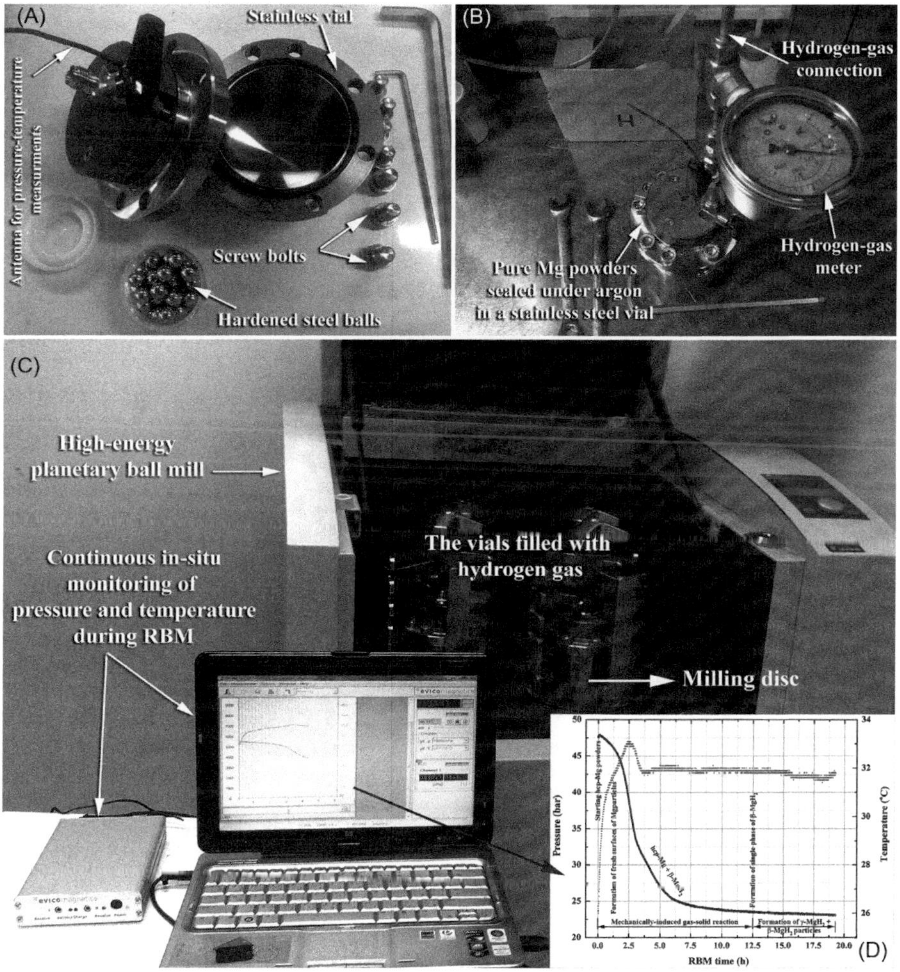

Figure 9.6 Photographs taken from KISR-EBRC/NAM Lab, Kuwait, show (A) the vial and milling media (balls) and (B) the setup performed to charge the vial with 50 bar of hydrogen gas. The photograph in (C) presents the complete setup of GST (supplied by Evico-magnetic, Germany) system prior to start the RBM experiment for preparing of MgH_2 powders, using Planetary Ball Mill P400 (provided by Retsch, Germany). GST system allows us to monitor the progress of RBM process, as indexed by temperature and pressure versus milling time (D).

9.4.3.1 Structural change of Mg powders upon RBM under hydrogen gas

XRD was used to monitor the structural changes of Mg powders upon RBM under 50 bar of hydrogen gas for different time. The XRD pattern of the initial pure hcp-Mg powder is displayed in Fig. 9.8A. Clearly, at the initial stage (0 h), the powders were polycrystalline hcp-Mg phase (JCPDS card no. 35-0821) with large grains, suggested by the sharp diffraction Bragg's peaks that refer to the hcp crystal structure

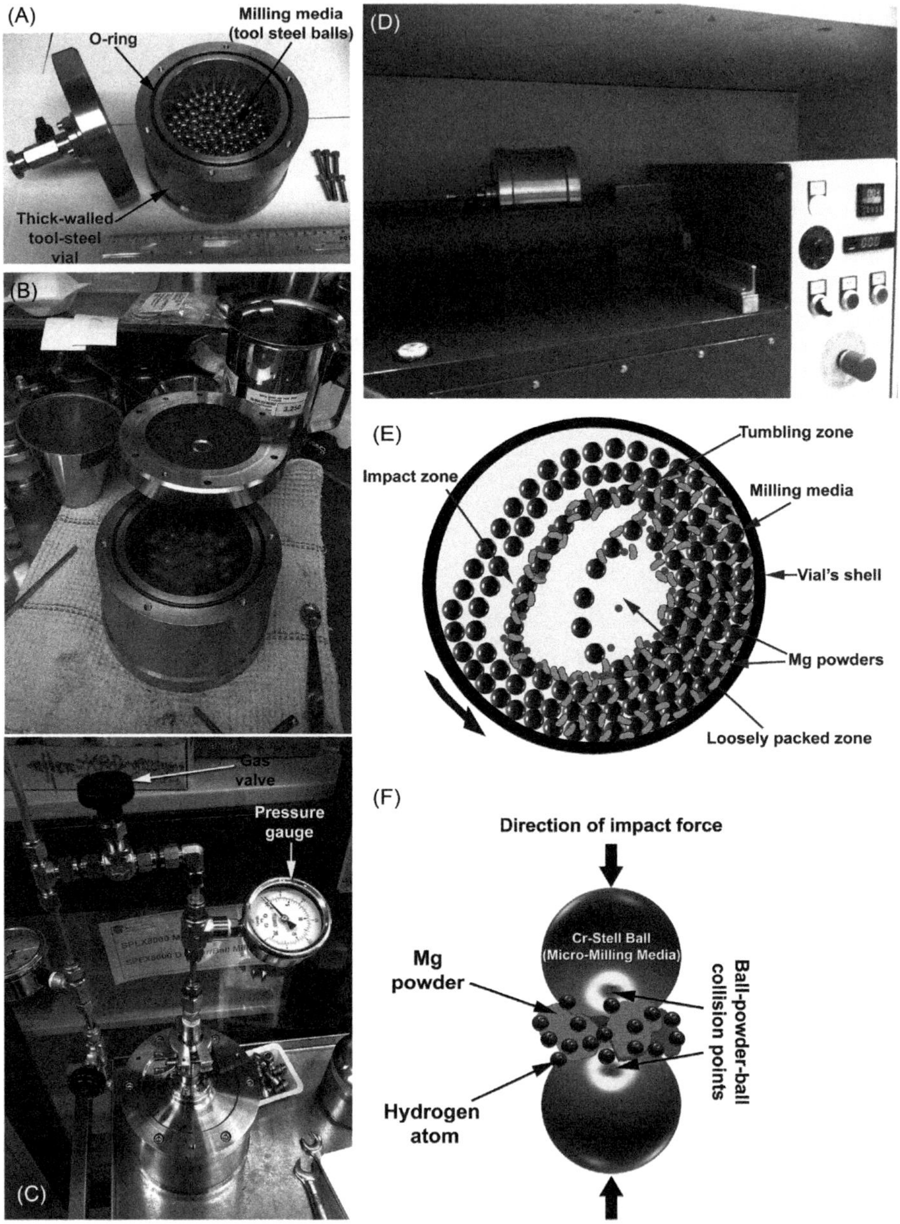

Figure 9.7 Photos taken from KISR-EBRC/NAM Lab, Kuwait, display setup of a lab-scale roller mill (1000 m in volume) showing (A) the milling tools including the balls (milling media and vial), (B) charging Mg powders in the vial inside inert gas atmosphere glove box, (C) evacuation setup and pressurizing hydrogen gas in the vial, and (D) ball milling processed, using a roller mill. Schematic presentations show the ball positions and movement inside the vial of a tumbler mall mill at a dynamic mode is shown in (E), where a typical ball-powder-ball collusion for a low energy tumbling ball mill is presented in (F).

(Fig. 9.8A). After 6 h of RBM (Fig. 9.8B), new Bragg's peaks appeared, suggesting the formation of a new crystalline phase. The major Bragg's peaks appeared at 2θ of 27.85, 35.72, 39.82, and 54.56 degree, as shown in Fig. 9.8B. Comparing these values with those reported for the tetragonal β-MgH_2 phase (JCPDS card no. 12-069735) confirmed the formation of significant molecular fractions of β-MgH_2 phase after 6 h of the RBM time coexisted with minor fraction of unprocessed Mg metal (Fig. 9.8B).

After 50 h of RBM time, the Bragg's peaks related to Mg crystals were disappeared, indicating the completion of gas-solid state reaction and the formation of β-MgH_2, as presented in Fig. 9.8C. Low-intensity Bragg peaks corresponding to γ-MgH_2 (PDF#: 00-035-1184) are shown in Fig. 9.8A, indicating β-γ-phase transformations.[63] Longer

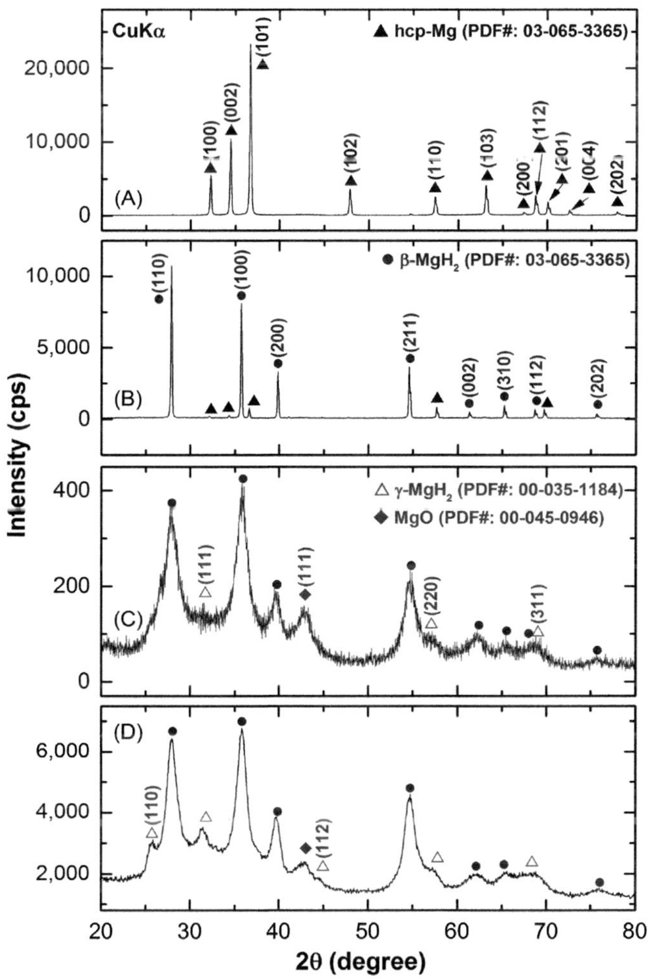

Figure 9.8 XRD patterns of Mg powders obtained after RBM for (A) 0, (B) 6, (C) 50, and (D) 100 h of RBM under hydrogen atmosphere.[63]

RBM time (100 h) led to further increasing in γ-phase, as indicted by the intensity increase seen in Bragg's peaks related to γ-MgH_2 (Fig. 9.8D). It has been suggested by several authors that increasing the milling time leads to increase the lattice imperfections introduced to β-MgH_2 powders, causing monotonic increase in the volume fractions of γ-MgH_2 phase.[63–65]

The bright field image (BFI) of the Mg powders obtained after 3 h of RBM is shown in Fig. 9.9A with the corresponding selected area diffraction pattern (SADP) (Fig. 9.9B). Obviously, the powders are composed of two phases. The first phase is hcp-Mg, and the second is tetragonal β-MgH_2 particles that are embedded into the soft metallic matrix of pure Mg. This would indicate that a considerable amount of Mg crystals reacted with hydrogen to yield the β-phase of MgH_2 with grain sizes ranging from 500 to 850 nm in diameter, as shown in Fig. 9.9A. The SADP in Fig. 9.9B shows the existence of two phases belonging to hcp-Mg and tetragonal β-MgH_2, respectively. No other phases could be detected at this resolution.

The dark field image (DFI) and the corresponding Miller-indexed SADP of the powders obtained after RBM time for 12.5 h are shown together in Fig. 9.9C and D, respectively. The powders after this stage of RBM have lenses-like structure morphology with an average grain size of 50 nm (Fig. 9.9C). The local structure of the powder shows a single phase of β-MgH_2 without any evidences for the existence of unprocessed hcp-Mg, as displayed in Fig. 9.9D.

The STEM-DFI image for two selected MgH_2 particles obtained after 25 h of RBM time is presented in Fig. 9.10A. The particles have nearly sphere-like morphology

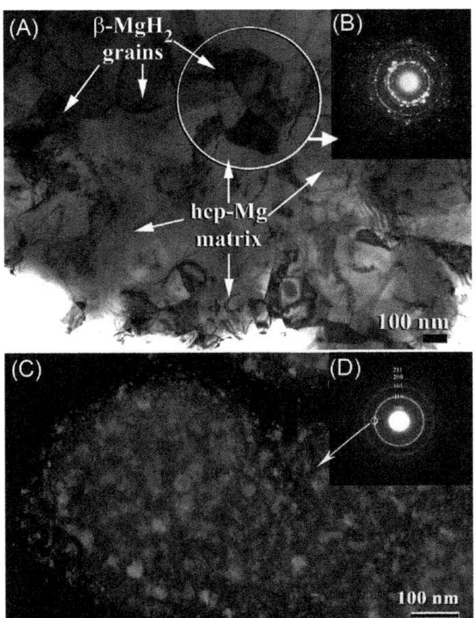

Figure 9.9 (A) BFI and (B) the corresponding SADP of Mg powders RBM for 3 h. The DFI and the corresponding SADP of the powders after milling for 12.5 h are shown in (C) and (D), respectively.[63]

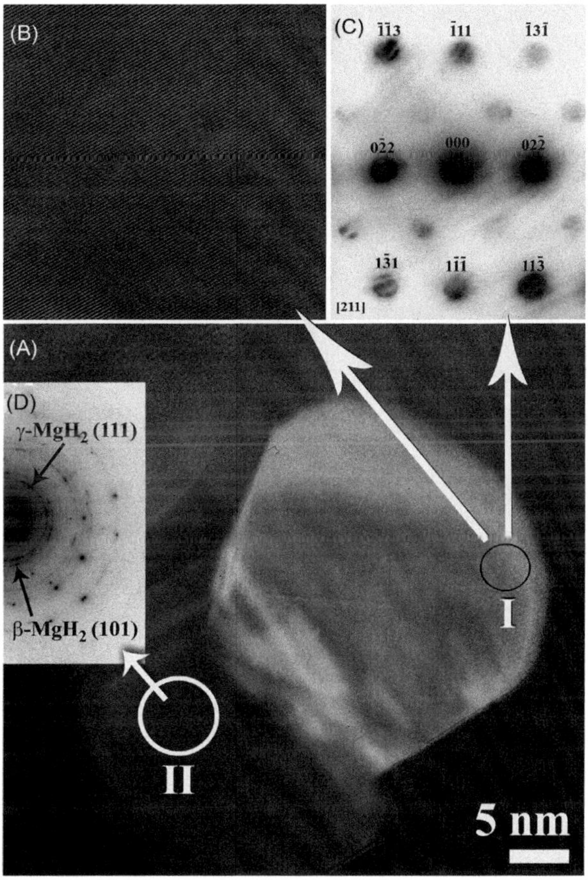

Figure 9.10 Image of the HRTEM of (A) scanning transmission electron microscope: dark field image (STEM:DFI) for two individual nanoparticles of MgH_2 obtained after RBM of Mg powders for 25 h. The lattice-fringe image of STEM-high-angle annular dark field (HAADF) of zone I (Fig. 10.7A) is shown in Fig. 10.7B. The indexed nano-beam diffraction (NBD) and the SADP presented in (C) and (D) are corresponding to zones I and II, respectively.[63]

with an average particle size of about 45 nm in diameter. Zone I of the first particle (Fig. 9.10A) was examined by STEM-high-angle annular dark field (HAADF) (Fig. 9.10B) and nano-beam diffraction (NBD) (Fig. 9.10C). The results indicate the formation of a single phase of β-MgH_2 and the absence of unprocessed Mg crystals, as indicated by the crystalline-fringe image (Fig. 9.10B) and the corresponding NBD of direction [211] (Fig. 9.10C). The Miller-indexed SADP (Fig. 9.10D), which is corresponding zone II in Fig. 9.10A, confirms the existence of two crystalline phases of β-MgH_2 (tetragonal) and metastable phase of γ-MgH_2 (orthorhombic).

The DFI for the powders obtained after 200 h of the RBM time is displayed in Fig. 9.11. The powders at this stage of milling are composed of nearly spherical fine grains with average size of about 2.8 nm. Comparing this image with that displayed in

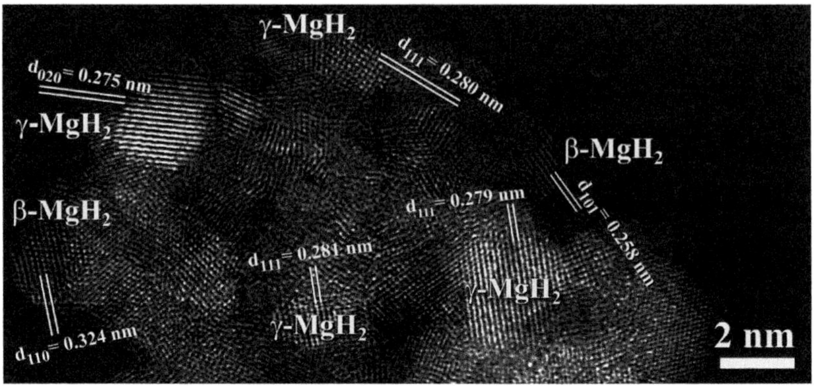

Figure 9.11 HRTEM/DFI for MgH_2 powders obtained after RBM for 200 h.[63]

Fig. 9.10A, it can be said that further milling (200 h) led to a significant refining in the grain sizes of MgH_2. It is worth mentioning that the powders of this stage of milling contain large volume fraction of γ-MgH_2 grains, as presented in Fig. 9.11. This implies continuity of lattice imperfections that led to β-γ phase transformations.

9.4.3.2 Morphological changes of Mg powders upon RBM under hydrogen gas

During the early stage of RBM the Mg powders were subjected to cold welding and tended to agglomerate to form larger particles with an average size of 860 μm in diameter, as displayed in Fig. 9.12A. As the RBM increased to 12.5 h, the agglomerated powder particles were disintegrated into smaller particles with an average particle size of 325 μm in diameter (Fig. 9.12B). The metallographic examination of the powders obtained after 12.5 h of RBM indicates the formation of composite system consisting of dark-gray spherical particles (less than 100 μm in diameter) embedded into the light-gray matrix presented in Fig. 9.12B. Those fine particles were probably MgH_2 and the light-gray matrix could be the unreacted Mg metal. After 25 h of RBM, the size of the powder particles was less than 50 μm in diameter, as shown in Fig. 9.12B.

Fig. 9.13A shows the backscattered electron (BSE) micrographs for the powders obtained after 200 h (A) and 500 h (D) of the RBM time. Dispersoids nanoparticles with an average particle size of less than 80 nm in diameter are embedded into the MgH_2 matrix as shown in Fig. 9.13A and D. The EDS dot mapping, which was used to investigate the nature of these particles, showed that the large gray particles were corresponding to Mg (Fig. 9.13B), where the small white nanoparticles are for Fe contamination (Fig. 9.13C) introduced to the powders upon using tool steel milling tools.[63] The percentage of the iron contamination in the sample that was ball-milled for 200 h reached to 2.35 wt.%.

Moreover, no undesirable reaction product could be identified, as suggested by the segregation of the elemental Mg and Fe in the milled powders, even after 500 h of the RBM time. The Fe contamination content increased to reach to a level of 3.76 wt.% upon increasing the RBM time to 700 h. This increasing in Fe content can be easily notified in Fig. 9.13F.

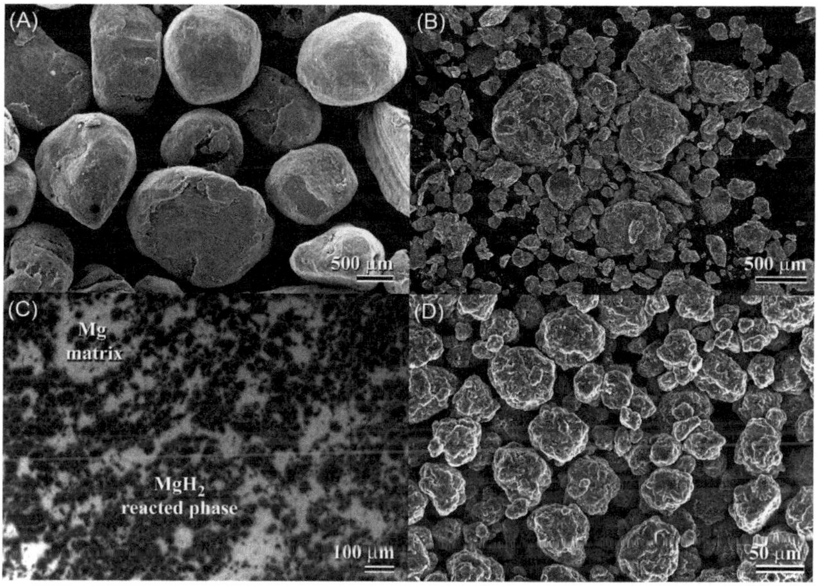

Figure 9.12 SEM micrographs of Mg powders milled under a hydrogen gas atmosphere for (A) 1, (B) 6, and (D) 12.5 h of RBM. The optical micrograph of the cross-sectional view for the powders milled for 6 h is shown in (C).[63]

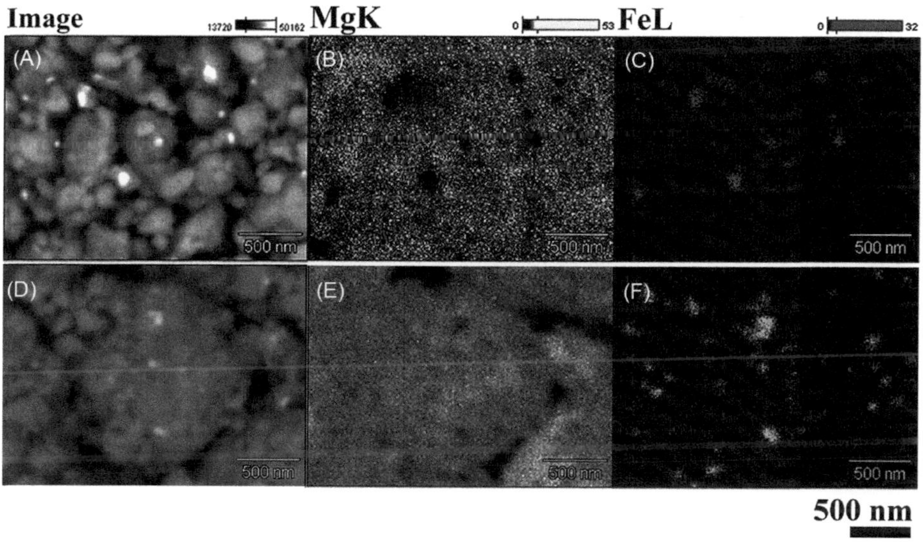

Figure 9.13 (A and D) BSE micrographs for the MgH_2 samples obtained after RBM time for 200 and 700 h, respectively. The corresponding EDS dot mapping images of elemental Mg and Fe for the 200 h sample are shown in (B) and (C), respectively. The EDS dot mapping images of elemental Mg and Fe for the 700 h sample are shown in (E and F), respectively.[63]

9.4.3.3 Thermal stability of MgH_2 powders obtained after different stages of RBM

Differential scanning calorimetry (DSC) was performed with a constant heating rate of 20 K/min in order to investigate the thermal stabilities of MgH_2 powders obtained after different RBM times. The DSC traces of the powders obtained after different RBM times are shown in Fig. 9.14. The onset temperature of the hydrogen desorption peak for β-MgH_2 phase obtained after 6 h of RBM time (Fig. 9.14A) appeared at a temperature of 714 K (441°C). The decomposition event took place through a single-endothermic peak, suggesting the existence of a single phase of β-MgH_2 in the ball-milled powders. The onset temperature for this endothermic reaction tends to shift to the low temperature side (707 K) upon milling the powders to 12.5 h of the RBM time (Fig. 9.14B), suggesting the destabilization of the formed β-MgH_2 phase. The DSC curve of the MgH_2 powders obtained after 25 h of the RBM time (Fig. 9.14C) revealed a shoulder-like endothermic peak appearing at an onset temperature of 692 K (419°C). This endothermic event vanished at an end-set temperature of about 750 K (477°C), as shown in Fig. 9.14C. Moreover, the onset temperature of the main endothermic peak did not show significant changes when compared with the powders obtained after 12.5 h of the RBM time.

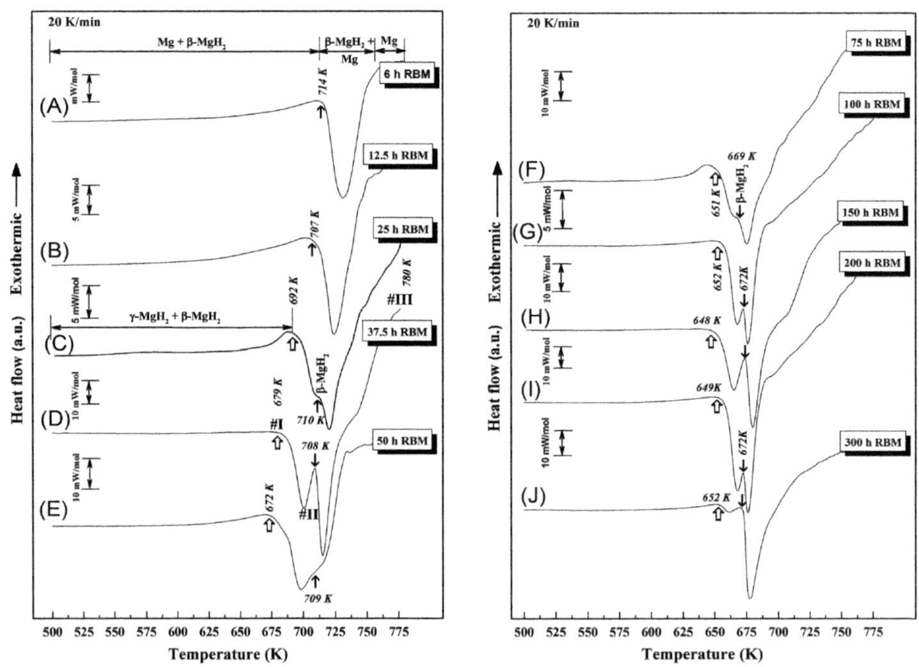

Figure 9.14 DSC curves of MgH_2 powders obtained after RBM for (A) 6, (B) 12.5, (C) 25, (D) 37.5, (E) 50, (F) 75, (G) 100, (H) 150, (I) 200, and (J) 300 h.[63]

After 37.5 h of the RBM time, the low- and high-temperature events clearly are seen into two separated peaks (Fig. 9.14D) appearing at onset temperatures of 679 K (409°C) and 708 K (435°C), respectively. The XRD pattern of the sample heated up to 679 K (406°C, just before the first endothermic event in Fig. 9.14D) shows the existence of two phases, that is, β-MgH_2 and γ-MgH_2, as shown in Fig. 9.15A. The XRD patterns of the sample #II, which was heated up to 708 K (Fig. 9.14D) and then quickly cooled down to the room temperature, reveals tetragonal structure of β-MgH_2 phase (Fig. 9.15B), implying γ-β phase transformation. In contrast to this, the XRD pattern (Fig. 9.15C) of the sample (#III) taken after the completion of the high-temperature endothermic peak (Fig. 9.14D) shows the existence of a single phase of hcp-Mg, indicating a complete decomposition of β-MgH_2 phase.

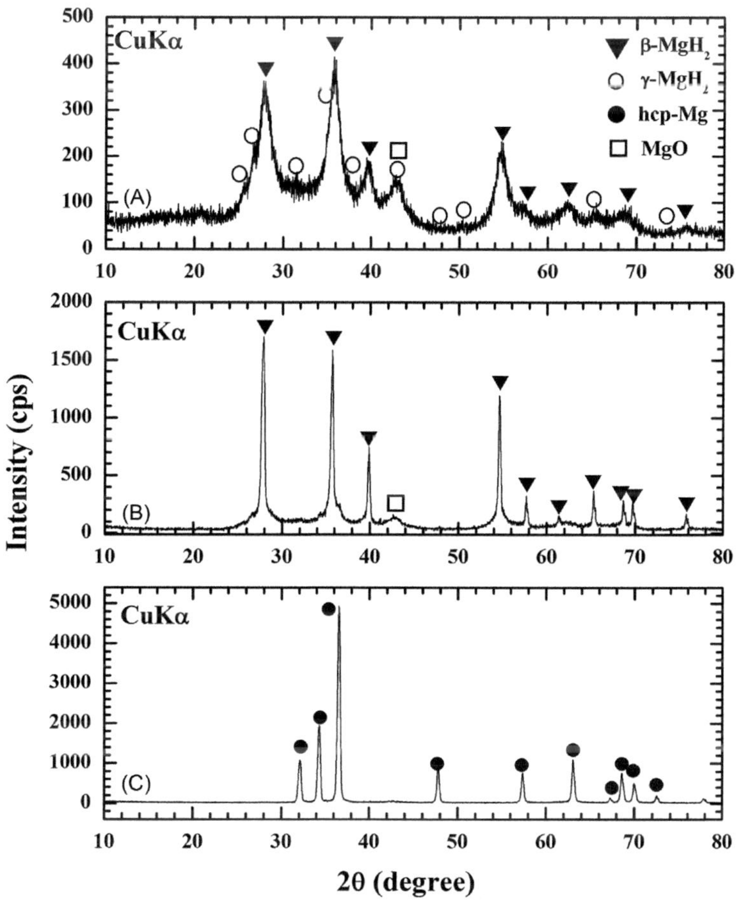

Figure 9.15 XRD patterns of MgH_2 powders obtained after 37.5 h of the RBM time and then heated up to (A) 679, (B) 708, and (C) 780 K in DSC under flow of an Ar gas atmosphere.

Fig. 9.14E shows the two separated endothermic peaks tending to be overlapping with each other for the sample obtained after 50 h of the RBM time. The onset temperatures of these two endothermic events were continuously shifted to the low-temperature side and recorded to be 670 K (397°C) and 709 K (336°C), respectively. Obviously, the area under the first endothermic peak showed to be relatively larger than that for the second endothermic reaction, indicating that the volume fraction of the existing phase in the milled powder was larger than β-MgH_2 phase. When the powders were successively subjected to further RBM time (75 h), the ratio between the formed two phases (γ-MgH_2 and β-MgH_2) was changed and the volume mole fraction of the high-temperature phase becomes larger, as indicated by the area under the second endothermic peak (Fig. 9.14F). Moreover, the onset temperatures of the two endothermic events shifted to the low-temperature side recorded as 650 K (377°C) and 669 K (396°C), respectively, suggesting a continuous destabilization of the formed hydride phases.

Again, and after 100 h of the RBM time (Fig. 9.14G), the two peaks were seen to split into individual events appearing at 652 K (379°C) and 672 K (399°C), respectively. This indicated a cyclic-β-γ phase transformations. The γ-MgH_2 phase, which was remarkably grown after 100 h of the RBM time tended to be more pronounced for the sample obtained after 150 and 200 h of the RBM time, as displayed in Fig. 9.14H and I, respectively. Further RBM (300 h) was seen to lead to a dramatic γ-β phase transformations, suggested by the obvious decrease in the area under the first endothermic peak, related to the decomposition of γ-MgH_2 phase, as presented in Fig. 9.14K.

In order to get more information about the thermal stabilities of the synthesized hydride phases and the role of RBM time on the destabilization of the β-MgH_2 phase, the apparent activation energy (E_a) of the powders was calculated using Kissinger's approach and plotted in Fig. 9.16. The E_a of β-MgH_2 phase obtained after 6 h of the RBM time showed high a value (161 kJ/mol), suggesting a high stability of this hydride phase. E_a decreased drastically upon further RBM time (6–37.5 h) to reach a lower value of 149 kJ/mol, as shown in Fig. 9.16. During the next stage of the RBM (37.5–50 h), the E_a values for both β-and γ-MgH_2 phases dramatically decreased with increasing the RBM time, suggesting a dramatic destabilization on the obtained hydride phased and an improvement on the hydrogen desorption kinetics. After 50 h, the calculated E_a values for β-and γ-MgH_2 phases are 139 kJ/mol and 125 kJ/mol, respectively. Further decreasing of E_a values performed upon RBM time for 100 h showed can be seen for both β-MgH_2 (133 kJ/mol) and γ-MgH_2 (124 kJ/mol). No significant improvements on the hydrogen desorption kinetics could be observed for the powders obtained upon further RBM times (100–300 h) where the E values for both β-and γ-MgH_2 phases were saturated at about 131 and 123 kJ/mol, respectively, as elucidated in Fig. 9.16. This significant decreasing of E was attributed to the monotonical decreasing of the grain size of the milled powders that led to increase in the specific Mg surface area and shortens the released pathways of hydrogen that were liberated from the Mg matrix.

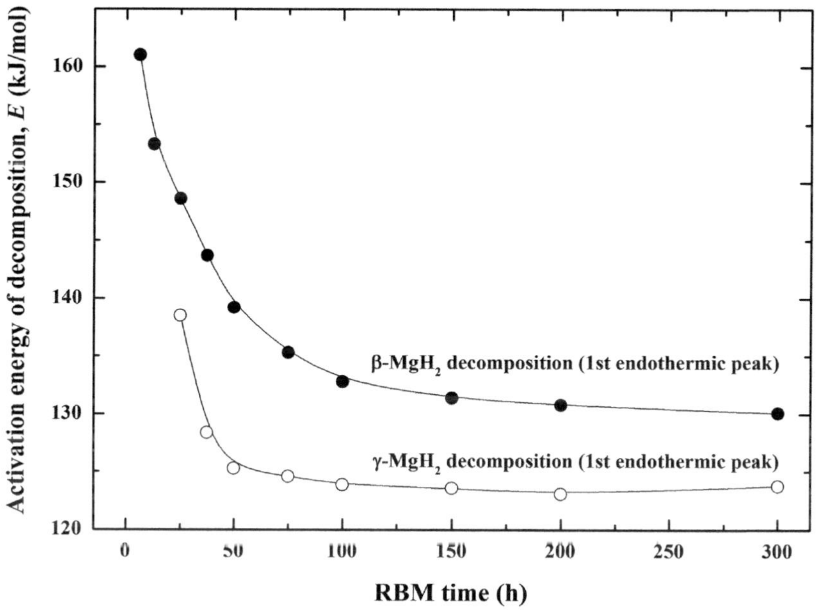

Figure 9.16 Effect of RBM on the apparent activation energy of γ-MgH_2 (open symbols) and β-MgH_2 (closed symbols) phases obtained after several stages of RBM.

9.4.3.4 *Effect of RBM time on the hydrogenation/dehydrogenation behavior of MgH_2*

The hydrogenation/dehydrogenation properties of MgH_2 are referred to the ability of metallic Mg to uptake (absorb) and release (desorb) hydrogen gas at certain hydrogen pressure and temperature. This property is always considered as the most important properties for evaluating any MgH_2 materials and to judge on their potential applications for fuel cell applications. The hydrogenation properties are indexed by volumetric and gravimetric density of hydrogen, temperature of hydrogen dissociation/decomposition, kinetics, and cyclic-lifetime (reversibility).

9.4.3.4.1 Pressure–composition–temperature (PCT)

There are different methods that can be used to determine PCT relationships of metal hydrides. One of the most popular methods is a volumetric approach, using Sievert-type apparatus. These measurements lead to construct PCT curve (known also as pressure–composition isotherm (PCI)), which is considered to be the most fundamental information for presenting the thermodynamic properties of any solid-state hydrides.

9.4.3.4.2 Hiden Isotherma

IMI system is one example of Sievert-type apparatus (Fig. 9.17). The apparatus consists of four main basic units known as reactor (calibrated volume reservoir),

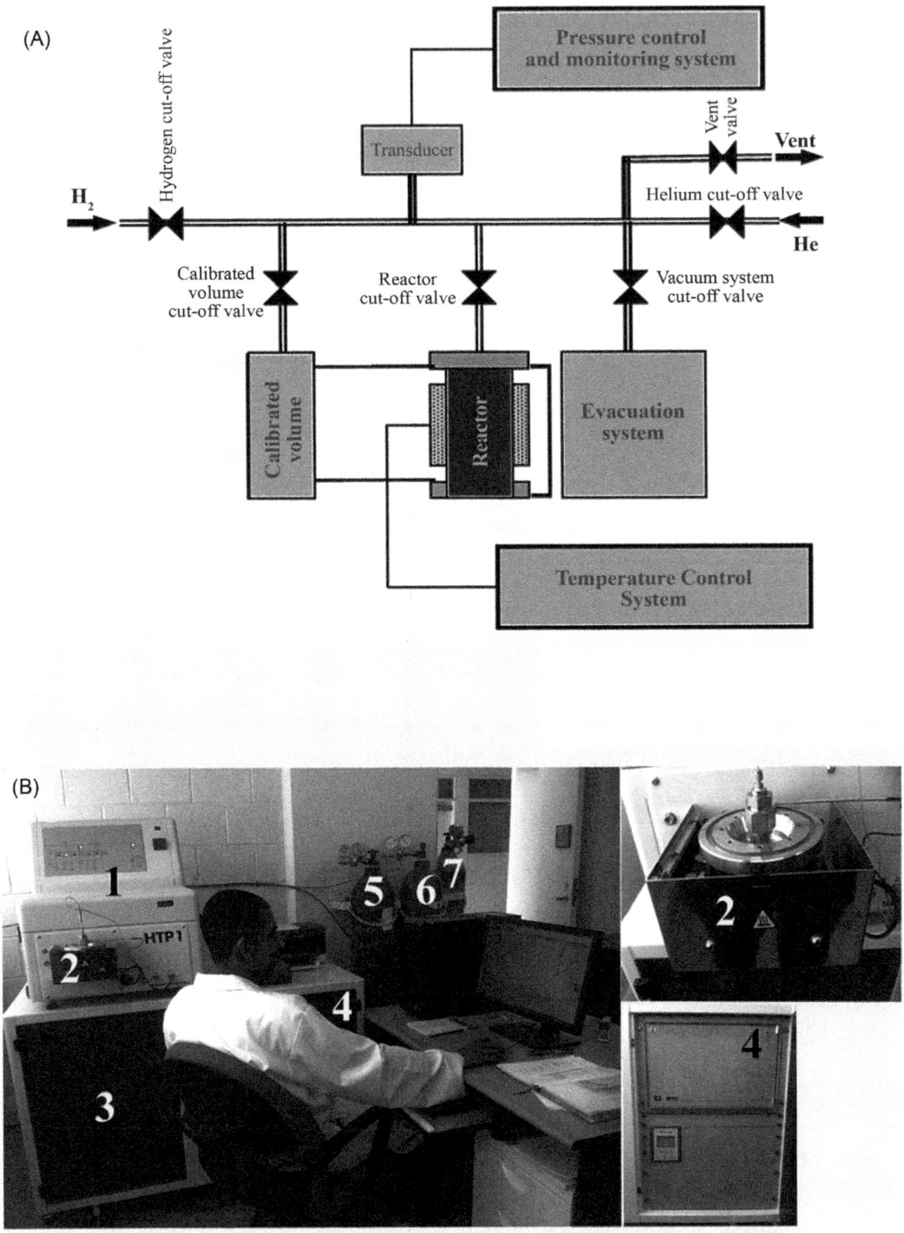

Figure 9.17 Schematic presentation of Sievert-type apparatus used for studying the hydrogenation properties of hydride materials, and photograph of the apparatus Sievert-type apparatus (Hiden Isotherma, IMI System, England, KISR-EBRC/NAM Lab, Kuwait).

temperature control unit system, which controls the temperature of the reactor, evacuation (vacuum) system, pressure control and monitoring unit system. There are different existing valves that control the gas inlet, outlet, and suction (vacuum) procedure. For PCT measurement, three types of high purified gas of hydrogen, helium, and compressed air are connected to the system. When hydrogen is supposed to be an ideal gas, the quantity of desorbed or absorbed hydrogen to Mg powders can be indexed by the number of hydrogen moles that can be calculated using the well-known ideal gas equation, which states that:

$$PV = nRT \tag{9.4}$$

where P = pressure, V = volume, n = number of H_2 moles, R = gas constant, and T = absolute gas temperature.

Before beginning the process of hydrogen absorption or desorption, the relation between pressure of hydrogen in a system and number of hydrogen moles at temperature T of the analyzed process can by expressed by

$$P_1V = n_1RT \tag{9.5}$$

However, after the desorption or absorption took place the relation in Eq. (9.5) can be written as:

$$P_2V = n_2RT \tag{9.6}$$

where $P_1 > P_2$ for absorption and $P_1 < P_2$ for desorption. Thus, both Eqs. (9.5) and (9.6) can be rearranged to give the number of moles of the absorbed (n_1) and desorbed (n_2) hydrogen gas, as follows:

$$n_1 = \frac{P_1V}{RT} \quad \text{and} \quad n_2 = \frac{P_2V}{RT} \tag{9.7}$$

The difference between n_1 and n_2 (Δn) can be estimated as follows:

$$\Delta n = n_1 - n_2 = \Delta P \frac{V}{RT} \tag{9.8}$$

where $\Delta = P_1 - P_2$.

Then, the mass of the absorbed or desorbed hydrogen gas (m_H) can easily be calculated from equation as follows:

$$m_H = 2.016 \Delta P \frac{V}{RT} \tag{9.9}$$

where 2.016 is the molecular mass of hydrogen.

A typical isotherm of a reversible MgH_2 nanocrystalline powders that were obtained after 200 h of RBM time is shown in Fig. 9.18. The measurements were achieved at

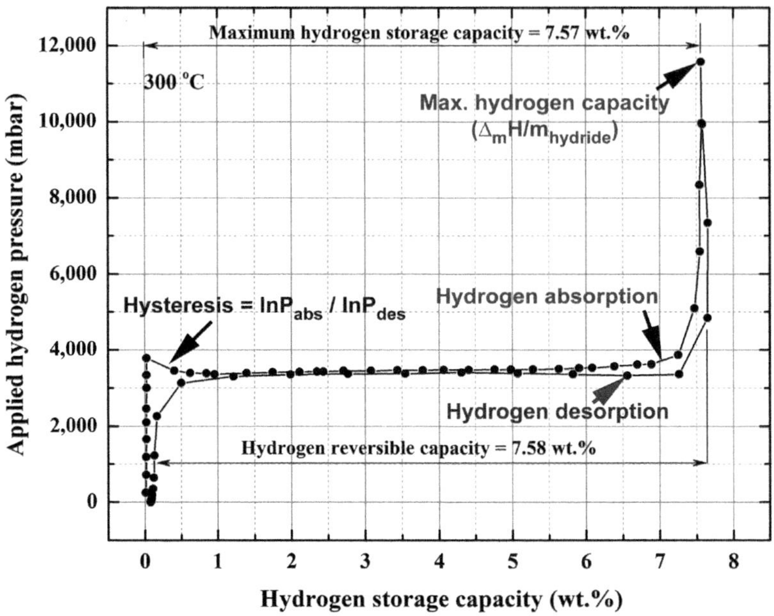

Figure 9.18 PCT curve measured at 300°C for MgH_2 powders obtained after 200 h of RBM.

a temperature of 300°C. The flat zone in the curve refers to the plateau pressure P and temperature T. The relationship between these two parameters is given by the so-called Van't Hoff equation:

$$\ln(p/p_o) = \frac{\Delta H}{RT} - \frac{\Delta S}{R} \tag{9.10}$$

where P_o = atmospheric pressure, ΔH and ΔS = enthalpy and entropy changes of hydrating/dehydrating reactions, respectively, T = absolute temperature, and R = gas constant.

Fig. 9.18 shows the PCT curve investigated at 300°C for MgH_2 sample obtained after 200 h of RBM time. The plateau region on the PCT curve was very flat, possessing only a minimal slope. The curve exhibited a pressure hysteresis with very low gab between the pressure needed for absorption (hydride formation), P_{abs} and the pressure required for hydride decomposition, P_{des} (Fig. 9.18). Moreover, the hydrogen storage capacity of the sample obtained after 200 h of RBM reached to a very closed value (7.57 wt.%) of the theoretical value (7.60 wt.%), as shown in Fig. 9.18. This is attributed to the role of the small grains of 200 h sample to facilitate short distance of hydrogen diffusion.

The hydrogen capacity in weight percent (H_c) of MgH_2 corresponding to any pressure is the ratio between the hydrogen mass (m_H) and the mass of MgH_2 (m_{MgH_2}), which can be calculated from the equation as follows:

$$H_c = \frac{m_H}{m_{MgH_2}} \times 100 \tag{9.11}$$

At higher gas pressure, the volume of hydrogen gas molecules driven from the ideal gas equation should be corrected by Van der Waals equation:

$$\left\{p + \frac{n^2 a}{V^2}\right\}(V - nb) = nRT \tag{9.12}$$

9.4.3.4.3 Experimental procedure

In order to obtain precious results, special care should be taken during sample preparation for Sievert-type apparatus. The careful steps for sample preparation are shown as follows and illustrated in Fig. 9.19:

- The bronze-type sample holder (reactor) should be cleaned up and dried in a drying oven at 150°C for 1 day and then brought out from the oven to cool to a room temperature (Fig. 9.19A),
- The samples, displaced (made of porous quartz wool), and the bottom part of the reactor were brought to the balance room (Fig. 9.19B). The empty reactor's bottom part is balanced firstly before charging the powder. Then, a specific weight of the sample (about 150–250 mg) is introduced into the reactor and balanced together. In order to avoid any loss of the sample during the evacuation steps in the apparatus, a small piece (about 2 mg in weight) of quartz wool might be placed onto the surface of the sample then all the system was balanced together (Fig. 9.19C)
- Knowing the weights of the empty reactor's bottom and quartz wool the net weight of the sample powder could be determined. Then the upper part of the sample holder was tightly sealed with its lower part (Fig. 9.19D),
- Before inserting the sample holder into the reactor part of the apparatus, a gasket should be firstly placed to avoid any leakage of the system during the experiments (Fig. 9.19E and F),
- Thus, the sample holder must be carefully loaded into the reactor chamber (Fig. 9.19G), where the thermocouple is inserted into the upper hole of the sample holder to know the temperatures of the system during the experiments (Fig. 9.19H),
- The system should be sealed tightly before starting the experiments (Fig. 9.19J),
- In all experiments, the samples must be activated before measurement starts. The activation process is usually taken place at temperature of 350°C with H_2-gas atmosphere at pressure of 35 bar (3.5 MPa) for 2–3 h at least. This activation step is necessary to clean up the surface of the sample powder and to break down the magnesium oxide (MgO) layer, which is usually formed upon handling the powder outside the Ar-glove box. The system can be then cooled down to room temperature, and the pressure was decreased to normal atmospheric pressure (1 bar),
- Then, the absorption-isotherm experiments can be started by setting the program at a desired heating rate. Once the sample reached to the set values of temperature and pressures of

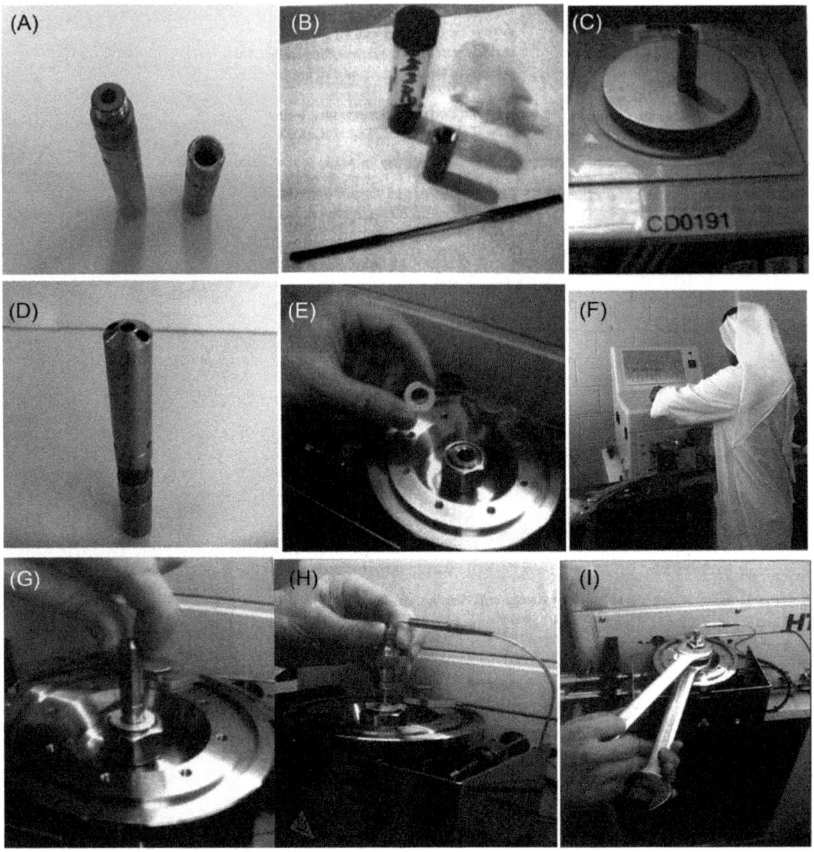

Figure 9.19 Photos present the several steps used to investigate the hydrogenation/dehydrogenation properties of as-prepared MgH_2 powders, using Sievert-type apparatus (Hiden Isotherma, IMI System, England, KISR-EBRC/NAM Lab, Kuwait).

300°C and 70 bar the desorption process took place by cooling down the system.The effect of applied temperature on pressure and hydrogen storage capacity when measured PCT of MgH_2 nanocrystalline powders obtained after 100 h of RBM time is shown in Fig. 9.20. Obviously, the sample measured at 100°C absorbed about 4. 48 wt.% H_2. This value was increased to 5.2 and 5.3 wt.% H_2 upon increasing the temperature to 125 and 150°C, respectively. According to Eq. (9.5), increasing the applied temperature leads to increase the hydrogen pressure. As a result, the number of H_2 molecules absorbed by Mg metal is increased. This is implied by the monotonical increase in the hydrogen storage capacity for the samples measured at higher temperature side (175–300°C). Since, Mg metal has poor kinetics of hydrogenation/dehydrogenation processes, application of high temperature (above 275°C) is required to enhance both processes.Another important application of Sievert-type apparatus

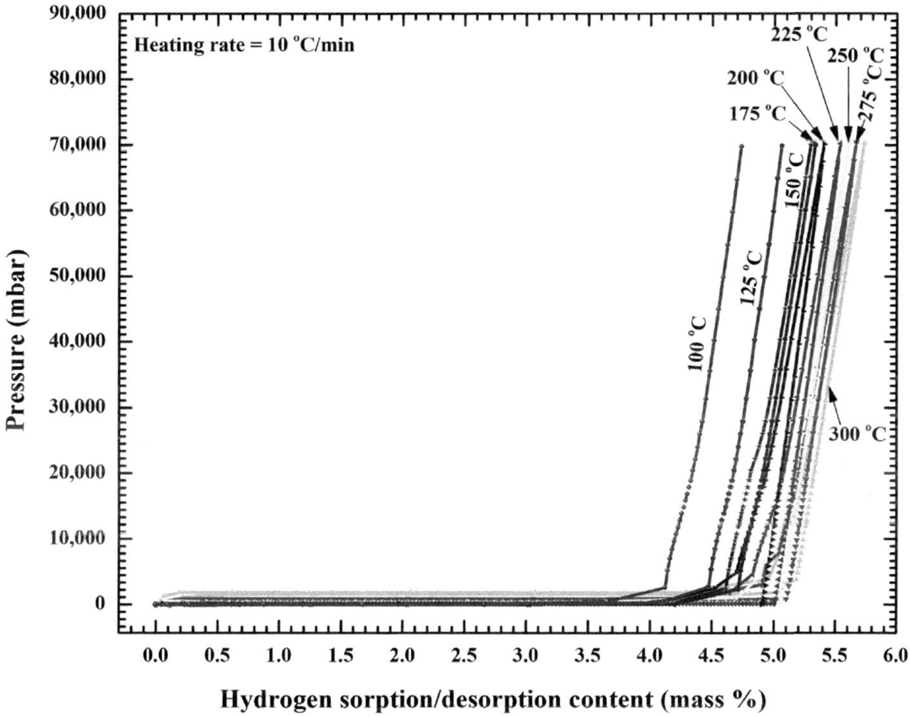

Figure 9.20 PCT curve measured at different temperatures for MgH_2 powders obtained after 100 h of RBM.

is the gas uptake and release kinetics at different temperatures and pressures. This test elucidates how the pressure of hydrogen in the reactor changes with changing the time. The test is also necessary to investigate the hydrogen mole fraction absorbed/desorbed by the sample. In addition, this application is very useful for measuring the hydrogenation/dehydrogenation cyclability (cycle-lifetime).

Fig. 9.21 shows the absorption (A and C) and desorption (B and D) kinetics of the sample obtained after 25 and 200 h of the RBM time, respectively. At 250°C, the 25 h sample absorbed about 5 wt.% H_2 within 5,000 s and saturated at 6.2 wt.% H_2 after about 32,000 s. When these values are compared with those obtained at 200°C (5 wt.% H_2/8120 s) and 225°C (5 wt.% H_2/6074 s), one can say that increasing the applied temperature leads to a moderate improve on the kinetics of hydrogen absorption seen with the temperature increase, as shown in Fig. 9.21A. The effect of temperature on the kinetics of desorption for the same sample is hardly seen, as displayed in Fig. 9.21B. The hydrogen released after 40,000 s at temperature of 200 and 225°C were 0.08 and 0.18 wt.%, respectively.

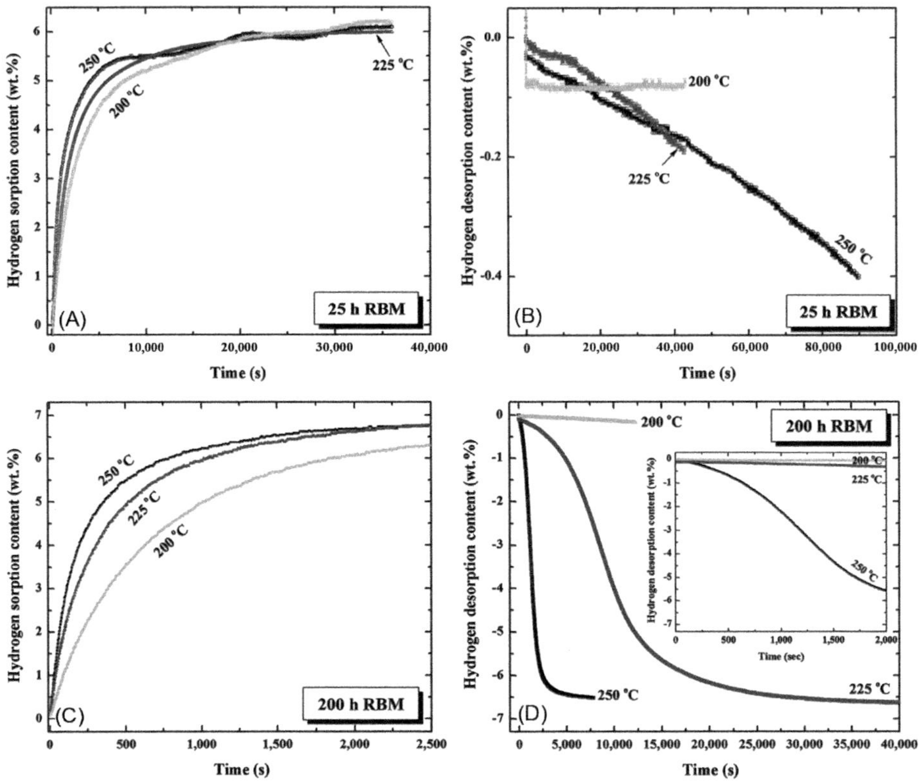

Figure 9.21 Effect of RBM time and applied temperatures on the (A and C) absorption and (B and D) desorption kinetics for MgH_2 powders, obtained after RBM time of (A and B) 25 and (C and D) 200 h.

In contrast to the 25 h samples, the powders obtained after 200 h of RBM showed significant improvement in the hydrogen uptake/release, as presented in Fig. 9.21C and D, respectively. At 250°C, the 200 h sample absorbed and released about 5 wt.% H_2 within 348 and 1657 s, respectively. The maximum hydrogen capacities for absorption and consequent desorption attained saturation values of about 6.8 wt.% within 3140 and 7400 s, respectively, as presented in Fig. 9.21. The kinetics of absorption decreased remarkably upon decreasing the temperatures to 225 and 200°C, as shown in Fig. 9.21C. At 225°C, the absorption kinetics reached to its saturation value of 6.9 wt.% within 3240 s. However, 35,207 s would be for complete desorption, as shown in Fig. 9.21D. At 200°C, further duration of time (5529 s) was necessary for complete absorption (~6.61 wt.%). However, the sample failed to release its absorbed hydrogen completely even after 12,394 s.[63]

Nanocrystalline MgH_2 powders obtained after 200 h of RBM possessed excellent cyclability, indexed by its hydrogen storage (7.5 wt.% H_2) stability with no obvious degradation on kinetics during a cycle-lifetime test, extended to 600 h, as presented in Fig. 9.22.[63] In order to investigate the structural changes of the sample

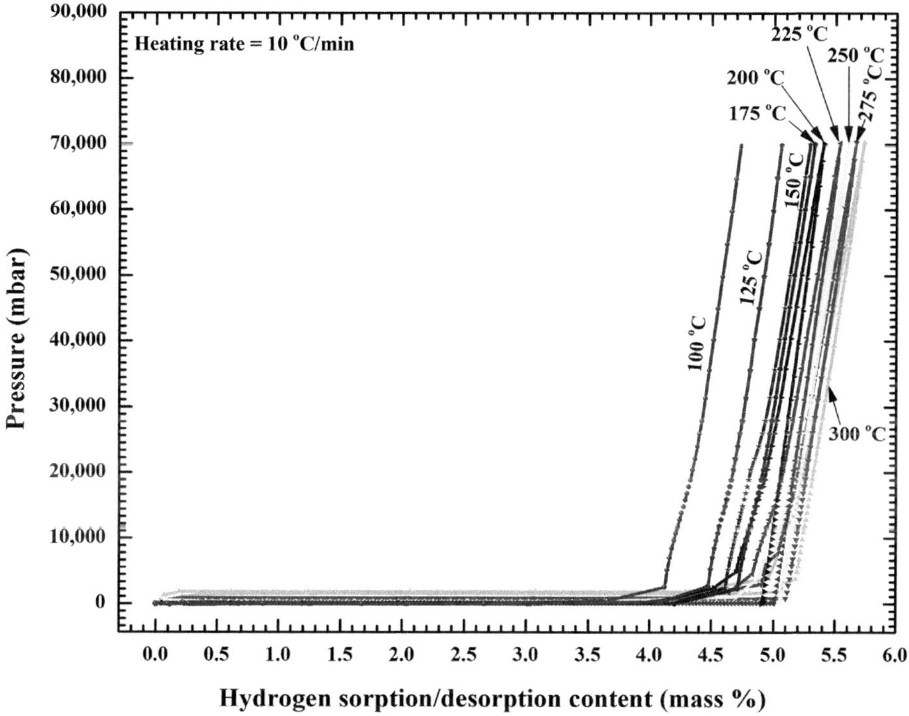

Figure 9.22 Hydrogen absorption and consequent desorption curves of 600 complete cycles of as-reacted ball-milled MgH_2 powders examined at 300°C under hydrogen pressure of 0/8 bar (absorption) and 200 mbar bar (desorption).

after performing cycle-lifetime test, XRD analysis was necessary. The XRD pattern of the sample taken after the completion of the 600 cycles of absorption/desorption (Fig. 9.22) showed sharp Bragg's peaks of single phase corresponding to hcp-Mg. This suggests a complete decomposition of MgH_2 phase.[63]

References

1. Key world energy statistics, Int. Energy Agency (2017).
2. Energy Information Administration (EIA). 2015. U.S. energy-related carbon dioxide emissions, 2014. Washinton, DC: U.S. Department of Energy. Online at http://www.eia.gov/environment/emissions/carbon/, accessed on December 10, 2018.
3. Jackson RB, et al. Environ Res Lett, 12: 110202 (2017).
4. David W.I.F. Faraday Discuss, 151: 399 (2011).
5. Latake PT, Pawar P, Ranveer AC. International Journal of Innovative Research and Creative Technology, 1: 333 (337).
6. Maryam S. International Journal of Hydrogen Energy, 42: 24927 (2017).

7. Wei Q., Qin F., Ma Q., She W. Carbon, 141: 542 (2019).
8. El-Eskandarany M. Sherif, Banyan M., Al-Ajmi F. RSC Adv., 8: 32003 (2018).
9. Marbán G., Valdés-Solís T. Int. J. Hydrogen Energy, 32: 1625 (2007).
10. Lanzia E., E. Verdolinia, Haščič. I. Energy Policy, 39: 7000 (2011).
11. Shashikala K. Functional Materials Preparation, Processing and Applications, 15: 607 (2012).
12. El-Eskandarany M. Sherif, Shaban E., and Alkandary A. Materials Today: Proceedings 5: 13726 (2018).
13. Jefferson, M., *Renewable Energy*, 31:571 (2006).
14. Satyapal, S., Petrovic, J., and Thomas, G., Scientific American, 296:80 (2007)
15. El-Eskandarany M. Sherif, Al-Matrouk H., Shaban E., Al-Duweesh A. Energy, 91: 274 (2015)
16. Moreno-Benito M., Agnolucci P., Papageorgiou L. G. Computers and Chemical Engineering, 102: 110 (2017)
17. Veziroğlu, T. N., and Sümer, S., Energy Conversion and Management, 49:1820 (2008)
18. Jain, I.P., Lal, C., and Jain A., International Journal of Hydrogen Energy, 35: 5133 (2010)
19. Neef, H. J., Energy, 34:327 (2010)
20. Walker G. Solid-state hydrogen storage: Materials and Chemistry, 1st Ed. Chapter 1, 2008. Woodhead Publishing Limited, New York.
21. Veziroğlu T. N., Sümer. S. Energy Conversion and Management, 49: 1820 (2008)
22. Schlapbach, L. and Züttel, A., *Nature*, 414:353 (2001)
23. Jain, I.P., Jain P, and Jain A., Journal of Alloys and Compounds, 503:303 (2010)
24. White C.M., R. Steeper R.R., Lutz A. International Journal of Hydrogen Energy, 31: 1292 (2006)
25. Das L. M. Hydrogen-fueled internal combustion engines, Compendium of Hydrogen Energy, Volume 3: Hydrogen Energy Conversion, Woodhead Publishing Series in Energy, 2016, Pages 177-217. DOI: https://doi.org/10.1016/B978-1-78242-363-8.00007-4
26. Parra D., Valverde L., Pino F. Javier, Patel K. M. Renewable and Sustainable Energy Reviews, 101: 279 (2019)
27. El-Eskandarany M. Sherif, Shaban E., Aldakheel F., Alkandary A., Behbehani M., Al-Saidi M. Scientific Reports, 7: 13296 (2018), www.nature.com/scientificreports, DOI:10.1038/s41598-017-13483-0
28. Robles J. O., De-León Almaraz S., Azzaro-Pantel C. Hydrogen as a Pillar of the Energy Transition. Hydrogen Supply Chains: Design, Deployment and Operation, Academic Press, 2018, Chapter 1, Pages 3-35. https://doi.org/10.1016/B978-0-12-811197-0.00001-4
29. Jorgensen, Scott W., Current Opinion in Solid State and Materials Science, 15: 39 (2011)
30. Sinigaglia T, Lewiski F, Martins MES, Siluk JCM. Int J Hydrogen Energy, 42: 24597 (2017)
31. Napoli G., Micari S., Dispenza G., Di Novo S., Antonucci V., Andaloro L. Int J Hydrogen Energy, 42: 28034 (2017)
32. Contestabile M. Batteries, Hydrogen Fuel Cells, and Bioethanol in Passenger Cars, Ethanol Science and Technology, ELSEVIER, 2019, Chapter 1, Pages 407-423.
33. Moriarty P., Honnery D. International Journal of Hydrogen Energy. 35: 12374 (2010)
34. Stetson N.T., McWhorter S., Ahn C.C. Compendium of Hydrogen Energy. Edited by Gupta, R. B., A. Basile, and T. N. Veziroğlu, Vol. 2, Chapter 1, 2016 (Oxford Elsevier).
35. Sherif S.A., Barbir F., Veziroglu T. N. Handbook of Hydrogen Energy, Mechanical and Aerospace Engineering, Edited by Sherif S.A., Goswami D. Y., Stefanakos E. K., Steinfeld A., Chapter 1, 2014, CRC Press, Pages 1-16.
36. Andreasen, A., Vegge, T., and Pedersen, A. S., J. Phys. Chem. B, 109:3340 (2005)

37. Züttel A. Materials for hydrogen storage. Materials Today, 6: 24 (2003)
38. El-Eskandarany M.S. Sci Rep, 6: 26936 (2016), DOI: 10.1038/srep26936
39. Sandi, G., *The Electrochem. Soc. Interface*, 13:40 (2004)
40. Chen, P., and Zhu, M., *Materials Today*, 11:36 (2008)
41. Sherif S., Barbir F., Veziroglu T. Journal of Scientific & Industrial Research, 62: 46 (2003)
42. Ouyang LZ, Dong HW, Peng CH, Sun LX, Zhu M. Int J Hydrogen Energy, 32: 3929 (2007)
43. Ahluwalia R.K., Hua T.Q., Peng J.-K. Int J Hydrogen Energy, 32: 3592 (2007)
44. Ismail M. Energy, 79: 177 (2015)
45. Aceves S.A., Petitpas G., Espinosa-Loza F., Matthews M.J., Ledesma-Orozco E. Int J Hydrogen Energy, 38: 2480 (2013)
46. Rusman N.A.A., Dahari M. International Journal of Hydrogen Energy, 411: 12108 (2016)
47. Dillon A.C., Jones K.M., Bekkedahl T.A., Kiang C.H., Bethune D.S., Heben M.J. Nature, 386: 377 (1997)
48. Angela L., Palph T. Yang. Journal of Catalysis, 206: 165 (2002)
49. Seung M. L., Kay H. A., Y. H. L., Gotthard S., Thomas F. Journal of American Chemistry Society, 123: 5059 (2001)
50. Balderas-Xicohténcatl R., Schmieder P., Denysenko D., Volkmer D., Hirscher M. Energy Technology 6: 510 (2017)
51. Graham T. Proc. R Soc. Lond., 17: 212 (1868)
52. Sakintuna, B., Lamari-Darkrim F., and Hirscher, Michael, International Journal of Hydrogen Energy, 32:1121 (2007)
53. Pistidda C., et al. Journal of Power Sources, 270: 554 (2014)
54. Hardian R., et al. Int. J. Hydrogen Energy, 43: 16738 (2018)
55. Polanski M., Nawra D., Zasada D. Journal of Alloys and Compounds, 776: 1029 (2019)
56. Stampfer J.F, Holley C.E, Suttle J.F. J Am Chem Soc, 82: 3504 (1960)
57. Ellinger F. H, Holley Jnr C. E., McInteer B. B., Pavone D, Potter R. M., Staritzky E., et al. J Am Chem Soc, 77: 2647 (1955)
58. Calka A. Appl Phys Lett., 59: 1568 (1991)
59. El-Eskandarany M. Sherif, Sumiyama K., Aoki K., Suzuki K., Mater Sci Forum, 88: 801(1992)
60. El-Eskandarany M. Sherif, Alkandary A., Aldakheel F., Al-Saidi M., Al-Ajmi F., Banyan M. RSC Adv., 8: 38175 (2018)
61. Valiev R.Z, Islamgaliev R.K., Alexandrov I.V. Prog. Mater Sci., 45: 103 (2000)
62. Amira S., Huot J. J. Alloy. Compd, 520: 287 (2012)
63. El-Eskandarany M Sherif, Shaban E., Al-Halaili B. Nanocrystalline Int J Hydrogen Energy 39:12727 (2014)
64. Xiao X., Liu Z., Saremi-Yarahmadi S., Gregory D. H. Phys. Chem. Chem. Phys., 18: 10492 (2016)
65. Shen C., Aguey-Zinsou K.-F. J. Mater. Chem. A, 5: 8644 (2017)

Mechanically induced-catalyzation for improving the behavior of MgH_2

10

In the previous chapter, we explained and discussed the most common method used for preparing nanocrystalline MgH_2 powders, using RBM technique. In addition, the characterization techniques used to investigate structural, morphology, thermal stability, and hydrogenation/dehydrogenation kinetics were elucidated. Unfortunately, and in spite of the attractive properties found in MgH_2 there are still major drawback issues restricting implementation of this attractive materials in real applications. In this chapter we will present and discuss the approaches taken place in the world to improve the kinetics behavior of MgH_2, using high-energy ball milling method. The figure shows (A) BFI image and corresponding (B) BFI image of nanocomposite $MgH_2/5Ni/5Nb_2O_5$ powders obtained after 50 h of mechanical ball milling. The corresponding BF-STEM images of zones I (A) and III (B) are shown in (C) and (D), respectively. The HAADF-STEM of zone II is shown in (E) together with the corresponding NBDP (inset of E). The corresponding NBDP of the indexed Ni particle located at the edge of MgH_2 matrix (B) is presented in (F) [1].

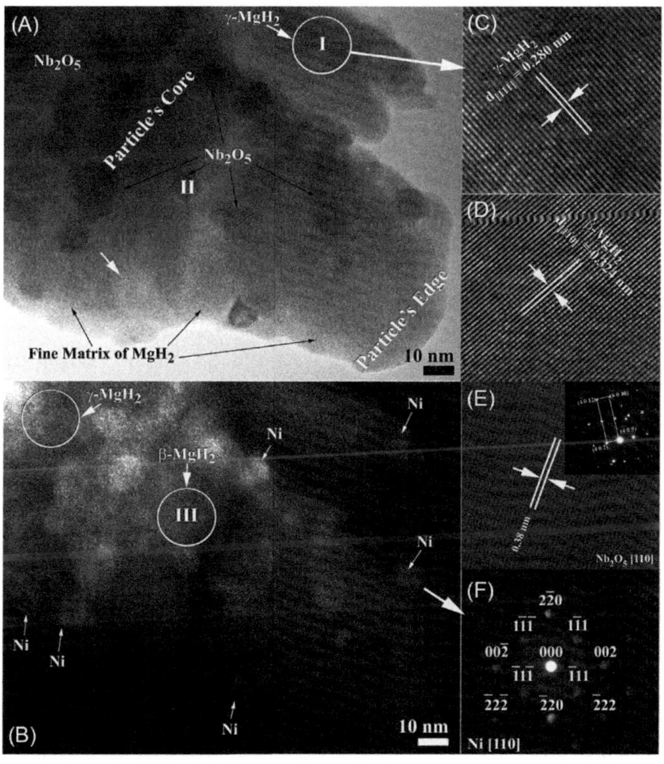

Mechanical Alloying. http://dx.doi.org/10.1016/B978-0-12-818180-5.00010-8

10.1 Introduction

Although MgH_2 solid-state hydrogen storage materials possesses excellent hydrogen storage capacity (7.60 wt.%), reversibility, cost effective, availability, and cyclability, there are major drawback issues restricting its implementation real applications [2–4]. MgH_2 in its pure form is very stable, possessing a large negative heat of formation ($\Delta H^{for} = -75$ kJ/mol.H_2) [5]. It has very slow kinetics of dehydrogenation at temperatures less than 400°C [6]. In addition, Mg metal is very sensitive to the surrounded atmosphere, particularly to oxygen. Coating the powder with MgO and $Mg(OH)_2$ films during materials handling usually hinders the hydrogen atoms to penetrate the hard ceramics coat to react with Mg surfaces. Removal of such films requires strict activation process of Mg under the application of high hydrogen pressure (above 35 bar) and high temperature (~ 350°C) [7]. Furthermore, the low thermal conductivity of MgH_2 system (0.4 W/mK [8]) is considered to be a serious problem since heat transfer is a very important factor required for both hydrogenation and dehydrogenation processes.

10.2 Scenarios for improving the behavior of MgH_2

10.2.1 Alloying elements for improving the hydrogenation/ dehydrogenation kinetics of Mg-based alloys

One practical solution proposed for enhancing the hydrogenation/dehydrogenation properties of MgH_2 is to melt pure bulk Mg with selected alloying elements such as Ni, Pd, and Nd to obtain less stable binary, ternary or multicomponent Mg-based alloy systems with lower heat of formation (ΔH^{for}). In many cases, the synthesized Mg-based alloy systems do not show attractive storage properties when compared with pure MgH_2. For example, alloying Mg with Ni to form binary Mg_2Ni lowers the ΔH^{for} of the metal hydride phase (Mg_2NiH_6) to -64.5 kJ/mol instead of -74.5 kJ/mol for pure MgH_2. However, the system shows a dramatic degradation on its hydrogen storage capacity (3.5 wt.%) with no significant decreasing in the decomposition temperature [9].

In the beginning of this century, Yamada et al. [10] proposed Mg-rich (90 at.%) systems of Mg-Pd, Mg-Nd and Mg-Pd-Nd to replace the traditional Mg_2Ni alloy. They reported that Mg-based system containing Pd show at 300°C pressure-composition-temperature (PCT) curves with three plateau-like regions and hydrogen absorbency of 5 wt.%. They pointed out that the hydrogenation/dehydrogenation of Mg-Nd system was influenced by a catalytic effect of the formed $NdH_{2.5}$ and NdH_3 phases that assisted hydriding and dehydriding of the Mg matrix. For Mg-Pd system, they investigated that disproportional reaction of Mg_6Pd to form Mg_5Pd_2 and MgH_2 retarded the overall reaction kinetics. In spite of their efforts dedicated to introduce a new Mg-based alloy system with advanced kinetics behavior, their proposed $Mg_{89}Pd_7Nd_4$ alloy required 50 min to absorb about 4 wt.% H_2 at 300°C. In addition, the dehydrogenation process of this system required 150 min at the same operating temperature to be completely achieved [10].

In 2009, Ouyang et al. [11] demonstrated a complete study of Mg_3Pr and $Mg_3PrNi_{0.1}$ hydride systems. Based on their results, both alloys showed reversible hydrogenation (290°C/35 bar H_2)/dehydrogenation (290°C/0.003 bar) cycles. Under these conditions, Mg_3Pr and $Mg_3PrNi_{0.1}$ exhibited storage capacities of 2.58 and 3.23 wt.%, respectively [11]. More recently, in situ formation of cycle stable $CeH_{2.73}$-MgH_2-Ni nanocomposites, from the hydrogenation of as-melt $Mg_{80}Ce_{18}Ni_2$ alloy, with excellent hydrogen storage performance has been proposed in 2014 by Ouyang et al. [12]. This nanocomposite system demonstrated reversible hydrogen storage capacity of 4.0 wt.%, at a low desorption temperature (232°C) with fast kinetics and long cycle life (500 cycles). Moreover, $CeH_{2.73}$-MgH_2-Ni nanocomposite exhibited a low apparent activation energy of 63 kJ/mol [12].

In 2011, indium (In) was used as an alloying element to form $Mg_{0.95}In_{0.05}$ solid solution, using mechanical alloying (MA) technique [13]. Fully reversible transformation in the Mg–In–H system with reduced hydrogen sorption reaction enthalpy was demonstrated. The hydrogen storage capacity of $Mg_{0.95}In_{0.05}$ solid solution measured at 30 bar/300°C showed a high value of 5.5 wt.% [13]. In a different proposed by the same school [14], elemental In was replaced by Sn with the same atomic percent (0.05) to form $Mg_{0.95}Sn_{0.05}$ solid solution, using MA technique. They reported that Mg–5 at.% Sn nanocomposite exhibited elevated plateau pressure and destabilized thermodynamics due to the introduction of large amount of interface energy in Mg/Mg_2Sn nanocomposite. However, the hydrogen storage capacity measured at 30 bar/323°C was lower (4.8 wt.%) than Mg-In system. The corresponding hydrogenation (50 bar/362°C)/dehydrogenation (0.003 bar/300°C) kinetics, for Mg–5 at.% Sn system were 170 and 130 min, respectively [14].

Further efforts dedicated for lowering the stability of MgH_2 system was achieved by Cao and his team by introducing elemental Al and Ti into Mg(In) solid solution to form $Mg_{85}In_5Al_5Ti_5$ alloy, using plasma milling technique [15]. They claimed that addition of Ti and in-situ synthesized MgF_2 improved the kinetics and the introduction of In as well as Al imparted enhanced thermodynamics to the $Mg_{85}In_5Al_5Ti_5$ system. In their study, the dehydrogenation enthalpy change and activation energy of this new system were 65.2 kJ/(mol H_2) and 125.2 kJ/mol, respectively. The hydrogen storage capacity this system measured at 30 bar/340°C was 5.5 wt.% [15].

Recently, nanocrystalline and amorphous microstructures prepared by rapid solidification (RS) processing are effective in improving the absorption/desorption kinetics of the Mg-based hydrogen storage alloys. It has been shown by Zhang et al. [16] that nanocrystalline and amorphous Mg-5Ni-3La (at.%) alloys show better absorption/desorption kinetics than that of the as-cast alloy due to the smaller grain size and more grain boundaries. The values of ΔH for the Mg-H_2 system of the nanocrystalline and amorphous alloys were −74.0 kJ/mol and −66.1 kJ/mol, respectively, which are significantly lower compared to that of the as-cast alloy (−77.7 kJ/mol) [16].

More recently, nanocrystalline hydrogen storage $La_{1.5}Mg_{0.5}Ni_{6.5}Co_{0.5}$ alloy obtained by mechanical alloying and heat treatment (850°C/0.5 h), has been demonstrated by Dymek [17]. The electrochemical charge/discharge multicycling of powder composite electrodes was carried out in view of material potential usefulness as electrode material for Ni/MH_x batteries. Partial substitution of Ni by Co in $La_{1.5}Mg_{0.5}Ni_7$

nanocrystalline alloy was prone to increase of electrode discharge capacity (up to 14%) and enhances its activation (strong decrease of N_{max} cycle and simultaneous increase of QNmax) [17]. On the other hand, Co addition worsens corrosion degradation behavior, but the corrosion resistance can be improved (30%) by the application of amorphous nickel coating. Finally, Co modified $La_{1.5}Mg_{0.5}Ni_7$ nanocrystalline alloy reveals 15% greater exchange current density of H_2O/H_2 system and improved hydrogen diffusivity compared to the reference, nanocrystalline Co-free $La_{1.5}Mg_{0.5}Ni_7$ hydrogen storage material [17].

10.2.2 Doping MgH_2 with catalysts

Since the mid of 1990s, enormous efforts have been devoted in order to improve the hydrogenation/dehydrogenation behaviors of MgH_2, using a wide variety of catalysts. In this section, we will present typical examples of several attempts that have been dedicated to improve the kinetics of MgH_2 for hydrogenation and subsequent dehydrogenation.

10.2.2.1 Metal and metal alloys

Because of their superior capability for hydrogen splitting (dissociation) and re-combination, transition metal (TM) groups have received great attention, particularly in 1990s and 2000s. However, the effect of pure metals, such as Ti, Zr, V, Nb, Fe, Co, Ni, Pd, Pt, on the kinetics reaction of MgH_2 have been intensively studied with the past 20 years. This catalyst group, specially Ni, Ti, Zr, and their alloys are still receiving great attentions. Liang et al. [18] reported the results obtained upon doping of MgH_2 powders with 5 wt.% of Ti, V, Mn, Fe and Ni. In their work, the as-received MgH_2 powders were individually high-energy ball milled under argon atmosphere with for 20 h. They pointed out that elemental Ti and V showed better catalytic effects for hydrogen absorption and desorption compared with Ni. Since then, many researchers have reported attractive results upon using wide range of pure metal.

Zaluska et al. reported the earliest study on doping of MgH_2 powders with 3-d elements, when they selected Pd, Ti, Fe, V, Zr, and Mn for catalyzing MgH_2 powders, either individually or as a catalytic mixture [19]. The catalyzed MgH_2 powders were prepared by high-energy ball milling composite powders in the range of 15 min to 20 h. Based on their finding, elemental V and Zr, as well as a composite Mn/Zr were the most effective agents used to improve the kinetics behavior of MgH_2. Some typical MgH_2/TM composite systems with their corresponding characteristics are listed in Table 10.1.

Meanwhile, intermetallic compounds and alloys have received a great attention since 1979, when Karty et al. have reported beneficial effect on hydrogenation and dehydrogenation of Mg upon doping MgH_2 with Mg_2Cu eutectic alloy [22]. They attributed the enhancement achieved in the kinetics reaction of MgH_2 to the dominant role of Mg_2Cu consists in minimizing the oxide inhibition by providing a clean unoxidized external surface at which the hydrogen gas readily adsorbs, dissociates, and diffuses into the interior where hydride formation occurs; essentially the reverse process occurs in dehydriding [23]. Since then Mg, was doped with alloys of different

Table 10.1 Effect of pure metal additives on the hydrogenation and dehydrogenation characteristics of MgH_2 powders.

Catalyst used	Dose	Doping approach	Hydrogenation/ Dehydrogenation properties	Apparent activation energy (kJ/mol)	References
NPs-Fe	10 wt.%	Ar- HEBM/4 h	**Absorption**/10 bar/350°C: 5.1 wt.%/9.1 min	86	[20]
NPs-Co	10 wt.%	Ar- HEBM/4 h	**Absorption**/10 bar/350°C: 4.3 wt.%/7.7 min	75	[21]
NPs-Ni	10 wt.%	Ar-HEBM/4 h	**Absorption**/10 bar/350°C: 5 wt.%/6.1 min	72	[21]
NPs-Cu	10 wt.%	Ar- HEBM/4 h	**Absorption**/10 bar/350°C: 4.6 wt.%/15.3 min	76	[21]
Zr	1 at%	H_2-RBM/300 rpm/20 h + annealing at 100°C/40 bar H_2/24 h	**Absorption**: 5.5 wt.% /4000 s/300°C. **Absorption** of nanocomposite: 3.5 wt.% H_2/6000 s/100°C. Dehydrogenation: 5.5 wt.% H_2/2000 s/350°C.	40	[22]

HEBM, High-energy ball milling; RBM, reactive ball milling; NA, not available.

concentrations, including Mg_2Ni [24,25], $Mg(La_2Mg_{17})$-$LaNi_5$ [26], Mg–FeTi [27], Mg–$ZrFe_{1.4}Cr_{0.6}$ [28], $LaNi_5$ [29], ZrCrNi [30], Ti-V based bcc alloys [31].

Dehouche et al. demonstrated a detailed useful study on the doping effect of activated $Zr_{100\text{-}x}Ni_x$ alloy powders on the uptake/release kinetics of MgH_2 powders [32]. They pointed out that catalyzing MgH_2 with such activated $Zr_{100\text{-}x}Ni_x$ alloys led to remarkable increasing in the kinetics behavior of MgH_2. Their results have shown that the best kinetics properties of MgH_2 was obtained upon doping with 10 wt.% of eutectoid $Zr_{47}Ni_{53}$, however, mixing with different compositions (Zr_9Ni_{11} and Zr_7Ni_{10}) led also to significant improvement but not as much as the eutectoid composition [32]. They elucidated that when MgH_2/0.2 mol% Zr_7Ni_{10} ground with activated carbon, further kinetics improvement was attained.

Table 10.2 lists selected intermetallic compounds and alloys that have been efficiently used as catalytic agents to enhance the kinetic behavior of MgH_2. Further information of different intermetallic compounds catalytic agents are presented in different comprehensive review articles [20, 33–35].

Table 10.2 Doping effect of intermetallic compounds and alloys on the hydrogenation and dehydrogenation behavior of MgH_2 powders.

Catalyst used	Dose	Doping approach	Hydrogenation/ Dehydrogenation properties	Apparent activation energy (kJ/mol)	References
$LaNi_5$	5 wt.%	H_2-HEBM/40 h	**Absorption/** 20 bar/285°C: 5.1 wt.%/2700 s **Desorption/** 285°C: 5.1 wt.%/1800 s	NA	[26]
Mg_2Ni	5 mol%	Ar- HEBM/3 h	**Absorption /** 20 bar/220°C: 5.5 wt.%/30 min **Desorption/** 220°C: 5.5 wt.%/60 min	NA	[25]
$Ti_{0.4}Cr_{0.15}Mn_{0.15}V_{0.3}$	20 wt.%	Ar- HEBM/3 h	**Absorption/** 20 bar/350°C: 6.5 wt.%/300 min	71.2	[31]
ZrCrNi	10 wt.%	Ar- HEBM	**Absorption/** 5 bar/300°C: 5.8 wt.%/30 min **Desorption/** 0.1 bar/300°C: 5.8 wt.%/45 min	NA	[30]
TiAl	14 wt.%	H_2-RBM/4 h	**Absorption/** 1 bar/100°C: 4 wt.%/30 min **Desorption/** 0.1 bar/240°C: 4 wt.%/7 min	65	[36]
$Ti_{0.05}Cr_{0.20}V_{0.75}$	10 wt.%	H_2-RBM/4 h	**Absorption/** 1 bar/25°C: 3 wt.%/75 min **Desorption/** 0.01 bar/240°C: 5.5 wt.%/5 min	71.2	[37]
FeTi	5 wt.%	Ar- HEBM/8 h	**Absorption/** 30 bar/300°C: 5.8 wt.%/800 s **Desorption/** 1.58 bar/300°C: 2.57 wt.%/110 min	NA	[38]
FeTiMn	40 wt.%	Ar- HEBM/40 h	**Absorption/** 40 bar/80°C: 4 wt.%/3600 s **Desorption/** 3 bar/240°C: 2.57 wt.%/7200 min	NA	[38]

HEBM, High-energy ball milling; RBM, reactive ball milling; NA, not available.

10.2.2.2 New approach for doping MgH_2 with pure metals

In parallel to the worldwide efforts paid to enhance the behavior of MgH_2 with using a long list of metallic catalysts, researchers in Kuwait Institute for scientific Research (KISR) in Kuwait have recently developed a new method so-called in-situ catalyzation approach [39]. In this method, Ni-balls were used as milling media during RBM of Mg under 50 bar of H_2. Upon high-energy ball milling for 6 h, Ni shots (~~65 nm) worn out from the balls and introduced into the MgH_2 matrix, as shown in Fig. 10.1. Increasing RBM time (12.5–25 h) led to monotonical decrease in Ni particle size and improvement of the Ni particles distribution in MgH_2 matrix (Fig. 10.1D, G). Toward the end of the processing time (50 h) nano-Ni powders (~5 nm) were homogeneously embedded in the matrix to for form a uniform nanocomposite powders with Ni concentration of 5.5 wt.%, as displayed in Figs. 10.1J-L.

MgH_2/5.5 wt.% Ni nanocomposite powders obtained after short RBM time (25 h) revealed low values of decomposition temperature (218°C), where the apparent activation energy (E_a) was 75 kJ/mol [40]. In addition, hydrogenation (Fig. 10.2A) /dehydrogenation (Fig. 10.2B) kinetics measured at moderate temperature (275°C) for

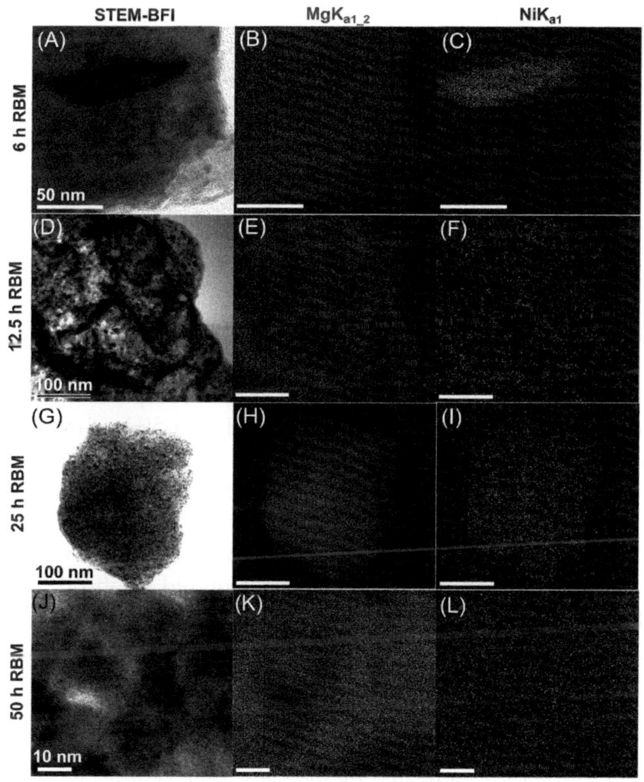

Figure 10.1 (A, D, G, J) STEM-BFIs, (B, E, H, K) Mg-Kα_{1-2}, and (C, F, I, L) Ni-Lα_1 of the powders obtained after different stages of RBM, using Ni-balls milling media [40].

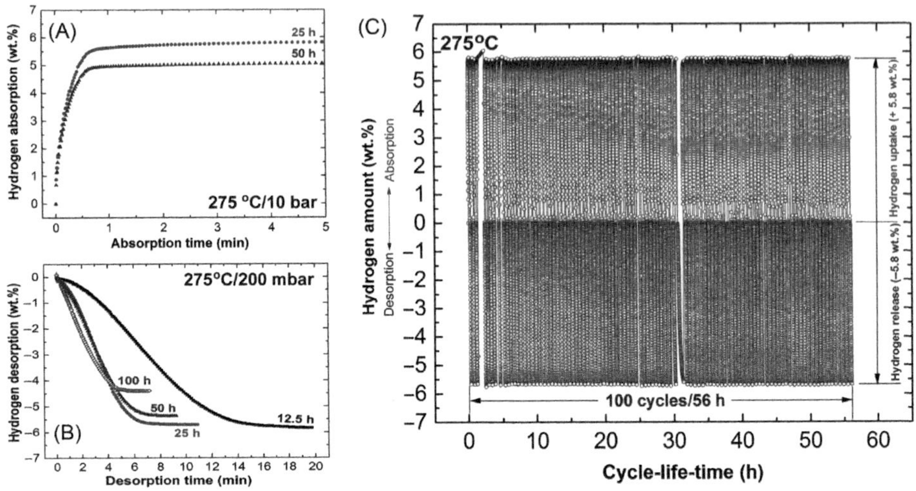

Figure 10.2 (A) Hydrogenation and (B) dehydrogenation kinetics examined at 275°C of MgH_2 powders obtained after different RBM time, using Ni-balls milling media. The Cycle-life-time conducted at 275°C for the sample obtained after 25 h of RBM time is presented in (C) [39].

25 h-sample to uptake and release 5.8 wt.% H_2 were 2.5 min and 8 min, respectively. The authors claimed that such fast kinetics behavior was attributed to the formation of homogeneous nanocomposite system, in which Ni nano-powders were adhered to all MgH_2 powders without exceptions [39]. The synthesized nanocomposite performed 100-continuous cycle-life time of hydrogen charging/discharging at 275°C within 56 h without failure or degradation, as elucidated in Fig. 10.2C.

More recently, Ti balls were used as milling tools upon RBM of Mg under hydrogen (50 bar) atmosphere [41]. Likewise, Ni-balls, Ti shoots were worn out from the balls and embedded in MgH_2 matrix to form uniform nanocomposite MgH_2/5.3 wt.% Ti powders after 50 h (Fig. 10.3A). The measured decomposition temperature and calculated E_a of this nanocomposite system obtained after 50 h of milling were 230°C and 72 kJ/mol, respectively [41]. The hydrogenation kinetics for the sample obtained after 50 h of RBM 50 h of RBM time (end product) was investigated under 10 bar of H_2 at applied temperatures of 100°C, 125°C, 150°C, and 175°C, as presented in Fig. 10.3B. The sample showed outstanding hydrogenation behavior at a very low temperature (100°C), indexed by the short time required to absorb 4.9 wt.% H_2 in 3 min (Fig. 10.3B). At 175°C, better storage capacity and kinetics were attained, as indicated by short time needed (3 min) to uptake 5.2 H_2, as shown in Fig. 10.3B.

The dehydrogenation kinetics behaviors for the sample obtained after 50 h of RBM time were investigated at different temperatures in the range between 175°C and 250°C (Fig. 10.3C). This sample possessed moderate dehydrogenation kinetics in a low temperature range (175–225°C), indicated by the short time (4 min) necessary to release −2.1 and −3 wt.% hydrogen, respectively, as displayed in Fig. 10.3C. Obviously, the kinetics of the sample was enhanced by increasing the dehydrogenation time to 7 min,

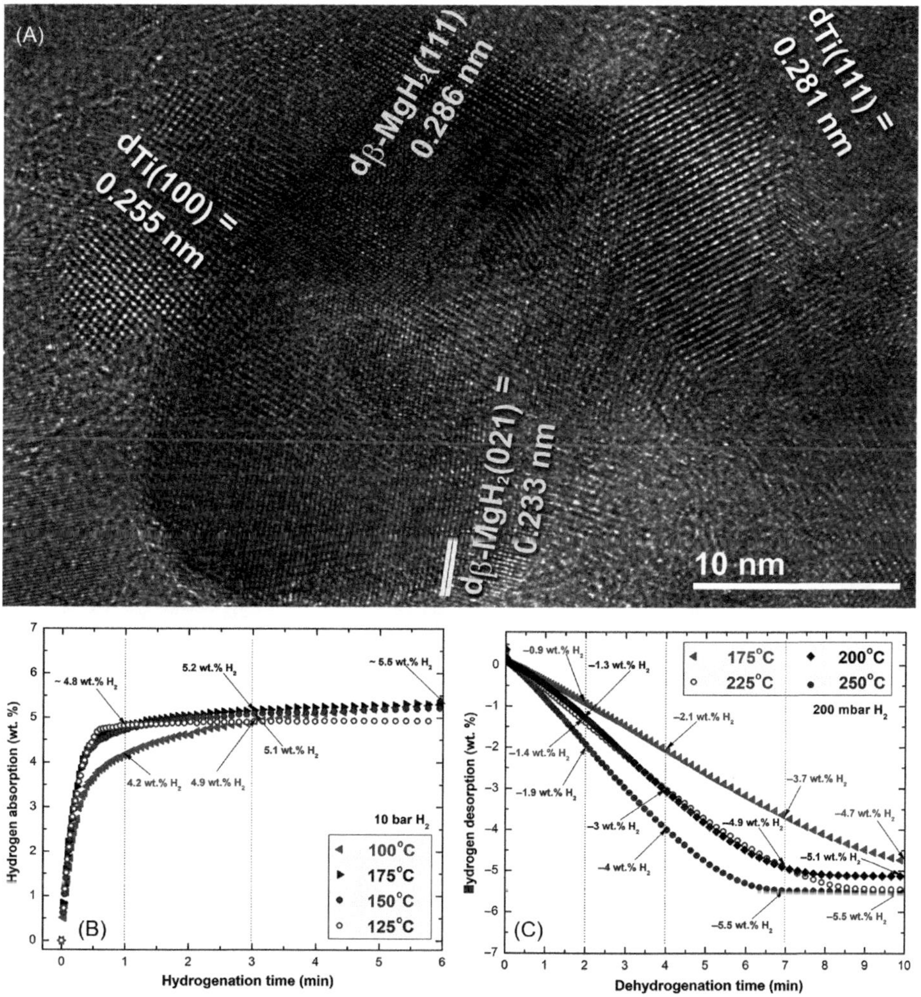

Figure 10.3 (A) FE-HRTEM image of MgH_2 powder containing about 5 wt.% Ti metallic particles after RBMing for 50 h, using Ti-balls milling media. The hydrogenation and dehydrogenation kinetics measured at different temperatures of 50 h-sample presented in (B) and (C), respectively [41].

where the hydrogen release increased with increasing applied temperature from −3.7 wt.% (175°C) to −4.9 wt.% (200°C and 225°C), respectively. The two samples measured at 225°C and 250°C reached to their saturation values of −5.1 and −5.5 wt.% H_2, respectively after 10 min, as indicated in Fig. 10.3C.

10.2.2.3 New intermetallic catalytic agents

Among intermetallic alloys used as catalytic agents for MgH_2 system, two binary systems of $Mn_{3.6}Ti_{2.4}$ [42] and $ZrNi_5$ [43] alloys showed excellent potential as catalysts.

The end-product of MgH_2/7 $Mn_{3.6}Ti_{2.4}$ nanocomposite powders obtained after 50 h of RBM were aggregated to form particles with apparent particle size of about 250 nm in diameter (Fig. 10.4A). These aggregates consisted of MgH_2 (Fig. 10.4B doped with nano-sized $Mn_{3.6}Ti_{2.4}$ dispersoids (Fig. 10.4C,D). At this final stage of ball milling, the $Mn_{3.6}Ti_{2.4}$ particles possessed lens-like morphology with particle size distribution laying in the range between 8 and 32 nm in diameter, as displayed in Figs. 10.4C,D.

Due to the good distribution of $Mn_{3.6}Ti_{2.4}$ nano-lenses in MgH_2 matrix, the end product of this system revealed good hydrogenation/dehydrogenation kinetics at moderate temperature of 275°C [42]. This is implied by the short time required to absorb and desorb ~5.3 wt.% H_2 within 2 min (Fig. 10.5A,B) and 7 min (Fig. 10.5C,D), respectively. Formation of ultrafine nano-scaled powder particles facilitated good hydrogen diffusion cyclability, as suggested by the long (1400 h) absorption/desorption cycle-life-time presented in Fig. 10.5E. However, a minor degradation (~0.3–0.4 wt.% H_2) in the hydrogen storage capacity was observed between 400 h (300 cycles) and 1400 h (1000 cycles), as elucidated in Fig. 10.5E. This minor degradation took place due to slight grain growth came off in the Mg/$Mn_{3.6}$ $Ti_{2.4}$ grains, as indicated in Fig. 10.5F,G [42].

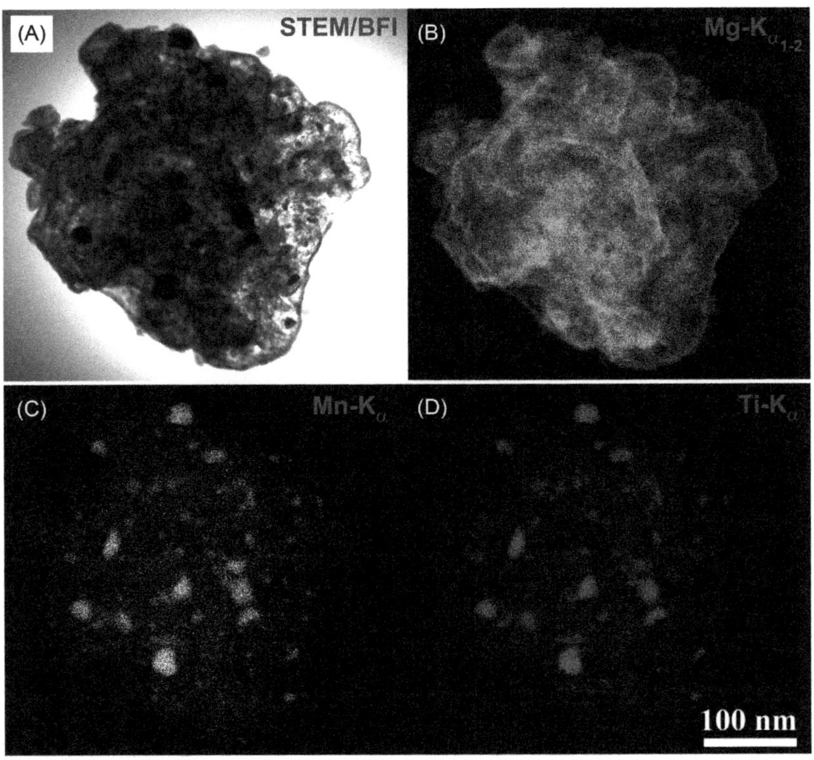

Figure 10.4 (A) STEM-BFI and the corresponding X-ray elemental mapping of (B) Mg, (C) Mn, and (D) Ti of nanocomposite MgH_2/7 wt.% $Mn_{3.6}Ti_{2.4}$ powders obtained after 50 h of milling time [42].

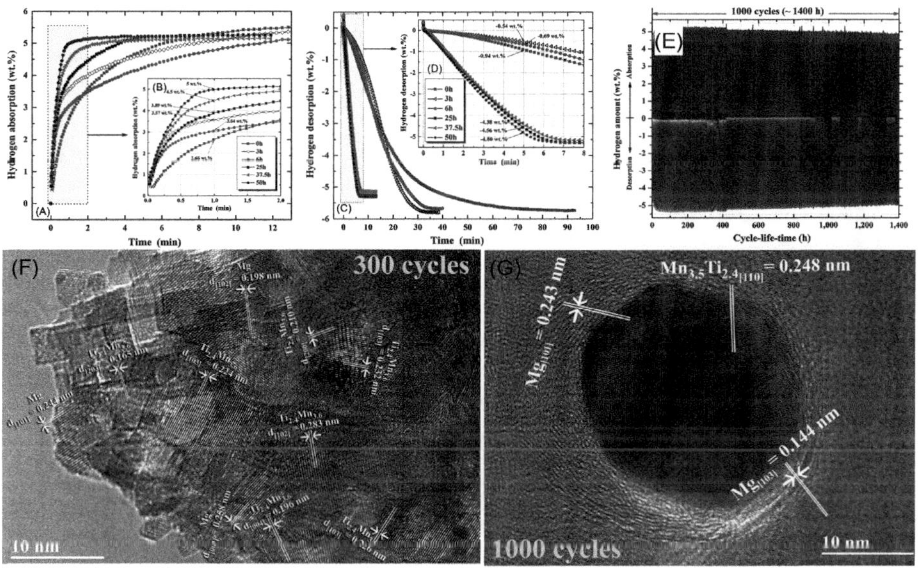

Figure 10.5 Effect of ball milling time on the (A) hydrogenation and (B) dehydrogenation kinetics of nanocomposite MgH_2/7 wt.% $Mn_{3.6}Ti_{2.4}$ powders. The hydrogenation and consequent dehydrogenation curves of 1000 complete cycles for the nanocomposite powders obtained after 50 h of milling time are presented in (C). The HRTEM images of the nanocomposite powders subjected to 300 and 1000 are presented in (F) and (G), respectively [42].

More recently, nanocomposite MgH_2/10 wt.% $ZrNi_5$ powders were fabricated by RBM of as-synthesized nanocrystalline MgH_2 doped with 10 wt.% of arc-melt $ZrNi_5$ powders for 50 h [43]. Fig. 10.6 displays the temperature and time effects on the hydrogen absorption (A, B) and consequence desorption (C, D) kinetics of nanocomposite MgH_2/10 wt.% $ZrNi_5$ powders obtained after 50 h of the ball milling time. The synthesized nanocomposite powders showed good potential for absorbing hydrogen gas in a wide temperature range (50–275°C) with different hydrogen capacity, as shown in Fig. 10.4A. After 0.5 min, the powders absorbed 4.6 wt.% H_2 at 250°C and 275°C (Fig. 10.6B). In contrast, at this absorption time, the samples examined at 150°C, 100°C, and 50°C revealed modest hydrogen gas uptake reached to 0.8, 0.1, and 0.0 wt.% H_2, respectively (Fig. 10.6B). Increasing the hydrogen gas uptake time to 1 min led to a magnificent improvement of the absorbed hydrogen gas concentration, reached to 5.1 and 5.3 wt.% at applied temperature of 250°C and 275°C, respectively as shown in Fig. 10.6B. Those samples examined 150°C and 100°C showed significant improvement, suggested by a dramatic increasing in the hydrogen gas absorbed that reached to 3.8 and 2.2 wt.%, respectively (Fig. 10.6B). A marginal improvement on the absorption kinetics could be observed for that sample measure at 50°C that absorbed only 0.2 wt.% H_2 after 1 min, as displayed in Fig. 10.6B. Increasing the absorption time to 30 min (Fig. 10.6A) did not lead to increase the absorbed amount of hydrogen for the sample examined at 250°C and 275°C that tended to be saturated at 5.3 wt.% H_2. An outstanding improvement in the hydrogen concentration (4.4 wt.%)

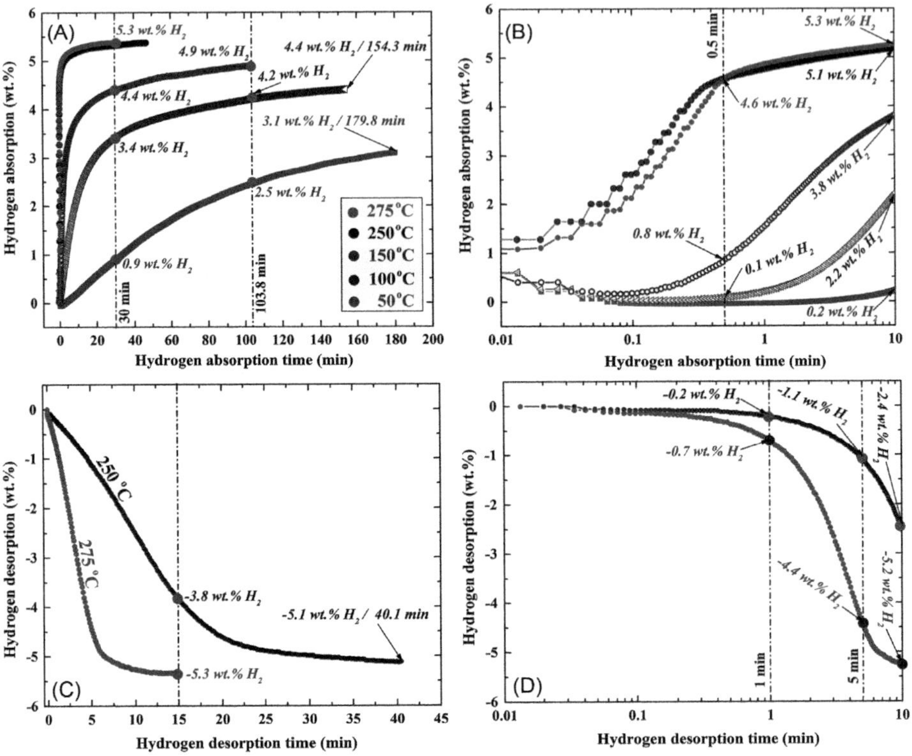

Figure 10.6 Temperature and time effects on the (A, B) hydrogenation, and (C, D) dehydrogenation kinetics of nanocomposite MgH_2/10 wt.% $ZrNi_5$ powders obtained after RBM for 50 h [43].

absorbed at 150°C could be attained after 30 min and tended to increase to a higher value (4.9 wt.%) after 103.8 min (Fig. 10.6A). After 30 min, the samples examined at 100°C and 50°C, absorbed about 3.4 and 0.9 wt.% H_2, as shown in Fig. 10.6A. The hydrogen concentrations of these samples reached to 4.2 and 2.5 wt.% after 103.8 min (Fig. 10.6A). About 180 min was required for the samples measured at 100°C and 50°C to absorb 4.4 and 3.1 wt.% H_2, as elucidated in Fig. 10.6A.

The corresponding desorption's kinetics of the nanocomposite powders investigated at 250°C and 275°C are shown in Fig. 10.6C,D. The sample examined at 275°C showed excellent desorption kinetics, indexed by the relatively short time (~5 min) required to release about −4.4 wt.% of hydrogen (Fig. 10.6D). This sample reached to a very closed saturation value (−5.2 wt.% H_2) after only 10 min, as shown in Fig. 10.6D. The sample measure at 250°C took about 5 and 10 min to release about −1.1 and −2.4 wt.% H_2, respectively (Fig. 10.6D). After 15 min of desorption time (Fig. 10.6C), the samples examined at 275°C and 250°C released about −5.3 and −3.8 wt.% H_2, respectively. About 40 min was required for the sample examined at 250°C to desorb its full hydrogen storage capacity (−5.1 wt.% H_2), as shown in

Fig. 10.6C. This hydrogen storage nanocomposite system revealed good cyclability (600 complete cycles), conducted at 275°C without serious kinetics degradation [43].

Metal oxides possess strong catalytic effect, leading to improve the hydrogenation/dehydrogenation kinetics of MgH_2. One of the earliest work was reported by Oelerich et al. [44] when they prepared different systems of nanocrystalline Sc_2O_3, TiO_2, V_2O_5, Cr_2O_3, Mn_2O_3, Fe_3O_4, CuO, Al_2O_3, SiO_2 powders, using high-energy ball milling. 5 wt.% of nanocrystalline oxides powders obtained after 20 h of milling, were individually mixed with MgH_2 powders and then high-energy ball milled for 100 h under Ar atmosphere. The results showed that all nanocrystalline oxides used in their study had a remarkable effect on improving the absorption/desorption kinetics of MgH_2 [44]. They reported that composite material containing Fe_3O_4 shows the fastest absorption kinetics at 300°C followed by V_2O_5, Mn_2O_3, Cr_2O_3, and TiO_2.

In 2007, Aguey-Zinsou et al. ball-milled MgH_2 with 17 wt.% Nb_2O_5 for 200 h, using a high-energy ball mill [45]. Based on their results, pure MgH_2 revealed fast hydrogen sorption kinetics (Table 10.3) after a long milling time (700–1000 h) with Nb_2O_5. They attributed this remarkable improvement to a so called "process control agent" in which Nb_2O_5 played the three roles as lubricant, dispersing and cracking agent [45]. Properties related to selected MgH_2/metal oxides composite systems are shown in Table 10.3.

It has been recently demonstrated by El-Eskandarany et al. [53] that intensive high-energy ball milling of MgH_2 powders doped with different wt.% of Nb_2O_5 for long milling time led to outstanding improvement on the kinetics of gas uptake and release [1]. In their experiments, the as-prepared MgH_2 powders obtained after 200 h of RBM were simply mixed firstly with 2.5, 5, and 10 wt.% of Nb_2O_5 nanopowders. The mixed powders were then high-energy ball milled under hydrogen for 50 h. The hydrogenation/dehydrogenation kinetics were examined for pure MgH_2, and MgH_2/x wt.% Nb_2O_5 (x; 2.5, 5, 10) powders in temperature range between 200–250°C, as displayed in Fig. 10.7 [1].

The absorption kinetics investigated at 200°C, 225°C and 250°C for pure MgH_2 nanocrystalline powders obtained after 200 h of RBM time (Fig. 10.7A) was faster when compared with the powders obtained after short milling time [1]. After this long milling run, the powders were subjected to dramatic grain refining associated together with severe lattice imperfections. This led to drastic powder disintegration and the formation of ultrafine grains ranging between 5 and 8 nm in diameter [1]. These favorable morphological properties led to enhance the uptake kinetics even at relatively low temperature of 200°C. For example, at this temperature the sample absorbed about 3.5 wt.% H_2 within 500 s and approached 6.3 wt.% H_2 after about 2500 s, as shown in Fig. 10.7A. When these values are compared with those obtained at 225°C (4.9 wt.% H_2/500 s and 6.8 wt.% H_2/2500 s) and 250°C (5.49 wt.% H_2/500 s and 6.81 wt.% H_2/2500 s), it can be realized that little increasing of applied temperature led to a significant improve on the kinetics of hydrogen absorption for nanocrystalline MgH_2 powders (Fig. 10.7A). Nanocrystalline MgH_2 powders obtained after 200 h of RBM possessed slow desorption kinetics at 200°C and 225°C, however, they showed moderate kinetics at 250°C, indicated by releasing of 6.51 wt.% H_2 after 7500 s, as presented in Fig. 10.7(B). Further comparison between the hydrogen storage properties

Table 10.3 Doping effect selected metal oxides on the hydrogenation and dehydrogenation behavior of MgH_2 powders.

Catalyst used	Dose	Doping approach	Hydrogenation/Dehydrogenation properties	Apparent activation energy (kJ/mol)	References
Nb_2O_5					
	17 wt.%	Ar-HEBM/ZrO_2 milling tools/200 h	**Absorption**/300°C: 6 wt.%/120 s **Desorption**/300°C: 6 wt.%/300 s 10 cycles	NA	[44]
	5 wt.%	Ar- HEBM/24 h MgH_2 + 20 min Nb_2O_5	**Absorption**/ 10 bar/300°C: 5.2 wt.%/78 s **Desorption**/0.1 kPa/300°C: 6 wt.%/300 s	NA	[45]
	10 wt.%	Ar- HEBM	**Absorption**/ 5 bar/300°C: 5.8 wt.%/30 min **Desorption**/ 0.1 bar/300°C: 5.8 wt.%/350 min	NA	[46]
	14 wt.%	H_2-RBM/ 4 h	**Absorption**/ 1 bar/100°C: 4 wt.%/30 min **Desorption**/ 0.1 bar/240°C: 4 wt.%/7 min	65	[47]
TiO_2					
	20 wt.%	H_2-RBM/1 h/8 bar	**Absorption**/20 bar/350°C: 3.8 wt.%/2 min **Desorption**/1 bar/350°C: 4.4 wt.%/8.5 min	NA	[48]
Fe_2O_3					
	10 wt.%	H_2-RBM/10 bar	**Absorption**/12 bar/320°C: 4.22 wt.%/10 min	NA	[49]
Cr_2O_3					
	1.2 wt.%	Ar-HEBM	**Absorption**/10 bar/300°C: 6.4 wt.%/5 min **Desorption**/0.25 bar/300°C: 6.4 wt.%/20 min	NA	[50]

Fe_3O_4@Graphene					
	5 wt.%		**Absorption**/15 bar/290°C: 6.2 wt.%/2.5 min 25 cycles	90.53	[51]
La_2O_3					
	12 wt.%	Ar-HEBM/ 20 h	**Absorption**/10 bar/300°C: 4.75 wt.%/8 min **Desorption**/0.3 bar/300°C: 6 wt.%/25 min	NA	[52]
V_2O_5					
	27	Ar-HEBM/ 20 h	**Absorption**/8.4 bar/300°C: 4 wt.%/1 min **Desorption**/0.003 bar/300°C: 4.7 wt.%/4 min	NA	[44]

HEBM, High-energy ball milling; RBM, reactive ball milling; NA, not available.

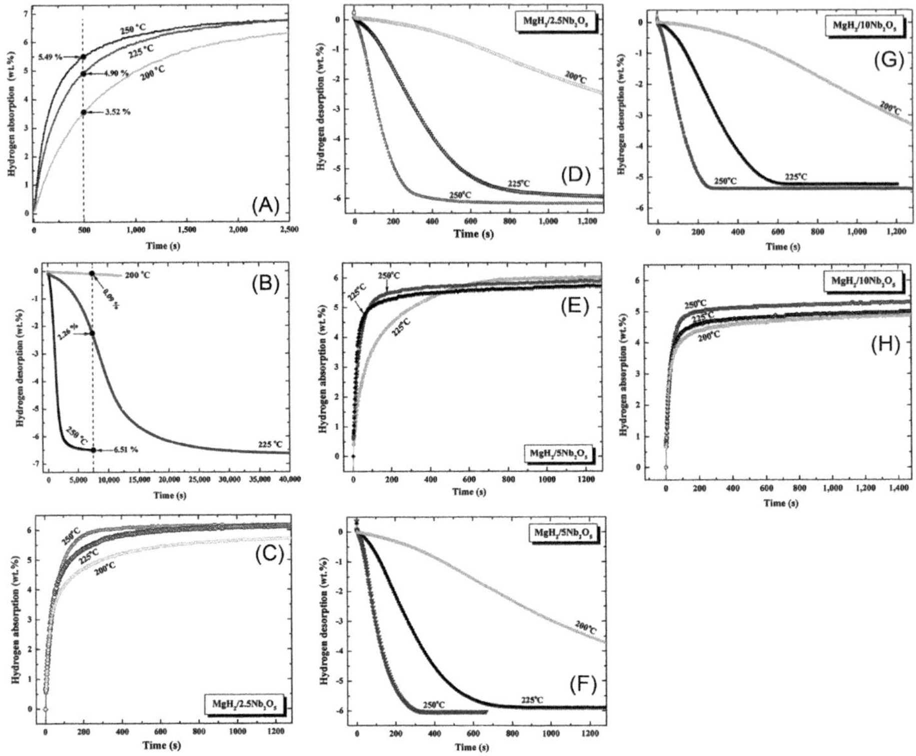

Figure 10.7 Effect of Nb_2O_5 doping dose and applied temperature on the (A, C, E, G) hydrogenation, and (B, D, F, H) dehydrogenation kinetics of MgH_2 powders [1].

of starting synthetic MgH_2 and MgH_2/x wt.% Nb_2O_5 (*x*; 2.5, 5, 10) is presented in Table 10.4.

The authors attributed the improvement achieved in kinetics behavior of MgH_2 powders upon mechanically induced doping with metal oxide to the micro-cracking agent media role played by Nb_2O_5. Moreover, it is believed that Nb_2O_5 can act as effective catalyst in recombination with hydrogen molecules and accelerate the promotion of gas-solid reaction [1].

10.2.2.4 Catalyzation with metal/metal oxide nanocomposite powders

The as prepared MgH_2 was doped with 5 wt.% Ni and 5 wt.% Nb_2O_5 and high-energy ball milled under H_2 for 50 h. The results showed that the hydrogenation/dehydrogenation kinetics of MgH_2 were enhanced with using binary Ni/Nb_2O_5 system, implied by the shorter time required for absorption when compared with MgH_2/5 Nb_2O_5 system [1]. This can be attributed to elemental Ni additive, which is well known as a hydrogen splitter element. The cycle-life-time of nanocomposite MgH_2/5Ni/5Nb_2O_5 powders is displayed in Fig. 10.8. This ternary nanocomposite system shows an excellent storage properties, indexed by high cyclic stability even after about 94 h (~180 cycles)

Table 10.4 Hydrogen storage properties of nanocrystalline MgH_2-based Nb_2O_5 nanopowders prepared by mechanically-induced doping process, using high-energy ball mill [1].

Catalyst used	Dose	Doping approach	Hydrogenation/Dehydrogenation properties
MgH_2			
		H_2-RBM/50 bar/200 h	**Absorption/** 200°C: 6.3 wt.%/10 bar/2500 s 225°C: 6.4 wt.%/10 bar/2500 s 250°C: 6.5 wt.%/10 bar/2500 s **Desorption/** 225°C: 6.4 wt.%/0.02 bar/40000 s 250°C: 6.5 wt.%/0.02 bar/7500 s
Nb_2O_5			
	2.5 wt.%	H_2-RBM/50 bar/50 h	**Absorption/** 200°C: 5.7 wt.%/10 bar/1250 s 225°C: 5.9 wt.%/10 bar/1000 s 250°C: 6.1 wt.%/10 bar/600 s **Desorption/** 200°C: 2.5 wt.%/0.02 bar/1250 s 225°C: 5.8 wt.%/0.02 bar/1250 s 250°C: 6.1 wt.%/0.02 bar/800 s
	5 wt.%	H_2-RBM/50 bar/50 h	**Absorption/** 200°C: 6.0 wt.%/10 bar/1250 s 225°C: 6.0 wt.%/10 bar/800 s 250°C: 6.0 wt.%/10 bar/600 s **Desorption/** 200°C: 3.7 wt.%/0.02 bar/1250 s 225°C: 6.0 wt.%/0.02 bar/1000 s 250°C: 6.0 wt.%/0.02 bar/280 s
	10 wt.%	H_2-RBM/50 bar/50 h	**Absorption/** 200°C: 4.8 wt.%/10 bar/1400 s 225°C: 5.0 wt.%/10 bar/1200 s 250°C: 5.4 wt.%/10 bar/500 s **Desorption/** 200°C: 3.3 wt.%/0.02 bar/1250 s 225°C: 5.3 wt.%/0.02 bar/700 s 250°C: 5.4 wt.%/0.02 bar/280 s
Ni/Nb_2O_5			
	5 wt.%/5 wt.%	H_2-RBM/50 bar/50 h	**Absorption/** 200°C: 5.2 wt.%/10 bar/250 s 225°C: 5.2 wt.%/10 bar/220 s 250°C: 5.4 wt.%/10 bar/190 s **Desorption/** 200°C: 5.4 wt.%/0.02 bar/2000 s 225°C: 5.5 wt.%/0.02 bar/500 s 250°C: 5.5 wt.%/0.02 bar/190 s

HEBM, High-energy ball milling; RBM, reactive ball milling; NA, not available.

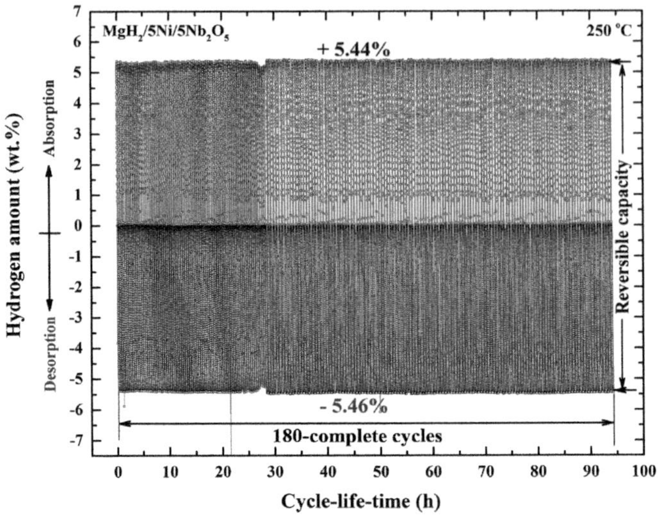

Figure 10.8 Hydrogen absorption and consequent desorption curves of 180 complete cycles of nanocomposite MgH_2/5 wt.% Ni/5 wt.% Nb_2O_5 powders examined at 250°C under hydrogen pressure of 10 bar (absorption) and 200 mbar (desorption) [1].

with no obvious degradation on the hydrogen storage capacity, which exhibited nearly constant absorption and desorption values of +5.44 and −5.46 wt.% H_2, respectively, as displayed in Fig. 10.8.

10.2.2.5 Catalyzation with titanium carbide nanopowders

Different compounds related to this group of refractory materials have had good role for improving the kinetics behavior of MgH_2. Many authors have pointed out that such hard spherical catalytic powders led to dramatic refining of MgH_2 powders, implied to a significant improvement on the hydrogen storage property of Mg. Micromilling media refer to those abrasive hard material powders such as TiC that when ball milled with MgH_2 powders ultrafine hydride powders were obtained [53]. A group in KISR, succeed to prepare ultrafine spherical $Ti_{55}C_{45}$ particles (~15 nm in diameter; Fig. 10.9A, using mechanically induced self-propagating reaction [54].

The sphericity and fineness of TiC allowed these hard powders to be embedded and occupied catalytic sites at the grain boundaries of Mg matrix, as presented in Fig. 10.9B. Metal hydride powders played vital role of hindering Mg grains from growing during practical use at moderate temperature of 275°C [55]. Accordingly, the end-product of nanocomposite MgH_2/5 wt.% TiC obtained after 50 h of ball milling revealed excellent absorption/desorption cyclability of 530 complete cycles with no degradation, as shown in Fig. 10.9 C [55]. In addition, during the absorption/desorption cycles the hard spherical TiC powders penetrated into MgH_2 surfaces to form groves, as shown in Fig. 10.9D. Formation of such grooves facilitated excellent hydrogen diffusion paths [53]. The as fabricated nanocomposite MgH_2/5TiC powders possessed

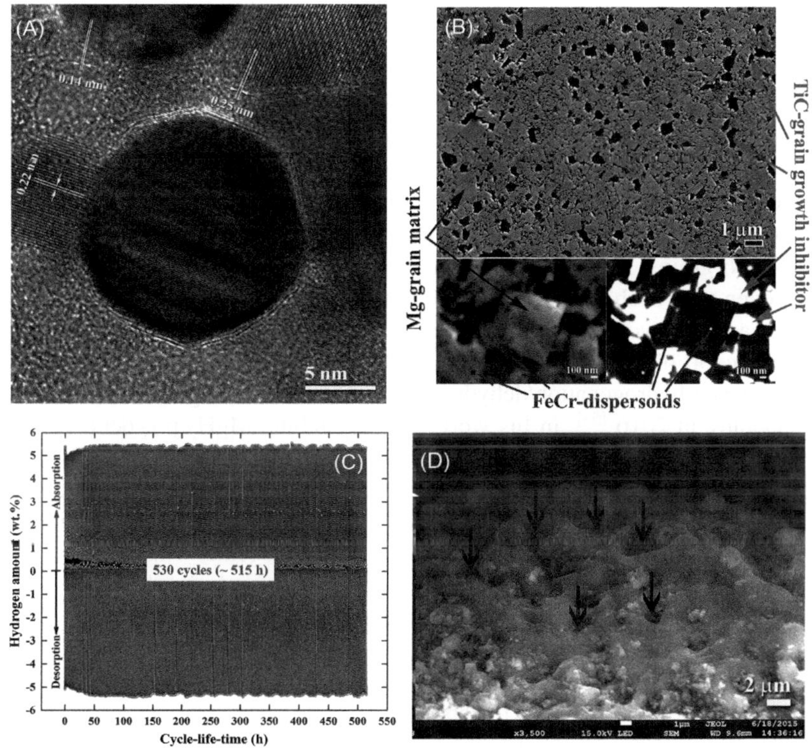

Figure 10.9 (A) FE-HRTEM image of as-prepared $Ti_{55}C_{45}$ particle, using mechanically induced self-propagating reaction [54], The low magnification FE-SEM, (B) high-magnification FE-SEM, and (C) backscattered electron (BSE) micrographs of the cross-sectional view for the nanocomposite powders obtained after 50 h of ball milling and then hot-pressed in vacuum at 525°C under an uniaxial pressure of 125 MPa. (A) Hydrogenation and consequent dehydrogenation curves of 697 complete cycles are shown together in (B). The hydrogenation and consequent dehydrogenation curves conducted at 275°C of 530 complete cycles performed within 515 h for nanocomposite MgH_2/5TiC powders obtained after 50 h of ball milling time is shown in (C). The FE-SEM micrograph of the powders obtained after completion of 530 cycles is presented in (D) [55].

superior hydrogenation /dehydrogenation characteristics, suggested by a low value of the activation energy (97.74 kJ/mol), and the short time required for achieving a complete absorption (6.6 min) and desorption (8.4 min) of 5.5 wt.% H_2 at moderate temperature of 275°C under a hydrogen gas pressure ranged from 0 bar to 8 bar [53].

10.2.3 Catalyzation with metastable phases of Zr-based nanopowders

The idea of utilizing Zr-based metastable phases for improving the hydrogenation/ dehydrogenation kinetics of MgH_2 system dates back to 2015, when El-Eskandarany et al. [56] doped the hydride phase with 10 wt.% of big-cube Zr_2Ni nanopowders. The results have shown that high-energy milling of MgH_2 with big-cube hard particles for 50 h led to superior storage properties of MgH_2 powders, as indicated by the short time required to absorb and desorb 5.1 wt.% H_2 within 100 s and 613 s, respectively [57]. At this temperature, the synthesized nanocomposite powders possessed excellent absorption/desorption cyclability of 2546 complete cycles within 1250 h [56]. However, a minor degradation (~0.5 wt.% H_2) in the hydrogen storage capacity was observed after 2546 h of the cycle-life-time, due to a minor grain growth came off in the metallic Mg and Zr_2Ni grains. It is worth mentioning that Zr_2Ni maintained its big-cube (fcc) structure even after such long cyclic processing time. This can indicate a good stability against the applied cyclic pressure and temperature.

The first report of using metallic glassy (MG) $Zr_{70}Ni_{20}Pd_{10}$ nanopowders as catalytic agent to enhance hydrogenation/dehydrogenation kinetics of MgH_2 was introduced by El-Eskandarany in 2016 [58]. In his work, nanocrystalline MgH_2 powders prepared by RBM technique were doped with a small mole fractions (5 wt.%) of MG-$Zr_{70}Ni_{20}Pd_{10}$ powders and then high energy ball-milled for 50 h under 50 bar of H_2. The HRTEM image taken near the edge of MgH_2/5 wt.% amorphous $Zr_{70}Ni_{30}Pd_{10}$ composite particle obtained after 50 h of RBM time is shown in Fig. 10.10A together with the corresponding NBDP (Fig. 10.10B). Overall, the composite powders obtained after this stage of milling consisted of continuous amorphous matrix (maze-like morphology shown in Fig. 10.10A) hosting ultrafine nanoclusters (~4 nm in diameter) of order-structure (related to MgH_2). It is worth mentioning that the MgH_2 grains were distributed into the metal glassy powders in a segregation fashion with the absence of agglomerates or aggregated grains, as displayed in Fig. 10.10A. The NBDP shows halo-diffraction pattern related to amorphous $Zr_{70}Ni_{30}Pd_{10}$ coexisted with spot-like pattern came from nanocrystalline γ- and β-MgH_2 phases oriented in different axial directions [59].

The end-product (50 h) of MgH_2/5 wt.% MG-$Zr_{70}Ni_{20}Pd_{10}$ nanocomposite powders consisted of fine powders in the range between 180 to 500 nm, as presented in Fig. 10.11A. The STEM-BFI and the corresponding STEM-DFI of a selected nanocomposite MgH_2/5 wt.% MG-$Zr_{70}Ni_{30}Pd_{10}$ powder particle are shown in Figs. 10.11A,B, respectively. The powders had nearly spherical-like morphology (Fig. 10.11B), containing fine lenses of less than 10 nm in diameter that were homogeneously distributed

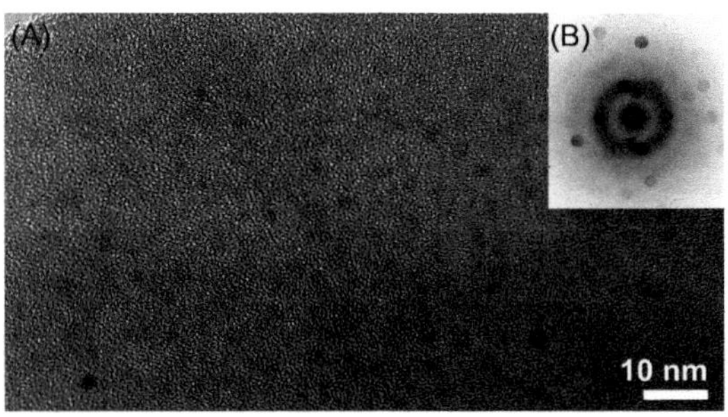

Figure 10.10 (A) HRTEM image of MgH_2 powders doped with 5 wt.% $Zr_{70}Ni_{30}Pd_{10}$ and milled for 50 h, (B) corresponding NBDP [58].

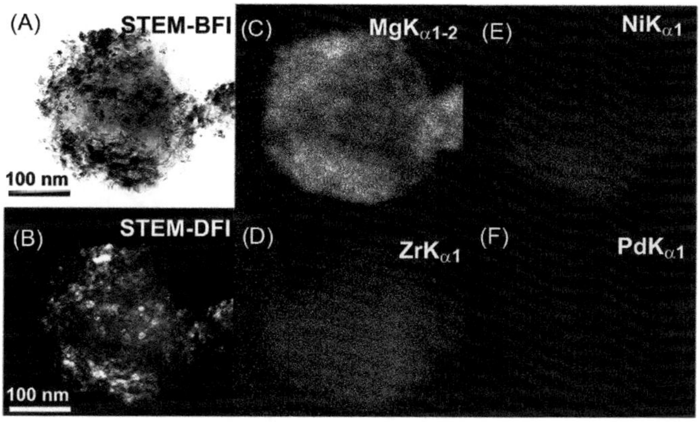

Figure 10.11 STEM-BFI and –DFI images for MgH_2 powders doped with 5 wt.% $Zr_{70}Ni_{30}Pd_{10}$ are displayed in (A, B), respectively. The X-ray-elemental mapping for Mg, Zr, Ni and Pd corresponding to (A) are presented in (C–F), respectively [58].

into the metallic glassy matrix (Fig. 10.11C,D). The fine spherical lenses were corresponding to Mg (Fig. 10.11C), where the elemental mapping of Zr, Ni, and Pd are presented in Fig. 10.11D–F, respectively. The EDS-analysis indicated excellent elemental distribution beyond the sub-micro level [59].

Fig. 10.12A,B show the pressure-composition-temperature (PCT) curves investigated at 200°C, 350°C for nanocomposite MgH_2/5 wt.% MG- $Zr_{70}Ni_{30}Pd_{10}$ powders obtained after 50 h of ball milling (Fig. 10.12A) and MgH_2 powders obtained after 6 h of RBM time (Fig. 10.12B), respectively. Successful hydride formation (complete absorption) of pure MgH_2 sample, high temperature (350°C) and pressure (P_{abs}, in the range between 200 mbar to 40 bar) were required to absorb about 4.7 wt.% H_2, as shown in Fig. 10.12B. This indicates that the sample requires the application of higher temperature to get its theoretical hydrogen storage capacity (7.6 wt.% H_2). Pure MgH_2 showed poor dehydrogenation behavior, indexed by exhibiting a clear pressure hysteresis with significant large gabs (~17 bar) between the pressure needed for absorption (hydride formation), P_{abs} and the pressure required for hydride decomposition, P_{des}, as presented in Fig. 10.12B. In addition, pure MgH_2 powders obtained after 6 h of RBM time failed to release its stored hydrogen content completely even after 12 h of the desorption time (Fig. 10.12B).

In contrast to pure MgH_2 sample, the nanocomposite powders showed an excellent PCT hydrogenation/dehydrogenation curves, indexed by the complete hydrogen sorption/desorption event at lower temperature (200°C) and pressure (50 mbar to 10 bar), as shown in Fig. 10.12A. At such relative low temperature and pressure, the nanocomposite sample reached to a higher value of hydrogen storage capacity (5.8 wt.%) when compared with pure MgH_2 (4.7 wt.% H_2). Moreover, the plateau region for the nanocomposite sample was very flat with negligible slope with the absence of multistep hydrogenation/dehydrogenation, as shown in Fig. 10.12A. In addition, the sample succeed to achieve complete desorption through an almost flat PCT curve with minimal difference value (~72 mbar) between P_{abs} and P_{des}, as shown in Fig. 10.12A.

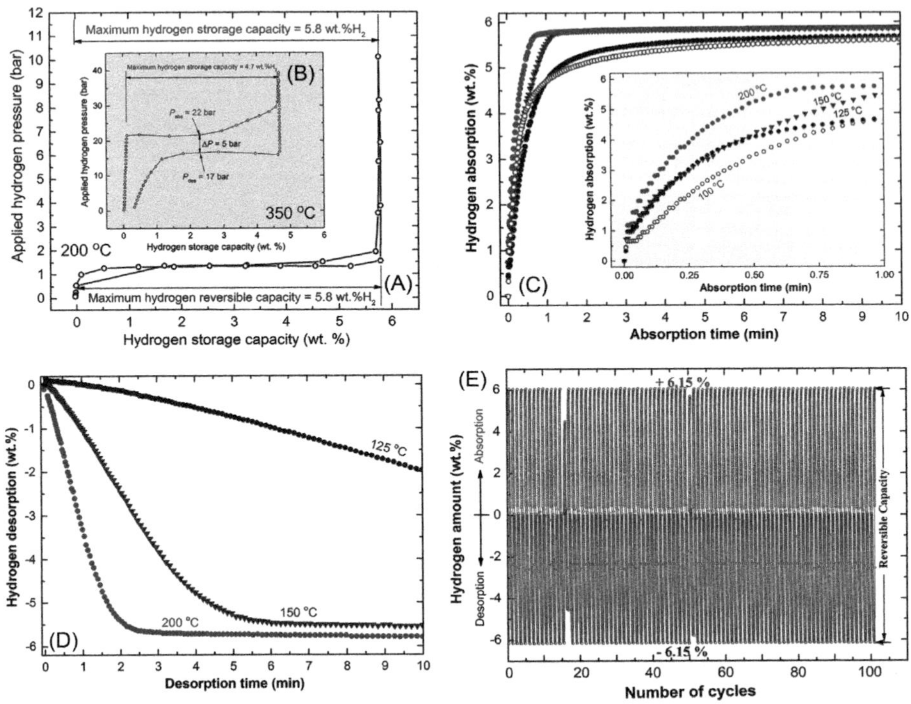

Figure 10.12 The PCT curves for the end-products of ball-milled MgH_2/5 wt.% metallic glass $Zr_{70}Ni_{20}Pd_{10}$ nanocomposite and MgH_2 powders are shown in (A,B), respectively. Whereas the PCT experiment conducted at 200°C for the nanocomposite powders (A), it was accomplished at 350°C for MgH_2 system (B). Hydrogenation/dehydrogenation kinetics of MgH_2/5 wt.% metallic glass $Zr_{70}Ni_{20}Pd_{10}$ nanocomposite and MgH_2 powders as a function of applied temperature, are presented in (C) and (D), respectively. Cycle-life-time test for the end product of nanocomposite powders conducted at 200°C is displayed in (E) [58].

Complete hydrogenation (Fig. 10.12C)/dehydrogenation (Fig. 10.12D) processes were successfully achieved with fast kinetics to charge/discharge about 6 wt.% H_2 at 200°C within 1.18 min/3.8 min, respectively [58]. In addition, this new nanocomposite system shows high performance of achieving continuous 100 hydrogen charging/discharging cycles without degradation, as elucidated in Fig. 10.12E.

10.2.3.1 Mechanism of enhancing MgH_2 kinetics upon doping with metallic glassy abrasive nanopowders

El-Eskandarany claimed that during the early stage of ball milling MgH_2 doped with abrasive MG-$Zr_{70}Ni_{30}Pd_{10}$ nanopowders (micro-milling media), the ball-powder-ball collusions (Fig. 10.13A) led to break down the large MgH_2 powder particles and assisted adherence of the fine powders onto the surface of MgH_2 particles (Fig. 10.13B) [58]. At this stage of milling, the hard nanopowders penetrated the oxide layer formed on

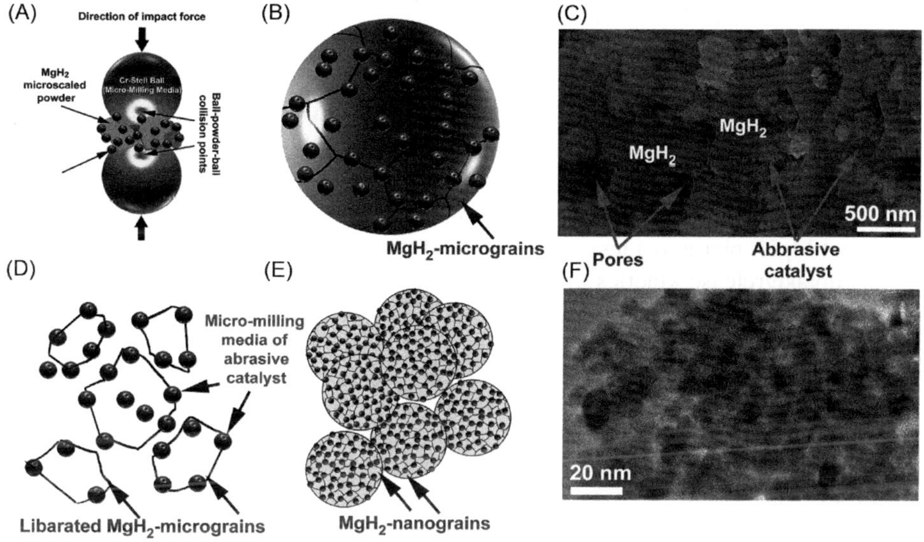

Figure 10.13 A schematic illustration presents a mixture of MgH_2 and metallic glassy $Zr_{70}Ni_{20}Pd_{10}$ powders that were charged into a stainless steel vial together with stainless steel balls. The large MgH_2 powders were subjected to sever mechanical deformation upon high-speed ball-powder-ball collusions (A), leading to disintegrated into smaller particles. The *red-colored spheres* (gray in print version) shown in the (B) represent the ultrafine metallic glassy powders adhered on the surface of the metal hydride powders (B). These hard spherical metallic glassy powders penetrated the surface of MgH_2 powders and created pores (C). Metallic glassy powders played important role as abrasive micro-milling media and led to liberate MgH_2 grains (D). Continuous disintegration was taking place with further increasing the milling time (E), where the end product possessed ultrafine morphological structure (F) [58].

the surfaces of MgH_2 powders to create micro-holes on their surfaces (Fig. 10.13C). Increasing the ball milling time led to "migrate" a large volume fraction of the abrasive milling media powders through the cavities and micro-channels created in the body of MgH_2 particles and located at the their grains. Hence, they break up the large MgH_2 grains along their weak grain boundary zones (Fig. 10.13D) and forming finer powders with large volume fraction of nanostructured grains (Fig. 10.13E).

Since the hydrogen diffusion is much faster along the grain boundaries when compared with inside grains, the hydrogenation/dehydrogenation kinetic behaviors of MgH_2 were gradually improved with increasing the number of "liberated" grains. During the last stage of ball milling, the role of ball-powder-ball collusions on achieving further refining of MgH_2 powder particles was almost absent since the size of the powders became ultrafine (less than 1 μm). Accordingly, the final refining process was attained by the nanosized abrasive milling media, which were homogeneously distributed within MgH_2 to form typical homogeneous nanocomposite powders (Fig. 10.13F).

10.3 Combination of cold rolling and ball milling for improving the kinetics behavior of MgH_2 powders

In spite of the beneficial effects obtained upon doping MgH_2 with catalytic agents, these additives always lead to significant decreasing of the hydrogen storage capacity of [60]. Apart from doping MgH_2 powders with catalysts, it has been experimentally demonstrated by some authors that changing the crystal structure of stable β-tetragonal MgH_2 phase to a less stable phase of γ-orthorhombic MgH_2 led to improve the gas uptake/release kinetics and decrease the hydrogenation temperature without drastic decreasing of the storage capacity [60–62]. β-to-γ phase transformations can be attained via severe plastic deformation, SPD [63] at ambient temperature by different approaches such as high-energy ball milling, HEBM [64], cold rolling, CRing [65], equal channel angular pressing, ECAP [66], and high-pressure torsion, HPT [67]. A common result of these employed techniques is the formation of nanocrystalline phase along with introducing high intensity defects, leading to increase of grain boundaries density. Presence of these defects in the Mg lattice leads to create nucleation points for hydrogenation, where existence of large number of grain boundaries assists fast diffusion pathways for hydrogen [68].

More recently, MgH_2 powders were charged and sealed in stainless steel (SUS304) tube (0.8 cm diameter and 20 cm length) inside a glove box under helium gas atmosphere, as shown in Fig. 10.14A. The tube contained MgH_2 powders were severely

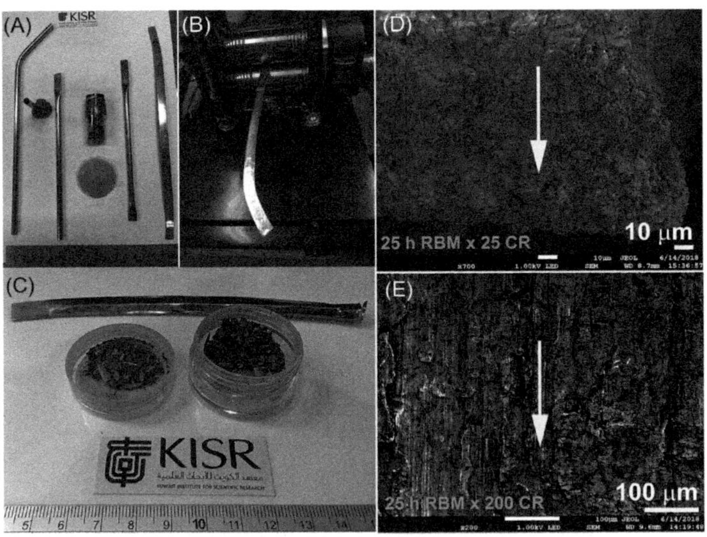

Figure 10.14 (A) The as RBMed MgH_2 powders obtained after 25 h of high energy ball milling under hydrogen were charged into SUS304 tubes and then were cold-rolled for 200 passes (B). (C) A photograph of the powders obtained after 25 passes. FE-SEM micrographs taken at accelerated voltage of 1 kV of MgH_2 powders obtained after 25 h of RBMing and then CRed for (A) 25, and (B) 200 passes [69].

cold rolled (CR) for different number of passes (1–200 passes), using two-drum type manual cold roller (11 cm wide × 5.5 cm rollers diameter), as displayed in Fig. 10.14B. After 25 passes the powders were agglomerated and forming large aggregates upon CRing for 25 passes (Fig. 10.14C,D). After 200 passes, micro-bands with thickness of 143 μm were developed as a result of cold working generated during CRing process, as shown in Fig. 10.14E. The as-CRed powders for 200 passes were then HEBMed under hydrogen gas for 50 h.

The XRD pattern of CRed MgH_2 powders obtained after 200 CR passes and then HEBMed for 50 h is displayed in Fig. 10.15A. Obviously, the Bragg peaks corresponding to γ-and β-MgH_2 phases were completely vanished and replaced by new Bragg peaks of unreported phase, appeared at scattering angles (2θ) of, 35.034, 40.584, 58.758, 70.115, and 73.868 (Fig. 10.15A). XRD analysis indicated that this discovered MgH_2 phase has face cantered cubic structure (fcc) of space group, Fm-3m(225). The lattice parameter (a_o) of this phase, calculated from (111) was 0.44361 nm [69].

The FE-HRTEM image of the as-CRed powders for 200 passes and then HEBMed for 50 h is shown in Fig. 10.15B. The image revealed Moiré fringe image for three intimated nanograins. Filtered-fringe images corresponding to Zone I is presented in Fig. 10.15B. The d-spacing related to fcc-MgH_2 (111) was calculated and found to be 0.2559 nm, where the corresponding a_o was calculated and found to be 0.4433 nm [69].

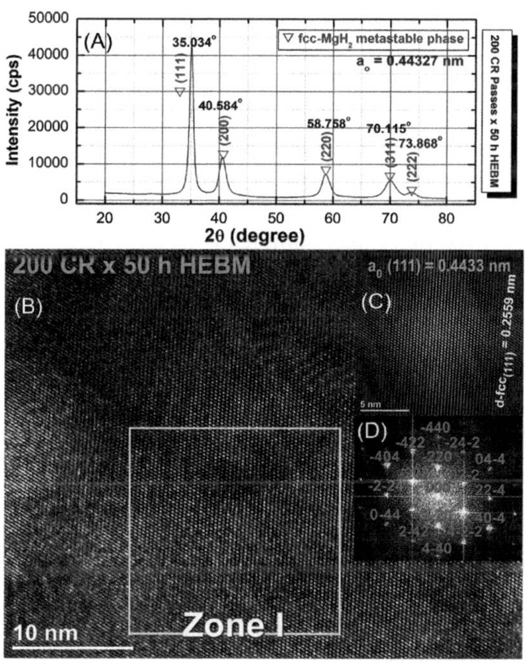

Figure 10.15 (A) XRD pattern of CRed MgH_2 powders obtained after 200 passes and then HEBMed for 50 h, (B) FE-HRTEM image, (C) filtered atomic resolution image of Zone I, and (D) FFT image of MgH_2 powders obtained after 25 h of RBM, consequently CRed for 200 passes and finally HEBMed for 50 h [69].

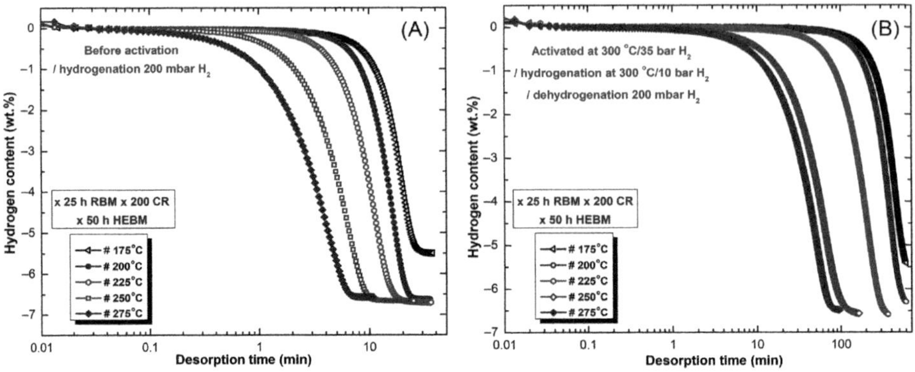

Figure 10.16 Dehydrogenation kinetics measured for 10 individual samples of MgH_2 powders obtained after 25 h of RBM, consequently CRed for 200 passes and then HEBMed for 50 h. The measurements were conducted under 200 mbar H_2 (A) without powders activation, and (B) after activated the powders at 300°C under 10 bar H_2 [69].

These values matched well with the XRD analysis. The FFT image, which is displayed in Fig. 10.15C, was taken from Zone I and oriented to [001] axis, where the corresponding FFT showed spot-electron diffraction pattern related to fcc-MgH_2 (111), as presented in Fig. 10.15D.

Fig. 10.16 displays the dehydrogenation kinetics behaviors of the CRed powder obtained after 200 passes and then HEBMed for 50 h (fcc-MgH_2 phase). The measurements were conducted in the temperature range of 175–275°C under 200 mbar (A) without powders activation, and (B) after activated the powders at 300°C under 10 bar H_2. In order to maintain the crystal structure of the powders (fcc-MgH_2 phase) obtained after this stage of milling, the desorption kinetics measurements measured without activation for five individual samples (Fig. 10.16A). All samples were successfully desorbed their hydrogen storage with different time scale. Generally, the fcc-MgH_2 powders showed advanced dehydrogenation characteristics over γ- and β-MgH_2 phases. At 175°C, the sample desorbed ~ 4 wt. % H_2 within 40 min (Fig. 10.16A). However, the powders desorbed about 6.5 wt. % H_2 within 22 min at 200°C, (Fig. 10.16A). Increasing the applied temperature to 225°C and 250°C, improved the dehydrogenation kinetics of fcc-MgH_2, indicated by the short time required to desorb 6.6 wt. % H_2 in 17 min and 10 min, respectively (Fig. 10.16A). Outstanding desorption kinetics was conducted at 275°C, when the sample desorbed 6.6 wt. % H_2 within 7 min, as elucidated in Fig. 10.16A. We should emphases that the XRD patterns for the examined samples at temperature range laid between 175°C to 225°C revealed presence of Bragg peaks corresponding to γ- and β-MgH_2 phases. However, the XRD pattern for those samples examined at higher temperatures (250°C and 275°C) revealed diffracted lines of hcp-Mg phase [69].

Fig. 10.16B displays the dehydrogenation kinetics conducted for the same, however, they were activated firstly at 300°C (far above the transformation temperature of fcc-MgH_2 to γ- and β-MgH_2 phases) under 25 bar H_2 [67]. Based on this activation step condition, all the samples lost their fcc-structure and completely transformed into γ- and β- MgH_2 phases. This led to a significant deduction of dehydrogenation kinetics that become very slow when compared with the same samples shown in Fig. 10.16A. For example, the sample examined at 175°C required more than 650 min to desorb less than 5.5 wt. % H_2 (Fig. 10.16B). The two samples examined at 200°C and 225°C desorbed about 6.4 and 6.6 wt. % H_2 within 600 and 390 min, respectively (Fig. 10.16B). Comparing the necessary desorption time (100 min) for the sample examined at 275°C to release 6.6 wt. % H_2 (Fig. 10.16B) with that time required (7 min) for fcc-MgH_2 sample (Fig. 10.16A) to desorb same hydrogen amount at the same temperature, we can realize that the kinetics desorption of fcc-MgH_2 phase is 14 times faster than γ- and β- MgH_2 phases [69].

References

1. El-Eskandarany M. Sherif, Shaban E., Al-Shemmiri A. International Journal of Hydrogen Energy, 39: 21097 (2014).
2. Jain I. P. Hydrogen the Fuel for 21st Century. Int. J. Hydrogen Energy, 34: 7368 (2009).
3. Varin R. A, Czujko T., Wronski Z. S. Nanomaterials for Solid State Hydrogen Storage. 1st ed. Chapter 1 (Springer Science+Business Media, LLC 2009).
4. Webb C. J., Journal of Physics and Chemistry of Solids, 2014; http://dx.doi.org/10.1016/j.jpcs.2014.06.014
5. Moretto P, Zlotea C, Dolci F, Amieiro A, Bobet JL, Borgschulte A, et al. Int J Hydrogen Energy, 38:6704 (2013)
6. Bogdanović B., Bohmhammel K., Christ B., Reiser A., Schlichte K., Vehlen R., et al. J Alloys Compd, 282: 841999.
7. El-Eskandarany M. Sherif, Al-Nasrallah E., Banyan M., Al-Ajmi F. Int. J. Hydrogen Energy, 27: 23382 (2018)
8. Camirand C.P. Thermochim Acta, 417: 1 (2004)
9. Xiong Y., Ba J., Qing W., Jing W. J. Plasma Fusion Res. Series , 10: 94 (2013)
10. Yamada T., Yin J., Tanaka K. Materials Transactions, 42: 2415 (2001)
11. Ouyang L. Z., Yang X. S., Dong H. W., Zhu M. Scripta Materialia, 61: 339 (2009).
12. Ouyang, L. Z., et al. J. Phys. Chem. C, 118: 7808 (2014)
13. Zhong H.C., Wang H., Liu J. W., Sun D. L., Zhu M. Scripta Materialia, 65: 285 (2011)
14. Zhong H.C., Wang H., Ouyang L. Z., Zhu M. Journal of Alloys and Compounds, 509: 4268 (2011)
15. Cao, Z., et al. Journal of Alloys and Compounds, 623: 354 (2015)
16. Zhang B., Lv Y., Yuan J., Wu Y. Journal of Alloys and Compounds, 702: 126 (2017)
17. Dymek M.,Nowak M, Jurczyk M., Bala H. Journal of Alloys and Compounds, 780: 697 (2019)
18. Liang G., Huot J., Boily S., Van Neste A., Schulz R. J Alloys Compd, 292: 247 (1999)
19. Zaluska A., Zaluski L., Ström–Olsen J.O. J Alloys Compd, 288: 217 (1999)
20. Sun Y., Shen C., Lai Q., Liu W., Wang Da-Wei, Francois K., Zinsou A. Energy Storage Materials, 10: 168 (2018)
21. Chen J., Xia G., Guo Z., Huang Z., Liu H., Yu X. J. Mater. Chem. A, 3: 15843 (2015)
22. Ding X., Li Y., Fang F., Sun D., Zhang Q. J. Mater. Chem. A 5: 5067 (2017)
23. Karty A., Grunzweig-Genossar J., Rudman P. S. J. Appl. Phys. 50:7200 (1979)
24. Boulet J. M., Gerard, N. J. Less-Common Met, 89: 151 (1981)

25. Zaluska A., Zaluski L., Ström-Olsen J.O. Journal of Alloys and Compounds, 289: 197 (1999)
26. Dutta K., Srivastava O. N., Journal of Materials Science, 28: 3457 (1993)
27. Mandal P., Dutta K., Ramakrishna K., K. Sapru K., Srivastava O. N. Journal of Alloys and Compounds, 184: 1 (1992)
28. Wang P., Zhang H. F., Ding B. Z., Hu Z. Q. Acta mater, 49: 921 (2001)
29. Fu Y., Groll M., Mertz R., Kulenovic R. Journal of Alloys and Compounds, 460: 607 (2008)
30. Agarwal S., Aurora A., Jain A., Jain I.P., Montone A. Int J Hydrogen Energy, 34: 9157 (2009)
31. Yu X.B., Yang Z.X., Liu H.K., Grant D.M., Walker G.S. Int J Hydrogen Energy 35: 6338 (2010)
32. Dehouche Z., Peretti H.A., Hamoudia S., Yooc Y., Belkacemi K. Journal of Alloys and Compounds, 455: 432 (2008)
33. Jain IP, Lal C, Jain A. International Journal of Hydrogen Energy, 35: 5133 (2010)
34. Rusman N.A.A., Dahari M. A review on the current progress of metal hydrides material for solid-state hydrogen storage applications International Journal of Hydrogen Energy, 411: 12108 (2016)
35. Ouyang LZ, Dong HW, Peng CH, Sun LX, Zhu M. Int J Hydrogen Energy 32: 3929 (2007)
36. Zhou C., Fang Z.Z., Ren C., Li J., Lu J. J. Phys. Chem. C, 117: 12973 (2013)
37. Ren C., Fang Z.Z., Zhou C., Lu J., Ren Y., Zhang X. J. Phys. Chem. C, 118: 21784 (2014)
38. Vijay R, Sundaresan R, Maiya MP, Srinivasa Murthy S, Fu Y, Klein HP, et al. J Alloys Compd, 384(1-2): 283 (2004)
39. El-Eskandarany M. Sherif, Shaban E., Ali N., Aldakheel F., Alkandary A. Method of synthesizing MgH_2/Ni nanocomposites. US 9,828,245 B1, Nov. 28, 2017.
40. El-Eskandarany M. Sherif, Shaban E., Ali N., Aldakheel F., Alkandary A. Sci. Rep., 6: 37335 (2016) doi: 10.1038/srep37335
41. El-Eskandarany M. Sherif, Alkandary A., Aldakheel F., Al-Saidi M., Al-Ajmi F., Banyan M. RSC Adv., 8: 38175 (2018)
42. El-Eskandarany M. Sherif, Al-Matrouk H., Shaban E., Al-Duweesh A. Int J Hydrogen Energy, 40: 10139 (2015)
43. El-Eskandarany M. Sherif, Shaban E., Al-Matrouk H., Behbehani M., Alkandary A., Aldakheel F., et al. Materials Today Energy 3: 60 (2017)
44. Oelerich W, Klassen T, Bormann R. Journal of Alloys and Compounds, 315: 237 (2001)
45. Aguey-Zinsou K.-F., Ares Fernandez J.R., Klassen T., Bormann R. International Journal of Hydrogen Energy, 32: 2400 (2007)
46. Hanada N., Ichikawa T., Hino S., Fujii H. J. Alloy. Compd., 420: 46 (2006)
47. da Conceição M.O.T., Brum M.C., dos Santos D.S., Dias M.L. Journal of Alloys and Compounds, 550: 179 (2013)
48. Wang P., Wang A.M., Zhang H.F., Ding B.Z., Hu Z.Q. J. Alloy. Compd., 313: 218 (2000)
49. Kwon I.-H., Bobet J.-L., Bae J.-S., Song M.-Y. J. Alloy. Compd., 396: 264 (2005)
50. Dehouche Z., Klassen T., Oelerich W., Goyette J., Bose T.K., Schulz R. J. Alloy. Compd., 347: 319 (2002)
51. Bhatnagar A., Pandey S. K., Vishwakarma A., Singh S., Shukla V., Soni P. K., Shaz M.A., Srivastava O.N. J. Mater. Chem. 4 (2016) 14761–14772, http://doi.org/10.1039/C6TA05998H
52. Gupta R., Agresti F., Russo S.L., Maddalena A., Palade P., Principi G. J. Alloy. Compd., 450: 310 (2008)
53. El-Eskandarany M. Sherif, Shaban E. Materials, 8: 6880 (2015)

54. El-Eskandarany M. Sherif, Al-Hazza A. Materials Characterization 97: 92 (2014)
55. El-Eskandarany M. Sherif, Shaban E., Alsairafi A. Energy, 104: 158 (2016)
56. El-Eskandarany M. Sherif, Al-Matrouk H., Shaban E., Al-Duweesh A. Energy, 91: 274 (2015)
57. El-Eskandarany M. Sherif, Al-Matrouk H., Behbehani M., Shaban E., Alkandary A., Aldakheel F., Al-Saidi M. Materials Chemistry and Physics, 203: 17 (2018)
58. El-Eskandarany M.S. Sci Rep 6, 26936 (2016) doi: 10.1038/srep26936
59. El-Eskandarany M. Sherif. Composition for Hydrogen Storage. US 9,533,884 B1, Jan. 3, 2017.
60. Xiao X., Liu Z., Saremi-Yarahmadi S., Gregory D. H. Phys. Chem. Chem. Phys., 18: 10492 (2016)
61. Edalati K. Kitabayashi K., Ikeda Y., Matsuda J., Li H., Tanaka I., Akiba E., Horita Z. Scripta Materialia, 157: 54 (2018).
62. Shen C., Aguey-Zinsou K.-F. J. Mater. Chem. A, 5: 8644 (2017)
63. Valiev R.Z, Islamgaliev R.K., Alexandrov I.V. Prog. Mater Sci., 45: 103 (2000)
64. El-Eskandarany M Sherif, Shaban E., Al-Halaili B. Int J Hydrogen Energy, 39:12727 (2014)
65. Amira S., Huot J. J. Alloy. Compd, 520: 287 (2012)
66. Jorge A. M., de Lima G. F., Triques M R M., Botta W. José, Langdon T.G. Int. J. Hydrogen Energy, 39: 3810 (2014)
67. Vajeeston P., Ravindran P., Kjekshus A., Fjellvåg H. Phys. Rev. Lett, 89: 175506 (2002)
68. Mine, Y., Tsumagari, T., Horita, Z. Scripta Mater, 63: 552 (2010)
69. El-Eskandarany M. Sherif, Banyan M., Al-Ajmi F. RSC Adv., 8: 32003 (2018)

Implementation of MgH_2-based nanocomposite for fuel cell applications

11

Storing hydrogen gas into cylinders under high pressure of 350 bar is not safe and still needs many intensive studies dedicated for tank's manufacturing. Liquid hydrogen faces also severe practical difficulties due to its very low density, leading to larger fuel tanks 3 times larger than traditional gasoline tank. Moreover, converting hydrogen gas into liquid phase is not an economic process since it consumes high energy needed to cool down the gas temperature to $-252.8°C$. One practical solution is storing hydrogen gas in metal lattice such as Mg powder and its nanocomposites in the form of MgH_2. There are two major issues that should be solved first. One related to MgH_2 in which its inherent poor hydrogenation/dehydrogenation kinetics and high thermal stability must be improved. Second is related to providing a safe tank. Here, we present the experimental results of using MgH_2-based nanocomposites as solid-state hydrogen source, used for an integrated Ti-tank/commercial proton exchange membrane (PEM)-fuel cell system. The photo shows an integrated hydrogen high-pressure storage Ti-tank manufactured filled with MgH_2/5.3 wt.% Ti powders and coupled with 40 W/4.5 A PEM fuel cell (Fig. 11.1).[1]

11.1 Introduction

Carbon dioxide (CO_2) emissions from fossil fuels and industry comprise ~90% of all CO_2 emissions from human activities.[2] Over the last 250 years, humans keep adding large amounts of CO_2 in Earth's atmosphere, which reached in May 2013 to 400 ppm for the first time in human history.[3] The continuous increase of such harmful emissions mainly resulted from uncontrolled growth in our energy demands that require huge consumption of fossil fuels. Hence, developing alternate energy sources has become a must for a sustainable common future on our planet. Of these alternative energy options, hydrogen energy has been considered as the most practical energy carrier[4,5] holding tremendous promise as a new clean energy option for the future.[6–8]

Because of its attractive properties of being convenient, safe, and versatile fuel source that can be easily produced from a renewable energy system and converted to a desired form of energy,[9] hydrogen has become a favorite energy carrier. Moreover, hydrogen has a very high calorimetric value, with a lower heating value of 120 MJ kg^{-1}, compared with petrol, which is approximately one-third of this at 43 MJ kg^{-1}. A key advantage of hydrogen in environmental point of view is that when burned, CO_2 is not produced.

Mechanical Alloying. http://dx.doi.org/10.1016/B978-0-12-818180-5.00011-X

Figure 11.1 Integrated hydrogen high-pressure storage Ti-tank manufactured filled with MgH_2/5.3 wt.% Ti powders and coupled with 40 W/4.5 A PEM fuel cell.[1]

Hydrogen fuel cells have a wide spectrum range of potential applications, operating from a few watts up to gigawatts. Hydrogen micro-fuel cells have great potential to power portable electronics such as laptops and mobile phones.[7] The superior energy density of a hydrogen polymer electrolyte membrane (PEM) micro-fuel cell opens a new horizon for manufacturing of longer battery lifetimes. For mobile applications, fuel cells could provide auxiliary power for electrical units such as air conditioning and refrigerators.[8]

11.2 Hydrogen reactors

During 1990s $LaNi_5$ alloy was considered as the most suitable alloy that can be used for hydrogen storage reactor system due to its fast uptake/release kinetic at low temperatures.[10–13] Developments achieved in MgH_2-based nanocomposite system that show fast hydrogenation/dehydrogenation kinetic have provided excellent opportunities for useful usages as hydrogen storage reactors. One of the earliest studies on employing nanostructured MgH_2 for pilot tank development was reported in 2007 by de Rangeo et al.[14] In their study, small-scale tank was developed and tested in different cooling conditions. At 280°C, the system absorbed 4.9 wt.% H_2. They have reported that it is essential to maintain the material below 370°C to avoid the rapid crystal growth of the Mg powders associated to a strong degradation of the kinetics of the hydriding/dehydriding reactions.[14] Same authors enhanced the hydrogen sorption of MgH_2 used in their first reactor[14] model by mixing with expanded natural graphite (ENG) and consolidated the powders into compacted disks.[15] They reported that incorporating ENG before compression drastically improves the thermal conductivity in the direction normal to compression axis. Moreover, the thermal conductivity increases linearly with ENG content, and can be adjusted to fulfill the loading time requirements.[15] Since then, different schools have introduced various studies focused on design and manufacturing of hydrogen reactor based on MgH_2 system.[16–18]

11.2.1 Bulk nanocomposite MgH_2/10 wt.% (8 Nb_2O_5/2 wt.% Ni) system

XRD pattern of Mg powders obtained after RBM for 200 h under 50 bar of H_2 atmosphere is displayed in Fig. 11.2A. The obtained powders revealed broad Bragg peaks corresponding to β (PDF # 03-065-3365)- and γ(PDF # 00-035-1184)-MgH_2 phases. Formation of metastable γ-phase upon long term of RBM was related to the destabilization of β-MgH2 phase into a less stable γ-phase.[19] The β-to-γ phase transformation at ambient temperature is attributed to the severe lattice imperfections and plastic deformations introduced to the powders during the high-energy ball-milling process.[20]

The FE-HRTEM image of this end product confirmed the formation of nanocrystalline MgH_2 grains (~17.3 nm in diameter) that did not reveal specific orientation, as

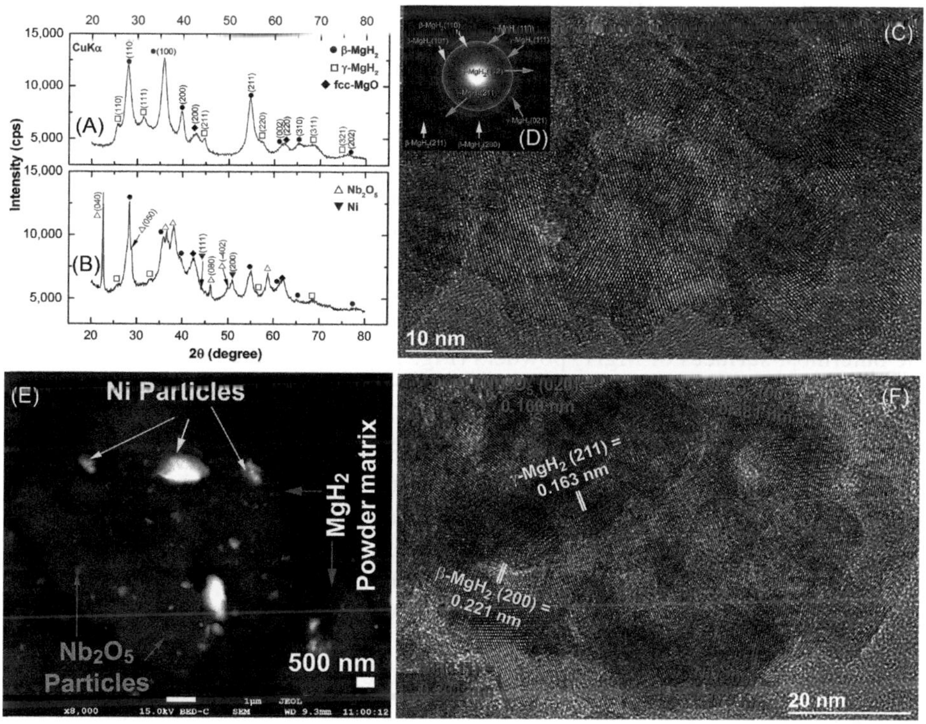

Figure 11.2 (A) XRD pattern of nanocrystalline MgH_2 powders obtained after 200 h of RBM time. The XRD pattern of nanocomposite MgH_2/8 wt.% Nb_2O_5/2 wt.% Ni powders obtained after 50 h of milling is shown in (B). The HRTEM image of as-synthesized MgH_2 powders is presented in (C) together with the corresponding SADP (D). The BES-FESEM micrograph and HRTEM image of nanocomposite MgH_2/8 wt.% of Nb_2O_5/2 wt.% Ni powders obtained after ball milling for 50 h is presented in (E) and (F), respectively.[1]

elucidated in Fig. 11.2C. The development of dislocation tangles into subgrain boundaries can be seen clearly in the figure. The dislocations presented inside the individual grains suggesting the nanostructure development attained by severe plastic deformation upon RBM process. The selected area diffraction pattern (SADP) displayed in Fig. 11.2D revealed Debye continuous rings corresponding to β-(tetragonal) overlapped with γ-(orthorhombic) MgH_2 phases. Such diffracted ring fashion refers to a random orientation of as-prepared nanocrystalline MgH_2 grains (Fig. 11.2D). In addition, the ring-like SADP consisted of numerous closely spaced spots, each from diffraction of a single crystallite, as presented in Fig. 11.2D. The absence of unprocessed metallic Mg crystallites from this end product suggests the compilation of mechanically induced solid-gas reaction via RBM process.

The XRD pattern of nanocrystalline MgH_2 obtained after 200 h of RBM and then mechanically ball-milled with a mixture of (8 wt.% Nb_2O_5/2 wt.% Ni) powders for 50 h is displayed in Fig. 11.2B. The sample disclosed the formation of nanocomposite powders, indicated by the presence of broad Bragg peaks related to the matrix and additive species of MgH_2, Nb_2O_5 (PDF # 00-030-0873), and fcc-Ni (PDF # 00-004-0580) phases (Fig. 11.2B). The as-prepared MgH_2 powder particles possessed porous morphology, as indicated by the pores and cavities presented in the back electron scattering (BES) FE-SEM micrograph of Fig. 11.2E. Formation of these pores yielded as a result of milling MgH_2 powders with abrasive Nb_2O_5 nanoparticles, which penetrated and embedded into the rather soft surfaces of MgH_2. Accordingly, several nano-dimensional pores were created in the metal hydride matrix (Fig. 11.2E). Moreover, the abrasive Nb_2O_5 fine particles acted as micro-milling media during RBM process led to disintegrate both of microscaled MgH_2 and Ni powders into submicron particles, as elucidated in Fig. 11.2E. It can be concluded from the BES observations that Nb_2O_5 (~50 nm in diameter) and Ni (0.5–1 μm) particles with their huge surface areas were homogenously adhered to the outermost surfaces of MgH_2 powders (Fig. 11.2E) to create numerous numbers of nanocatalytic sites beyond the nano level.

The atomic resolution FE-TEM image taken at the near-edge of nanocomposite MgH_2/8 wt.% Nb_2O_5/2 wt.% Ni particle obtained after 50 h of RBM time is shown in Fig. 11.2F. This examined zone consisted of ultrafine nanodimensional grains with sizes ranged between 6 and 8 nm in diameter, as elucidated in Fig. 11.2F. The MgH_2 matrix composited of β- and γ-phases, as indexed by the Miller fringes for different randomly oriented nanograins (Fig. 11.2F). The MgH_2 matrix hosted the additive powders of Nb_2O_5 and Ni grains, implied by the Moiré-fringes corresponding to each of species (Fig. 11.2F).

DSC experiments conducted under a flow of He gas were performed at constant heating rate to investigate the effect of RBM time and the additives of Nb_2O_5 and Ni powders on the thermal stability of MgH_2 powders. The DSC trace of pure MgH_2 powders obtained after 200 h of RBM revealed an endothermic event observed at a peak temperature (T_p) of 455°C (Fig. 11.3A). This event was related to the decomposition of MgH_2 phase, as confirmed by the XRD pattern of the sample taken after the completion of this endothermic event. To investigate the effect of additives (Nb_2O_5 and Ni) and ball-milling time on the thermal stability of MgH_2 powders, MgH_2/8 wt.%

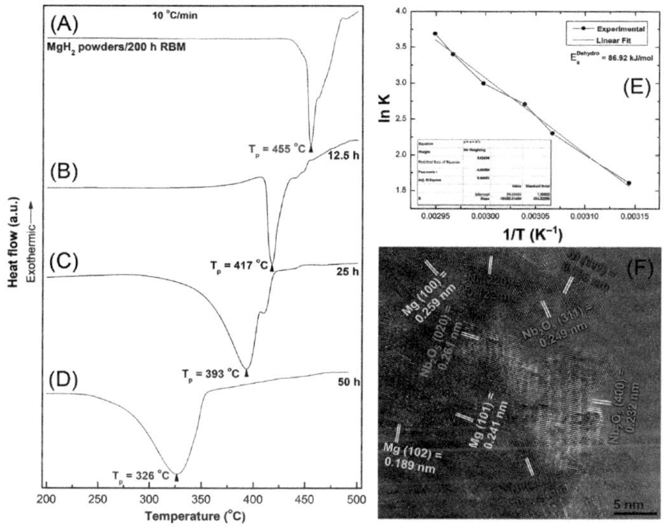

Figure 11.3 (A) DSC curve of as-synthesized nanocrystalline MgH_2 powders obtained after 200 h of RBM time. The DSC curve of MgH_2 doped with a mixture of 8 wt.% Nb_2O_5/2 wt.% Ni powders and then ball-milled for 12.5, 25, and 50 h (end product) are displayed in (B), (C), and (D), respectively. The apparent activation energy (E_a) calculated for hydrogen desorption, using Arrhenius approach, is presented in (E). HRTEM image of the powders taken after completion of the DSC experiment is shown in (F).[1]

Nb_2O_5/2 wt.% Ni powders were milled for different milling time (12.5–50 h). After 12.5 h of milling time, T_p of MgH_2 was significantly decreased by 38°C, as presented in Fig. 11.3B. The decomposition peak tended to shift again to the low temperature side (393°C) after 25 h of milling, as displayed in Fig. 11.3C. The decomposition temperature for MgH_2/8 wt.% Nb_2O_5/2 wt.% Ni powders obtained after milling for 50 h showed significant improvement, suggested by the low T_p (326°C), as shown in Fig. 11.3D. It should be emphasized that further ball milling did not lead to remarkable decreasing on the T_p value that remained without notable change (324–323°C) after ball milling for 75–100 h.

In order to realize the effect of doping the metal hydride powders with 8 wt.% Nb_2O_5/2 wt.% Ni powders after milling for 50 h on the apparent activation energy of dehydrogenation (E_a), DSC experiments were conducted with different heating rates (5, 10, 15, 20, 30, and 40°C/min). The E_a corresponding to each heating run was calculated according to the Arrhenius equation. The E_a value was determined by measuring the corresponding T_p of decomposition peaks related to different k and then plotting ln(k) versus 1/T_p (Fig. 11.3E). E_a was then obtained from the line slope presented in Fig. 11.3E. The E_a of the nanocomposite sample obtained after 50 h of ball milling showed a low value (86.92 kJ/mol), as elucidated in Fig. 11.3E.

The FE-HRTEM image of the sample taken after completion of the endothermic event shown in Fig. 11.3F indicated disappearance of MgH_2 and the presence of metallic hcp-Mg coexisted with Nb_2O_5 and Ni phases (Fig. 11.3F). No reacted phases (e.g., Mg_2Ni, Mg_2NiH_4, $Nb_{100-x}H_x$) could be detected in this sample, implying that Nb_2O_5 and Ni powders acting as heterogeneous catalytic agents.

van't Hoff approach for indirect measurement of the enthalpy change of hydrogenation and dehydrogenation. The pressure–composition–temperature (PCT) relations of MgH_2 powders obtained after 200 h of RBM time and then ball-milled with a mixture of 8 wt.% Nb_2O_5/2 wt.% Ni powders were volumetrically investigated by Sievert's approach at different temperatures of 225, 250, 275, 300, 325, and 350°C. Within this temperature range, the powders possessed fair PCT hydrogenation/dehydrogenation characteristics, as displayed in Fig. 11.4. This can be realized by the near pressure values required for absorption (P^{abs}) and desorption (P^{des}) processed at 225, 250. 275, 300, and 325°C. Moreover, the powders manifested outstanding single-step hydrogenation/dehydrogenation behavior with negligible slope, as displayed in Fig. 11.4.

The hydrogen equilibrium pressure measurements were used in the present study to investigate the heat of hydrogen absorption, using van't Hoff equation:

$$\ln\left(\frac{P_{eq}}{P_0}\right) = -\frac{\Delta H}{RT} + \frac{\Delta S}{R} \tag{11.1}$$

where P_{eq} is the hydrogen pressure under equilibrium state at a given specific temperature, T, P_0 is a reference pressure of 1 bar, R is the gas constant (0.0083145 kJ/mol), ΔH is the molar enthalpy of metal hydride formation (MgH_2), and ΔS is the entropy of absorption. Thus, ΔH can be estimated from plotting the natural log of each P_{eq} point versus the corresponding $1/T$, as shown in Fig. 11.5A. In the present work, the calculated ΔH of hydrogenation for MgH_2/8 wt.% Nb_2O_5/2 wt.% Ni powders obtained after 50 h of ball-milling time was –72.74 kJ/mol.

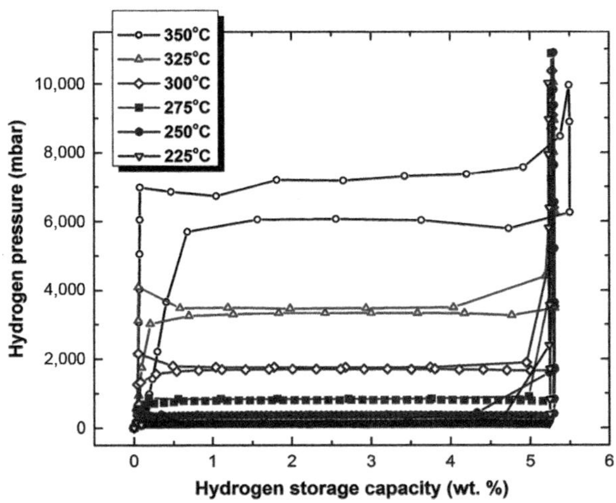

Figure 11.4 Pressure–composition–temperature (PCT) curves measured at different temperatures for nanocomposite MgH_2/8 wt.% Nb_2O_5/2 wt.% Ni powders obtained 50 h of RBM time.[1]

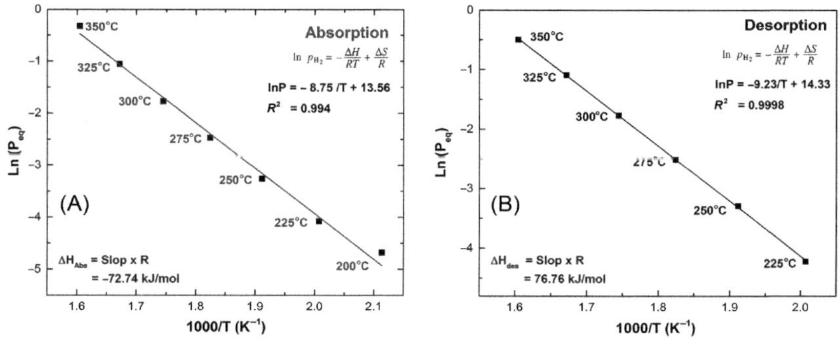

Figure 11.5 van't Hoff plot of MgH_2/8 wt.% Nb_2O_5/2 wt.% Ni ternary system of (A) hydrogenation and (B) dehydrogenation.[1]

The strength of Mg-H bonds, which can be expressed by the enthalpy of decomposition, was also calculated by van't Hoff approach, using the equilibrium dehydrogenation pressure in the PCT measurements. The van't Hoff plot illustrating the relationship between ln(P) and $1/T$ for the decomposition of MgH_2 phase in the nanocomposite powders obtained after 50 h is shown in Fig. 11.5B. ΔH was calculated from the slope of the curve presented in Fig. 11.5B and found to be 76.76 kJ/mol.

Measuring the kinetic behavior of hydrogen absorption/desorption is one of the most important characterizations required to evaluate the hydrogen storage materials and their potential usage in real applications. One objective of the present study is to improve the kinetics of both absorption and desorption for as-prepared MgH_2 upon mixing with a mixture of 8 wt.% Nb_2O_5/2 wt.% Ni powders. The hydrogenation kinetic behavior investigated at different temperatures in the range between 200 and 250°C, using 10 bar H_2 gas pressure for both nanocrystalline MgH_2 and nanocomposite MgH_2/8 wt.% Nb_2O_5/2 wt.% Ni powders are presented in Fig. 11.6A and B, respectively. The as-prepared MgH_2 nanocrystalline powders required a very long time (20 min) to absorb almost 5 wt.% H_2 at 200°C, as shown in Fig. 11.6A. At this temperature, however, a significant improvement in the hydrogenation kinetic was attained upon doping MgH_2 with 8 wt.% Nb_2O_5/2 wt.% Ni powders and milling for 50 h. This is indicated by the shorter time (~2 min) needed to absorb 5 wt.% H_2 (Fig. 11.6B).

Increasing the applied absorption temperature to 225°C led to enhance the absorption behavior of pure MgH_2 powders, as suggested by the rather shorter time (10 min) required to uptake 5 wt.% H_2 (Fig. 11.6A). The effect of such slight increasing of the temperature on the kinetic of hydrogenation process was manifested for nanocomposite MgH_2/8 wt.% Nb_2O_5/2 wt.% Ni system, implied by the notable reduction of the hydrogenation time (~1 min) required to uptake 5 wt.% H_2 (Fig. 11.6B). Further increasing of the hydrogenation temperature (250°C) led to a moderate improvement of the hydrogen uptake kinetics for pure MgH_2, as advocated by the shorted time (~8 min) needed to absorb 5 wt.% H_2. Increasing the absorption time showed a monotonical improvement on the hydrogen storage capacity, where MgH_2 powders

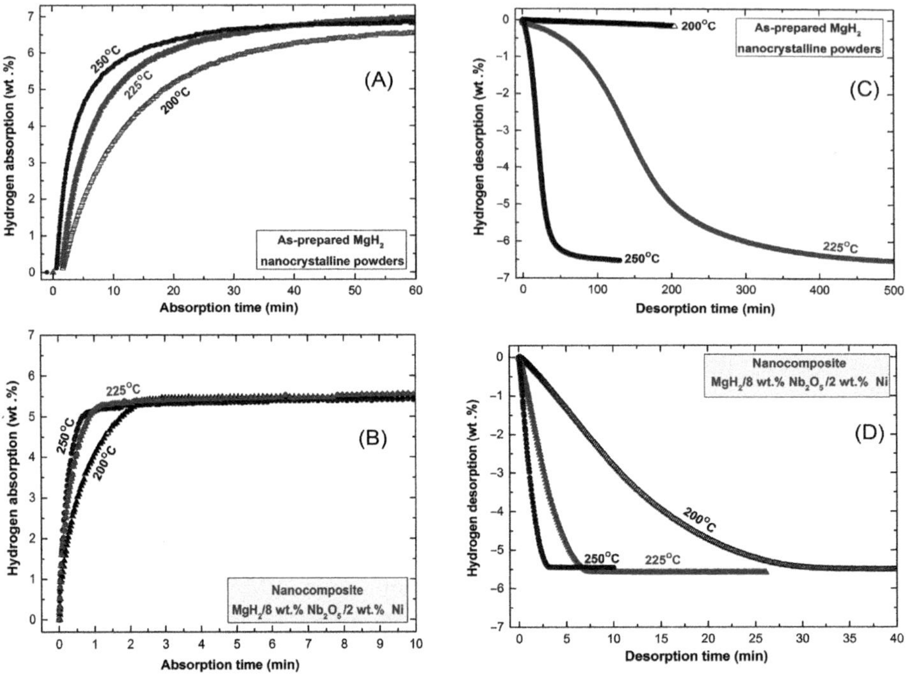

Figure 11.6 Kinetic of hydrogenation investigated at different temperature under 0–8 bar H_2 for (A) nanocrystalline MgH_2, (B) nanocomposite MgH_2/8 wt.% Nb_2O_5/2 wt.% Ni powders. The corresponding kinetics of dehydrogenation achieved at 0.2 bar H_2 are presented in (C) and (D), respectively.[1]

absorbed about 6.5 wt.% H_2 after 30 min (Fig. 11.6A). The powders tended to attain a saturated value of 6.86 wt.% H_2 after 50–50 min, as displayed in Fig. 11.6A. At this applied temperature (250°C), superior uptake kinetics behavior was achieved for nanocomposite MgH_2/8 wt.% Nb_2O_5/2 wt.% Ni powders. The powders at this moderate temperature were capable to absorb 5 wt.% H_2 within ~1 min, as presented in Fig. 11.6B. The storage capacity of these nanocomposite powders saturated at 5.25 wt.% H_2 after 3 min, where they did not reveal further storage improvement even after 10 min, as elucidated in Fig. 11.6B.

The kinetic of hydrogen releasing for the MgH_2 powders was investigated before and after doping procedure. At low temperature (200°C) the powders showed very slow hydrogenation kinetics, indicated by the very long time (~210 min) required to desorb only 1.8 wt.% H_2 (Fig. 11.6C). In contrast when MgH_2 was doped with 8 wt.% Nb_2O_5/2 wt.% Ni powders the dehydrogenation kinetics showed significant improvement, indicated by the rather short time (22.5 min) needed to desorb 5 wt.% H_2, as presented in Fig. 11.6D. In addition, at this temperature (200°C) the nanocomposite powders succeed to full discharging their hydrogen storage capacity (5.5 wt.% H_2) within 35 min (Fig. 11.6D).

Pure MgH_2 nanocrystalline powders revealed marginal improvement on the kinetic of dehydrogenation upon increasing the applied temperature to 225°C. This is implied by the rather short time to release 5 wt.% H_2 within 205 min, as displayed in Fig. 11.6C. However, longer time (500 min) was necessary to enhance the powders for desorbing their nearly hydrogen storage capacity (6.5 wt.% H_2), as displayed in Fig. 11.6C. Doping MgH_2 with the catalytic agent of 8 wt.% Nb_2O_5/2 wt.% Ni powders showed extreme fast dehydrogenation kinetic, suggested by the very short time (8 min) required to complete releasing of hydrogen storage capacity of 5.5 wt.% H_2, as displayed in Fig. 11.6D. The undopped MgH_2 powders showed notable dehydrogenation improvement upon increasing the hydrogen discharging temperature to 250°C. At this temperature, MgH_2 powders released 5 wt.% H_2 within 41 min and achieved a complete discharging of 6.5 wt.% H_2 after 134 min, as displayed in Fig. 11.6C. Application of such a moderate temperature (250°C) had an obvious effect on enhancing the dehydrogenation kinetic of nanocomposite MgH_2/8 wt.% Nb_2O_5/2 wt.% Ni powders, when the material succeeds to discharged its full hydrogen storage capacity (5.5 wt.% H_2) in a very short time (~2.6 min), as elucidated in Fig. 11.6D.

However, mechanical doping MgH_2 with 8 wt.% Nb_2O_5/2 wt.% Ni powders showed a major beneficial effect on the hydrogen uptake and releasing kinetics, doping the hydride phase heavy atomic mass materials such as Nb_2O_5 and Ni powders, which they do not react with H_2 gas led to a drastic decrease in the hydrogen storage capacity of the system by about 1 wt.%.

The capability of nanocomposite MgH_2/8 wt.% Nb_2O_5/2 wt.% Ni powders to achieve continuous cyclic hydrogenation/dehydrogenation processes was investigated at 225°C. This test is necessary to realize the ability of the fabricated nanocomposite system on maintaining their hydrogen storage capacity and fast hydrogenation/dehydrogenation kinetic without serious degradation or failure. In the present study, a small piece (~350 mg) of nanocomposite sample obtained after cold pressing the powders milled for 50 h was continuously subjected to hydrogenation/dehydrogenation cycles conducted at 225°C under a hydrogen gas pressure ranging between 10 bar (hydrogenation) and 200 mbar (dehydrogenation) for 400 h (~2 weeks). In this experiments, the powders were firstly activated by applying cyclic hydrogen gas sorption/desorption under pressure of 35 bar at 350°C for 10 continuous cycles. This treatment is necessary for surface cleaning of the powders and to break down the oxide phase (MgO) formed on the powder surfaces during handling the sample outside of the glove box. Fig. 11.7A shows the hydrogen absorbed/desorbed cycles achieved continuously for 400 h at a temperature of 225°C. No remarkable degradation in the hydrogen storage capacity (5.25 wt.% H_2) could be detected even after 400 h, as shown in Fig. 11.7A. Moreover, the kinetic of hydrogenation/dehydrogenation processes remaining nearly constant without failure or serious decay. It can be concluded that doping MgH_2 with a mixture of 8 wt.% Nb_2O_5/2 wt.% Ni and ball milling for rather long time (50 h) allowed the catalytic agent particles to occupy the surface of MgH_2 powders with uniform distribution. Thus, controlling the morphological properties including the shape, size, and the composition homogeneity is considered as the most important factor that controls the hydrogenation/dehydrogenation kinetic behavior and performance of hydrogen storage materials.

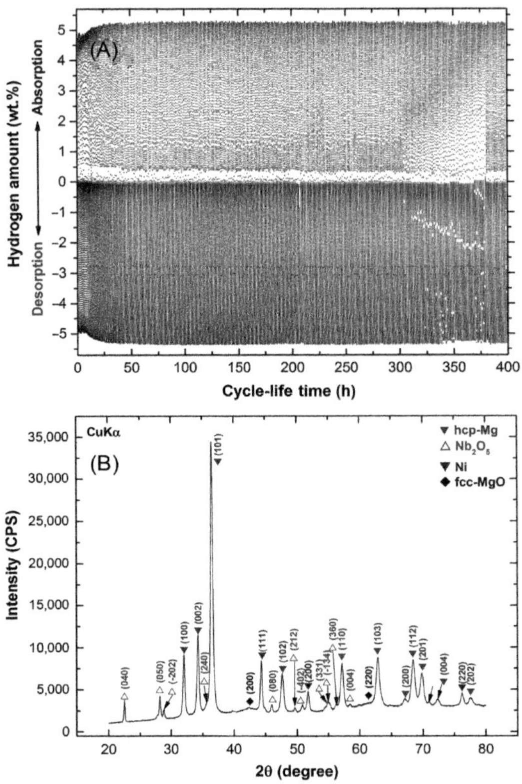

Figure 11.7 (A) Cycle-lifetime achieved at 225°C for as-cold pressed nanocomposite sample obtained after 50 h of RBM time examined. The XRD pattern of the powders obtained after completion the cycle-lifetime test is presented in (B).[1]

The XRD pattern of the sample taken after 400 h of cycle-lifetime is presented in Fig. 11.7B. The sample revealed a domain structure of hcp-Mg metal coexisted with Nb_2O_5 and Ni, as displayed in Fig. 11.7B. Neither niobium hydride(s) nor Mg_2NiH_4 phases could be detected, implying that Nb_2O_5 and Ni can be considered as heterogeneous catalysts, in which they enhanced the hydrogenation/dehydrogenation kinetic without entering into a chemical reaction with MgH_2 matrix.

11.2.1.1 Implementation of nanocomposite MgH_2/8 wt.% Nb_2O_5/2 wt.% Ni green compacts for fuel cell applications

For the purpose of the present study, the as-prepared nanocomposite powders obtained after 50 h of milling were cold-pressed at ambient temperature into disk buttons of about 3.25 cm in diameter. In the consolidation experiments, an amount of ~30 g of as-prepared MgH_2/8 wt.% Nb_2O_5/2 wt.% Ni nanocomposite powders obtained after 50 h of milling were charged into a tool steel die (Fig. 11.8A) without adding any binders. The powders were then subjected to uniaxial load of 20 ton, using a

Figure 11.8 Preparations procedure used for fabrication of nanocomposite MgH_2/8 wt.% $Nb_2O5/_2$ wt.% Ni disks. (A) About 30 g of as-prepared nanocomposite powders were charged into a 3.25-mm-diameter tool steel powder die and sealed with upper/bottom bunches, and plunger in a glove box filled with He gas atmosphere. The system was then subjected to an axial vertical load of 20 ton at ambient temperature for 50 h, using 100-ton cold press (B). The as-cold pressed solid-hydrogen nanocomposite system had relative density of ~68.7% (C). Three cold-pressed nanocomposite bulk samples with a total weight of about 87 g were placed into high-pressure hollow vessel made of pure titanium metal (hydrogen tank) of 8 cm high × 6.2 cm outer diameter × 5.4 cm inner diameter (D, E, and F). The system was sealed with eight hexagonal bolts (hexagonal head cap screws) and connected to high-pressure gate ball valve with a 20-bar pressure gauge inside a glove box filled with He gas (G).[1]

commercial 100 ton-cold press (Fig. 11.8B). The compacts obtained after this single consolidation step (Fig. 11.8C) had average relative density of 68.7%. In order to avoid the swelling of the green compacts upon introducing H_2 at high pressure (~20 atm), these green compacts were subjected to recycle loading from 0 to 25 ton for 30 cycles and left under 30 ton for 24 h. The as-cycled disks possessed higher relative density of

~92%, as displayed in Fig. 11.8D. A simple customized made hydrogen storage Ti-metal tank was used to host the three nanocomposite disks, as presented in Fig. 11.8E and F. The system was then placed in He-atmosphere glove box to seal the buttons in the Ti-tank under, using copper metal gasket. A pressure gauge (20 bar) and a high-pressure ball gate valve system were perfectly installed into the tank cap's to monitor the hydrogen gas released and charged (Fig. 11.8G). The reason of using powder cold pressing came from our preliminary results of preparing several bulk samples, using hot pressing and spark plasma sintering (SPS) techniques, when the powders were consolidated at temperature ranged between 530 and 550°C to obtain nearly full dense disks with relative densities of ~99.5%. Our preliminary results showed that when the consolidated buttons had rather modest relative density (~92%), successful hydrogen desorption/sorption processes were attained. In contrast, those full-dense disks of the same powders consolidated by hot pressing/SPS techniques consolidated at temperatures far above the decomposition temperature of our prepared MgH_2 (327°C) led to decompose the metal hydride phase into Mg metal and hydrogen gas. Moreover, significant mole fractions of Ni that is existed in the catalytic agent mixture with Nb_2O_5 powders preferred to react with Mg metal to form an undesired Mg_2Ni phase.

It should be notified that consolidation of the nanocomposite powders at such high temperature showed several drawbacks such as: (1) applying such high temperature led to enhance the grain growth of the nanocrystalline powders. Accordingly, the hydrogenation/dehydrogenation kinetic became very slow upon increasing the hydrogen diffusion distance associated with the grain growth process of the hot-pressed powders, (2) however, increasing the density of the consolidated samples led to improve the mechanical properties of the compacts, but at the same time it decreased the volume fraction of the pores beyond the micro-level. Disappearance of such desirable pores led to slow down the hydrogenation/dehydrogenation kinetic and restrict the hydrogen diffusion process, and (3) application of high consolidation temperature led to enhance the reaction between Mg and Ni to form Mg_2Ni alloy when hydrogenated Mg_2NiH_4 phase was yielded. Formation of this undesired phase led to decrease the hydrogen storage capacity of the system from 5.5 wt.% to 4.2 wt.%. In addition, this significant molar fraction of Ni metal preferred to react with Mg and leading to decrease the concentration of elemental Ni in Nb_2O_5/Ni system. As a result, decreasing the concentration of Ni metal, which is known as an excellent hydrogen splitter agent, led to slow down the hydrogenation kinetic of MgH_2/Nb_2O_5/Ni system.

The solid-hydrogen Ti-tank hosting three MgH_2/8 wt.% Nb_2O_5/2 wt.% Ni green compacts was then evacuated to 10^{-5} bar, using rotary/molecular pump (Fig. 11.9A). A Cu-jacket heater and hot plate (Fig. 11.9B) were used as heating sources required to decompose MgH_2. The tank was placed into the Cu-jacket heater and wrapped well with a jacket temperature insulator (Fig. 11.9C). The end of the ball gate valve was connected to a rotary pump interfaced with 40W/4.5A proton exchange membrane (PEM) fuel cell (Fig. 11.9C and D). This rotary pump was necessary for assisting evacuation of the hydrogen released upon heating the system and to control the path of H_2 gas released toward the PEM-FC (Fig. 11.9C and D).

The hydrogen released from the nanocomposite disks upon heating the storage system to 225°C was used to feed the PEM-fuel cell with H_2. Fig. 11.10A presents the

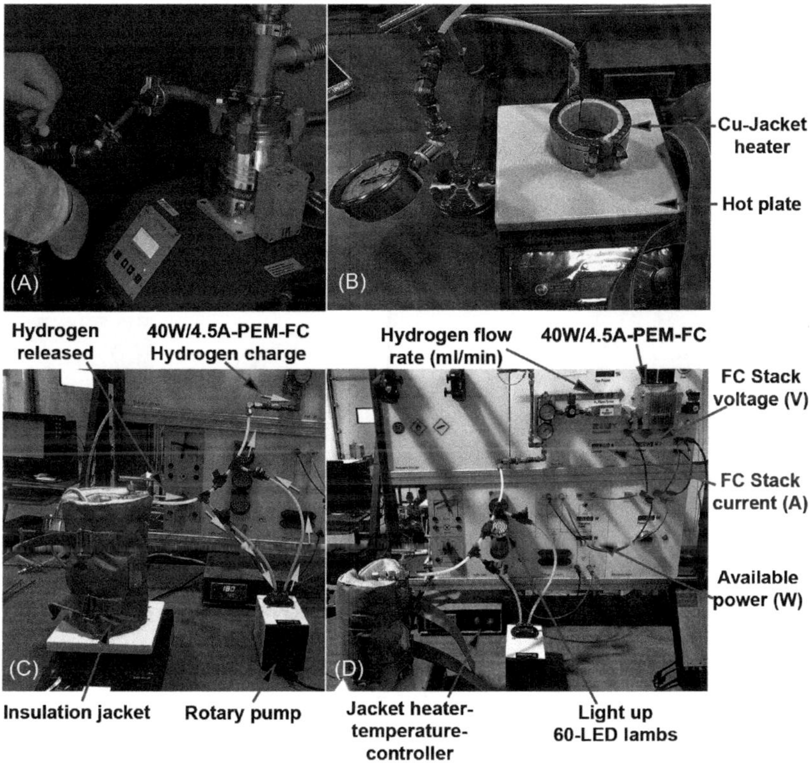

Figure 11.9 Solid-hydrogen storage tank integrated with 40W/4.5A PEM-FC system. (A) The Ti-tank containing the nanocomposite disks was evacuated with rotary/molecular pump, (B) temperature controlled hotplate, and Cu-jacket heater, (C) Ti-tank hydrogen storage system filled with MgH_2/8 wt.% Nb_2O_5 /2 wt.% Ni nanocomposite disks, and (D) the solid-hydrogen tank was coupled with 40 W/4.5 A PEM-FC.[1]

experimental setup utilized for charging the battery of a cell phone device through 5V voltage regulator. The PEM-fuel cell system is controlled and operated with a software where the data output corresponding to hydrogen flow rate, voltage, and current were obtained and stored. During the first few min (0.033 min) of the battery charging time, conducted at 225°C the MgH_2 powders were completely decomposed, where the pressure of released hydrogen saturated at a value of 20 bar, as indicated in Fig. 11.10C. To avoid any reaction between the Mg powders and the hydrogen gas released and stored inside the Ti-tank, heating process of the system was immediately stopped. Then the ball gate valve was then opened where the rotary pump facilitated H_2 succession forward the PEM-fuel cell. The monotonical decreasing of H_2 pressure inside the Ti-tank upon increasing the battery charging time implies a continuous hydrogen gas delivered to the PEM-fuel cell. It should be mentioned that using a rotary pump led to deliver the hydrogen gas to the fuel cell at a constant rate of ~160 mL/min, as shown in Fig. 11.10B. Providing the fuel cell with such a constant rate of hydrogen gas flow

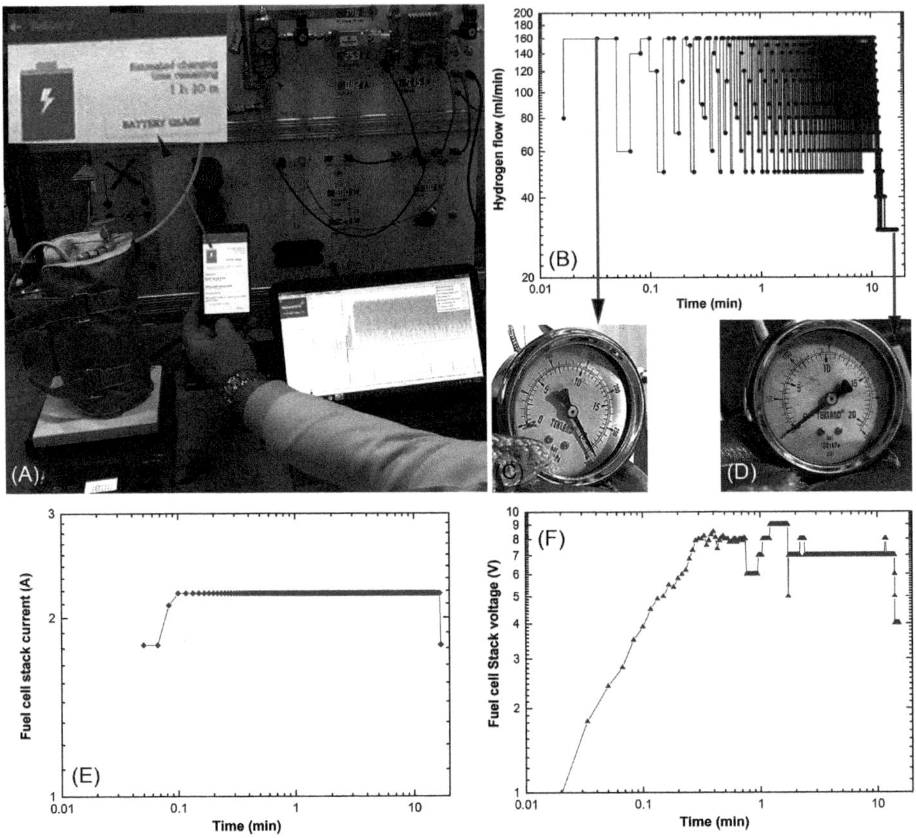

Figure 11.10 Charging of battery for a cell phone device via solid-hydrogen storage tank integrated with 40W/4.5A PEM-FC system. Hydrogen storage system filled with MgH_2/8 wt.% Nb_2O_5/2 wt.% Ni nanocomposite disks coupled with 40W/4.5A PEM-FC to charge battery of a cell phone device is presented in (A). The hydrogen flow during the cell-phone's charging process is displayed in (B) as a function of battery charging time. (C) and (D) show the hydrogen pressure of Ti-tank at the beginning and final charging stages, respectively. Correlation between charging time and fuel cell-stack current and voltage are presented in (E), and (F), respectively.[1]

leads to generate constant values of current (2.2A) and voltage (7.3 V), as displayed in Fig. 11.10E and F, respectively. After 18 min, the pressure inside the Ti-tank reached to 1 bar Fig. 11.10D, indicating the completion of hydrogen releasing and end of the battery charging process.

In order to investigate the effect of performing 50-hydrogen charging/discharging cycles during PEM-FC experiment on the hydrogenation/dehydrogenation kinetics behavior of the nanocomposite system, the Ti-tank was opened in a He-atmosphere glove box after completion of the battery charging. We found that the upper cold-pressed nanocomposite disk suffering from swelling and the existence of numerous

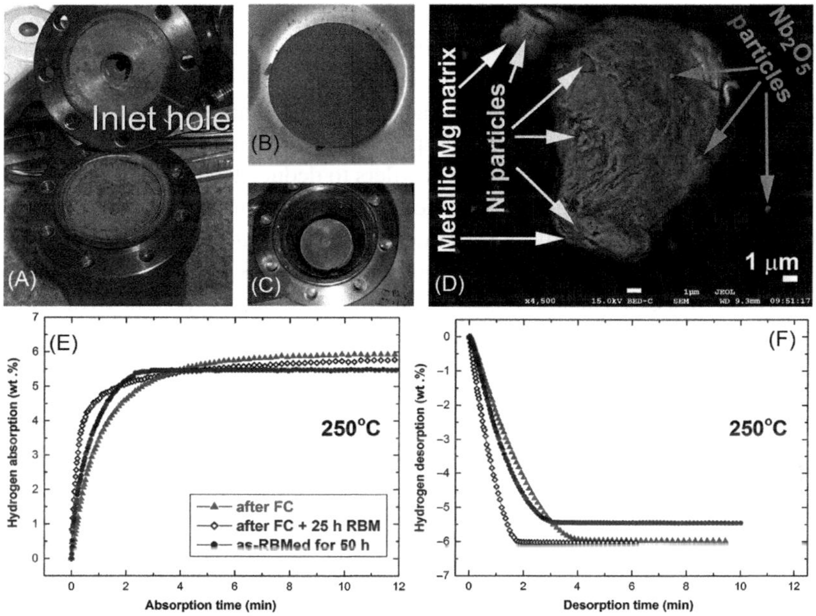

Figure 11.11 Effect of charging a cell phone's battery on the morphology and kinetics of nanocomposite disks. (A) The nanocomposite disk placed on the top inside Ti-tank (upper disk) after completion 50 cycles of charging a cell phone's battery. The photos corresponding for the middle and lower disks are displayed in (B) and (C), respectively. FE-SEM micrograph of an aggregated powder taken from the upper disk is shown in (A). The hydrogenation kinetics of (E) hydrogenation and (B) dehydrogenation of the samples obtained after 50 h RBM, and after processing in FC experiment for 50 cycles. The sample processed for 50 cycles in FC experiment and then ball-milled for 25 h is presented as well.[1]

cracks, as displayed in Fig.11.11A. In contrast, the middle (Fig. 11.11B) and lower (Fig. 11.11C) disks placed far from the hydrogen inlet hole shown in Fig. 11.11A did not possess any serious defects. Hence, the swelling happened in the upper disk may be attributed to its close position with the internal hole, where hydrogen introduced at a high pressure (20 atm). The FE-SEM micrograph of a sample taken from the upper disk after the last dehydrogenated cycles is shown in Fig. 11.11D. Obviously, the powders composited of aggregated metallic Mg matrix reinforced with Nb_2O_5 and Ni particles, as elucidated in Fig. 11.11B.

In part to investigate the effect of powder growth on the hydrogenation and dehydrogenation kinetics of the system, powder samples taken from the upper disk were subjected to hydrogen absorption/desorption kinetics analysis. Surprisingly, the hydrogen storage capacity measured for the sample taken after completion of FC experiments was higher (~6 wt.%) than the original one obtained after 50 h of milling (5.25–5.5 wt.%), as displayed in Fig. 11.11E and F. This can be attributed to the surface cleaning achieved for the powders during the FC experiments. Particle size growing of this sample, however, led to slow down the kinetic of absorption and desorption, as elucidated in Fig. 11.11E and F, respectively.

To improve the hydrogenation/dehydrogenation kinetic of the as-used powders obtained after FC experiments, the upper disk was removed from the Ti-tank inside the He-glove box and then gently ground into coarse powders, using agate pestle and mortar. The powders were then charged into the reactive ball mill under 20 bar of hydrogen gas and mounted on the roller mill for 25 h. Reprocessing the materials used for FC experiments had obvious effect on refining the powders to deduce their particle sizes.

To realize the effect of reprocessing the as-used nanocomposite powders obtained after RBM for 25 h on their hydrogenation/dehydrogenation behaviors and the possibility of reuse again, hydrogen absorption/desorption kinetic examinations were tackled. The reprocessed nanocomposite powders revealed excellent hydrogenation kinetic when compared with the original powders used for FC experiment, as indicated by the very short time (1.4 min) required to absorb 4.5 wt.% H_2 (Fig. 11.11E). More importantly, these treated powders showed the best hydrogen storage capacity (~6 wt.%) and dehydrogenation kinetic (~2 min) when compared with the other two samples (as-RBMed for 50 h, and after FC), as shown in Fig. 11.11F. It can be then concluded that the as-prepared and then processed nanocomposite powders for FC experiments can be easily treated to refine their grains and improve both of their hydrogen storage capacity and hydrogenation/dehydrogenation kinetic.

References

1. El-Eskandarany M. Sherif, Alkandary, Abdullah, Aldakheel, Fahad, Al-Saidi, Mariam, Ajmi, Fahad, Banyan, Mohammad, RSC Adv., 8: 38175 (2018)
2. Jackson, RB, et al., *Environ Res Lett*, 12: 110202 (2017)
3. Latake, PT, Pawar, P, Ranveer, AC., *International Journal of Innovative Research and Creative Technology*, 1: 333 (2015)
4. Chen, J, Fu, H, Xiong, Y, Xu J, Zheng, J, Li, X., *Nano Energy*, 10: 337 (2014)
5. Nowotny, J., et al., *Renewable and Sustainable Energy Reviews*, 81: 2541 (2018)
6. Lin, H.J,, et al., *Nano Energy*, 9: 80 (2014)
7. Schlapbach, L., Züttel, A., Nature, 414: 353 (2001)
8. Walker, G. Solid-state hydrogen storage: Materials and chemistry, 1st Ed. Chapter 1 (Woodhead Publishing Limited, New York, 2008).
9. Stetson, N.T., McWhorter, S. & Ahn, C.C. Compendium of Hydrogen Energy. Edited by Gupta, R. B., Basile, A. & Veziroglu, T. N. Vol. 2, Chapter 1 (Elsevier, Oxford, 2016).
10. Pons, M., Dantzer, P., *International Journal of Hydrogen Energy*, 19: 611 (1994)
11. Jemni, A., Nasrallah, S. Ben, *International Journal of Hydrogen Energy*, 20: 43 (1995)
12. Jemni, A, Nasrallah, S. Ben, Lamloumi, J., *International Journal of Hydrogen Energy*, 24: 631 (1999)
13. Demircan, A., Demiralp, M., Kaplan, Y., Mat, M.D., Veziroglu, T.N., *International Journal of Hydrogen Energy*, 30: 1437 (2005)
14. de Rangeo, P., et al., Journal of Alloys and Compounds, 446-447: 52 (2007)
15. Chaise, A., et al., International Journal of Hydrogen Energy, 34: 8589 (2009)
16. Khandelwal, A., et al., *International Journal of Hydrogen Energy*, 35: 3565 (2010)
17. Capurso, G., et al., *International Journal of Hydrogen Energy*, 509S: S646 (2011)
18. Garrier, S., et al., *International Journal of Hydrogen Energy*, 36: 9719 (2011)
19. Varin, R.A., Czujko, T., Wronski, Z., *Nanotechnology*, 17: 38 (2006)
20. El-Eskandarany, M. S., Shaban, E., Al-Halaili, B., *Int J Hydrogen Energy*, 39: 12727 (2014)

Utilization of ball-milled powders for surface protective coating

12

This chapter introduces selected ball-milled feedstock systems, used for surface coating by means of high velocity thermal spray coating and cold spray techniques (the photo presents the robotic system of a cold spray equipment provided by Startack Co. Ltd., housed in the Nanotechnology Laboratory, Energy and Building Research Center (EBRC), Kuwait Institute for Scientific Research (KISR) (Fig. 12.1)).

12.1 Introduction

Surface engineering is focused on the design and modifications of the bulk materials' surfaces (substrates) to provide specific physical, chemical, and engineering properties that are not inherently existed in the original bulk materials. Improving the wear-, oxidation-, and corrosion- and resistances, friction coefficients, bio-inertness, electronic properties, and thermal insulations are some of those properties that can be successfully enhanced with surface treatments. Improving the surface properties can be achieved by metallurgical, mechanical, or chemical approaches. Coating, which is simply defined as a single or multilayered materials deposited artificially on the surface of bulk object (substrate) made of another material, is a well-known process used in part to obtaining some required technical or decorative properties or to protect the material from expected chemical and physical interactions with its surrounded environment. Coatings are widely used in automotive systems, boiler components, and power generation equipment, chemical process equipment, aircraft engines, pulp and paper processing equipment, bridges, rollers and concrete reinforcements, orthopedics and dental, land-based and marine turbines, ships.[1]

There are different approaches and technologies that can be employed to deposit the appropriate surface protective layer with different thickness, ranging from few micrometer (below 10–20 μm) to more than 30 μm and up to several millimeters. In general, the coating process can be classified based on the media to the following two categories:

- Wet coating approach, such as electroplating, electroless plating, and hot-dip galvanizing methods,
- Dry coating approach, such as brazing, weld overlays, vapor deposition (physical vapor deposition (PVD, Fig. 12.2A), chemical vapor deposition (CVD, Fig. 12.2B), thermal spray techniques, and most recently cold spray technique.

Fig. 12.3 shows the correlation between the principal coating processes and the typical coating thicknesses attainable.

Mechanical Alloying. http://dx.doi.org/10.1016/B978-0-12-818180-5.00012-1

Figure 12.1 Robotic system of a cold spray equipment provided by Startack Co. Ltd., Japan, housed in the Nanotechnology Laboratory, Energy and Building Research Center (EBRC), Kuwait Institute for Scientific Research (KISR).

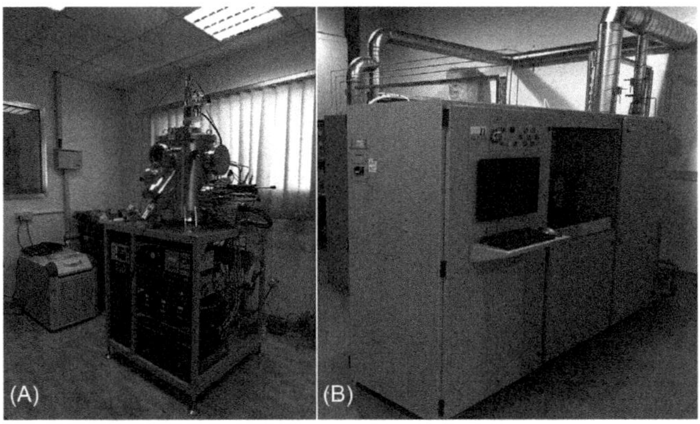

Figure 12.2 Typical electron beam physical vapor deposition (EBPVD, manufactured by Blue Wave Semiconductors, USA) and chemical vapor deposition (CVD, manufactured by First Nano, USA) equipment used in coating processes. (The equipment is housed in the Nanotechnology Labs, EBRC-KISR.)

12.2 Thermal spraying

Thermal spraying technology comprises a group of coating approaches and processes that can be roughly divided into two groups according to the types of feedstock materials as; metallic and nonmetallic groups. Apart from thermal spray coating, and in 1990s new technology so called cold gas-dynamic spray technology, or cold spray appeared. As its name indicates, cold spray is in which the high energy gas flow is

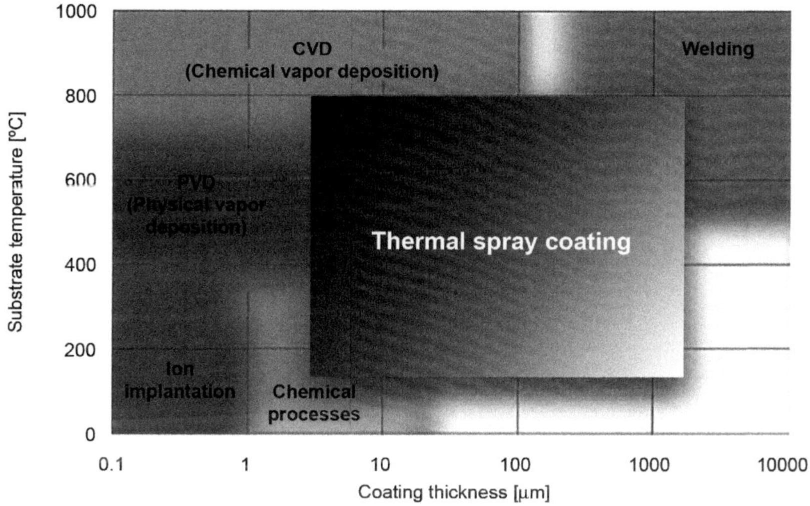

Figure 12.3 Correlation between the coating approaches, applied temperatures, and coating thicknesses of selected coating processes.

produced by a compressed relatively cold gas ($T < 800°C$) expanding in a nozzle, perhaps is considered as the most promising technology that maintains the size of sprayed nanocrystalline powders in their original nano-dimension range (below 100 nm in diameter). The feedstock materials of thermal spray processes can be in the form of powders, rods, or wires. Pawlowski[2] proposed a more useful commercial classification based on the material properties, as shown in Fig. 12.4.

12.2.1 Combustion-based processes

As was previously discussed, thermal spray techniques can be classified based on the applied temperatures and energy source into two groups, namely combustion or electric, and the cold-spray process. The combustion methods can be divided into two subgroups: low velocity and high velocity methods. The applied temperatures in these groups are similar because of using the same fuel sources of gas(es) (or liquids).

12.2.1.1 High velocity oxygen thermal spraying (HVOF)

High HVOF technique, which is a relatively new method of thermal spray coating family, has received great attention as one of the most powerful techniques used to deposit fine powder particles on different types of material substrates. The fuels used in this process could be propylene, propane, kerosene (liquid), acetylene, and hydrogen. Whereas oxygen is used as an oxidant agent, an atmospheric air is utilized to pressurize the combustion chamber.

Likewise, the other thermal spray techniques (except cold spray) HVOF is not a solid-state process since the coating materials are deposited on the surface of the substrate in a molten or semi-molten state to tailor the desired coat. The size of the

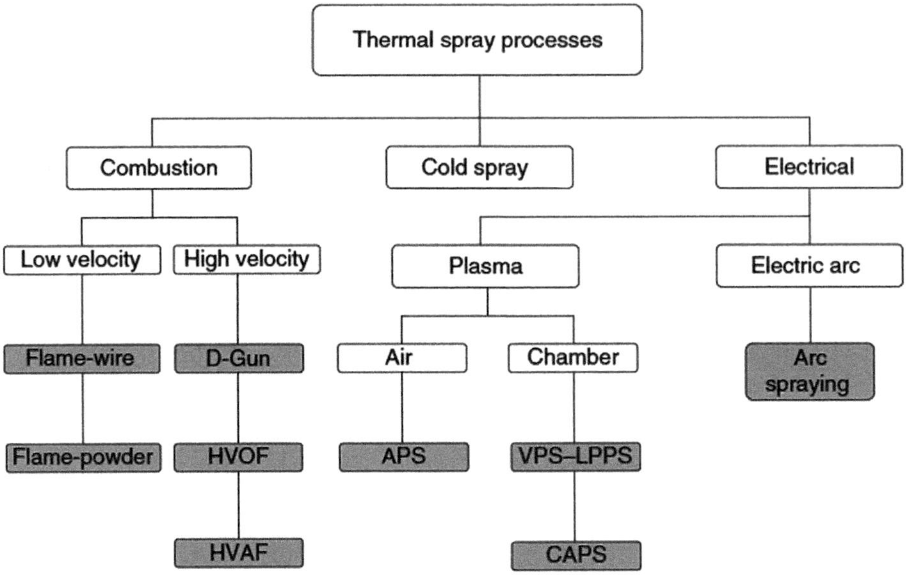

Figure 12.4 Types of thermal spraying processes and the corresponding techniques. (*After Guilemany and Nin*[3]).

powders used as feedstock materials is considered as critical factor for successful coating process. The powders should enjoy particle sizes in the range between 10 and 40 μm. Coarse powders that consist of large particles above 40 μm usually lead to non-melted particles in the structure of the deposited layer while fine particles (above 10 μm) burn or degrade completely before reaching the substrate.

The HVOF uses a supersonic jet, setting it apart from conventional flame spray, the speed of particle impact on the substrate is much higher, resulting in improved coating characteristics. The mechanism differs from flame spraying by an expansion of the jet at the exit of the gun (Fig. 12.5A and B). Fuel gases of propane, propylene, acetylene, hydrogen, and natural gas can be used, as well as liquid fuels such as kerosene. The powder coating materials are heated to reach to their molten state and accelerated by employing such a high temperature, high-velocity gas stream toward a substrate (Fig. 12.5C). Such stream of molten powder impacts on the substrate object in the form of a stream of droplets in the high-energy gas stream. The droplets flatten or deform on the substrate and generate lamellae called "splats." The piling up of multiple layered splats forms the coating. One advantage of HVOF method is the dwell times required that lead to minimize degradation of the starting feedstock powders. In addition, the short processing time can overcome/minimize the interaction between the molten powder stream with the surrounding atmosphere.

12.2.1.2 Utilization of ball-milled powders as feedstock materials for HVOF

There are wide spectrum of powder production methods employed to produce feedstock materials suitable for HVOF technique such as fusion and crushing methods, atomization, spray drying, self-propagating high temperature synthesis (SHS), and

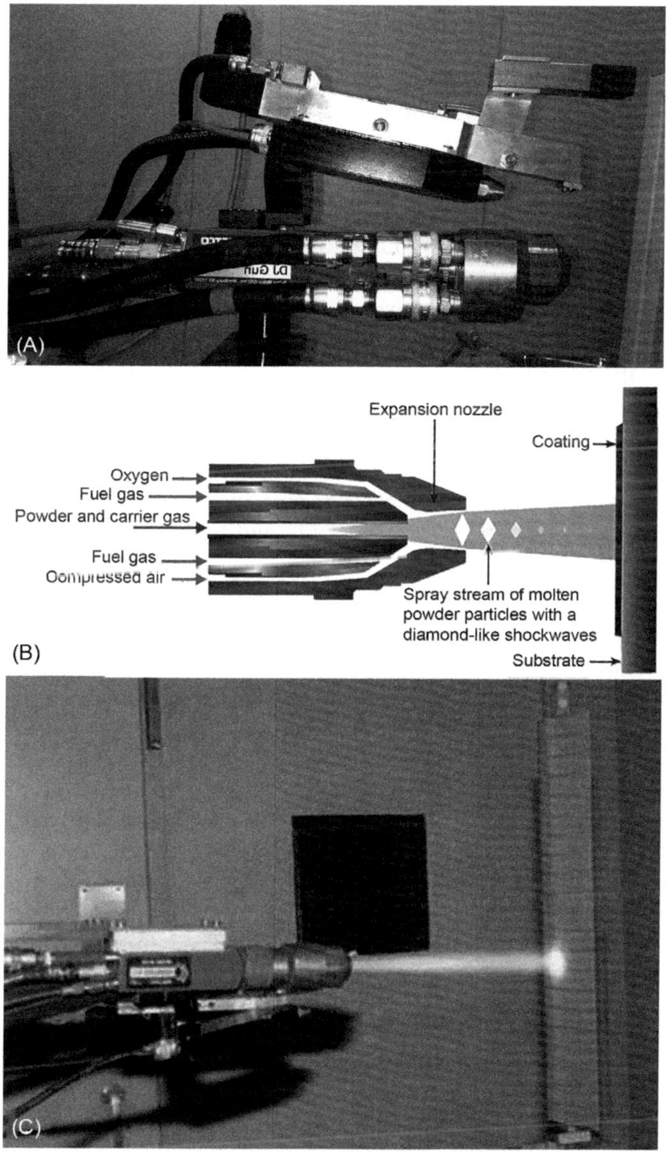

Figure 12.5 (A) Photo of the gun for HVOF, (B) schematic diagram of the HVOF process, and (C) photo of the gun during spraying process of HVOF using an equipment provided by Sulzer, Germany, and housed in the Nanotechnology Labs, EBRC-KISR.

sol–gel technique. Apart from the traditional and commercial material powders used in HVOF, using of nanostructured powders as feedstock-HVOF materials are continuously increased due to their enhanced wear-resistant properties and hence they become new hot-research topic of thermal spray. Mechanical alloying/milling have been considered as the most powerful methods for preparing different types of nanocrystalline (metallic and nonmetallic) and nanocomposite (metal matrix and ceramic matrix) powders with particle sizes in the range recommended for using in HVOF. Some typical examples of mechanically alloyed powders used to tailor protective coating via HVOF will be presented in this section.

12.2.1.2.1 HVOF reactive spraying of mechanically alloyed Ni–Ti–C powders

Horlock et al.[4] mechanically alloyed a mixture of Ni(Cr) prealloyed powder and Ti and C to fabricate nanocomposite powders, using planar-type ball mill. Their results showed that metastable Ni–Ti–C powders were obtained with the nominal composition 50wt.%Ni–40wt.%Ti–10wt.%C with a particle size range of 8–38 μm. They employed HVOF spraying to deposite approximately 250 μm thick onto mild steel substrates. Based on the XRD results, they suggested that a self-propagating high-temperature synthesis (SHS) reaction had occurred in the powder particles during the spraying process and the major phases presented in the coating, which had splat like morphology, were nanoscale TiC (~50–200 nm) and a Ni-rich solid solution coexisted with small volume fractions of NiTi, TiO_2, and $NiTiO_3$.

12.2.1.2.2 HVOF of nanostructured Cr_3C_2-Ni20Cr coatings

Da Cunha et al.[5] used wet high-energy ball milling for preparations of Cr_3C_2-25(Ni20Cr) powders. They used liquid hexane as milling control agent to reduce the average particle size. In addition, they milled the powders under nitrogen gas atmosphere and found that this led to the formation of uniform nanoscale powders with an average particle size of 11 mm in diameter. The as-ball-milled Cr_3C_2-25(Ni20Cr) powders under nitrogen gas for 4 h were used as feedstock for HVOF on AISI 310 specimens 50 × 20 × 2 mm to tailor a coat with an average thickness of 200 μm (Fig. 12.6). Their results showed that the microhardness of as-HVOF nanocrystalline coating is higher (1124.5 VHN) when compared with those coating where microscale powders were used (810 VHN).

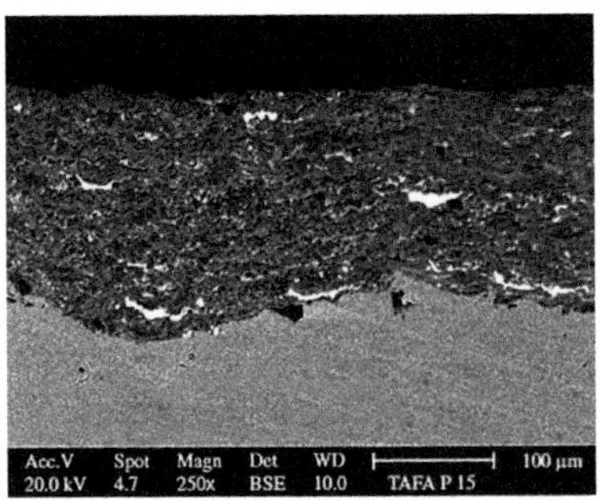

Figure 12.6 SEM micrograph of the cross-sectional view of HVOF coating of nanostructured Cr_3C_2-25(Ni20Cr) on AISI 310 substrate. (*After Da Cunha et al.*[5])

12.2.1.2.3 HVOF of nanocrystalline iron aluminide

It has been suggested by Hu et al.[6] that refining the grain size of the starting powders to the nanocrystalline level led to improve the strength of the coating while enhancing ductility and toughness. In their study, nanocrystalline iron aluminide intermetallic compounds powder synthesized by mechanical alloying were used as feedstock materials for HVOF. The coating contained essentially a fairly disordered FeAl phase coexisted with nanocrystalline grain size. The results showed that the hardness of deposited nanocrystalline coating is higher than that of a conventional coating and bulk material, whereas the elastic modulus is lower than that of bulk material.

12.2.1.2.4 High-feed-milled HVOF sprayed WC-Co coatings

More recently, Tillmanna et al.[7] have reported that textured surface patterns were obtained upon HVOF sprayed WC-Co coatings with nanosized WC particles, using a HFM post process. They have investigated that that the HFM post process leads to refine the WC crystallite size as a result of increase of its internal strain, which both can be attributed to plastic deformations during HFM.

12.2.1.2.5 HVOF sprayed diamond reinforced bronze coatings

Das et al.[8] suggested that the hardness and wear resistance of soft bronze metal were significantly improved upon reinforcing with 10 wt.% of diamond powders, using ball-milling approach. The composite powders were used as feedstock for HVOF process. Based on their experimental results, the hardness and elastic modulus were increased by 35% and 15%, respectively, upon addition of diamond in bronze coatings. Moreover, Diamond–Bronze composite coatings offered up to 65% increase in wear resistance.

12.2.2 Cold spray process

A new comer, named cold gas-dynamic spray process or simply called as cold spray, which was developed at the Institute of Theoretical and Applied Mechanics of the Russian Academy of Sciences in Novosibirsk,[9] joined the thermal spray family in the mid-1980s. Some other names includes cold gas spray, micro cold spray, cold gas dynamic, spray, kinetic spray, and supersonic particle deposition for the same process are used.[10] In contrast to the conventional thermal spray approach, cold spray procedure is a solid-state process, in which the powder feedstock remains in its solid form without melting. This relatively new process takes place with supersonic acceleration of the powder by a carrier gas through a convergent divergent nozzle and impact onto the desired substrate with a high velocity (300–1200 m/s), as elucidated in Fig. 12.7. Fig. 12.8 displays digital photos of a typical cold spray system, composing of (1) high-pressure air compressor to compress the air in a high-pressure tank (Fig. 12.8A), (2) a dust collector unit is installed outside of the building (Fig. 12.8B) to collect the fine sprayed powders suspended in atmosphere of the spray room (Fig. 12.8C) during the coating process, and (3) gas control unit with a gas mixer is used to control the gas mixture (e.g., helium and nitrogen gas) with the desired flow rate and pressure (Fig. 12.8D). The indoor cold spraying process is fully automated and controlled by

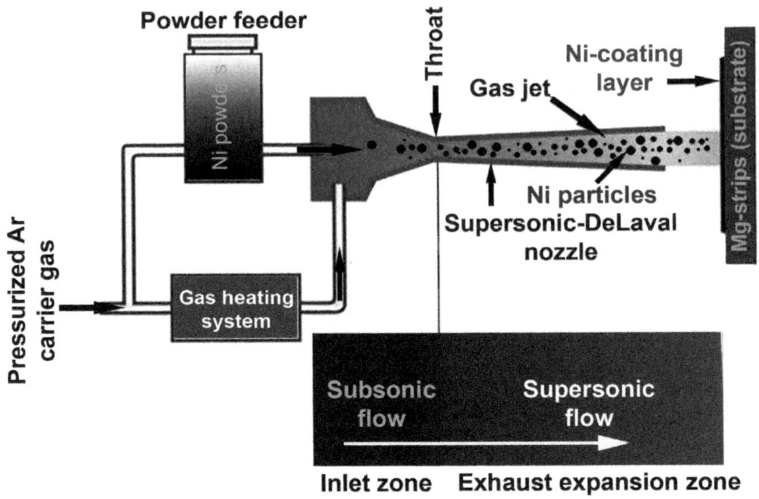

Figure 12.7 Schematic of cold spray process.

central process control unit, using a software to select the desired operating parameters (Fig. 12.8E). Fig. 12.8F displays the components installed in the spray room. The main components are powder feeders and a robot, which carries the gun during the coating process. The gun with the expansion nozzle can be oriented in a horizontal or vertical positions, as shown in Fig. 12.9A and B, respectively. An optional low-pressure cold spray system, which is used for outdoor coating, can be also installed and manually operated inside the room (Fig. 12.8F). Horizontal (Fig. 12.8G) and vertical (Fig. 12.8H) turntables used for coating the outer and inner parts of the objects (e.g., pipelines) can be installed inside the spray room.

As was discussed in the previous section (12.2.1.1), in thermal spraying combustion, molten or semi-molten powder particles are deposited onto a substrate material to form a coating. Unlike in the thermal spray combustion-based methods, the cold spray approach is achieved at low temperature (in the range between 50° and 900°C), being far below than the melting points of the feedstock powders. Because cold spray method is a solid-state process achieved at low processing temperatures, it is suitable process for many of the oxygen-sensitive powders (e.g., aluminum, copper, or titanium) used for coating applications. Moreover, it may open a new horizon in utilizing the mechanically alloyed/milled metastable powder materials such as nanoscaled, amorphous, and metallic glassy powders as cold spray feedstock materials in surface protective coating applications. It is expected that the as-cold sprayed nanocrystalline powders can maintain their original metastable phase and nano-dimensions (1–100 nm in diameters) with a minimum grain growth.

Scientific publications include books, and review articles presented and discussed the fundamental, process and applications of cold spray in different area have been monotonically increased since 1990 (Fig. 12.10).

Figure 12.8 Photos taken from the Nanotechnology Laboratory, Energy and Building Research Center (EBRC), Kuwait Institute for Scientific Research (KISR), present the components of a cold spray system (provided by Startack Co. Ltd., Japan). The system consists of high-pressure air compressor to compress the air in a high-pressure tank (A), a dust collector (B), spray room (C), gas control unit with a gas mixer (D), and central process control unit (E). The component of the system installed inside the spray room including the powder feeder, a robot, and low-pressure cold spray system is shown together in Fig. 12.7F, which carries the gun during the coating process. (The gun with the expansion nozzle can be oriented in a horizontal or vertical positions, as shown in Fig. 12.8A and B, respectively, which is used for outdoor coating that can be also installed and manually operated inside the room (Fig. 12.7F). Horizontal and vertical turntables are displayed in Fig. 12.7G and H, respectively.

12.2.2.1 Advantages

Among the thermal spraying coating process, cold spray method enjoys wide advantages that make it a unique approach and desirable technique for nanocrystalline and amorphous materials. Some of these advantages can be summarized as follows:

The coatings can exhibit wrought-like microstructures with near theoretical density values:

- Well defined small spray trace (in the range between 1 and typically 1–25 mm^2) allowing precise control on the substrate area of deposition;
- The impact velocity gained by the sprayed powder particles upon applying the jet with high velocity reaching to 1200 m/s leads to subjecting the particles under compressive stresses and this yield to sever plastic deformation on the deposited powders. As a result, thickness

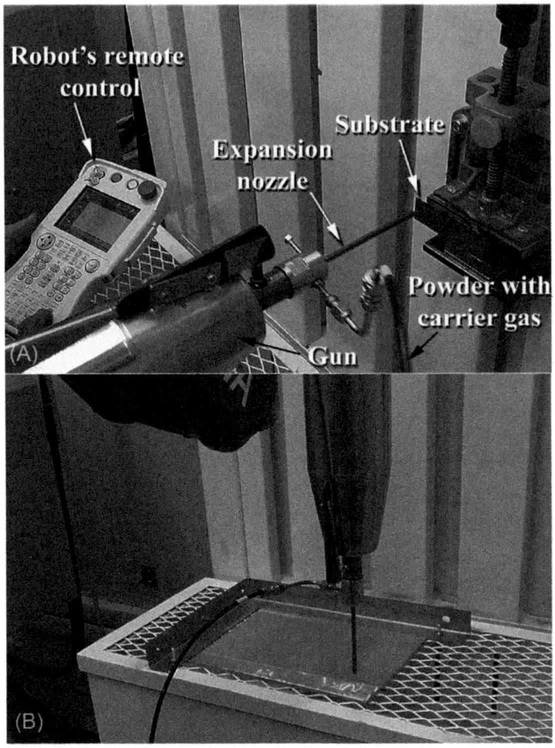

Figure 12.9 The gun with the expansion nozzle of a cold spraying system can be oriented using a robot system in horizontal (A) or vertical (B) positions.

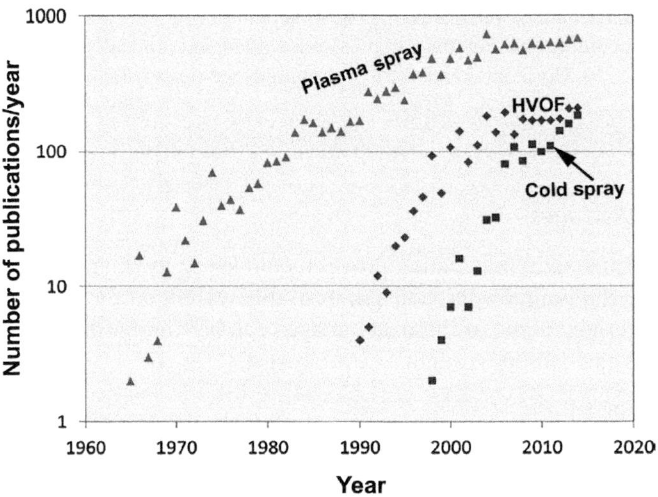

Figure 12.10 Collected data from Scopus representing the number of review articles and books of coating technology published in the period between 1965 and 2018.

of the as-coated materials produced with such severe compressive stresses can be ultra-thick layers reaching in the range between 5 and 50 μm without adhesion failures;

- The cold-sprayed coating materials possess uniform concentration of the sprayed powder particles in the jet with a minimal varying concentration limit;
- The temperature sensitive powder particles such as copper (Fig. 12.11A and B), aluminum, nanocrystalline, amorphous, and metallic glassy powders can be used safely as feedstock materials without sever phase transformation and/or grain growth after the coating process;
- Applications of medium range temperature leads to use sensitive materials (e.g., glass or polymers) as substrates (Fig. 12.11C);
- Low thermal load on materials (typically: substrate 50–250°C, feedstock sprayed powders 50–900°C);
- Cold spray can be used to coat substrate objects of different sizes by spraying soft metallic powders such as Cu (Fig. 12.11A) or hard materials such as WC-Co-based nanocomposites (Fig. 12.11D);
- High deposition rate up to 14 kg/h; and
- Low porosity coating.

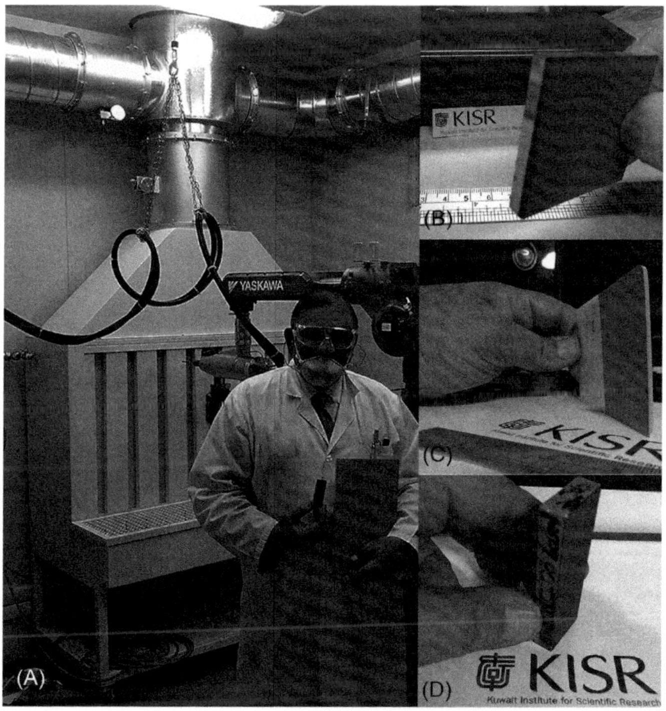

Figure 12.11 Combinations between particles' temperature, velocity, and size lead to a successful cold spraying processing for depositing wide range of (A) and (B) pure metals such as Cu onto a variety of substrate materials including steel (A and B) or polymers (C). Cold spraying can be used also for deposition of hard composite materials such as WC-Co-based nanocomposite powders. (The photo presents the author in the cold spray laboratory with the cold-sprayed substrate samples achieved by a cold spray equipment provided by Startack Co. Ltd., housed in the Nanotechnology Laboratory, Energy and Building Research Center (EBRC), Kuwait Institute for Scientific Research (KISR).)

12.2.2.2 Mechanism

However, the coating mechanism in thermal spraying combustion approach is understood and can be summarized as the occurrence of local adiabatic shear instabilities at the particle-substrate and particle-particle zones (interfaces) upon subjecting to thermal softening, the full mechanism of cold spraying process is still not completely understood.[8] The principles of cold gas spraying have been described in more details elsewhere[11–16] and are schematically summarized in Fig. 12.7. It is agreed that the possible mechanism can be explained as follows:

- The particle and carrier gas velocities are increased upon heating (without using combustion) to temperatures far below the melting points of the sprayed powders.
- Impact the powders with a high-velocity at temperatures far below the melting point of the feedstock powders, the solid powder particles undergo to severe plastic deformation.[13]
- The solid powder particles sprayed at such high velocity hit the substrate and hence an intimate conformal contact is developed. It is assumed that the kinetic energy of the particles with high contact pressure promotes bonding with the substrate surface and develops the coating formation.[15,16]

12.2.2.3 Cold spraying of metastable powders obtained by mechanical alloying

The worldwide interest on using amorphous and/or nanocrystalline layers on metal substrates is due to the expected improving on the expected surface performance such as wear and corrosion resistance for the bulk material objects. Cold spraying technique opened a new horizon for using metastable phases (e.g., nanocrystalline and metallic glassy powders) for using as feedstock materials in spraying process at temperature well below their phase transformations and grain growth. Mechanical alloying and mechanical milling techniques has been considered as the most powerful and cost effectiveness tool used to produce the feedstock nanocrystalline powders for thermal spraying process.

The possibility of employing mechanical milling for preparing nanocrystalline aluminum 5083 powders starting from commercially available atomized material powders was reported by Ajdelsztajn et al.[17] In their experiments, the mechanically milling process was achieved, using an attritor mill under liquid nitrogen. The end product obtained after 8 h of milling was nanocrystalline powders with grain size in the range of 20–30 nm. They used the mechanically milled powders as feedstock powders for cold spraying process, using a room temperature inert carrier gas of helium. The stagnation inlet pressure was set at 1.7 MPa, while the injection pressure was set to provide a pressure difference of 80 kPa at the injection site. They reported[14] that a successful spraying process was achieved for deposition of the mechanically nanostructured powder onto pure Al substrate with a coating thickness of 270 μm, as displayed in Fig. 12.12A.[17] The as-cold sprayed coating, which consisted of nanocrystalline grains (Fig. 12.12B), was porous free with an excellent interface with the substrate material. In addition, they found that the rough surface of the cryomilled powders provided higher surface area and a larger number of irregularities that increases

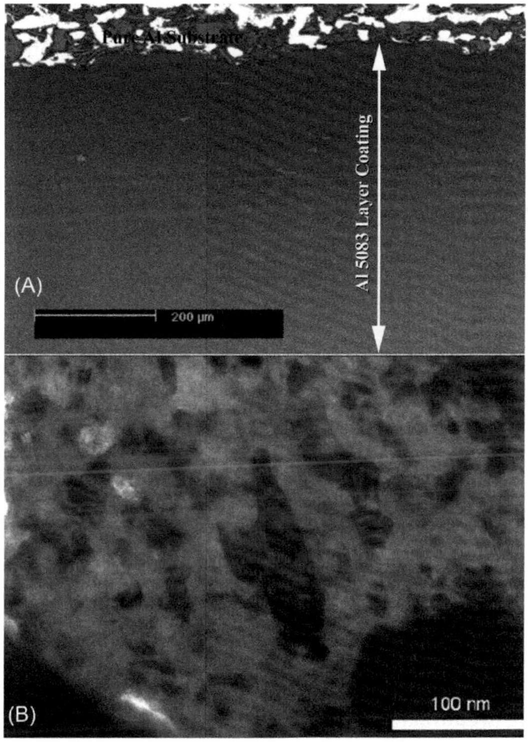

Figure 12.12 SEM image of the cross-sectional view (A) and bright-field TEM image (B) of the nanocrystalline cold-sprayed Al 5083 coating onto pure Al substrate. (*After Ajdelsztajn et al.*[17])

the possibility of mechanical interlocking. The coating hardness value, which was 261 HV (300 g), is higher than any other values reported for Al 5083 consolidated with other techniques.

Another example of using ball-milling technique for preparing the nanocrystalline Al-based alloys (Al 6061 and Al 5083 alloys) feedstock material powders utilized for cold spray process is given by Hall et al.[18] They pointed out that cold spraying led to refine the microstructure of an ultra-fine-grained powder and consolidate it to create a homogenous nanocrystalline (20–40 nm) microstructure. They suggested that the grain refinement observed in Al was due to cold spraying process that leads to a severe plastic deformation of the sprayed powders.

Li et al.[19] used planetary ball mill for preparing uniform Al5356/TiN composites powders starting from gas-atomized Al5356 and commercial TiN powders. The as-ball-milled composite powders were then cold sprayed onto pure Al substrate. The cold spray process was achieved using a high-pressure air as at a pressure of 2.7 MPa and a temperature of about 510 C. Argon gas was used as powder carrier gas with a pressure of about 3 MPa. The reported that a uniform composite layer was successfully formed with a microhardness higher than that of pure Al5356 coating. They attributed such a high hardness values of the coating deposit to the ball-milling process and the existence of TiN grains in the matrix of Al5356 alloy. Moreover, their results were shown that the coefficient of frication and wear rate decreased by about 14 and 50 times than that of pure Al5356 deposit, respectively.

Nickel titanium alloy powders fabricated by mechanical alloying process was employed to deposit $Ni_{50}Ti_{50}$ alloy by cold spraying.[20] It was observed that a self-propagating high-temperature combustion synthesis (SHS) phenomenon occurs during cold spraying of the mechanically alloyed powders. The carrier gas temperature igniting the SHS reaction during cold spraying decreased with the increase of ball-milling time. The ignition of SHS reaction led to partial melting of Ni/Ti powder. It was found that the melted particle fraction which resulted from SHS reaction led to splashing of spray particles off substrate on impact, which significantly decreased the deposition efficiency.[20]

Nanocrystalline Cu powders (5–40 nm) were prepared by cryomilling for 12 h and used as feedstock materials for spraying onto Al6061 substrate, using cold spraying process.[21] The cold spraying process was achieved using nitrogen as a powder carrier gas at a temperature of 300°C with accelerating gas stagnation pressure of 1.5 MPa. The nanoindentation results showed that the hardness value of sprayed Cu coating reached to 3.3 GPa, which was higher than that of its coarse grained counterpart.

More recently, Rokni et al.[22] employed electron microscopy and nanoindentation to investigate the relationship between microstructure and nanohardness of nanocrystalline 5083 Al alloy powder before and after being deposited by cold spraying process. Microstructural investigations observed the presence of nano grains in the powder microstructure, ranging from 20 to 80 nm and with a typical grain size of 40–50 nm. It was also revealed that the nanocrystalline structure of the powder is retained after cold spraying. As a result, almost no change in nanohardness was indicated between the powder and the particles interior in the cold-sprayed layer. However, hardness was substantially higher in some regions in the cold-sprayed layer, which was attributed to the particle–particle interfaces or other areas with very small nanograin size.

List et al.[23] have reported the possibility of using metallic glassy powders of FeCoCrMoBC obtained by gas atomization technique as feedstock powders in cold spraying process. The powders were sprayed onto different substrate materials such as $AlMg_3$ alloy, pure Cu, pure Ti and SUS 3161. The XRD patterns (Fig. 12.13) showed that the cold sprayed coatings contain more than 98% of amorphous structures, similar to phase contents of the original feedstock powders.[19] They found that the coating microstructures show a porosity of about 2.5%. In addition, the coating hardness was measure and found to be more than 1100 HV 0.3, similar to that of cast Fe-based bulk metallic glass. In their study, applying higher temperatures and deformation rates on harder substrates are beneficial for shear banding and bonding, leading to higher deposition efficiency.[23]

Mechanically alloyed $Cu_{50}Ti_{20}Ni_{30}$ metallic glassy alloy powders, proposed for antibacterial protective coating applications in medical and food sectors were investigated for the first time by El-Eskandarany and Al-Azmi[24] in 2017. In their study, mechanical alloying process was utilized for synthesizing of metallic glassy $Cu_{50}Ti_{20}Ni_{30}$ alloy powders starting from the elemental alloying, using a low energy

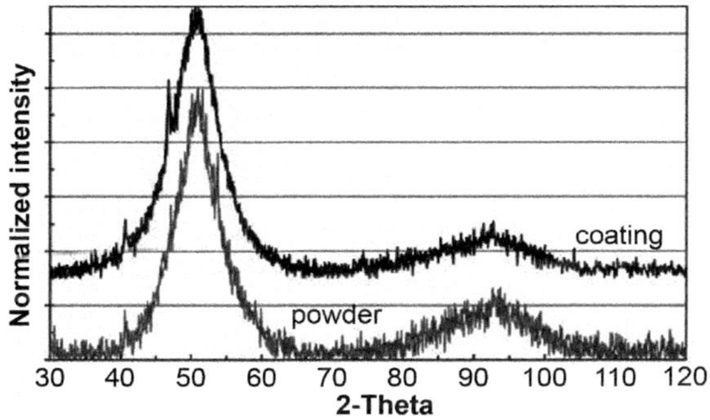

Figure 12.13 XRD patterns of FeCoCrMoCB bulk metallic glass (coat) cold-sprayed onto SUS 3161 substrate under a gas pressure of 4 MPa and a gas temperature of 900°C. The XRD pattern of the original FeCoCrMoCB metallic glassy powders is presented below to that pattern. (*After List et al.*[23])

ball-milling technique. The metallic glassy powders obtained after 100 h of ball milling had an average particle size of 1.7 μm in diameter and possessed excellent homogeneous distribution of the alloying elements, as shown in Fig. 12.14. Moreover, this end product possessed excellent thermal stability, indexed by a relatively high glass transition temperature (358.3°C) with a wide supercooled liquid region extended up to 61°C (Fig. 12.15A and B). This amorphous phase tended to crystallize into Ti_2Cu and $CuTiNi_2$ ordered phases through two overlapped crystallization temperatures at 419.3 and 447.5°C, respectively (Fig. 12.15B and C). In a part of employing the fabricated metallic glassy powders in useful applications in medical and food industries, the powders of the end product (100 h) were employed as feedstock materials to double-face coating the surface of SUS 304 substrate (Fig. 12.16), using cold spraying

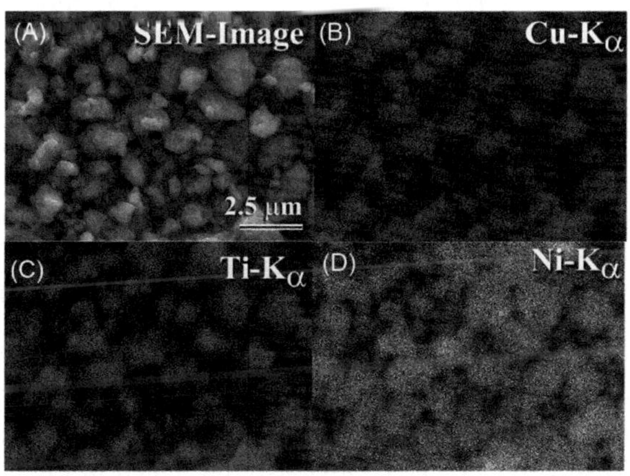

Figure 12.14 (A) SEM micrograph and the corresponding X-ray elemental mapping of (B) Cu, (C) Ti, and (D) Ni of mechanically alloyed $Cu_{50}Ti_{20}Ni_{30}$ powders obtained after 100 h of the ball-milling time.[25]

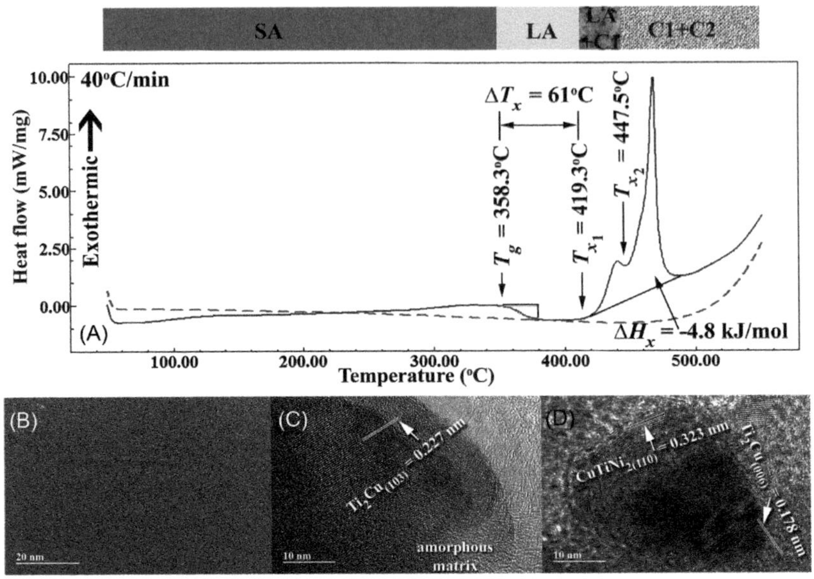

Figure 12.15 DSC thermogram of the powders obtained after 100 h of the ball-milling time. The regions of solid amorphous (SA), liquid amorphous (LA), LA + crystalline phase 1 (LA + C1), and crystalline phases C1 + C2 are schematically elucidated in the bar above the figure. The glass transition temperature (T_g), supercooled liquid region (ΔT_x), crystallization temperature (T_x) of phases 1 (T_{x1}) and 2 (T_{x2}), and the enthalpy change of crystallizations (ΔH_x) are indexed in the figure. The HRTEM images for the samples heated in the DSC up to 400°C, 420, and 560°C are shown in (B), (C), and (D), respectively.[25]

process.[25] The coating process was achieved under helium gas pressure at temperature laid in the supercooled liquid region (400°C). The as-cold sprayed metallic glassy material maintained its original amorphous structure with existence of small volume fraction of ordered-Ti_2Cu phase, as displayed in Fig. 12.17. In addition, the coating glassy material possessed excellent morphological and elemental distribution with the absence of any compositional degradation within the microscale, as investigated by SEM/EDS technique (Fig. 12.18). The nanoindentation experiments indicated that this metallic glassy coating material had an extraordinary high nanohardness value of 3.1 GPa (Fig. 12.19).

Moreover, it showed a high resistance to wear with a low value of the coefficient of friction ranging from 0.45 to 0.45, as experimentally investigated (Fig. 12.20). The macrograph taken for the worn surfaces of $Cu_{50}Ti_{20}Ni_{30}$ coated SUS304 specimen after achieving the wear test is shown in Fig. 12.21A. The indexed circular symbol taken at the bottom zone of the wear track presented in Fig. 12.21A refers to the area examined by FE-SEM (Fig. 12.21B). Obviously presented in Fig. 12.21B, the wear track zone achieved at the coated metallic glassy layer consists of large bright shots with irregular shapes and were embedded into the remained coating layers. In order to understand the nature of these bulk shots materials, X-ray elemental mapping was

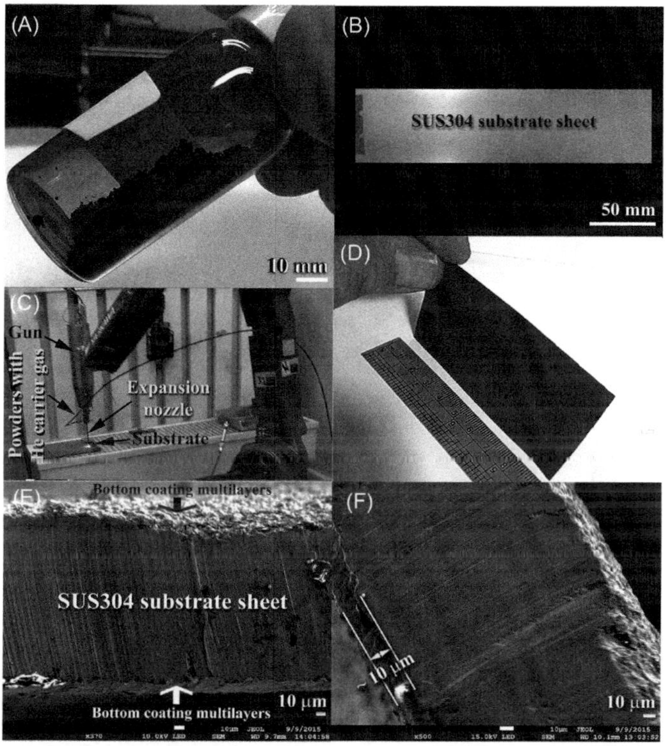

Figure 12.16 (A) as-prepared $Cu_{50}Ti_{20}Ni_{30}$ metallic glassy powders obtained after 200 h of MA time, (B) as-surface treated sheet strip of SUS304 alloy with a dimension of ~20 mm length × 4 cm width, (C) cold spray coating processes using the as-prepared metallic glassy powders as feedstock coating materials, and (D) SUS304 substrate coated with $Cu_{50}Ti_{20}Ni_{30}$ metallic glassy powders. The low-magnification FE-SEM micrographs for the elevation and side views of the metallic glassy coated/SUS304 are shown in (E) and (F), respectively.[25]

performed to identify the elemental distribution of W (Fig. 12.21C), C (Fig. 12.21D), Ni (Fig. 12.21E), Cu (Fig. 12.21F), and Ti (Fig. 12.21G). Basically, the coating material consisted of the alloying elements of Ni, Cu, and Ti, as shown in (C), (F), and (G) of Fig. 12.21. The large grains embedded at the metallic glassy coating layer surface in the shallow wear grooves were related by those shots released from the WC/Ni ball during the wear test, as suggested by Fig. 12.21C–E. However, the microhardness of the ball (pin) used against the metallic glassy coated SUS304 (disk) has a high value of about 14.37 GPa, it severely worn during the wear test. This is implying high hardness values of the disk materials (metallic glassy coat) and its high resistance. Fig. 12.21H shows higher magnification FE-SEM micrograph of a selected zone taken from the groove existed Fig. 12.21B. The micrograph displays two individual zones with different elevations. The upper zone possessed large cracked grains morphology and consisted of WC/Ni composite, indicated by the X-ray elemental distribution presented in Fig. 12.21I–K. Moreover, small sized WC/Ni shots were embedded into the lower

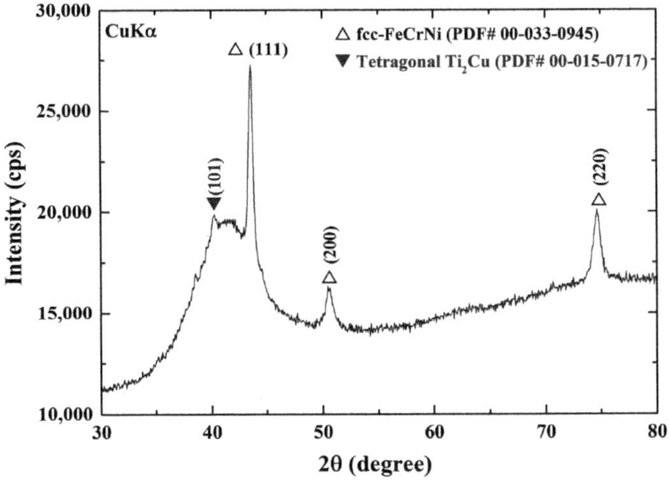

Figure 12.17 XRD pattern of cold spray metallic glassy $Cu_{50}Ti_{20}Ni_{30}$ powders coated SUS304.

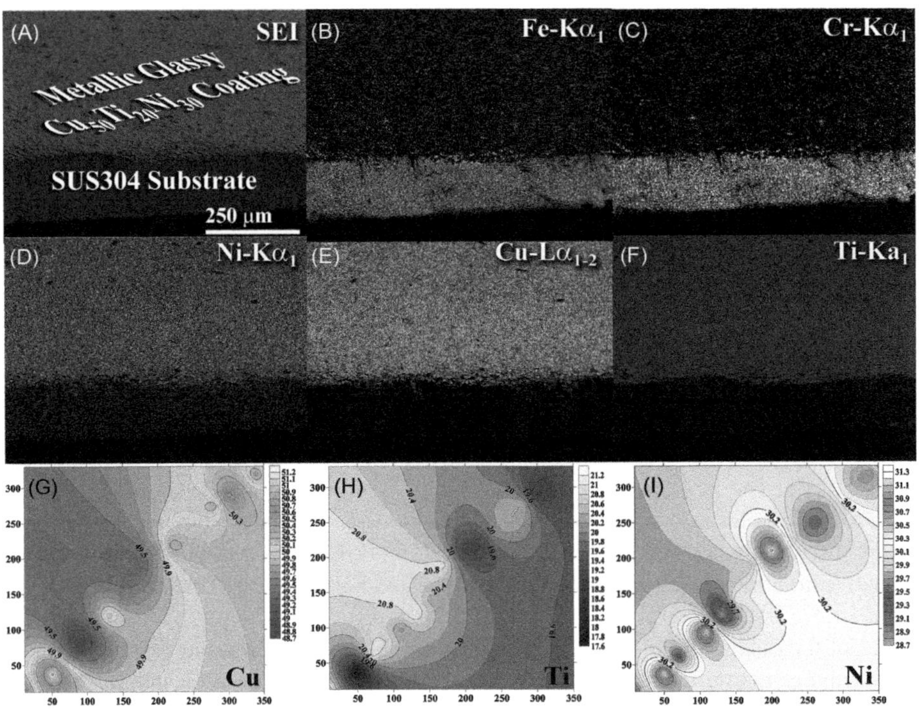

Figure 12.18 (A) Scanning electron image (SEI) and the corresponding EDS elemental mapping of (B) Fe, (C) Cr, (D) Ni, (E) Cu, and (F) Ti of cold-sprayed metallic glassy $Cu_{50}Ti_{20}Ni_{30}$ powders coated SUS304. The elemental EDS-isochemical contour maps of Cu, Ti, and Ni alloying elements located at in (A) are shown in (G), (H), and (I), respectively.[25]

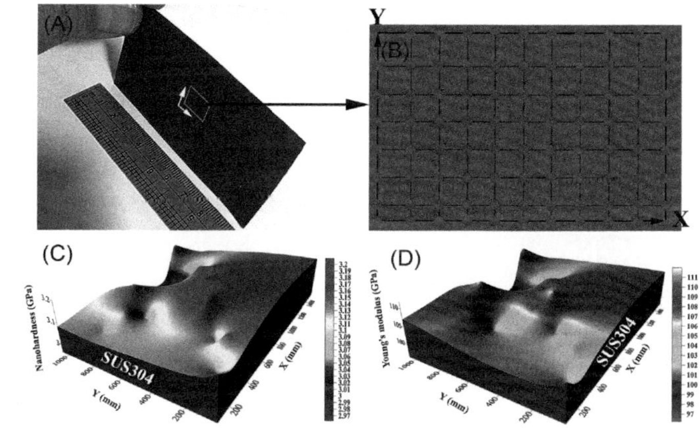

Figure 12.19 (A) A strip portion (~9 × 5 cm) of cold spray metallic glassy $Cu_{50}Ti_{20}Ni_{30}$ powders coated SUS304. The small squared-portion shown in the middle of the strip in (A) refers to the sample selected for performing a nanoindentation test, under a load of 400 mN and is presented in a different scale in (B). The test was achieved at the *red*-circular symbols shown in (B). The corresponding nanohardness and Young's modulus results are shown in isometric projections in (C) and (D), respectively.[25]

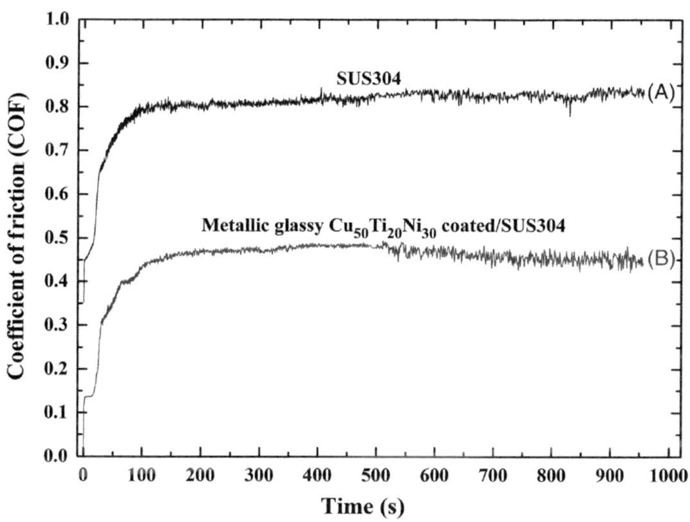

Figure 12.20 Dynamic COF curves of (A) uncoated SUS304 substrate, and (B) cold spray metallic glassy $Cu_{50}Ti_{20}Ni_{30}$ powders coated SUS304. The wear test was performed with sliding wear pin-on disk testing technique, using a tool steel ball under a load of 100 N at ambient temperature for 950 s.[25]

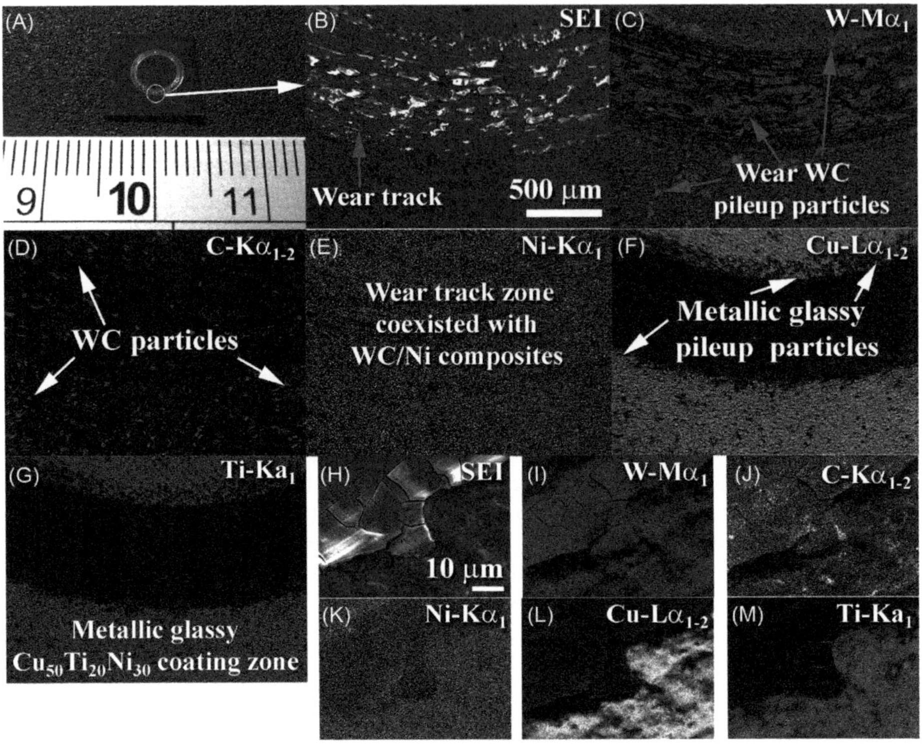

Figure 12.21 (A) A macrograph for the $Cu_{50}Ti_{20}Ni_{30}$ coated SUS304 specimen after achieving the wear test. The corresponding FE-SEM and X-ray mapping for W, C, Ni, Cu, and Ti for the indexed circular zone shown in (A) are presented in (B–G), respectively. A higher magnification FE-SEM micrograph and the corresponding X-ray elemental mapping for a selected worn surface zone shown in (B) are presented in (H–M).[25]

zone, as suggested by the X-ray elemental mapping shown in Fig. 12.21. This lower zone is corresponding to the metallic glassy coated SUS 304, as suggested by the elemental distribution of Ni (Fig. 12.21K), Cu (Fig. 12.21L), and Ti (Fig. 12.21M). It showed be notified in Fig. 12.21H that a large volume fraction of the metallic glassy coating phase was existed under beneath the large-grained WC/Ni, as suggested by the presence of the alloying elements in X-ray mapping shown in Fig. 12.21K–M.

Fig. 12.22 shows the inhibitory effect of coated substrate against *Escherichia coli* biofilm formation incubated for 24, 48, and 72 h. High bacteria count was observed on the negative control substrate SUS304, only Cu coated substrate showed no growth as shown in Fig 12.22A. Although the inhibitory effect of Cu on biofilm formation is not well known, it is thought that Cu ion released from Cu plays an important role in alter the protein structure of bacterial cell wall.

The substrates coated with $Cu_{50}Ti_{20}Ni_{30}$ remarkably inhibited colony formation of *E. coli* compared with SUS304 as shown in Fig. 12.22A. These results were statistically significant ($p < 0.05$). Moreover, results suggest that $Cu_{50}Ti_{20}Ni_{30}$ coated

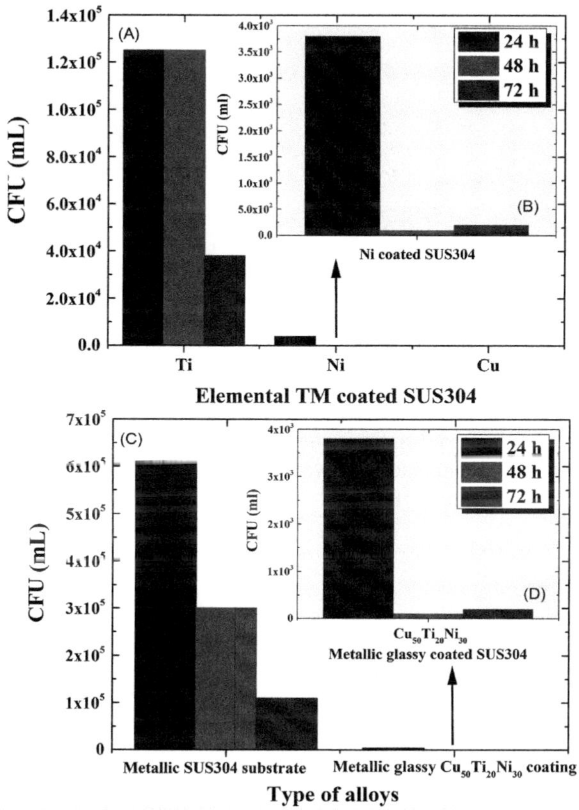

Figure 12.22 Effect of (A and B) elemental metal and (C and D) metallic alloys cold-sprayed coatings on biofilm formation by *E. coli* (ATCC 25922) after 24, 48, and 72 h.[25]

materials will be very effective against biofilm formation in addition, results showed the improvement of antimicrobial effect of Ni when combined with Cu and Ti. In this work, the pure Cu-100% significantly decreased biofilm formation. However, antimicrobial activity was also well obtained with only Cu-50%. That may suggest synergic effect between Cu, Ti, and Ni.

A different attractive application of cold spray for catalyzation of cold-rolled Mg strips by Ni powder particles has been recently reported by El-Eskandarany et al.[26] In their study, the plastically deformed Mg-strips obtained after 300 passes of cold rolling were exposed to a stream of supersonic Ni particles carried by high-velocity jet (500 m/s) of argon gas (Fig. 12.23). Accordingly, the pelted Ni particles were able to penetrate the Mg-substrate strips and created numerous micropores hosting the Ni particles to compose catalytic active sites (Fig. 12.24). The results showed that increasing the number of cold spray coating times (1–3 times) led to increase the Ni concentration from 1.48 wt.% to 5.28 wt.%. Increasing the number of Ni layers (three layers) engaged with the formation of homogeneous layer coating had excellent synergetic

Figure 12.23 (A) The as-CR Mg strips (1) were mounted and fixed by clips (2) on stainless steel (SUS304) sheet, (B) two-movable jaws (3) were used to fix the sample that was aligned to be perpendicular to the nozzle (5) of cold spray (CS) gun (6). A photo presenting Mg-strips after coating with Ni powders for 3 times is shown in (C). The as-coated Mg-strips were wrapped with balance papers and placed between two-jaw type vise for 16 h (D), and then CR for 10 times to ensure straightening (E).[26]

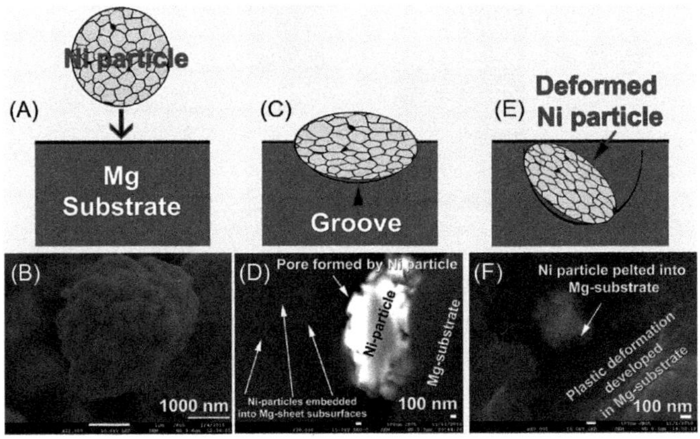

Figure 12.24 (A) CS experimental setup, and (B) schematic illustration of CS process of Mg-substrate doped by supersonic Ni powder particles. The subsonic and supersonic zones presented inside the nozzle gun are indicated in (B).[26]

effect in improving the hydrogen storage behavior of Mg-strips. This was indicated by the capability of bulk Mg/5.28 wt.% Ni to absorb 6.1 wt.% hydrogen within 5.1 min at a low temperature (150°C) under moderate hydrogen pressure (10 bar), as presented in Fig. 12.25A. The formed MgH_2 strips catalyzed with 5.28 wt.% Ni possessed good dehydrogenation characteristics, implied by their capability to release 6.1 wt% hydrogen in 11 min at 200°C under 200 mbar of hydrogen pressure (Fig. 12.25B).

12.2.2.4 *Cold spraying of metal matrix reinforced with carbon nanotubes (CNTs)*

Recently, cold spray has been employed to fabricate dense and thick CNTs-reinforced metal matrix composite (MMC) such as ball-milled Cu-CNT,[27] Cu-CNT-AlN,[28]

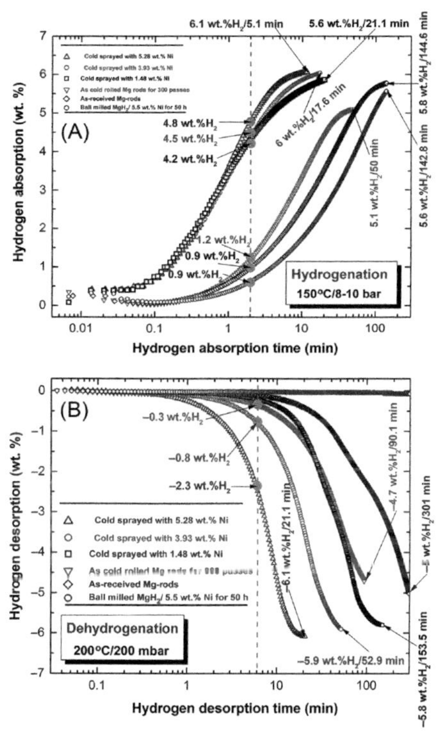

Figure 12.25 (A) Hydrogenation and (B) dehydrogenation kinetics of as-received Mg-rods, CR Mg-rods for 300-passes, CR Mg-rods for 300-passes, and then CS with Ni powders for 1-, 2-, 3 times, and ball-milled MgH_2/5.5 wt.% Ni obtained after 50 h. The hydrogenation and dehydrogenation kinetics measurements were conducted under 150°C/10 bar and 200°/200 mbar, respectively.[26]

Cu-CNT-SiC,[29] and Al-CNT[30] systems. The fabricated coating composite systems possessed homogeneous distribution of the reinforcement CNTs, as implied by SEM and EDS analysis (Fig. 12.26).

Although ball-milling technique has excellent capability for preparing homogeneous CNT-MMC-based materials, but it shows as serious technical problem related to the damage taking place in CNT during the milling process, as pointed out by Cho et al.[31] Above that, employing the high-velocity impact with cold spray coating procedure led to further fracture of CNT, as suggested by Bakshi et al.[30]

12.2.2.5 Cold spraying of metal matrix reinforced with diamond powders

Due to its extraordinary high hardness, diamond and nanodiamonds (NDs) have become favorable reinforcement powders/nanopowders, ball-milled with MMCs powder to improve their mechanical properties. However, diamond-reinforced MMCs can be prepared in bulk form with traditional methods that require the application of extremely high temperatures, such as thermal spray,[32] and pressure infiltration techniques,[33] cold spray method shows beneficial processing advances over these

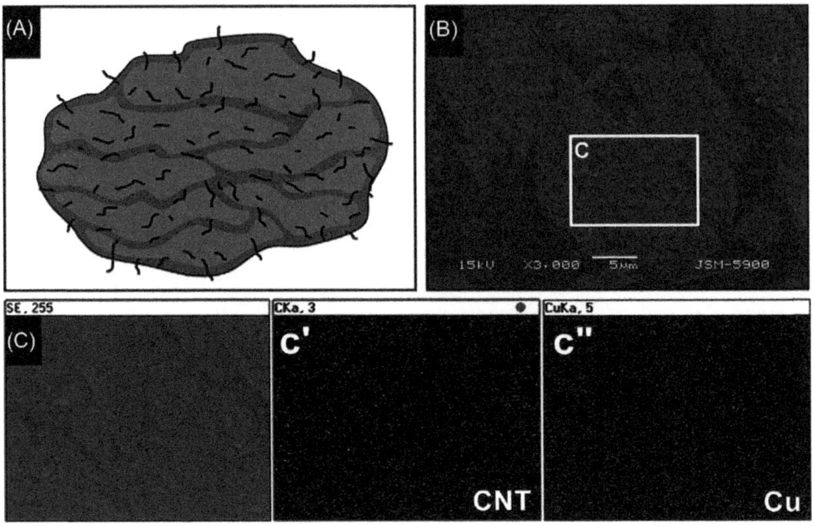

Figure 12.26 Scanning electron image of the cross-sectional view and EDX mapping of a ball-milled CNT-reinforced Cu-MMC powder.[27]

common methods. This is implied by the very low temperature used in cold spray, which protect the composite system risk of high-temperature-induced diamond phase graphitization and simultaneously, retain high diamond contents as pointed out by Yao et al.[34]

12.2.2.6 Cold spraying of metal matrix reinforced with tungsten carbide

Since the World War II, tungsten carbide (WC) has been receiving great attention due to its superior hardness and stiffness that making it desirable hard material. WC powders have been widely used in industrial applications as cutting tools, and for surface protective coating. One way to control the brittleness of WC is to mix the powders with certain volume fractions of metallic Co or with Ni powders, using ball-milling technique. Composite WC-Co can find wide range of application as surface protective coating in oil and water industries. It had been pointed out that cold spray of WC-Co powders led to fabrication of fully dense coat with the absence of pores.[35] In contrast to the other thermal spray coating techniques, cold spray process that is taking place with the absence of decarburization and phase transformation.[36] Accordingly, cold-sprayed WC-Co coatings possesses higher Vickers hardness with lower wear rate when compared by those coating with the same composition obtained by high velocity oxy-fuel technique.[35,37]

12.2.2.7 Applications of cold spray coating feedstock powders

In fact, cold spray is more than a coating process, it can be also used as a manufacturing of near net shape components and for repairing defects in manufacture components and parts.[38,39] Cold spray coating, using ball-milled powders as feedstock materials can find wide range of industrial applications that are probably difficult to be completely covered in this section. Of these, we will give some examples of the powder families, as shown hereafter.

- *Ni-based alloy powders:* This class of materials reveals excellent resistance to wear and corrosion, especially at high temperature. They are mainly employed for coating the machinery and parts in aerospace, automobile, as well as medical devices. In addition, they are used for piston extrudes and wearing plates.
- *Co-based alloy powders:* This family finds excellent applications used for improving the wear resistance and impact strength for the equipment and devices used in the offshore sector such as gate valves, and in the chemical processing plants.
- *Cu-based alloy powders:* Are the best choice for using as coating materials for improving the electrical conductivity in those applications that need to enhance their electrical and thermal conductivity.
- *Ti- and hard Cr-based alloy powders:* Based on their superior corrosion resistance they are used in aerospace and medical applications that are require to have extreme resistance to corrosion. They are used in piston rings, aircraft engine, and hydraulic pistons and cylinders.

References

1. Turunen, E., Varis, T., Gustafsson, T.E., Keskinen, J., Falt,T., and Hannula, Simo-Pekka, *Surface Coatings & Technology*, 200:4987 (2006)
2. Pawlowski, L., *The Science and Engineering of Thermal Spray Coatings*, John Wiley, New York. (1995)
3. Guilemany, J.M., and and Nin, J., Thermal spraying methods for protection against wear, in Surface coatings for protection against wear, Edited by Mellor, B.G., (2006) Woodhead Publishing Limited, Cambridge, England.
4. Horlock, A. J., Sadeghian, Z., McCartney, D. G., and Shipway, P. H., *Journal of Thermal Spray Technology*, 14:77 (2005)
5. Da Cunha, Cecilio Alvares et al., *Materials Research*, 11:137 (2008)
6. Hu, W., Kato, T., and Fukumoto, M., *Materials Transactions*, 44:2678 (2003)
7. Tillmanna, W., Hagen L., Stangier D., Paulus M., Metin T., Sakrowski R., et al. *Surface & Coatings Technology*, 374: 448 (2019).
8. Das, P., Paul, S., and Bandyopadhya, P.P. Diamond and Related Materials, 93: 15 (2019)
9. Alkhimov, A.P., Kosarev, V.F., and Papyrin, A.N., *Dokl. Akad. Nauk SSSR*, 315:1062 (1990)
10. Assadi H., Kreye, H., Gärtner, F. Klassen, T., *Acta Materialia*, 116: 382 (2016)
11. Dickinson, M. E., and Yamada, M., *Nanoscience and Nanotechnology Letters*, 2:348 (2010)
12. Lima, R. S., Karthikeyan, J. C., Kay, M., Lindemann, J., and Berndt, C. C., *Thin Solid Films*, 416:129 (2002)
13. Lima, R. S., Kucuk, A., Berndt, C. C., Karthikeyan, J., Kay, C. M., Lindemann, J., *Journal of Materials Science Letters*, 21:1687 (2002)
14. Stoltenhoff, T., Kreye, H., and Richter, H., *J Thermal Spray Tech*, 11:542 (2002)
15. Assadi, H., Gartner, F., Stoltenhoff, T., and Kreye, H., *Acta Materialia*, 51:4379 (2003)
16. Kim, Y. S, Kim, K. T, Kim, B. T., and Bae, J. I., *Key Engineering Materials*, 353-358:848 (2007)

17. Ajdelsztajn, L., Jodoin, B., Kim, George E., and Julie M. Schoenung, *Metallurgical and Materials Transactions A*, 36:659 (2005)
18. Hall, A. C., Brewer, L. N., and Roemer, T. J., *Journal of Thermal Spray Technology*, 17:352 (2008)
19. Li, W.-Y., Zhang, G., Zhang, C., Elkedim, O., Liao, H., and Coddet, C., *Journal of Thermal Spray Technology*, 17:316 (2008)
20. Zhou,Y., Li, Chang-Jiu, Yang, Guan-Jun, Wang, Hong-Duo, and Li, Geng, *Intermetallics*, 13:2154 (2010)
21. Liu, J., Cui, H., Zhou, X., Wu,X., and Zhang, J., *Met. Mater. Int.*, 18:121 (2012)
22. Rokni, M. R., Widener, C. A., Nardi, A. T., and Champagne, V. K., *Applied Surface Science*, 305:797 (2014)
23. List, A., Gartner, F., Schmidt, T., and Klassen, T., *Journal of Thermal Spray Technology*, 21:2012 (2009)
24. El-Eskandarany, M. Sherif, and Al-Azmi, Ahmed, Metallic Glassy Alloy Powders for Antibacterial Coating, US 9,609,874 B1, April 4, 2017
25. El-Eskandrany, M. Sherif, and Al-Azmi, Ahmed, *Journal of Mechanical Behavior of Biomedical Materials*, 56:183 (2016)
26. El-Eskandarany, M. Sherif, Banyan, M., Al-Ajmi, F., *Int J Hydrogen Energy*, 44: 16852 (2019)
27. Pialago, E.J.T., Park, C.W., *Appl. Surf. Sci.*, 308:63 (2014)
28. Pialago, E.J.T., Kwon, O.K., Kim, M.S., Park, C.W., *J. Alloy. Comp.*, 650: 199 (2015)
29. Pialago, E.J.T., Kwon, Park, C.W., *Ceram. Int.*, 41: 6764 (2015)
30. Bakshi, S.R., Singh, V., Balani, K., McCartney, D.G., Seal, S., Agarwal, A., *Surf. Coat. Technol.*, 202: 5162 (2008)
31. Cho, S., Takagi, K., Kwon, H., Seo, D., Ogawa, K., Kikuchi, K. et al., *Surf. Coat. Technol.*, 206: 3488 (2012) 30]
32. Yang, L., Li, B., Yao, J., Li, Z., *Mater.*, 58: 139 (2015)
33. Dong, Y., Zhang, R., He, X., Ye, Z., Qu, X., *Mater. Sci. Eng. B* ,177: 1524 (2012)
34. Yao, J., Yang, L., Li, B,. Li, Z., *Appl. Surf. Sci.*, 330: 300 (2015)
35. Dosta, S., Bolelli, G., Candeli, A., Lusvarghi, L., Cano, I.G., Guilemany, J.M., Acta Mater. 124 (2017) 173
36. Couto, M., Dosta, S., Guilemany, J.M., *Surf. Coat. Technol.*, 268: 180 (2015)
37. Couto, M., Dosta, S., Fernández, J., Guilemany, J.M., J. *Therm. Spray Technol.*, 23: 1251 (2014)
38. Pathak, S., and Saha G. C., *Coatings*, 7: 122 (2017)
39. Fauchais P.L. et al., Thermal Spray Fundamentals: From Powder to Part, Chapter 18, Springer Science+Business Media, New York, 2014

Mechanically induced solid-state amorphization

13

Mechanical alloying process has been employed for preparing homogeneous amorphous alloy powders at room temperature. One advantage of this technique appears to be that there is a very large composition range for which amorphous metallic alloys may be fabricated. The figure shows the micrograph taken by high-resolution transmission electron microscope and the corresponding diffraction pattern of metallic glassy $Zr_{65}Al_{7.5}Ni_{10}Cu_{12.5}Pd_5$ alloy powders fabricated by high-energy ball milling technique.

13.1 Introduction

Ionic and covalent glasses are often formed when the corresponding melt fails to crystallize during relatively slow cooling, of the order of 1 K/s. Naturally occurring oxide glasses, such as obsidian, were first known to man who, as early as 70,000 BC, used them to make tools. Around 5,000 BC the Phoenicians discovered oxide-glass making.

The field of glassy metals or metallic glass has seen enormous development during recent years. For the uninitiated, the notion of a metallic glass may be rather unusual, since one would primarily associate glass with transparent and insulating window material. However, the term glass is nowadays almost unanimously used for an amorphous substance, which is obtained by cooling the corresponding melt. Metallic glasses are new substances with exciting properties, which are of interest not only for basic solid-state physics, but also for metallurgy, surface chemistry, and technology. Metallic glasses have properties, which are quite different from solid metals making them promising candidates for technical applications. The following is a list of some of the important properties that can be found in metallic glasses;

- high mechanical ductility and yield strength
- high magnetic permeability
- low coercive forces
- unusual corrosion resistance
- temperature-independent electrical conductivity

Since the first pioneer investigation for preparing amorphous metallic alloys during the 1950s,[1] almost all amorphous materials are exclusively synthesized by one of the following techniques;

- rapid solidification of melts or vapors[2]
- atomic disordering of crystalline lattices[3]
- solid-state amorphization reaction between pure elements[4,5]
- solid-state transformations from metastable[6]

Mechanical Alloying. http://dx.doi.org/10.1016/B978-0-12-818180-5.00013-3

Several different methods for the preparation of amorphous alloys have been discovered since 1983. Yeh et al.[7] reported that the dissolution of hydrogen gas in crystalline Zr_3Rh results in amorphous $Zr_3RhH_{5.5}$. It was shown that chemical energies could be used to drive a crystal-to-amorphous transformation in the solid state. The hydrogen atoms, being small, can easily diffuse in crystalline intermetallics of large unit cells, allowing for the reaction to occur at a temperature below the crystallization temperature of the amorphous hydride. So far, this method has been used for preparing various amorphous alloys.[8,9]

Another kind of solid-state amorphization reaction was reported by Herd et al.[4] They reported that metals can diffuse at low temperatures into amorphous semiconductors, such as tellurium, selenium, and silicon. Schwarz and Johnson gave the first example of two pure crystalline metals reacting to form a single-phase amorphous alloy.[5] In this experiment, thin films of pure gold and lanthanum, a few tenths of a nanometer in thickness, were fully reacted at 343 K within a few kiloseconds. Two requirements are proposed for the solid-state amorphizing reaction:

- The two reacting metals must have a large negative heat of mixing.
- The two metals must have vastly different diffusivities in each other and in the amorphous alloy to be formed.

The first condition ensures that a thermodynamic driving force for the reaction exists. The second condition ensures that the amorphous alloy will form in preference to crystalline intermetallics, which have lower free energies. This kinetic selection of the reaction path is possible because one species diffusing in the other and in the amorphous alloy is sufficient for the solid-state amorphizing reaction. The micro-mechanism of the solid-state amorphizing reaction in multilayers was studied by Rutherford back-scattering spectroscopic (RBS) marker experiments on a Ni/Zr diffusion couple[10] and transmission electron microscopy (TEM) on Co/Zr.[11] These experiments reveal that the smaller atom in the diffusion couple is the moving species. The growth rate of the amorphous layer, which is formed at the interface of the two atomic species, turned out to be determined by the diffusion coefficient through this amorphous layer. Recently, Suryanarayana and Inoue have shown different examples of amorphous and metallic glassy materials mainly formed by rapid solidification technique.[12]

13.2 Fabrication of amorphous alloys by mechanical alloying process

A different technique for synthesizing amorphous materials at temperatures below the crystallization temperature was suggested in 1979 by White[13] when he milled an elemental mixture of Nb and Sn powders. This technique, called mechanical alloying (MA), was then established by Koch et al.[14] for the synthesis of amorphous $Ni_{60}Nb_{40}$ powders by milling a mixture of pure elements. In the typical experiments for the preparation of an amorphous $Ni_{60}Nb_{40}$ alloy, elemental powders of Ni and Nb were milled for about 40 ks in air or in helium (using a Spex Mixer/Mill Model 8000 and steel balls as a milling media). The XRD patterns for the end product of amorphous

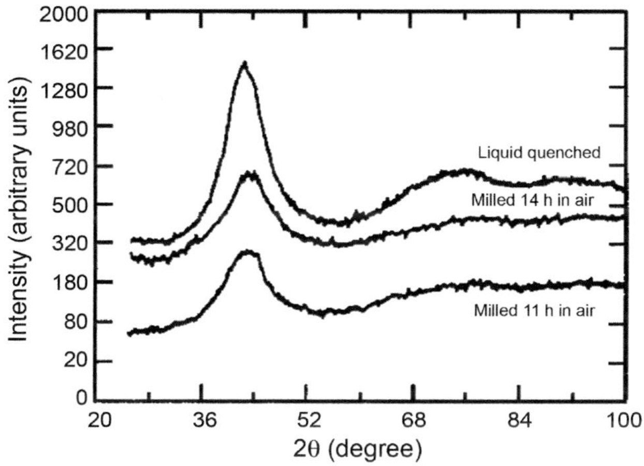

Figure 13.1 XRD patterns of amorphous $Ni_{60}Nb_{40}$ alloys prepared by liquid quenching, MA (14 h) in air, and MA (11 h) in helium.
Source: After Koch et al.[13]

$Ni_{60}Nb_{40}$ alloy powders of as-milled elemental powders in air and helium, prepared by melt spinning, are shown together in Fig. 13.1.

The positions of the maximum of the principal peak in the patterns of the alloys obtained by milling under helium and of the quenched $Ni_{60}Nb_{40}$ are similar ($2\theta = 42.6°$). For the samples that were milled in air, this maximum is at a somewhat higher angle ($2\theta = 43°$).

The exothermic crystallization peaks of amorphous $Ni_{60}Nb_{40}$, using different preparation techniques, measured by differential scanning calorimetry (DSC), are not the same (Fig. 13.2). The onset of the first crystallization peak of the melt-spin and the

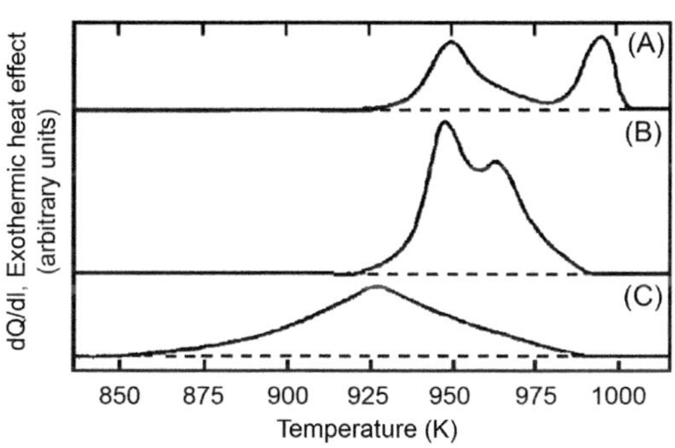

Figure 13.2 DSC plots (heating rate 40 K/min) of amorphous $Ni_{60}Nb_{40}$ prepared by (A) liquid quenching, (B) MA in helium, and (C) MA in air.
Source: After Koch et al.[13]

in-helium MA $Ni_{60}Nb_{40}$ are at T_x= 920 K. The onset of the second crystallization peak of the helium MA is about 30 K lower than when rapidly cooled. In contrast, the in-air MA $Ni_{60}Nb_{40}$ has one broad crystallization peak with an onset at 850 K. These differences can be explained by the different oxygen concentrations in the alloys. Milling in air resulted in an oxygen concentration of about 3.4 wt %, while milling in helium resulted in a concentration of 0.46 wt % oxygen. These oxygen impurities lower the crystallization temperatures.[14]

The effect of iron contamination content on the crystallization temperature, T_x, has also been studied by El-Eskandarany et al.[15] Fig. 13.3 illustrates the DTA curves for (A) rod-milled (RM) and (b) ball-milled (BM) $Al_{30}Ta_{70}$ powders that were milled in an argon gas atmosphere. However, both the RM and BM amorphous alloys are crystallized through single sharp exothermic peaks; the onset of the final RM amorphous $Al_{30}Ta_{70}$ alloy (1440 ks) is 1216 K [Fig. 13.3(A)], higher than that for the BM $Al_{30}Ta_{70}$ alloy (1170 K) after the same milling time. This difference is attributed to the existence of high iron concentrations in the BM amorphous powders (2.0 at. %) compared with that for the RM $Al_{30}Ta_{70}$ alloy powders (less than 1 at. %).[15]

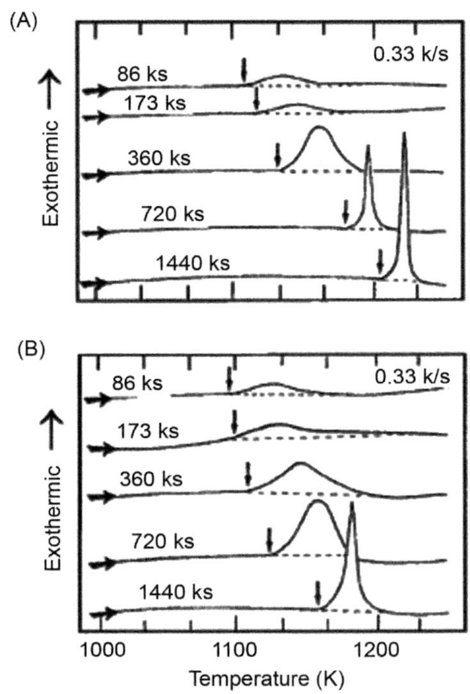

Figure 13.3 The DTA curves for mechanically alloyed $Al_{30}Ta_{70}$ powders as a function of the (A) rod-milling, RM, and (B) ball-milling, BM, times.
Source: After El-Eskandarany et al.[14]

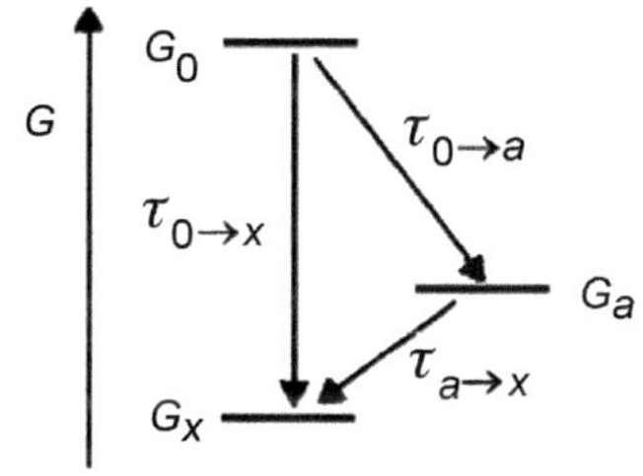

Figure 13.4 Basic principles of glass formation by a solid-state reaction.
Source: After Johnson.[17]

13.3 Crystal-to-glass transition

A free-energy diagram can successfully predict the products that are a result of solid-state amorphization reaction, SARR, for the initial and final products evaluated at the reaction temperature.[5,16] The free enthalpy of the equilibrium crystalline state (G_x) is always much lower than that of the amorphous state (G_a) for metallic systems below the melting temperature (T_m). The amorphous state is a metastable state, that is, an energy barrier exists which prevents amorphous metals from spontaneous crystallization.[17] In order to synthesize an amorphous metallic phase via SARR, it is necessary to create an initial crystalline state (G_0) with a high free enthalpy as proposed by the model illustrated in Fig. 13.4.[18]

Starting from this initial state, G_0, the free enthalpy of the system can be lowered either by the formation of the metastable amorphous phase or by the formation of the crystalline intermetallic phase. Energetically favored is, of course, the crystalline equilibrium phase, but the kinetics of the phase formation decides which phase is, in fact, formed. To evaluate this, the time scales of the possible reactions must be examined. The formation of the amorphous phase is possible and likely if the formation reaction for the amorphous phase is much faster than that for the crystalline phase:

$$\tau_{0-a} << \tau_{0-x}$$

(τ_{i-j} is the characteristic time scale of the reaction). During this reaction the amorphous phase should not crystallize:

$$\tau_{0-a} << \tau_{0-x}$$

It means that the reaction temperature, T_r, must be well below the crystallization temperature, T_x.

13.3.1 *The metastable phase diagram*

Amorphization by mechanical alloying occurs by an interdiffusion reaction at relatively low temperature in the milled powders. The powder particles are repeatedly

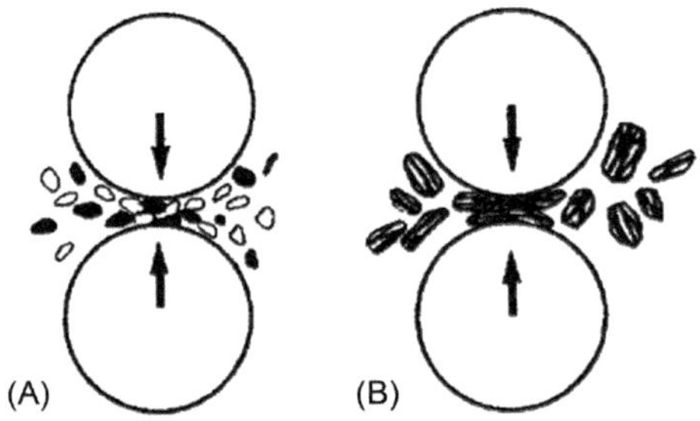

Figure 13.5 Schematic presentation of the mechanical alloying process. (A) At the beginning of the milling and (B) after some collision events.
Source: After Schultz.[16]

cold welded by the colliding balls [Fig. 13.5A].[17] Thus, the composite powder particles with a characteristically layered microstructure (Fig. 13.6)[19] are formed (Fig. 13.5A).[17]

During this interdiffusion reaction, intermetallic crystalline phases do not form, although these are thermodynamically stable phases. The metastable phase diagram can be deduced using free-energy curves.[20] Fig. 13.7 shows the thermodynamically stable phase diagrams of a hypothetical A-B alloy and the free energy curves at a temperature T_r. The free-energy curves show extended primary phases α and β. The phase boundaries are obtained from the double-tangent construction at the free-energy curves. At the temperature T_r, the free-energy curve for the liquid or (assuming that T_r is below the glass transition temperature, T_g) the amorphous phase is always above the composed curve for the stable state. Therefore, the amorphous phase is not present

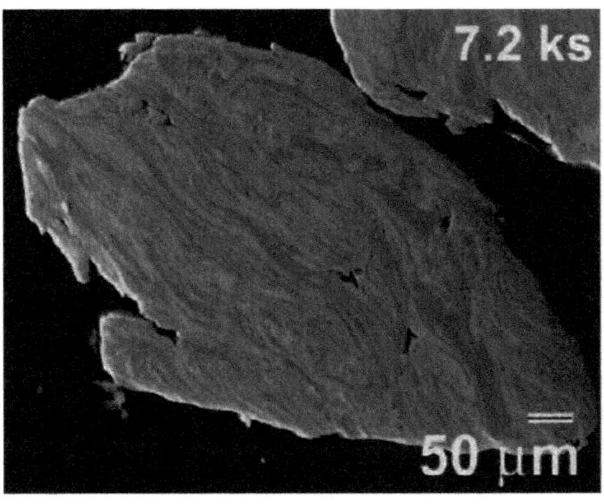

Figure 13.6 Cross-sectional view of mechanically alloyed $Al_{50}Zr_{50}$ powders after 86 ks of milling.
Source: After El-Eskandarany et al.[18]

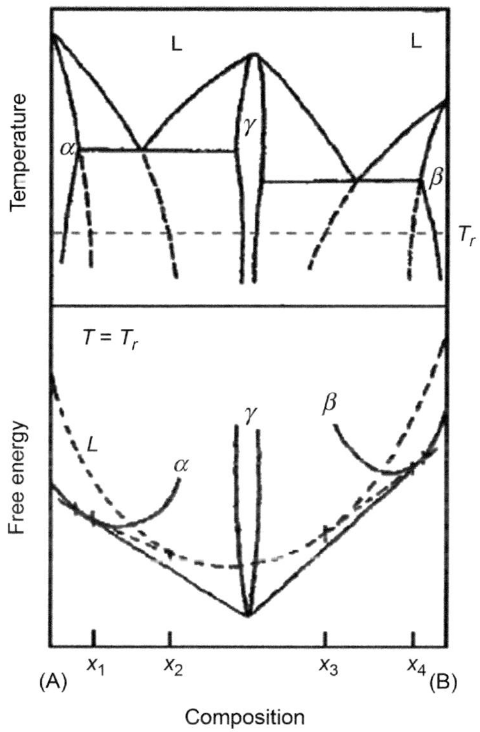

Figure 13.7 Free-energy curve and the resulting stable *(solid line)* and metastable *(broken line)* phase diagram.
Source: After Schwarz.[19]

in the phase diagram at T_r. If we consider the experimental fact that during the amorphization reaction the γ phase does not form, we must ignore the free-energy curve for the γ phase in Fig. 13.7. The free-energy curve for the amorphous phase is then lowest in the central part and the metastable phase diagram must be constructed using the free-energy curves for α and β, and by applying the double-tangent construction to these curves (broken lines in Fig. 13.7). As a result, we find extended solubilities for α and β, a broad glass-forming range in the central part and the two two-phase regions as α plus amorphous and β plus amorphous. Contrary to melt spinning, where glass formation would be expected in two composition ranges close to the two eutectics and not at the composition of the γ phase, glass formation by mechanical alloying preferentially occurs in the central part of the phase diagram, that is, at the composition of the high-melting intermetallic phase. The eutectic compositions of the stable phase diagram are not involved.

The modified phase diagram in Fig. 13.7 resembles a very deep eutectic diagram with the phases α and β in which the eutectic temperature is below the glass transition temperature. Therefore, amorphization by mechanical alloying can also be regarded as a eutectic melting.[18]

13.4 Mechanism of amorphization by mechanical alloying process

The amorphization process during ball milling or rod milling has been proposed by several authors for the following binary systems:

- Ni-TM (TM: Ti,[21–24] Zr,[25,26] and Nb[27])
- Fe-TM (TM: Zr,[28] and Nb[29])
- Al-TM (TM: Ti,[30,31] Zr,[30,32] Hf,[33,34] Nb,[30,35,36] and Ta[37–41])

As a typical example, we will describe, in detail, the formation of amorphous Al_xHf_{100-x}.[34]

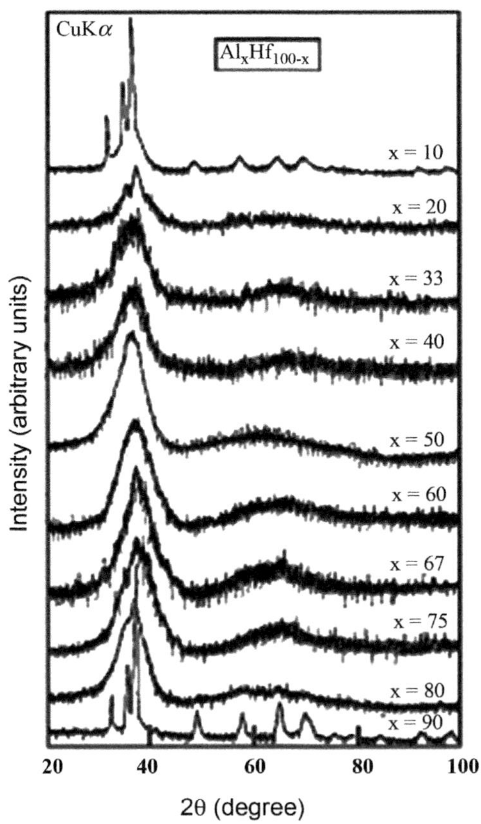

Figure 13.8 The typical XRD patterns of the end-products (720 ks of the MA time) for rod-milled Al_xHf_{100-x} powders.
Source: After El-Eskandarany.[33]

10.4.1 Structural changes with the milling time

10.4.1.1 X-ray analysis

Fig. 13.8 displays the XRD patterns of the end-product of mechanically alloyed Al_xHf_{100-x} powders at different atomic concentrations of Al, x. Obviously, a wide amorphization range ($33 \leq x \leq 75$) can be obtained after rod milling the elemental diffusion couples for 720 ks of the MA time. For $x = 20$ and 80, the final products are a mixture of the amorphous phase coexistent with unprocessed elemental Al and/or Hf crystals. No amorphous phases could be detected in the rich sides of the reactant materials (when $x = 10$ or 90) and the end products here are nanocomposites Al/Hf powders.

In order to understand the effect of MA time on the total structure of the rod-milled powders, different samples were taken after selected MA times for x-ray analyses. Fig. 13.9 presents the correlation between the MA time and the powder's structural changes, taking $Al_{50}Hf_{50}$ alloy as a typical example. In spite of the intact mixture of polycrystalline fcc-Al and hcp-Hf [Fig. 13.9A], at the early and the intermediate stages of MA time (0 –173 ks), the Bragg peaks for Al and Hf reflections become wider and their intensities are drastically decreased [Fig. 13.9B and C]. During the next stage of MA (360 ks), these peaks become broader, with clear diffuse haloes,

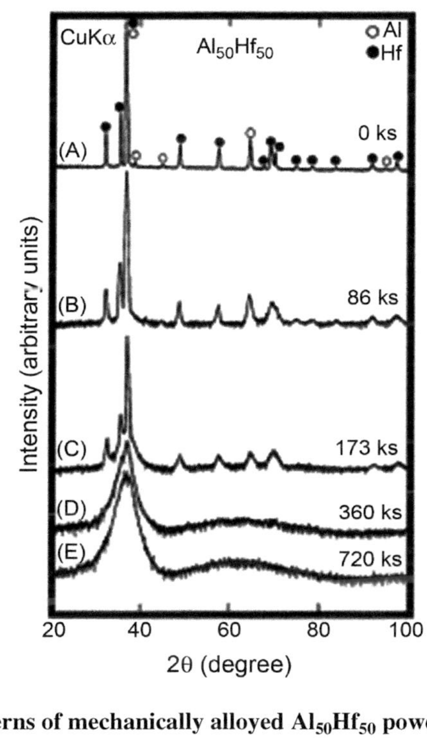

Figure 13.9 XRD patterns of mechanically alloyed $Al_{50}Hf_{50}$ powders after selected rod-milling processing times.

Source: After El-Eskandarany.[33]

suggesting the formation of an amorphous phase coexisting with Al and Hf crystals, as illustrated in Fig. 13.9D. At the end of rod-milling time (720 ks), a single homogeneous amorphous phase is yielded, characterized by diffuse haloes and smooth peaks, as displayed in Fig. 13.9E.

10.4.1.2 TEM observations

TEM analyses have been used to understand the local structure of the rod-milled powders, including the heterogeneity and homogeneity of the powders during the different stages of MA.

Fig. 13.10 presents the bright field image, BFI [Fig. 13.10A], and the selected-area diffraction pattern, SADP [Fig. 13.10B], for rod-milled $Al_{50}Hf_{50}$ powder particles after 6 ks of MA time. Neither amorphous nor metastable (e.g., solid solution) phases could be detected after this stage of milling, and the powders are just polycrystalline mixtures of coarse cells of Al and Hf, indicated by the sharp spot patterns as shown in Fig. 13.10B. The BFI and the corresponding SADPs of rod-milled powders at the intermediate stage of MA time (173 ks) are shown together in Fig. 13.11. Numerous faults with grain boundary fringes and heavy dislocations in the boundary appear clearly in the powders as a result of the shear stress generated by the milling media (rods), as displayed in Fig. 13.11A. A remarkable significant heterogeneity in the local structure of the powders can be distinguished [Fig. 13.11A]. According to this variety in the local structure, two different morphological regions (I and II) are chosen as SADPs and presented in Fig. 13.11B and C, respectively. A fine structure is observed at the right upper edge of the particle (region I), as presented in Fig. 13.11A. The corresponding SADP [Fig. 13.11B] shows a clear halo pattern, indicating the formation

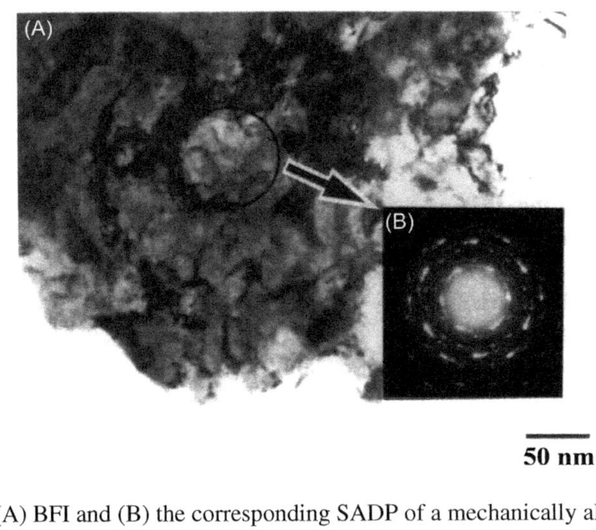

Figure 13.10 (A) BFI and (B) the corresponding SADP of a mechanically alloyed $Al_{50}Hf_{50}$ powder at the early stage of the rod-milling time (6 ks).
Source: After El-Eskandarany.[33]

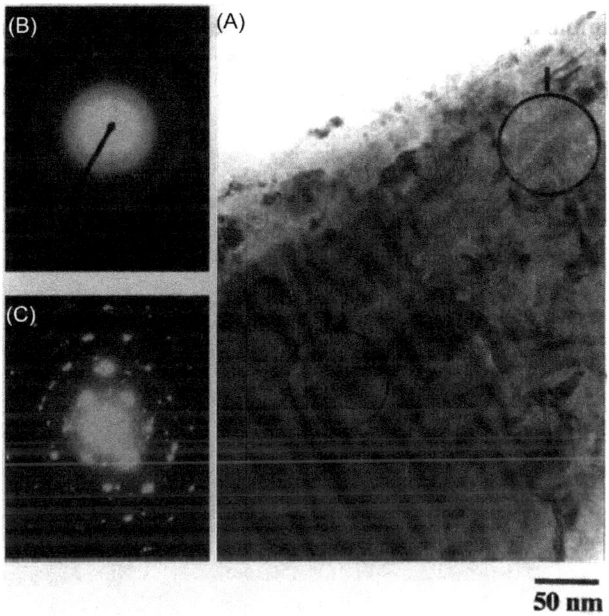

Figure 13.11 (A) BFI and (B and C) the corresponding SADPs of two selected zones (I and II) of a mechanically alloyed $Al_{50}Hf_{50}$ powder at the intermediate stage of the rod-milling time (173 ks).
Source: After El-Eskandarany.[33]

of an amorphous phase in the rod-milled powders at this stage of milling. In region II, however, several defects with grain boundary movement are clearly visible near the center and at the lower edge of the micrographs [Fig. 13.11A]. The oriented particle has twin and nanotwin boundaries on unprocessed Al and Hf, suggested by the Laue spots presented in Fig. 13.11C. The BFI and the corresponding SADP of the end product (720 ks) of $Al_{50}Hf_{50}$ powders are shown together in Fig. 13.12. Overall, the sample appears to have a homogeneous fine structure with no dominant facet structure [Fig. 13.12A]. Moreover, the SADP [Fig. 13.12B] shows a typical halo-pattern of an amorphous phase in good agreement with the XRD pattern that is presented in Fig. 13.9E.

EDS measurements, using an electron beam of 5 nm, have been used to determine the local composition and the degree of homogeneity for the end-product of the amorphous alloy. For this purpose, the sample that is shown in Fig. 13.13 was classified into a rectangular grid containing 50 circular regions (with a diameter of about 50 nm). Each region was divided into 10 zones and the content of Al in each zone was analyzed. Since such measurements were achieved over several days, the examined sample was kept under vacuum in the TEM apparatus until the completion of all of the analyses. The results of these experiments were used to construct an isochemical contour map of Al in the examined sample of Fig. 13.12A, using a contour interval of 0.2, as displayed in Fig. 13.13. Overall, the sample appears to have very close values of Al

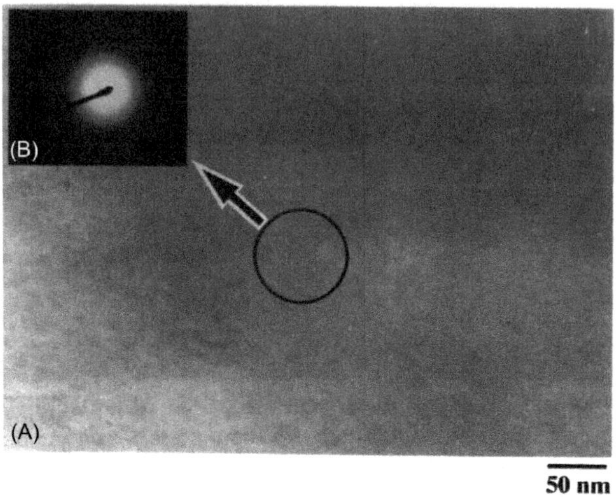

Figure 13.12 (A) BFI and (B) SADP of a mechanically alloyed $Al_{50}Hf_{50}$ powder at the final stage of the rod-milling time (720 ks).
Source: After El-Eskandarany.[33]

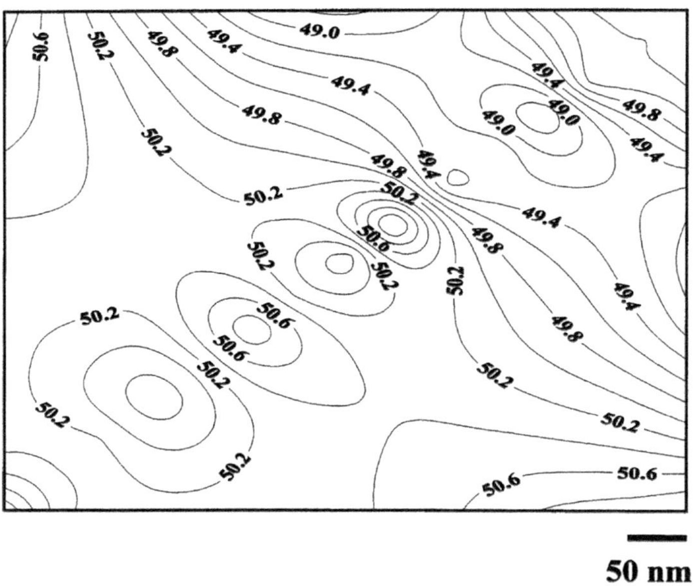

Figure 13.13 Isochemical contour map of Al concentration (at. %) of the sample shown in Fig. 13.12A.
Source: After El-Eskandarany.[33]

and varies slightly from 49.0 to 50.6 at.%, suggested by the slight gradient change of the Al isochemical line. This suggests the formation of a single homogeneous phase of $Al_{50}Hf_{50}$ alloy powder.

13.4.2 Morphology and metallography changes with the milling time

Detailed SEM observations were performed to understand the topology and metallography of the rod-milled powders after the different stages of MA. Fig. 13.14 shows the SEM micrographs of rod-milled $Al_{50}Hf_{50}$ powders after selected MA times. After a few kiloseconds of milling (22 ks), the starting ductile powders of Al and Hf tend to form globe-shaped agglomerate composite particles (nearly 700 μm in diameter), as shown in Fig. 13.14A.

The metallographical examinations of the cross section for the polished and etched particles that were milled for 86 ks show that the powders contain many intimate layers of elemental Al and Hf (Fig. 13.15). Contrary to the numerous thin layers of the equiatomic particles [Fig. 13.15B], the number of layers in both Hf and Al rich sides [Fig. 13.15A and C] are still fewer and wider in thickness. Further milling (173 ks) enhances the shear forces generated by the rods, leading to the formation of finer powders (about 25 μm in diameter) with cabbage leaf-like morphology, as presented in Fig. 13.14B. After 360 ks of MA time, the powders are dramatically decreased in diameter (about 2 μm or less) and tend to have nearly spherical-like morphology,

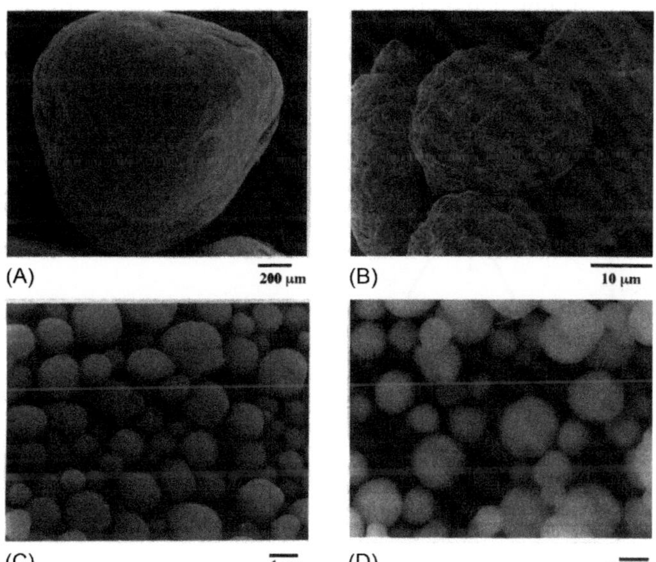

Figure 13.14 Detailed SEM micrographs of mechanically alloyed $Al_{50}Hf_{50}$ powders after (A) 22 ks, (B) 173 ks, (C) 360 ks, and (D) 720 ks of the rod-milling time.
Source: After El-Eskandarany.[33]

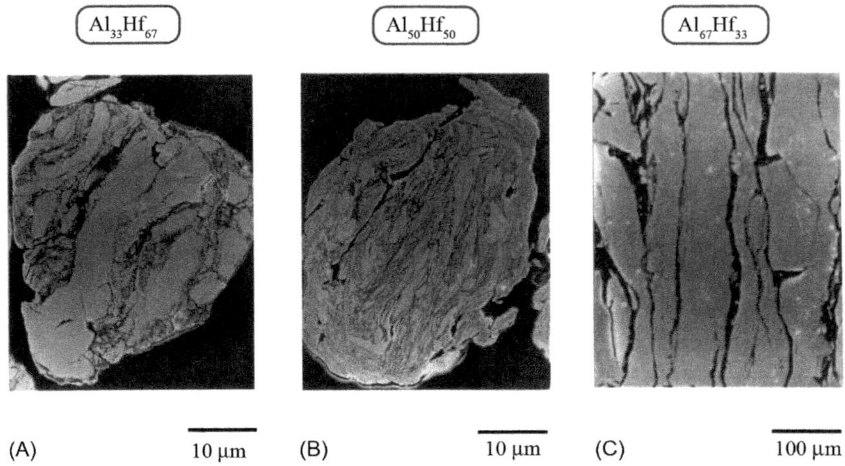

Figure 13.15 SEM micrographs of the cross-sectional view for mechanically alloyed (A) $Al_{33}Hf_{67}$, (B) $Al_{50}Hf_{50}$, and (C) $Al_{67}Hf_{33}$ after 86 ks of the rod-milling time.
Source: After El-Eskandarany.[33]

as shown in Fig. 13.14C. The powder of the end-product (720 ks) comprises homogeneous and smooth spheres with an average size of about 0.75 μm in diameter, as presented in Fig. 13.14D.

Figs. 13.16 and 13.17 summarize the SEM observations of rod-milled powders at different stages of milling. Obviously, the MA process performed by the rod-milling

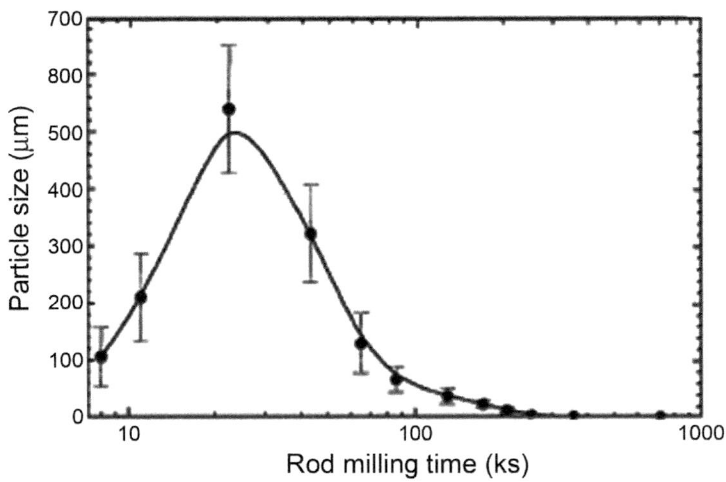

Figure 13.16 Influence of rod-milling time on the particle size distribution of mechanically alloyed $Al_{50}Hf_{50}$ powder.
Source: After El-Eskandarany.[33]

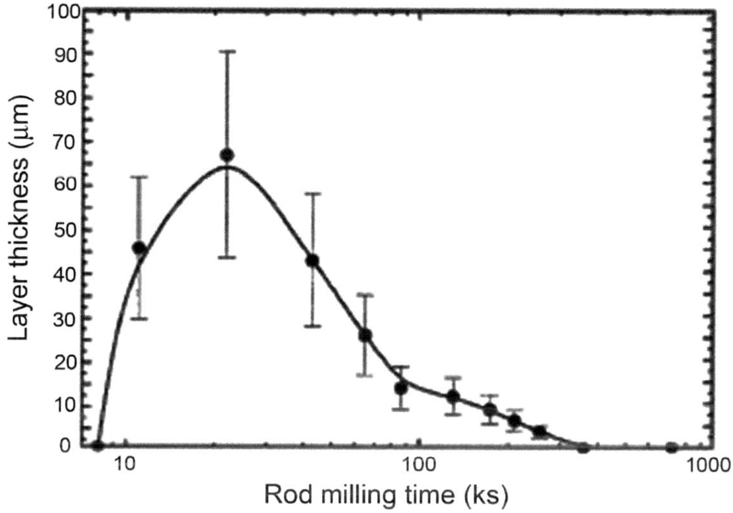

Figure 13.17 Effect of the rod-milling time on the layer thickness distribution of mechanically alloyed $Al_{50}Hf_{50}$ powders.
Source: After El-Eskandarany.[33]

technique can be classified into three stages that is to say, early or agglomeration, intermediate or disintegration and final or homogenization stages. At the agglomeration stage (0–22 ks) the alloying elemental powders of Al and Hf are agglomerated and grow in size, as a result of repeated cold welding. During this stage of milling, the powders vary widely in size from 150 to about 680 μm (Fig. 13.16). This, accompanied by an increase in the layer thickness and the number of layers for the individual particles, is presented in Fig. 13.17. During the subsequent disintegration stage (43–86 ks), the agglomerated powder particles are subjected to a continuous disintegration with fragmentation to form relatively finer powders with sizes less than 100 μm in diameter. This stage of milling provides very fine Al/Hf composite powder particles with narrow size distribution, as shown in Fig. 13.16. Furthermore, the layer thickness of the particles is dramatically decreased, indicating an increase in the interfaces between the reactant diffusion couples of Al and Hf, as displayed in Fig. 13.17. The final or so-called homogenization stage (360–720 ks) refers to the last stage of MA, in which all the particles are uniform in shape and size without any metallographical details (layered-structure morphology), suggesting the completion of the solid-state reaction and the formation of an amorphous phase.

13.4.3 Thermal stability

13.4.3.1 Amorphization process

The typical DTA thermograph of rod-milled $Al_{50}Hf_{50}$ powders is presented in Figs. 13.18 and 13.19 after selected MA times. These measurements were made at a constant heating rate of 0.33 K/s under an argon gas atmosphere. All the samples were

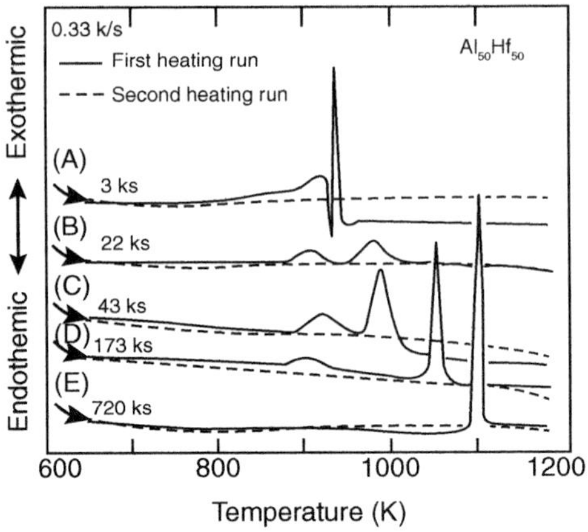

Figure 13.18 DTA curves of mechanically alloyed $Al_{50}Hf_{50}$ powders after (A) 11 ks, (B) 22 ks, (C) 43 ks, (D) 173 ks, and (E) 720 ks of the rod milling time.
Source: After El-Eskandarany.[33]

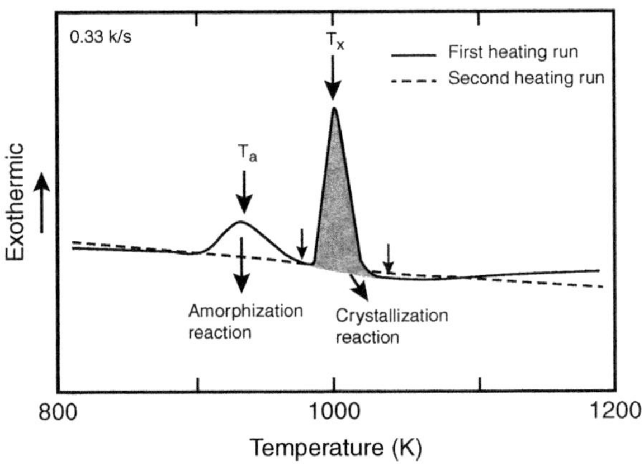

Figure 13.19 DTA thermogram associated with the amorphization reactions (the plain low temperature peak) and crystallization reactions (the shady high temperature peak) for mechanically alloyed $Al_{50}Hf_{50}$ powder after 86 ks of the rod milling time.
Source: After El-Eskandarany.[33]

heated to 1200 K (first run) and cooled to about 400 K. Then, second heating runs *(dashed lines)* were performed, in order to establish the base line.

After 11 ks, the scan reveals two reactions: the first reaction, observed at about 940 K, is endothermic and occurs due to the melting of pure Al in the starting reactant mixture of $Al_{50}Hf_{50}$ powders, and the second reaction (exothermic), occurs at about 980 K and is attributed to a partial reaction of the unprocessed materials of Al and Hf powders [Fig. 13.18A]. After 22 ks, 43 ks [Fig. 13.18B and C], and 86 ks (Fig. 13.19) of MA time, however, these reactions are replaced by a new reaction that takes place exothermally. The first exothermic reaction appears at a relatively low temperature (about 900 K) and does not change remarkably by changing the milling time. In spite of the first reaction, the second one takes place at a relatively high temperature and its position shifts remarkably to the high temperature side by increasing the MA time. The origin of these reactions, exemplified by the rod-milled 86 ks sample (Fig. 13.19), has been investigated by XRD, TEM/EDS, and metallographical examinations. The sample that was rod milled for 86 ks of the MA time is heavily dislocated and contains polycrystalline grains of Al and Hf, as suggested by the BFI and its corresponding SADP that is presented in Fig. 13.20A.

The metallographical examination of this sample reveals an intimate layered-structure of the diffusion couples of Al and Hf, as presented in Fig. 13.21A. These layers disappear upon heating the sample to 980 K (just above the first exothermic peak in Fig. 13.19), suggesting the formation of a single phase, as displayed in Fig. 13.21B. Moreover, the BFI of this sample shows the formation of fine and featureless images, and a clear halo pattern, as presented in Fig. 13.20B. Consequently, it is concluded that the first exothermic reaction occurs due to a crystalline-amorphous phase transformation conducted by a so-called *TASSA* reaction. The XRD patterns of the sample that was annealed just above the second exothermic reaction[47] show the formation of a crystalline phase, indicating that the high temperature peak resulted from crystallization of the amorphous phase(s) in the milled powder. As the milling time increases, the

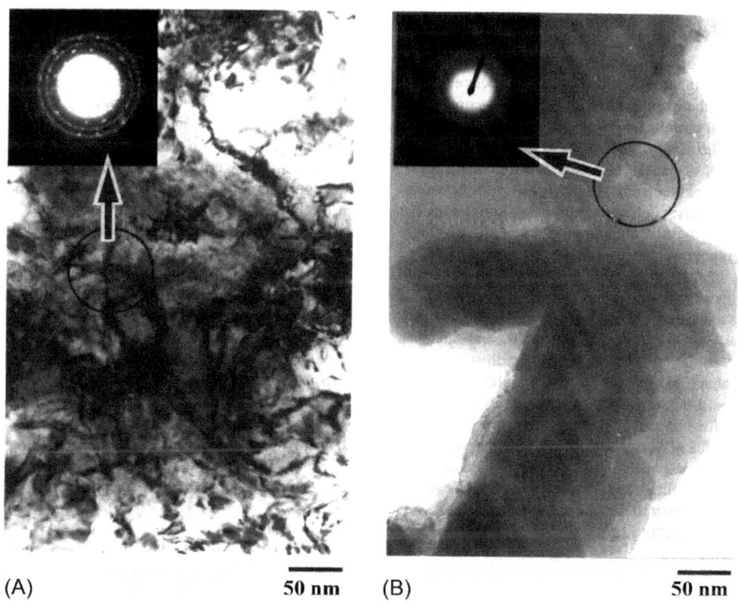

Figure 13.20 BFIs and the corresponding SADPs of $Al_{50}Hf_{50}$ powders after (A) rod milling for 86 ks, and (B) rod milling for 86 ks and then heating to 980 K in a DTA under Ar gas atmosphere.
Source: After El-Eskandarany.[33]

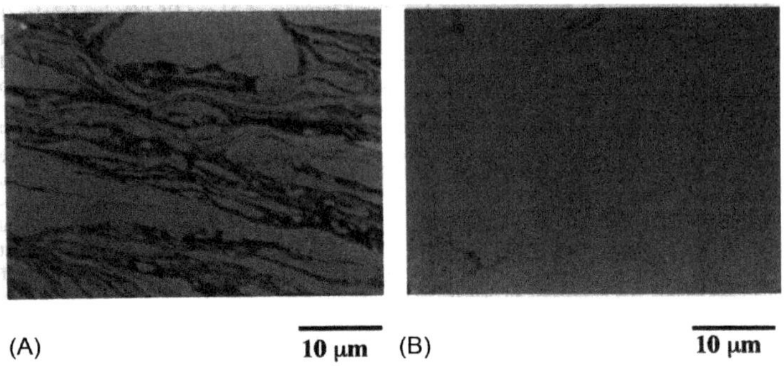

Figure 13.21 SEM micrographs of the polished surface of $Al_{50}Hf_{50}$ after (A) rod milling for 86 ks, and (B) rod milling for 86 ks and then heating to 980 K in a DTA under Ar gas atmosphere.
Source: After El-Eskandarany.[33]

crystallization peak moves to the high temperature side, becoming more pronounced and sharp; the amorphization peak, however, is hardly seen, as displayed in Fig. 13.18.

In order to understand the local composition and the degree of homogeneity of the amorphous phase that formed via the *TASSA* process, EDS analyses have been performed for the sample milled for 86 ks and then annealed at 980 K. Fig. 13.22 shows (A) the BFI and (B) the corresponding isochemical contour map of Al in the examined sample. The sample has been classified into 30 circular regions with a diameter of about 50 nm for each examined region. The results, which were obtained from the local analyses of at least 300 points (10 for each region, using an electron beam of 5 nm), were employed to construct the isothermal lines of Al concentration in the examined sample with a contour interval of 1 at. %, as shown in Fig. 13.22B. In contrast to the results presented in Fig. 13.13, the Al concentration of this sample varies widely from region to region and even within the individual region itself, suggested by the dramatic changes of the values for the isochemical lines (35 to 48 at. %) and the narrow distance between the contour lines. It is worth noting that at this stage of milling (86 ks), a considerable amount of the Al powder is still stacked on the milling tools due to the cold welding effect. Annealing such material leads to the formation of heterogeneous amorphous alloys so that the sample is either poor in Al content *(shaded area)* or approaches near equiatomic values *(plain area)*. It can be concluded that the amorphous alloy formed via *TASSA* is heterogeneous in composition, containing several compositions of amorphous Al_xHf_{100-x} powders.

The DTA measurements allow us to obtain the approximate values of the enthalpy of amorphization (heat release of the crystalline-to-amorphous phase transformation), ΔH_a, and the enthalpy of crystallization (heat release of the amorphous-to-equilibrium crystalline phase transformation), ΔH_x. The amorphization temperature, T_a, and the crystallization temperature, T_x, for amorphous Al_xHf_{100-x} powders have been investigated as well.

In the present study, the term $\Delta H_a TASSA$ is used to refer to the heat formation of the amorphous phase (enthalpy of amorphization) via *TASSA* that was directly measured

(A)

(B)

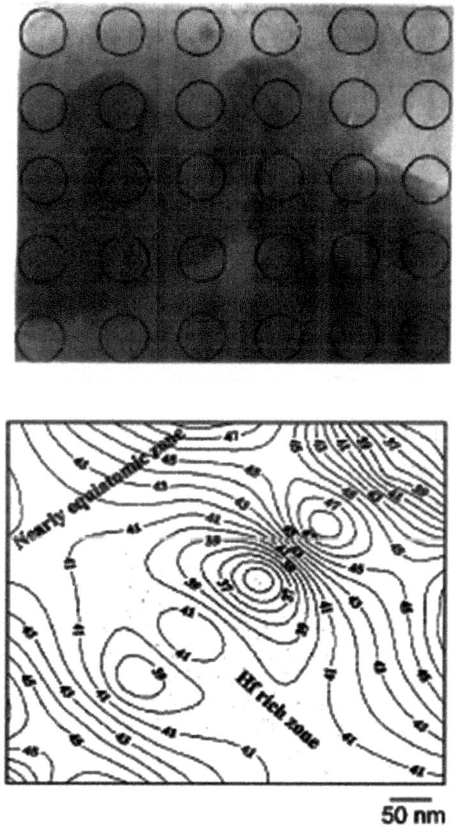

Figure 13.22 (A) BFI and (B) the corresponding isochemical contour map of Al concentration (at. %) of $Al_{50}Hf_{50}$ powder that was rod-milled for 86 ks and then heated to 980 K in a DTA under Ar gas atmosphere.
Source: After El-Eskandarany.[33]

from the area under the amorphization peaks and plotted in Fig. 13.23 as a function of the rod-milling time of selected Al_xHf_{100-x} powders. In all the compositions, ΔH_a^{TASSA} decreases drastically to approach minimum values and then increases dramatically to zero. Compared to the alloy of the equiatomic composition ($x = 50$), the ΔH_a^{TASSA} for the alloys at the Hf-rich side ($x = 20$) and Al-rich side ($x = 80$) take longer times to appear and also to vanish. This is attributed to the fact that the starting composite particles at the rich elemental sides contain such very thick elemental metallic layers that *TASSA* can not be performed between them.[43] Increasing the rod-milling time leads to refinement of the layers and, hence, *TASSA* takes place at the interfaces of these fine reactant layers.

In the present study, the term ΔH_a^{MDSSA} is used to refer to the heat formation of an amorphous phase via *MDSSA*. However, ΔH_a^{MDSSA} is not a measurable value; it can

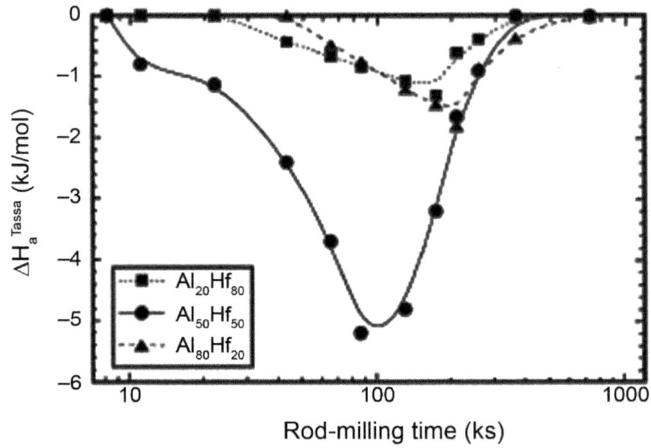

Figure 13.23 Dependence of rod-milling time and the Al concentration on the heat formation of amorphization via TASSA, $\Delta H_a{}^{TASSA}$ of selected mechanically alloyed Al_xHf_{100-x} powders.
Source: After El-Eskandarany.[33]

be estimated only during the final stage of milling (when the amorphization peaks are absent, i.e., $\Delta H_a{}^{TASSA} = 0$), using the following relation:[44]

$$\Delta H_{for} = \Delta H_a + \Delta H_x \tag{13.1}$$

Here ΔH_{for} is the reported value of the enthalpy of formation for an intermetallic compound at specific concentrations of Al(x) based on Miedema's model,[45] ΔH_a is the total heat formation of amorphous phases formed by *TASSA* and *MDSSA* processes, and the term ΔH_x is the sum total of enthalpy of crystallization for the amorphous phases caused by the two processes, *TASSA* and *MDSSA*.

Thus, ΔH_a can be given as follows:

$$\Delta H_a = \Delta H_a^{TASSA} + \Delta H_a^{MDSSA} \tag{13.2}$$

Here, $\Delta H_a{}^{TASSA}$ and $\Delta H_a{}^{MDSSA}$ are the heat of formation of amorphous phases formed by *TASSA* and *MDSSA*, respectively. Similarly, ΔH_x may be also written as

$$\Delta H_x = \Delta H_x^{TASSA} + \Delta H_x^{MDSSA} \tag{13.3}$$

Thus, Eq. (13.1) can be written as

$$\Delta H_{for} = \left(\Delta H_a^{TASSA} + \Delta H_a^{MDSSA}\right) + \left(\Delta H_x^{TASSA} + \Delta H_x^{MDSSA}\right) \tag{13.4}$$

During the final stage of milling (the stage in which the amorphous phase is formed due to the *MDSSA* process, that is, when the amorphization reaction peaks are absent), both ΔH_a^{TASSA} and ΔH_x^{TASSA} equal zero. Hence, the relation of Eq. (13.4) can be rewritten as

$$\Delta H_{for} = \Delta H_a^{MDSSA} + \Delta H_x^{MDSSA} \tag{13.5}$$

The value of ΔH_a^{MDSSA} can then be estimated directly from the following relation:

$$\Delta H_a^{MDSSA} = \Delta H_{for} - \Delta H_x^{MDSSA} \tag{13.6}$$

In fact, this value is very difficult to estimate at the intermediate stage of milling in which the amorphization peaks are present. At the intermediate stage, measured value of ΔH_x is caused by the crystallization of two amorphous phases resulting from *TASSA* and *MDSSA* together.

The correlation between the minimum values of ΔH_a^{TASSA} and ΔH_a^{MDSSA} with the Al content are presented in Fig. 13.24 as a function of the Al concentration, x. For all x, ΔH_a^{MDSSA} is larger than ΔH_a^{TASSA}, especially at the Al and Hf rich sides.

The amorphization ratios, $\Delta H_a^{MDSSA}/\Delta H_a^{TASSA}$, for amorphous Al_xHf_{100-x} alloy powders are listed in Table 13.1. Since the alloys of the rich sides of Al ($x \geq 75$) and Hf ($x \leq 33$) form composite particles with extremely thick layer structures (Fig. 13.15), the *TASSA* process takes place slowly. Thus, only a small mole fraction of amorphous phase(s) can result upon annealing these composites. Accordingly, the value of ΔH_a^{TASSA} for these alloys is lower than those alloys that are laid near the equiatomic composition ($33 \leq x \leq 75$), as previously presented in Fig. 13.24 and Table 13.1. Hence, the $\Delta H_a^{MDSSA}/\Delta H_a^{TASSA}$ ratio has large values (more than 14) for the rich Al and

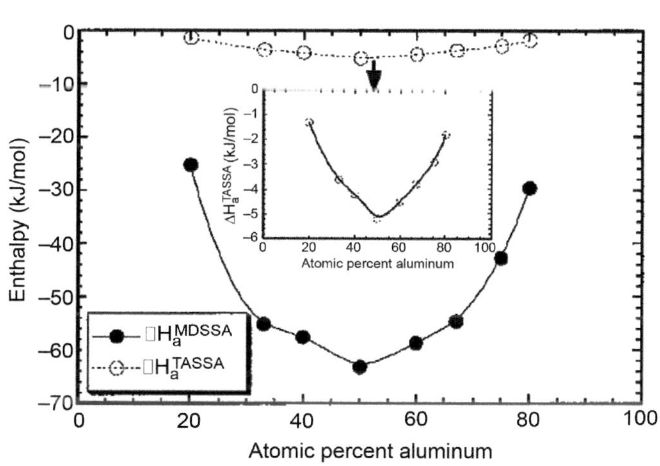

Figure 13.24 Correlation between the heat formation (enthalpy change) of amorphization via TASSA, ΔH_a^{TASSA} and MDSSA, ΔH_a^{MDSSA} for mechanically alloyed Al_xHf_{100-x} powders. The ΔH_a^{TASSA} of Al_xHf_{100-x} powders is shown inset in the figure in another scale.

Source: After El-Eskandarany.[33]

Table 13.1 The heat formation of amorphization via ***MDSSA***, ΔH_a^{MDSSA} and ***TASSA***, ΔH_a^{TASSA}, and the relative enthalpy of amorphization ratio, $\Delta H_a^{MDSSA}/\Delta H_a^{TASSA}$, for amorphous Al_xHf_{100-x} alloy powders.

Al concentration, *x* (at. %)	ΔH_a^{MDSSA} **(kj mol^{-1})**	ΔH_a^{TASSA} **(kj mol^{-1})**	$\frac{\Delta H_a^{MDSSA}}{\Delta H_a^{TASSA}}$
20	–25.2	–1.3	19.4
33	–55.1	–3.6	15.3
40	–57.6	–4.2	13.7
50	–63.2	–5.2	12.2
60	–58.7	–4.5	13.0
67	–54.6	–3.8	14.4
75	–42.8	–2.9	14.8
80	–29.6	–1.8	16.4

Source: After El-Eskandarany.[34]

Hf alloys, suggesting that the amorphous phases for these alloys are formed mainly due to the *MDSSA* process, which means long milling intervals are necessary.

The values of ΔH_a^{MDSSA} were estimated using Eq. (13.6), whereas, ΔH_a^{TASSA} was measured directly using DTA.

13.4.3.2 Crystallization process

The crystallization characteristics of selected Al_xHf_{100-x} alloy powders indexed by the crystallization temperature, T_x, and the enthalpy change of crystallization, ΔH_x, are presented as a function of the rod milling time, in Figs. 13.25, 13.26, respectively. It

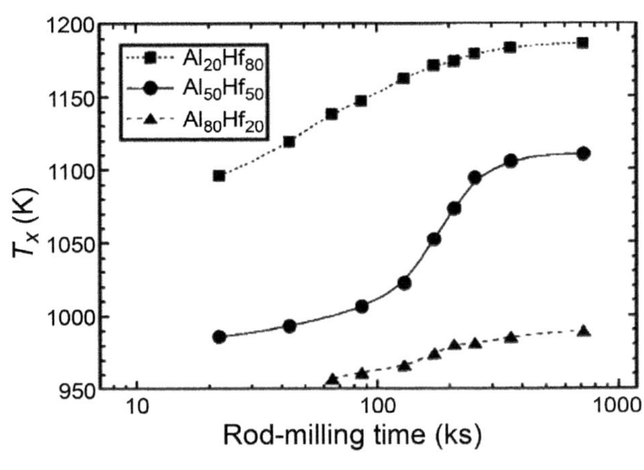

Figure 13.25 Effect of rod-milling time on the crystallization temperature, T_x, of mechanically alloyed Al_xHf_{100-x} powders.
Source: After El-Eskandarany.[33]

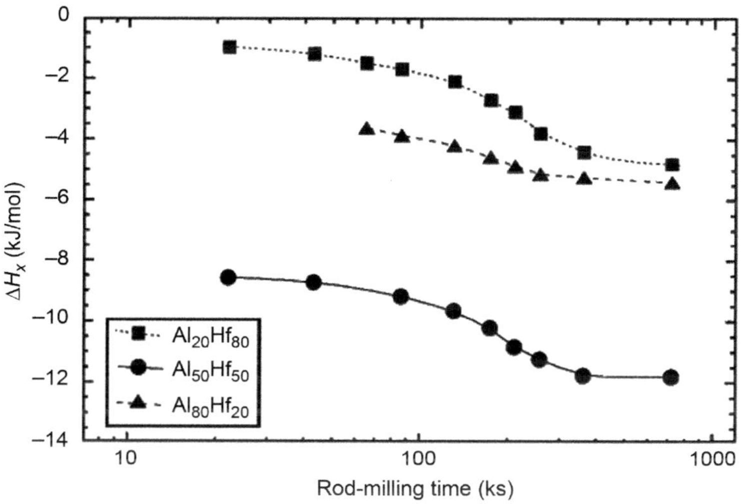

Figure 13.26 Effect of rod-milling time on the enthalpy change of crystallization, ΔH_x, of mechanically alloyed Al_xHf_{100-x} powders.
Source: After El-Eskandarany.[33]

can be seen in Fig. 13.25 that the T_x increases monotonically with an increase in the milling time during the early stage in which the amorphization process occurs only due to the *TASSA* process. This may suggest a gradual compositional change in the milled powders. It then increases dramatically during the intermediate stage in which the amorphous phases are formed due to the amorphization via *TASSA* and *MDSSA* together. This increase in the value of T_x indicates a drastic compositional change in the amorphous phases. During the final stage of milling, however, the amorphous phases are formed by the *MDSSA* process only so that the T_x approaches saturation value even after 720 ks of the MA time.

Further information on crystallization reactions is given by ΔH_x and presented in Fig. 13.26 as a function of the rod-milling time. During the first few kiloseconds of milling, the value of ΔH_x decreases monotonically, indicating a gradual increase in the volume fraction of the formed amorphous phases. A rapid increase in the volume fraction of the formed amorphous phase is suggested by the drastic decrease of ΔH_x during the intermediate stage of milling. Toward the end of the milling process, ΔH_x tends to have saturated values, suggesting the formation of homogeneous amorphous phases.

Fig. 13.27 illustrates T_x *(solid line)* of amorphous Al_xHf_{100-x} alloy powders after the final stage of milling (720 ks). In the figure, the melting points, T_m, of Al_xHf_{100-x} alloys based on the phase relation diagram[46] are also presented *(dotted line)*. The dependence of T_x on T_m can be noted. In other words, T_x increases approximately in parallel with increases in the T_m. Table 13.2 shows the melting points, T_m, and the crystallization parameters (T_x, T_m/T_x ratio, ΔH_x) for amorphous Al_xHf_{100-x} alloy powders prepared by the MA method using the rod-milling technique. Obviously, the shown T_m/T_x ratio is about 0.5, in good agreement with the metallic glasses.[47–49]

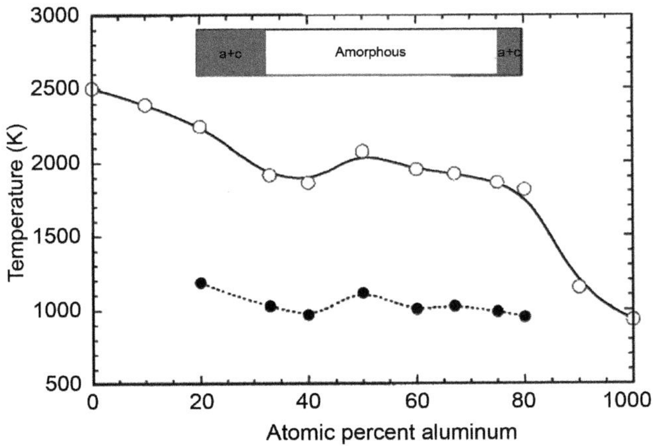

Figure 13.27 Correlation between the melting points, T_m *(dotted line)*, and the crystallization temperatures, T_x *(solid line)*, of Al_xHf_{100-x} powders.
Source: After El-Eskandarany.[33]

Table 13.2 The melting point, T_m,[46] crystallization temperature, T_x, T_m/T_x, and the enthalpy change of crystallization, ΔH_x, for amorphous Al_xHf_{100-x} alloy powders prepared in the present work.

Al concentration, x (at. %)	**T_m (K)**	**T_x (K)**	$\frac{T_x}{T_m}$	**ΔH_x (kJ mol^{-1})**
20	2245	1186	0.528	–4.8
33	1917	1028	0.536	-9.9
40	1863	970	0.521	–10.4
50	2073	1115	0.538	–11.8
60	1953	1007	0.516	–11.3
67	1923	1025	0.533	–10.4
75	1863	989	0.531	–8.2
80	1815	952	0.524	–5.4

Source: After El-Eskandarany.[34]

13.4.3.3 Mechanism

Amorphous Al_xHf_{100-x} alloy powders with wide amorphization ranges ($33 \leq x \leq 75$) have been synthesized by *TASSA* and/or *MDSSA* processes using the rod-milling technique. In the *TASSA*, the heat treatment of the composite powder particles containing a well-developed structure of fresh layers of metallic Al and Hf enhances the solid-state diffusion between these thin layers. This leads to a speed up in the rate of diffusion at the clean Al/Hf boundaries so that the free energy changes drastically from a nonstable phase (starting reactant materials) to a more stable phase (amorphous). In the *MDSSA* process, however, the solid-state reaction takes place in the same manner as occurs in the *TASSA* process, except for a lower diffusion rate and a longer MA time. This is attributed to the milling temperature, which is assumed to be far below that in *TASSA* (980 K). Thus, the interdiffusion between Al and Hf layers occurs slowly. Although the two processes seem to be the same with different time-scales, they produce amorphous alloys, which differ widely in thermal stability.

13.4.3.3.1 Amorphization via *TASSA* process: the early stage of milling

This duration of MA time (0–43 ks) refers to the first stage of milling, in which the amorphization process occurs only due to *TASSA* process. During the first few kiloseconds of this stage (0–22 ks), almost all the initial materials of Al and Hf are agglomerated to form powder particles containing coarse layers of the metallic diffusion couples. Further milling (22–43 ks) leads to the formation of composite particles containing numerous narrow layers of Al and Hf in good arrangement as a result of the shear force generated by the rods. These intensive shear forces, accompanied by cold working, create naturally multilayered composite particles that are typical for sputtered or evaporated diffusion couples.[5] In this stage, the as-rod-milled powders are large grained mixtures of polycrystalline Al and Hf with sharp grain boundaries. It is worth saying that the formation of the amorphous phase during this stage of milling occurs only due to *TASSA*, which is conducted in DTA. During the DTA measurements of the alloy powders at this stage, two clear separate exothermic peaks appear. The first exothermic reaction occurs at about 920 K and the second reaction occurs well above this temperature. The first exothermic peak refers to *TASSA* in which the free energy of the composite Al/Hf particles decreases from the initial state to the amorphous state formed via *TASSA*. The second exothermic peak occurs due to the crystallization of the amorphous phase that formed from the first reaction. The free energy then decreases again from the amorphous state to the equilibrium crystalline state. The DSC allow us to measure the enthalpy of amorphization via *TASSA*, ΔH_a^{TASSA}, directly.

Once the thickness of Al and Hf layers in the composite particles decreases by increasing the MA time, the number of these fresh layers (which became very narrow) increases. The increment of these layers plays a very important role for the amorphization by way of *TASSA*. The ΔH_a^{TASSA} increases drastically by increasing these layers, suggesting a dramatic increase in the volume fraction of the amorphous phase. It is worth noting that when the powder particles have no layers of the starting elements, heating them in the DTA does not lead to the formation of any amorphous phases and ΔH_a^{TASSA} becomes zero.

During the *TASSA* process, both T_x^{TASSA} and ΔH_x^{TASSA} increase drastically, suggesting a dramatic compositional change and an increase in the volume fraction of the amorphous alloy in the rod-milled powder. The crystallization process at this stage of milling involves only the crystallizing of the amorphous phase formed by *TASSA* process. It has been shown in a previous work[31] that, when the initial materials for rod-milling technique are intermetallic powders, the amorphization process or so-called mechanical disordering, MD, takes place without compositional change. This is because the final product of the milled powders has the same composition as the starting intermetallic compound powders. It has already been established[22] that the amorphization in the MD process requires an increase in the free energy of the compound from the lowest level (most stable phase) to a higher level (less stable phase).[50] This can be performed in the ball and/or rod mills by making the crystalline compounds store energy in the form of chemical disorder and point and lattice defects,[51] which is a typical example for the *MDSSA* process. Thus, the DTA curves for MD powders[51] have shown that the exothermic peaks which are related to the *TASSA* process are absent and the processes are carried out through a single stage. During all the stages of MD process, ΔH_a refers to the heat formation of an amorphous phase by the *MDSSA* process only, that is, ΔH_a^{MDSSA}. This value can be easily estimated at any MD processing time using Eq. (13.1).

In fact, *TASSA* is an interesting process in that an amorphous alloy can be formed simply by heating the layered-composite particles to 980 K without further milling. This phenomenon has also been shown in mechanically alloyed Ni-Ti,[22] Al-Nb,[52] Al-Ta,[50] Al-Ti,[31] and Al-Zr[32] binary systems. This process takes place in the same manner as solid-state amorphization in artificial metallic multilayered thin film.[54,55] One merit of the *TASSA* process is to shorten the milling time and this may lead to the formation of amorphous alloys with low contamination contents. However, the formed amorphous phase via *TASSA* method is rather heterogeneous, as indicated by broad crystallization peaks. Moreover, this amorphous phase has a low thermal stability, suggested by the low values of ΔH_x^{TASSA} and T_x^{TASSA}. The thermal stability and homogeneity of the amorphous phase can be improved by further milling times (see the following sections).

13.4.3.3.2 The intermediate stage of milling: the role of amorphization via *TASSA* and *MDSSA* processes

In parallel with the *TASSA* process, a *MDSSA* begins to take place at the fresh surfaces of Al/Hf layers with further MA time (86–173 ks). Thus, the amorphization reaction at this stage of milling occurs as a sum of two effects, the first is *TASSA* and the second is *MDSSA*. At the beginning of this stage, the elemental layers of Al and Hf of the composite powders become very thin and have excellent arrangement with numerous numbers of interfaces between the diffusion couples. Thus, the value of ΔH_a^{TASSA} decreases dramatically, being larger than ΔH_a^{MDSSA}. In other words, the value of ΔH_x^{TASSA} is larger than ΔH_x^{MDSSA}. Toward the middle of this stage, the elemental layers of Al and Hf gradually disappear and the amorphization process occurs mainly due to *MDSSA*. Thus, the values of ΔH_a^{MDSSA} and ΔH_x^{MDSSA} become larger than ΔH_a^{MDSSA} and ΔH_x^{TASSA}. Hence, in the amorphous phases, which form by way of *TASSA* plus *MDSSA* crystallizing together through a single broad exothermic peak (overlapped peak), it is very difficult to estimate the values of ΔH_a^{MDSSA}, ΔH_x^{MDSSA}, and ΔH_x^{TASSA} during this stage of MA. At the end of this stage (360 ks), the powder particles no longer have layered-structure morphology and the amorphization process starts to occur only due to *MDSSA*. Thus, the effect of the *TASSA* process in the amorphization route can be neglected.

13.4.3.3.3 The final stage of milling: the role of amorphization via *MDSSA* process

During this last stage of milling (360–1400 ks), the amorphization process occurs only due to the effect of *MDSSA*. Thus, both ΔH_a^{TASSA}, and consequently, ΔH_x^{TASSA}

become negligible. The free energy change of amorphization at this stage can be directly changed from the nonstable phase (starting material) to a more stable phase (amorphous). In fact, the formed amorphous phase via the *MDSSA* process is very homogeneous, and unprocessed Al and/or Hf atoms in the final product of the alloy powders are not detected. In addition, the alloyed powders possess excellent morphological properties such as homogeneous shapes (spherical-like morphology) with fine and smooth surface relief with uniform size (less than 1 μm in diameter). During this process, the amorphous alloys crystallize through single, sharp, exothermic peaks, suggesting that the amorphous phase formed is single and homogeneous in composition. Due to the crystallization process, the free energy changes from the amorphous phase to the most stable phase (crystalline). Compared with *TASSA*, the amorphization reaction via the *MDSSA* process occurs fast and much more homogeneously. This is demonstrated by the nearly constant values of T_x^{TASSA} and ΔH_x^{MDSSA}. One disadvantage of the *MDSSA* process is the long processing time required to change the free energy from the initial state to the amorphous state. This usually causes a rather high contamination of iron and oxygen.

13.5 The glass-forming range

As was mentioned in the preceding sections, the glass-forming range can be predicted using different models. It is worth noting that glass formed by mechanical alloying differs widely from rapid quenching; accordingly the glass-forming range for both processes differs.

Amorphization by mechanical alloying has been studied in a wide composition range for Ni-Ti,[21] Cu-Ti,[56] Fe-Zr,[57] Ni-Zr,[57–60] Co-Zr,[60,62] Ni-Nb,[63] Al-Nb,[35,64] Al-Ti,[31] Al-Hf,[34] and Al-Ta.[43,65] In most cases, the XRD diffraction patterns have been used to determine the glass and/or amorphous-forming range. For example, Fig. 13.28 shows the XRD patterns of the end product (1080 ks) for mechanically alloyed Al_xTa_{1-x}. At this stage of milling, clear halo-diffraction patterns are shown, suggesting the formation of an amorphous phase with a wide amorphization-range of $0.10 < x < 0.8$.[65]

Other useful and more reliable techniques are also used to determine several physical properties of the formed amorphous phases, such as the superconducting transition temperature (Fig. 13.29), crystallization temperature, T_x (Figs. 13.27, 13.30), enthalpy change of crystallization, ΔH_x (Figs. 13.26, 13.31), and crystallization temperature to melting point ratio, T_x/T_m (Fig. 13.32), which are qualitative properties depending on composition within the homogeneity range of the amorphous phase and being constant in the two-phase region.

Table 13.3 lists some typical metallic amorphous binary systems that have been obtained by way of mechanical alloying and their range of amorphization.

The formation of enthalpy of crystalline compounds can be estimated using Miedema's semi-empirical model.[70] This model can be applied to estimate the formation enthalpy of an amorphous alloy with chemical short-range order.[71] The

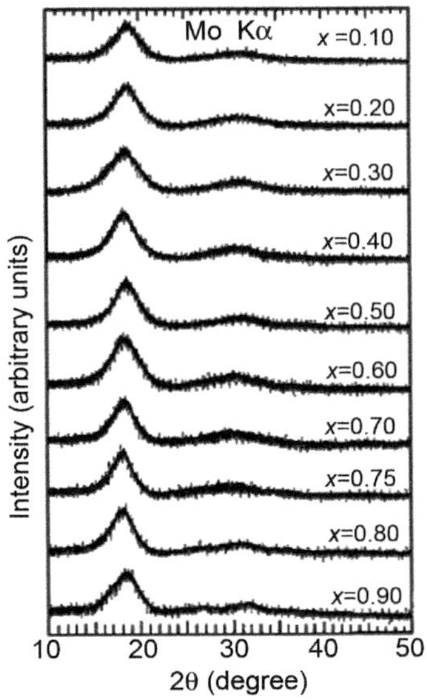

Figure 13.28 The XRD patterns of mechanically alloyed Al_xTa_{1-x} after 1080 ks (end product) of ball-milling processing time.
Source: After El-Eskandarany.[64]

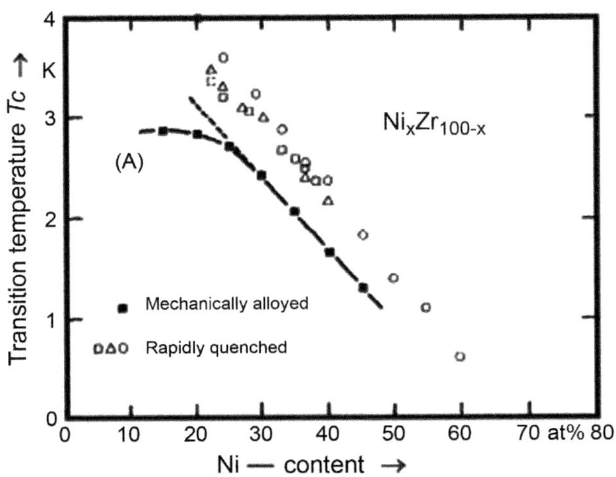

Figure 13.29 Superconducting transition temperature, T_c, of mechanically alloyed and rapidly quenched Ni_xZr_{100-x} as a function of Ni content.[65–67]

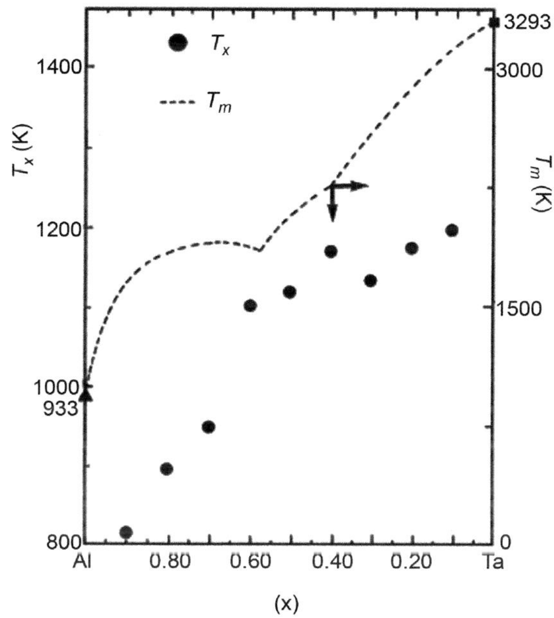

Figure 13.30 Correlation between the melting points, T_m *(dotted line)*, and the crystallization temperatures, T_x *(closed symbols)*, of mechanically alloyed Al_xTa_{1-x} after 1080 ks (end product) of ball-milling processing time.
Source: After El-Eskandarany.[68]

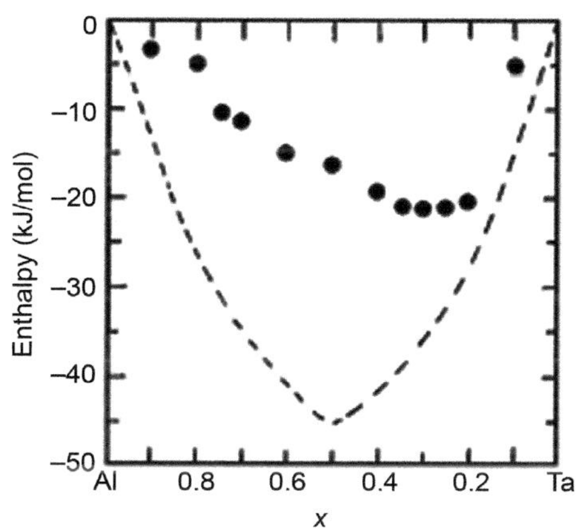

Figure 13.31 The enthalpy change of crystallization, ΔH_x, of mechanically alloyed Al_xTa_{1-x} after 1080 ks (end product) of ball-milling processing time.
Source: After El-Eskandarany.[64]

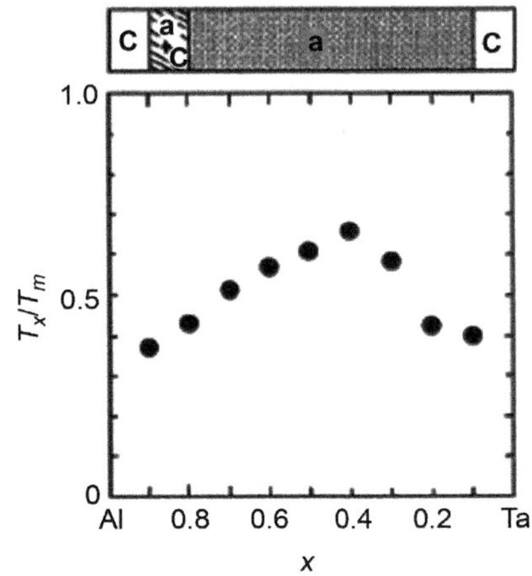

Figure 13.32 Dependence of T_x/T_m ratio of Al content, x, of mechanically alloyed Al_xTa_{1-x} after 1080 ks (end product) of ball-milling processing time.
Source: After El-Eskandarany.[64]

formation enthalpy of an alloy of a transition metal, A, in a transition metal, B, is given by the model as:

$$\Delta H_f = xA \int_B^A \Delta H_{sol}^{A\,in\,B} \tag{13.7}$$

where $\Delta H_{sol}^{A\,in\,B}$ is the solution enthalpy of A in B per mol A, x is the atomic fraction and $\int_B^A$ is the degree by which an A atom is surrounded by B atoms. For ordered crystalline alloys, $\int_B^A$ is defined as;

$$\int_B^A = C_B^S\left[1+8\left(C_A^S C_B^S\right)^2\right] \tag{13.8}$$

with C_B^S the surface concentration of B atoms:

$$C_B^S = \frac{x_B V_B^{2/3}}{x_A V_A^{2/3} + x_B V_B^{2/3}} \tag{13.9}$$

For amorphous alloys with chemical-short-range order, the order parameter is given as:[70]

$$\int_B^A = C_B^S\left[1+5\left(C_A^S C_B^S\right)^2\right] \tag{13.10}$$

For the formation enthalpy of the alloys, a weighted average of the formation enthalpies of A in B and B in A is taken:

$$\Delta H_f = x_A x_B \left(\int_B^A \Delta H_{sol}^{A\,in\,B} + \int_A^B \Delta H_{sol}^{B\,in\,A}\right) \tag{13.11}$$

Table 13.3 Glass and/or amorphous-forming range of selected metallic mechanically alloyed binary systems and the corresponding mill devices.

System	Milling device	Glass and/or amorphous-forming range	References
Ag-Cu	Spex mill	$Ag_{50}Cu_{50}$	[71]
Al-Fe	Planetary mill	$50 \leq Al \leq 80$	[72]
Al-Hf	Rod mill	$20 \leq Al \leq 80$	[34]
Al-Nb	Planetary mill	$Al_{50}Nb_{50}$	[63]
	Rod mill		[36]
Al-Ta	Ball mill Rod mill	$10 \leq Al \leq 80$	[69]
		$10 \leq Al \leq 75$	[43]
Al-Ti	Spex mill	$12.5 \leq Al \leq 26$	[73]
	Rod mill	$20 \leq Al \leq 75$	[31]
Al-Zr	Spex mill	$17.5 \leq Al \leq 40$	[74]
	Rod mill	$Al_{50}Zr_{50}$	[75]
Au-La	Vibratory mill	$Au_{55}La_{45}$	[76]
B-Fe	High-energy ball mill	B = 50, 60	[77]
Co-Nb	Attritor	$Co \leq 85$	[78]
Co-Ti	Planetary mill	$Co_{50}Ti_{50}$, $Co_{75}Ti_{25}$	[79,80,81]
Co-V	Planetary mill	$40 \leq Co \leq 67$	[82]
Co-Zr	Planetary mill	$27 \leq Co \leq 92$	[62]
		$75 \leq Co \leq 90$	[83]
Cr-Nb	Planetary mill	$25 \leq Cr \leq 80$	[84]
Cu-Hf	Planetary mill	$30 \leq Cu \leq 70$	[85]
Cu-Ta	Planetary mill	$Cu_{30}Ta_{70}$	[86]
Cu-Ti	Planetary mill	$10 \leq Cu \leq 87$	[56]
Cu-V	Planetary mill	$Cu_{50}Ta_{50}$	[87]
Cu-Zr	Spex mill	Cu = 40, 50, 60	[88]
Fe-Nb	Vibratory mill	$Fe_{52}Nb_{48}$	[29]
Fe-Si	Planetary mill	$Fe_{75}Si_{25}$	[89]
Fe-Ti	Spex mill	$Fe_{60}Ti_{40}$	[90]
Fe-Zr	Planetary mill	$20 \leq Fe \leq 80$	[28,91]
Fe-W	Ball mill	$Fe_{50}W_{50}$	[92]
Mn-Nb	Planetary mill	$45 \leq Mn \leq 55$	[84]
Mn-Zr	Planetary mill	$Mn_{60}Zr_{40}$	[93]
Nb-Ni	Spex mill	$20 \leq Ni \leq 83$	[94]
Ni-Ti	Spex mill	$35 \leq Ni \leq 80$	[95]
Ni-V	Planetary mill	$Ni_{40}V_{60}$	[96]
Ni-Zr	Planetary mill	$27 \leq Ni \leq 83$	[59]
Pd-Ti	Vibratory mill	$15 \leq Pd \leq 85$	[97]
Pd-Zr	Vibratory mill	$45 \leq Pd$	[98]
Si	Planetary mill	Si amorphous phase	[99]
Si-Ti	Planetary mill	$70 \leq Si \leq 80$	[100]
V-Zr	Vibratory mill	$V_{29}Zr_{71}$	[101]

Eqs. (13.10) and (13.11) are used to calculate the formation enthalpy of an amorphous alloy from the crystalline elements. This enthalpy (in J/mol) is estimated following Miedema and Niessen:

$$\Delta H = CT_m \tag{13.12}$$

where T_m is the melting temperature (in K) of the element and C is a constant for which an average value of 3.5 J/(mol K) will be used. From the difference in formation enthalpy of amorphous and crystalline phases calculated by Eqs. (10.8), (10.10), (10.11) and (10.13) the crystallization enthalpy of an amorphous phase can be estimated.

13.6 Amorphization via mechanical alloying when ΔH^{for}= Zero; mechanical solid-state amorphization of $Fe_{50}W_{50}$ binary system

The production of amorphous FeW alloys that have a heat of formation, ΔH^{for}, equal to zero[102] is difficult by melt quenching, sputter deposition, and thermal evaporation methods and few reports have been published on its amorphous formation.[103–105] This is due to the differences in the melting points and vapor pressure between Fe and W. The possibility of amorphous phase formation was suggested earlier by Boldrick et al.[106] when they milled pure W powders in a stainless steel ball mill. Because of the hardness of W powders, contamination and later alloying with Fe from abrasion of the stainless steel balls result in the amorphous phase. An amorphous-like Fe-W coexisting with unprocessed W powders was obtained after 180 ks, using a high-energy ball mill.[105]

In 1997, El-Eskandarany et al.[92] reported the formation of a single phase of $Fe_{50}W_{50}$ amorphous alloy powders via MA process using a low-energy ball mill operated under purified argon gas atmosphere at room temperature.

13.6.1 Structural changes with the milling time

The XRD patterns of mechanically alloyed $Fe_{50}W_{50}$ powders after selected MA times are presented in Fig. 13.33. In contrast to the intact mixture of polycrystalline bcc-Fe and bcc-W powders [Fig. 13.33A], after 86 ks of MA time [Fig. 13.33B], the Bragg peaks of α-W reflections become weak and broad, whereas, the peak positions do not change. On the other hand, the peak positions of α-Fe, taking Fe (110) as a typical example, shift to the low angle side. This suggests a solid-state diffusion of W atoms in the Fe matrix to form bcc Fe-W solid solution. The number of the W atoms that migrated to the Fe lattice increase with increasing MA time. Obviously, the intensity of the Bragg peaks for α-W decreases drastically after 173 ks of MA time, as shown in Fig. 13.33C. In parallel, the Bragg peaks of α-Fe shift to the low angle side and become broad. After 360 ks of MA time [Fig. 13.33D], no remarkable change in the peak position of α-Fe can be noted, indicating the completion of the solid-state reaction. After 720 ks of the

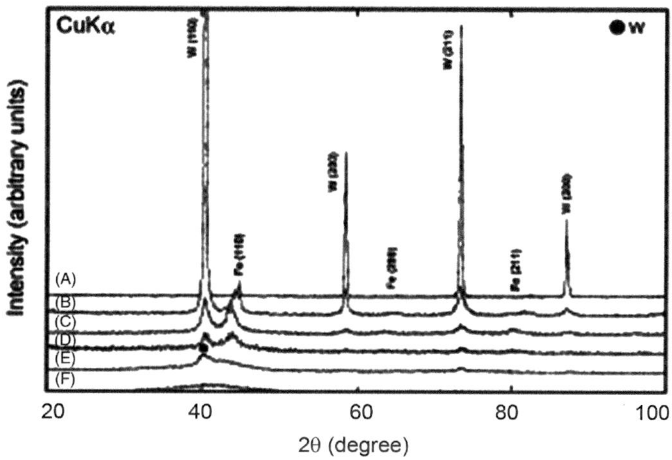

Figure 13.33 XRD patterns of mechanically alloyed $Fe_{50}W_{50}$ powders after (A) 0 ks, (B) 86 ks, (C) 173 ks, (D) 360 ks, (E) 720 ks, and (F) 1440 ks of MA time.
Source. After El-Eskandarany et al.[91]

MA time, the Bragg peaks of α-Fe surprisingly disappear, while a broad halo peak which overlaps those of α-W appears, as shown in Fig. 13.33E. This indicates the formation of amorphous FeW coexisting with a small fraction of unprocessed W powders. After 1440 ks of MA time, a solid-state amorphization reaction takes place between the amorphous phase of FeW and the W powders. The amorphous phase is characterized by a diffuse and smooth halo, as displayed in Fig. 13.33F. We emphasize that this amorphous phase is stable against mechanical deformation and does not change to any other phase(s) even after milling for a longer time, up to 2400 ks.

Fig. 13.34 displays the lattice parameter (a_0) of α-Fe as a function of the MA time. The a_0 monotonically increases with increasing the MA time. Increasing the MA time enhances the solid-state reaction so that the W atoms begin to migrate to the Fe lattice and this leads to a monotonical expansion in the value of a_0 for pure Fe. After 540 ks of MA time, the a_0 of α-Fe has a value of about 0.2915 nm, which is larger than that for pure Fe (0.2866 nm). This solid-solution phase transforms to a metastable phase (amorphous) after 720 ks of MA time, as presented in Fig. 13.33.

The TEM/EDS technique has been used to detect the local structural and compositional variations in mechanically alloyed $Fe_{50}W_{50}$ powders after several stages of milling. Fig. 13.35 summarizes the local (nanoscale) EDS analyses of mechanically alloyed $Fe_{50}W_{50}$ powders during several stages of milling. After 4 ks of MA time, the particles are either rich or poor in Fe content. After 22 ks of MA time, the solubility of W in Fe increases, and the particles have a very wide compositional distribution. At the intermediate stage of milling (22–173 ks), the particles are heterogeneous in composition. After 720 ks of MA time, the number of grains that have an average composition of $Fe_{70}W_{30}$ become predominant. Toward the end of the MA time (1440 ks), the composition of the powder particles approach the nominal composition, indicating the formation of a homogeneous alloy.

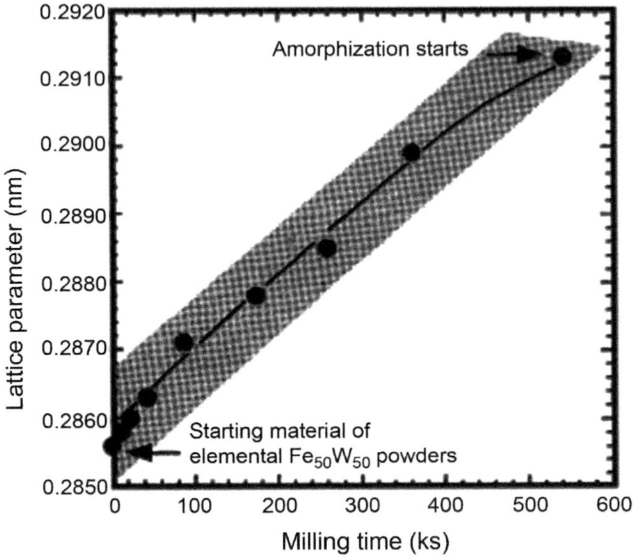

Figure 13.34 Correlation between the milling time and the lattice parameter, a_0, of a bcc Fe phase during milling of $Fe_{50}W_{50}$ powders.
Source: After El-Eskandarany et al.[91]

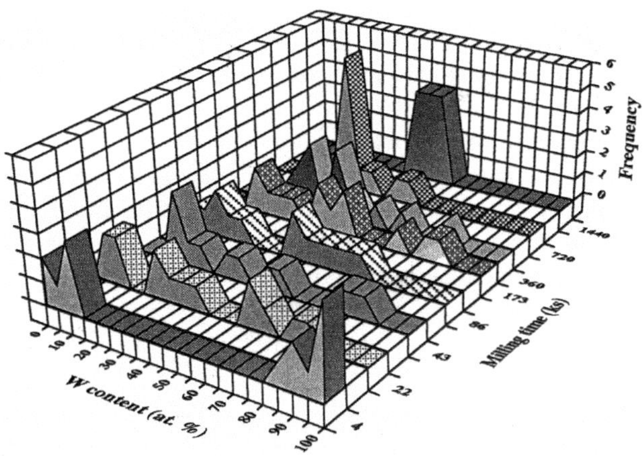

Figure 13.35 Correlation between the MA time, W content, and frequency of mechanically alloyed $Fe_{50}W_{50}$ powders.
Source: After El-Eskandarany et al.[91]

13.6.2 Magnetic studies

The magnetization of mechanically alloyed $Fe_{50}W_{50}$ powders at room temperature is plotted as a function of the MA time in Fig. 13.36. The rapid decrease in the magnetization during the early and the intermediate stages of milling suggests a dramatic decrease of pure Fe particles. Further milling (173–720 ks) enhances the solubility of

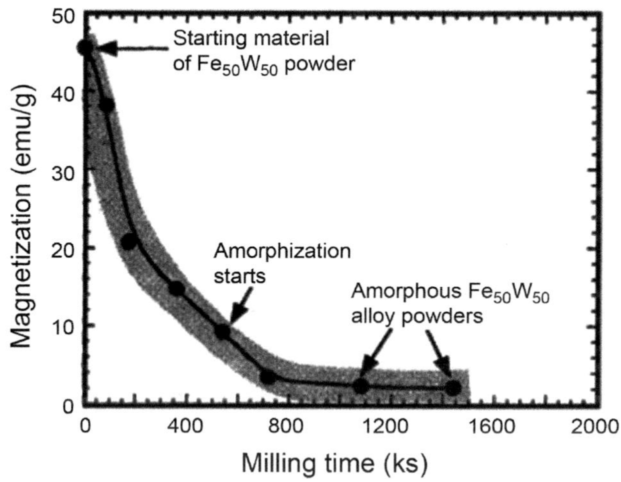

Figure 13.36 Dependence of magnetization, measured at room temperature for mechanically alloyed $Fe_{50}W_{50}$ powders, on the MA time.
Source: After El-Eskandarany et al.[91]

W in Fe lattice and this is translated to a monotonical decrease in the magnetization, as shown in Fig. 13.36. No dramatic change in the magnetization could be observed in samples milled for longer times (1080–1440 ks) and magnetization becomes constant at about 1.5 emu/g and still shows a saturation behavior, although FeW amorphous alloys are paramagnetic.[104] We suppose that this finite magnetization is attributable to the presence of Fe and Ni fine grains that came from the milling tools (balls and vial). It has also been reported[107] that the magnetization of mechanically alloyed FeV, surprisingly, increases toward the end of the MA time. Such increments have also been attributed to the presence of Fe contamination that came from the milling tools.

13.6.3 Thermal stability

The typical DTA thermogram of mechanically alloyed $Fe_{50}W_{50}$ powders is presented in Fig. 13.37 after selected MA times. The measurements were made at a constant heating rate of 0.67 K/s under flow of argon gas. All the samples were heated to 1300 K (first run) and cooled to about 300 K. Then, second heating runs (dashed lines) were performed, in order to get a base line. After 360 ks of MA time, an exothermic reaction takes place widely at about 580 K, suggesting amorphous-crystalline transformation, as shown in Fig. 13.37A. This broad exothermic peak shifts to the higher temperature side (about 785 K) and becomes pronounced after 720 ks of the MA time, as presented in Fig. 13.37B. After 1440 ks of the MA time, this crystallization reaction takes place at higher temperature, characterized by a single sharp exothermic peak centered at about 870 K, as shown in Fig. 13.37C.

The crystallization characteristics of amorphous $Fe_{50}W_{50}$ alloy powders indexed by the crystallization temperature, T_x (temperature of the exothermic crystallization reaction), and the enthalpy change of crystallization, ΔH_x (total area under the

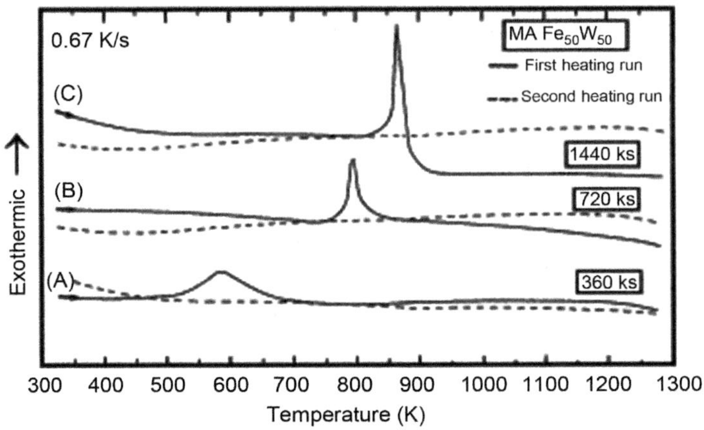

Figure 13.37 DTA curves of mechanically alloyed $Fe_{50}W_{50}$ powders after (A) 360 ks, (B) 720 ks, and (C) 1440 ks of MA time.
Source: After El-Eskandarany et al.[91]

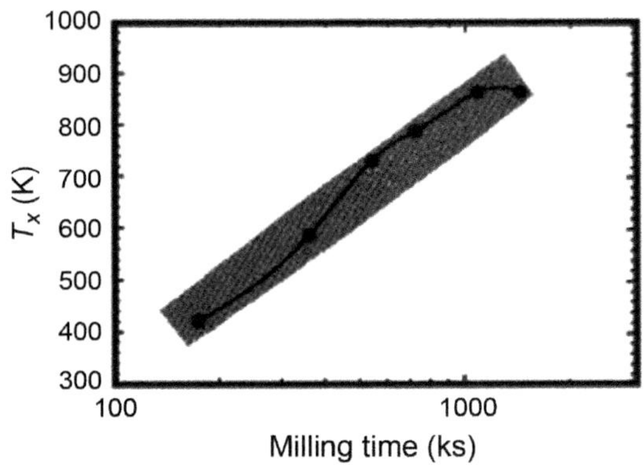

Figure 13.38 Effect of the ball milling time on the crystallization temperature, T_x, of mechanically alloyed $Fe_{50}W_{50}$ powders.
Source: After El-Eskandarany et al.[91]

exothermic crystallization reaction), are presented as a function of the MA time in Figs. 13.38, 13.39, respectively. It can be seen in Fig. 13.38 that the T_x of amorphous $Fe_{50}W_{50}$ alloy increases dramatically during the early and intermediate stages of milling. This is attributed to a continuous change in the composition of the amorphous phase. During the final stage of milling, the T_x approaches a saturation value of nearly 860 K, suggesting the formation of a single-phase amorphous alloy.

Further information on the crystallization reaction for amorphous $Fe_{50}W_{50}$ alloy powders is given by ΔH_x and presented in Fig. 13.39 as a function of the MA time. During the early and intermediate stages, the value of ΔH_x decreases monotonically

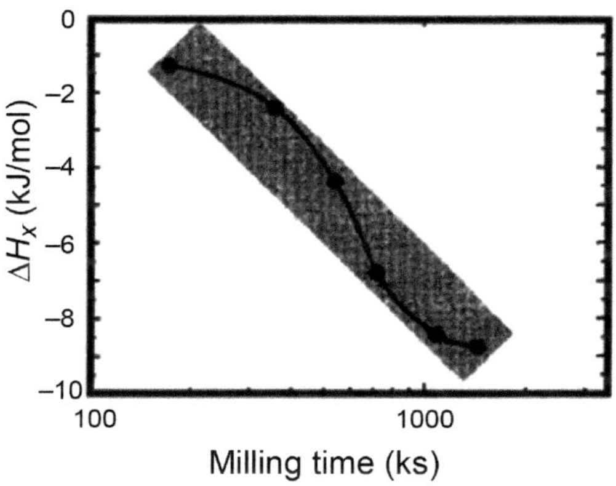

Figure 13.39 Effect of ball milling time on the enthalpy change of crystallization, ΔH_x, of mechanically alloyed $Fe_{50}W_{50}$ powders.
Source: After El-Eskandarany et al.[91]

with increasing MA time. At the end of the milling process (1080–1440 ks) ΔH_x saturates at a value of about −9 kJ/mol, suggesting that the milled powders contain a single phase of amorphous $Fe_{50}W_{50}$ alloy. These calorimetric studies clearly demonstrate that the amorphous phase becomes chemically and topologically homogeneous with increasing MA time.

13.6.4 Mechanism

As described in the previous section, a single phase of amorphous $Fe_{50}W_{50}$ powders has been synthesized by the milling of elemental Fe and W in a tumbling-ball mill. In this section, we discuss the amorphization process of the Fe-W binary system via MA. According to the characteristic properties of the alloy powders, the milling process is classified into the following three stages.

13.6.4.1 The stage of composite FeW powder particles formation

During MA time of 0–360 ks (the first stage), the initial powders of Fe and W are agglomerated to form large particles containing many layers of the reactant elements. In contrast to some of the other metallic binary systems,[35,36] annealing these multilayered composite powders does not lead to the formation of any amorphous phases. This is attributed to the null value of the heat of formation for the Fe-W binary system. Accordingly, no chemical driving force to lower the free energy is a priori present in this system. At the beginning of this stage (0–86 ks), the milled powders contain large grains of polycrystalline Fe and W. During the next duration of MA time (86–360 ks), the powders are subjected to heavy deformations, leading to a large amount of lattice and point defects. During this stage, the W atoms start to migrate to the Fe lattice to form a bcc Fe-W solid solution with finer grains. Towards the end of this stage (360 ks), a small fraction of an amorphous phase is formed by a mechanically driven solid-state amorphization (*MDSSA*). This amorphous phase is heterogeneous in its composition and structure: the local structure and composition differ widely from one particle to another and even within the particle itself. Moreover, the crystallization reaction takes place at low temperature and the exothermic peak of the amorphous phase formed at this stage is broad, indicating the low stability of this amorphous phase.

An empirical rule for the amorphous phase formation is proposed by the classical elastic continuum model. A large size difference in constituent elements makes substitutional atom packing unstable[108] and gives rise to the formation of an amorphous phase.[109] By calculating the local volume strain energy in an A matrix caused by introducing a B solute with a different atomic size for the $Fe_{50}W_{50}$ binary system, the critical concentration of A primary solid solution and an amorphous phase, $x_B^{\min}$ is expressed as:[109]

$$x_B^{\min} = 0.1V_A / \left|V_B - V_A\right| \tag{13.13}$$

where V_A and V_B are the atomic volumes of A and B, respectively. The $x_B^{\min}$ values are 0.38 for the Fe-rich primary solid solution and 48 for the W-rich primary solid solution. The present $Fe_{50}W_{50}$ alloy just satisfies this condition.

The forming abilities of an amorphous phase and a solid solution have also been discussed in terms of an alloy formation enthalpy, ΔH^{for}, and a difference in outer electron concentration of constituent elements, Δn:[111] an amorphous phase is favorable as ΔH^{for} is negative and the absolute values of ΔH^{for} and Δn become smaller. Since $\Delta H^{for}= 0$[111] and Δ=–2 for Fe-W alloys, this empirical relation predicts a wide range of Fe-W crystalline solid solutions in contrast to the present experimental results. However, random solid solutions such as an amorphous phase are stabilized by a free energy gain of the mixing entropy, because the metallic powder is mechanically milled at ambient temperature. In this context, the milled Fe-W is a ternary system of Fe, W, and defects.

13.6.4.2 The stage of formation of FeW solid solution

This stage of MA (360–540 ks) refers to the second stage of milling in that the number of W atoms that migrate to the Fe lattice are increased. Increasing the MA time enhances the solid-state reaction between the diffusion couples of Fe and W. This reaction is supported by a remarkable expansion of the lattice parameter of Fe accompanied by a drastic decrease in the magnetization. The T_x of the amorphous phase of this stage increases drastically, suggesting a continuous change in the composition. Moreover, the value of ΔH_x decreases monotonically with increasing MA time, suggesting an increase in the volume fraction of the amorphous phase against the crystalline phase of the milled powders. Toward the end of this stage, the alloy powders contain fine grains of bcc Fe-W solid solution coexistent with the residual W powders.

13.6.4.3 The stage of amorphous FeW formation

During this last stage of milling (360–1440 ks), the bcc Fe-W solid solution formed during the early and the intermediate stages transforms completely to an amorphous phase. The rapid diffusion process is enhanced by the several defects piled up in the milled powders. The amorphous phase coexists with a small fraction of unprocessed W powders. The W powders react with the FeW amorphous phase to produce a single amorphous phase of FeW powders. The local structures and compositions of these particles are homogeneous, as indicated by the EDS analyses. As the W content in the amorphous FeW matrix increases, both T_x and ΔH_x increase.

13.7 Special systems and applications

13.7.1 Amorphous austenitic stainless steel

Traditionally, steel and steel alloys are prepared by conventional melting and casting techniques. In 1994, El-Eskandarany and Ahmed[112] proposed an attractive technique for preparing austenitic stainless steel ($Fe_{74}Cr_{18}Ni_8$) by rod milling elemental powders of Fe, Cr, and Ni, at room temperature. The end product (after 1800 ks of milling) was a single amorphous alloy (Fig. 13.40). The thermal stability indexed by the crystallization temperature, T_x, of this amorphous alloy shows that the amorphous austenitic stainless steel is very stable up to temperatures as high as 1240 K. At this temperature (Fig. 13.41), the amorphous $Fe_{74}Cr_{18}Ni_8$ is transformed into an order phase of fcc $Fe_{74}Cr_{18}Ni_8$ (the typical structure of austenitic stainless steel), as shown in Fig. 13.42.

One significant potential technological attraction of this method of fabrication is its simplicity. Moreover, the end product of amorphous steel powder and/or annealed stainless steel (fcc structure) can be used as a powder for high corrosion resistant

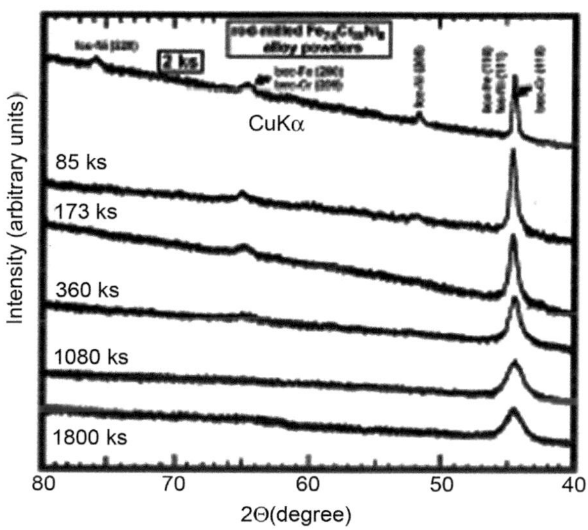

Figure 13.40 Effect of the rod-milling time on the structure of mechanically alloyed $Fe_{74}Cr_{18}Ni_8$ powders.
Source: After El-Eskandarany and Ahmed.[111]

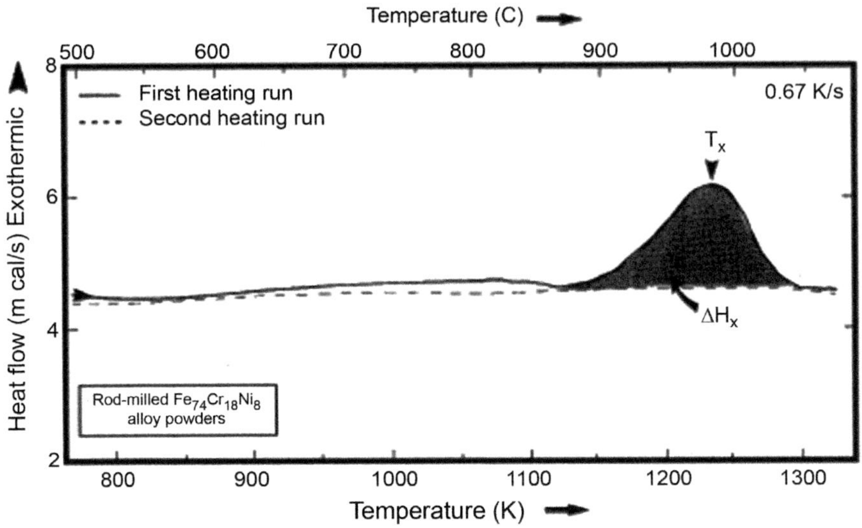

Figure 13.41 DTA curve of the end product for amorphous $Fe_{74}Cr_{18}Ni_8$ alloy powders as rod milled for 1800 ks.
Source: After El-Eskandarany and Ahmed.[111]

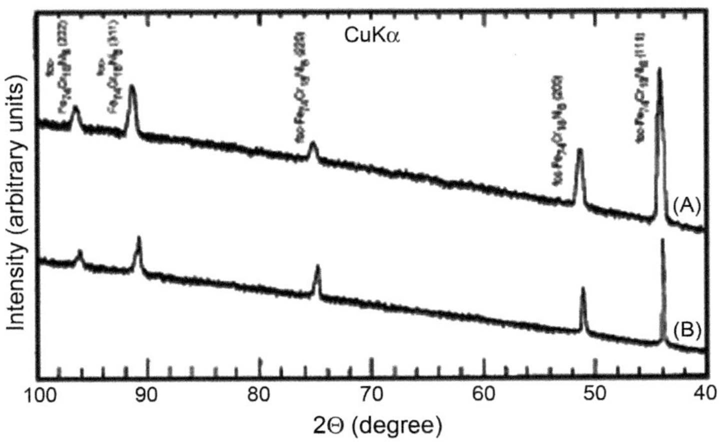

Figure 13.42 XRD patterns of (A) amorphous $Fe_{74}Cr_{18}Ni_8$ alloy powders as rod milled for 1800 ks, then heated to 1300 K during DTA measurements and (B) standard sample of austenitic stainless steel.
Source: After El-Eskandarany and Ahmed.[111]

coatings, or can be consolidated by hot pressing, or HIPing technique, thereby offering the prospect of manufacturing bulk metallic alloys made from amorphous (below crystallization temperature) and/or crystalline (above the crystallization temperature) steel phases by a simple and powerful technique.

13.7.2 Fabrication amorphous $Fe_{52}Nb_{48}$ special steel

Niobium is one of the most important alloying elements for the steel industry. It has a very strong chemical affinity for carbon and is used as a grain-refiner. The mechanical properties of steel, such as yield strength, tensile strength, and creep strength, are monotonically raised by the addition of elemental Nb. Unfortunately, the maximum solubility of Nb in α-Fe was reported to range from 0.095 to 0.330 at. % Nb at 873 to 1123 K, respectively.[113] Moreover, iron-niobium alloys with high Nb content (>45 at. %) cannot be obtained easily by the traditional method of liquid metallurgy, due to the remarkable gap differential in the melting points between pure Fe (1809 K) and Nb (2742 K) metals. This poor solubility and the difficulties in using the liquid metallurgy method restrict the fabrication of these technological alloys that could find a wide range of applications.

In 1999, El-Eskandarany et al.[29] proposed a room temperature ball-milling technique for fabrication of $Fe_{52}Nb_{48}$ amorphous alloy, using the mechanical alloying method. A single amorphous phase is formed after 144 ks of milling elemental Fe and Nb powders in a high-energy ball mill. The XRD patterns of mechanically alloyed $Fe_{52}Nb_{48}$ powder are presented in Fig. 13.43 after selected MA times. In spite of the intact mixture of polycrystalline bcc Fe and Nb powders that were taken after 7 ks of MA time [Fig. 13.43A], the Bragg peaks of Nb become broadened after 18 ks of milling [Fig. 13.43B]. Moreover, the intensities of the Bragg peaks for elemental Fe are markedly decreased, indicating a partial solubility of α-Fe in the Nb matrix. Toward

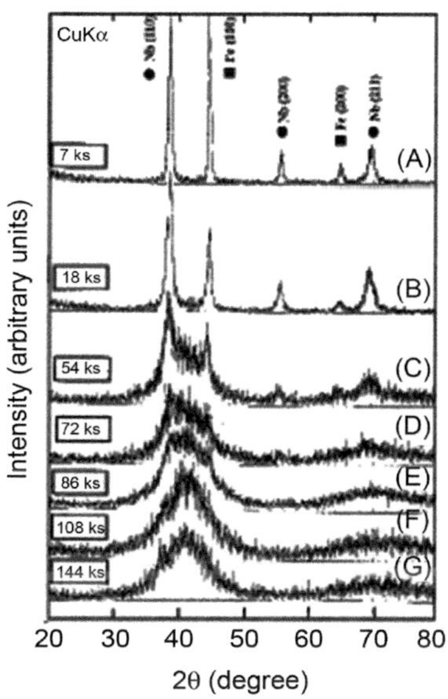

Figure 13.43 XRD patterns of mechanically alloyed $Fe_{52}Nb_{48}$ alloy powder after selected ball-milling time.
Source: After El-Eskandarany et al.[28]

the end of this early stage of MA (54 ks), a broad peak is overlapping the major Bragg peaks of Nb(110) and Fe(110), suggesting the existence of a major amorphous phase [Fig. 13.43C]. This first halo peak becomes more pronounced at the beginning of the intermediate stage of MA (72 ks), suggesting an increase in the amorphous phase and a decrease in the unreacted powders, as displayed in Fig. 13.43D. All the minor Bragg peaks of the elemental starting materials [Nb(200), Fe(200) and Nb(211)] have completely disappeared, and a second diffuse halo pattern has appeared [Fig. 13.43E]. At this stage of milling, a small mole fraction of metallic Nb and Fe powders is coexisting with the obtained amorphous phase, as suggested by the presence of Nb(110) and Fe(110) reflections in the first halo pattern. These peaks are hardly seen after 108 ks of MA time (final stage), and the material is completely in an amorphous phase, as shown in Fig. 13.43F. At the end of the milling (144 ks), a single homogeneous amorphous phase is obtained, characterized by smooth diffuse halos, as displayed in Fig. 13.43G.

This single amorphous phase transforms into an order phase (μ phase) upon heating at 1088 K (crystallization temperature, T_x) with enthalpy change of crystallization, ΔH_x, of −8.3 kJ mol^{-1}. The end product of this amorphous phase can be consolidated far below or well above its crystallization temperature to obtain either bulk amorphous or nanocrystalline full dense compacts.

13.7.3 Fe-Zr-B system

This important commercial system has been studied by L. Schultz et al.[17,114] In their experiments, elemental Fe, Zr, and submicrons of amorphous B powders were milled together, and the milled powders, during the early stage of milling, show a layered microstructure of Fe and Zr. The undeformed amorphous boron particles are caught by the colliding Fe and Zr particles and are embedded in the Fe/Zr interfaces. Further milling leads to refinement of this layered microstructure until, finally, the Fe and Zr layers react to form amorphous Fe-Zr (Fig. 13.44).

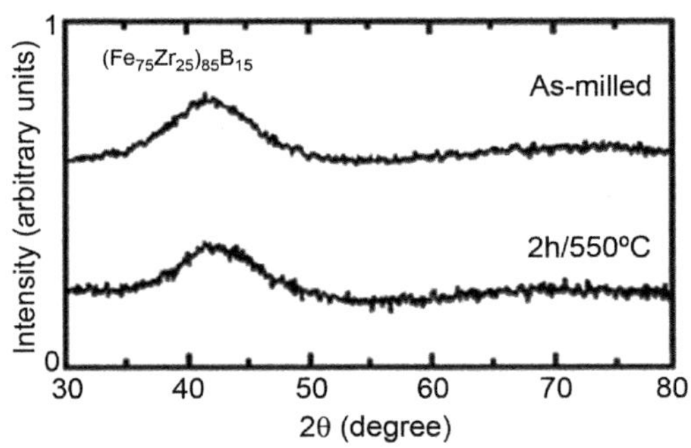

Figure 13.44 XRD patterns of $(Fe_{0.75}Zr_{0.25})_{85}B_{15}$ samples in the as-milled state and after 7.2 ks annealing at 728 K.
Source: After L. Schultz.[16]

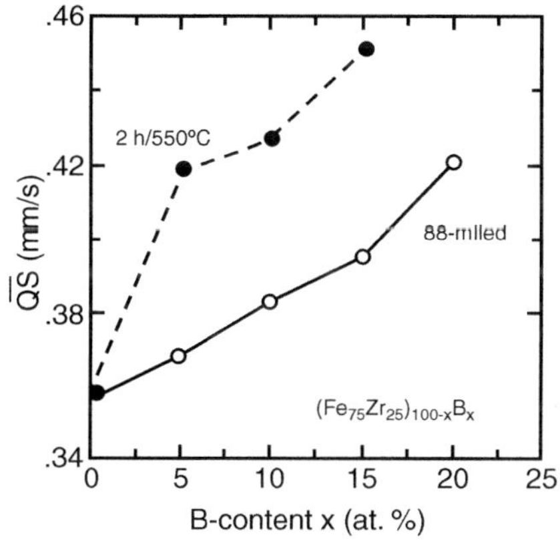

Figure 13.45 Average quadrupole splitting, QS, vs. the boron content for the as-milled and the reacted (annealed samples, 7.2 ks, 728 K) state.
Source: After L. Schultz.[16]

An additional solid-state reaction process (7.2 ks at 728 K) enables the boron to diffuse into the Fe-Zr, forming amorphous Fe-Zr-B (Fig. 13.44). The boron addition to amorphous Fe-Zr also increases the crystallization temperature.[17]

The Fe-Zr-B samples have also been studied by Mössbauer spectroscopy.[115] Fig. 13.45 shows the average quadrupole splitting, $\overline{QS}$, vs. the boron content for the as-milled and the reacted (annealed samples, 7.2 ks, 728 K) state. The $\overline{QS}$ increases during annealing because of the boron addition. However, the increase of $\overline{OS}$ with boron content for the as-milled samples shows that there is some boron [~ 5 at. % B for the $(Fe_{0.75}Zr_{0.25})_{85}B_{15}$ sample] dissolved in the amorphous Fe-Zr during mechanical alloying.

13.8 Difference between mechanical alloying and mechanical disordering in the amorphization reaction of $Al_{50}Ta_{50}$ in a rod mill

13.8.1 Background

There are only two categories of amorphization by ball milling and/or rod milling, that is to say, mechanical alloying (MA) and mechanical grinding (MG). The terms mechanical grinding,[116] or mechanical milling[117] are usually used to express the crystal-to-amorphous transformation without compositional change. Here, we are going to use mechanical disordering (MD)[50] as a more suitable terminology, which reflects the mechanism of crystal-to-amorphous transformation in much better understanding. As mentioned previously, the MA process synthesizes amorphous alloy powders by reacting elemental crystalline powders and/or intermetallic compound powders, with a negative heat of formation.[118] In the MD process, however, crystalline alloy

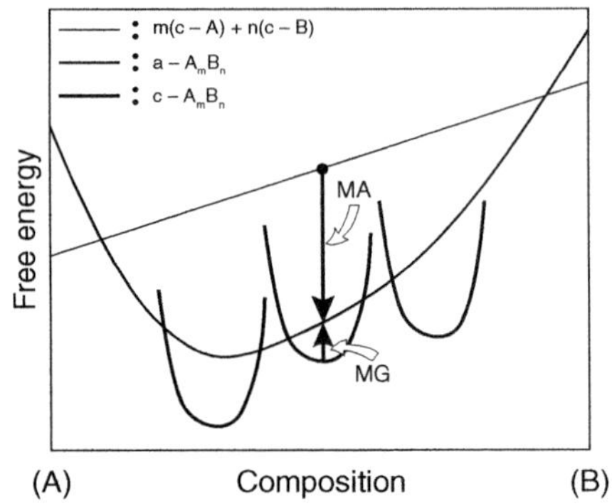

Figure 13.46 Thermodynamical processes of mechanical alloying (MA) and mechanical grinding (MG).
Source: After Suzuki.[118]

or compound powders are transformed into the amorphous solid state by relaxing the short-range order without compositional changes.[118] The MA and MD processes are reactions going in thermodynamically opposite directions of each other, as illustrated in Fig. 13.46,[118] which is a schematic free-energy diagram for the phases involved in the mechanical alloying and mechanical disordering processes. In the figure, *A* and *B* are the starting materials (before MA or MD), with three compositions (*A*-rich, equiatomic, and *B*-rich). The starting elemental crystalline state (*c*), the amorphous state (*a*), and the order state are assumed to have the same composition. The initial state is just a mixture of crystals of pure *A* and *B* so that the free energy of this state (point 1 in Fig. 13.46) is located along the straight line that is joining the free energies of the pure elemental powders of *A* and B. Point 2 is the free energy of the single amorphous phase (a-AB) that is formed by milling the elemental powders of *A* and *B* for a certain period of time. Point 3 is the free energy of the single crystalline intermetallic alloy with the same composition. The mechanical solid-state amorphization reaction takes place at a constant temperature that is sufficiently high enough to allow for the mechanical interdiffusion of *A* in *B* and/or *B* in *A*, but too low for promoting the nucleation and/or growth of the crystalline alloy. This kinetic selection of the reaction path is possible in binary systems in which the elements have largely different diffusivities in each other and in the amorphous phase.[5,21]

13.8.2 Procedure

In the MA process, elemental powders of aluminum (−325 mesh, 99.999%) and tantalum (−150 mesh, 99.99%) were mixed in a glove box under a purified argon gas atmosphere to give the desired average composition. On the other hand, the starting material for the MD process was prepared by arc-melting nominal amounts of 99.999% pure aluminum and 99.999% pure tantalum in an argon gas atmosphere (O_2 less than 0.2 ppm), purified with a titanium getter. The ingot was remelted five times, re-weighed, and checked metallographically. After the ingot had been crushed to powder using a stainless steel (SUS 304) mortar and pestle, the powder was classified in diameter to less than 50 μm, using screen analysis, and then analyzed, using the ICP emission method. The MA and MD processes were performed by the rod-milling technique.

13.8.3 Structural changes with milling time

The XRD patterns of mechanically alloyed and mechanically disordered $Al_{50}Ta_{50}$ are shown as a function of rod-milling time in Fig. 13.47A and B, respectively. As shown in Fig. 13.47A, the intensities of the Bragg peaks for pure Al and Ta powders decrease simultaneously with increases in the MA time. At the intermediate stage of MA (720 ks) all of the minor Bragg peaks from elemental Al and Ta crystals decrease with increasing MA time. After 86 ks of milling, the Bragg peaks from elemental fcc Al(200), (220) and (311) reflections disappear. In addition, the intensities of the Bragg peaks from elemental bcc Ta decrease quickly. During the intermediate stage (360–720 ks)

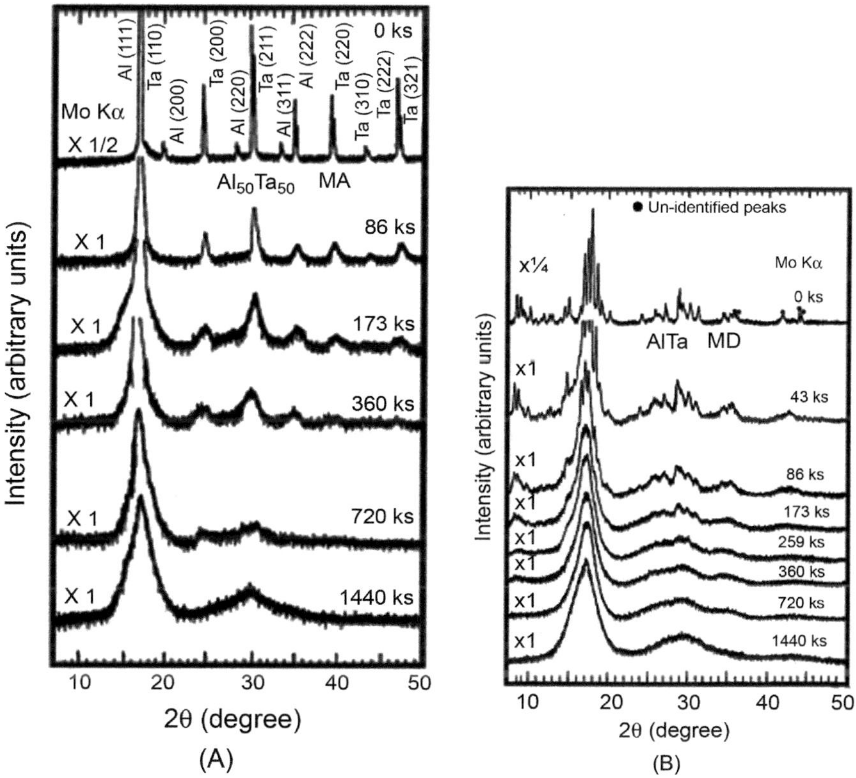

Figure 13.47 XRD patterns of $Al_{50}Ta_{50}$ alloy powders as a function of (A) MA time and (B) MD time.
Source: After El-Eskandarany et al.[49]

of MA time, the major Bragg peaks from pure Al(111) and Ta(110) and (211) reflections become wider, indicating the formation of an amorphous phase. Furthermore, the other Bragg peaks of pure Al and Ta have almost disappeared. Toward the end of the MA time (1440 ks), a homogeneous amorphous phase is formed, characterized by broad and smooth peaks.

As presented in Fig. 13.47B, the starting powders of an AlTa intermetallic compound have an XRD pattern with sharp Bragg peaks. The intensities of these peaks decrease by increasing the MD time. A halo pattern with no sharp Bragg peaks was obtained after 360 ks of MD time. At the final stage of milling (1440 ks), a homogeneous amorphous phase was formed, because small angle scattering at 2θ less than 10 degrees completely disappeared.

13.8.4 Morphological changes with milling time

Fig. 13.48 shows the scanning electron micrographs of the MA and MD $Al_{50}Ta_{50}$ powders after selected rod-milling times. The typical starting powders at 0 ks of MA and MD are shown in Fig. 13.48A and B, respectively. At the starting stage, the MD particles are blocky while the MA particles have a flake-like morphology. After 22 ks of MA time, the powder particles tend to agglomerate in size extending to more than 300 μm in diameter. In contrast, after 22 ks of MD, the particles' size is reduced to

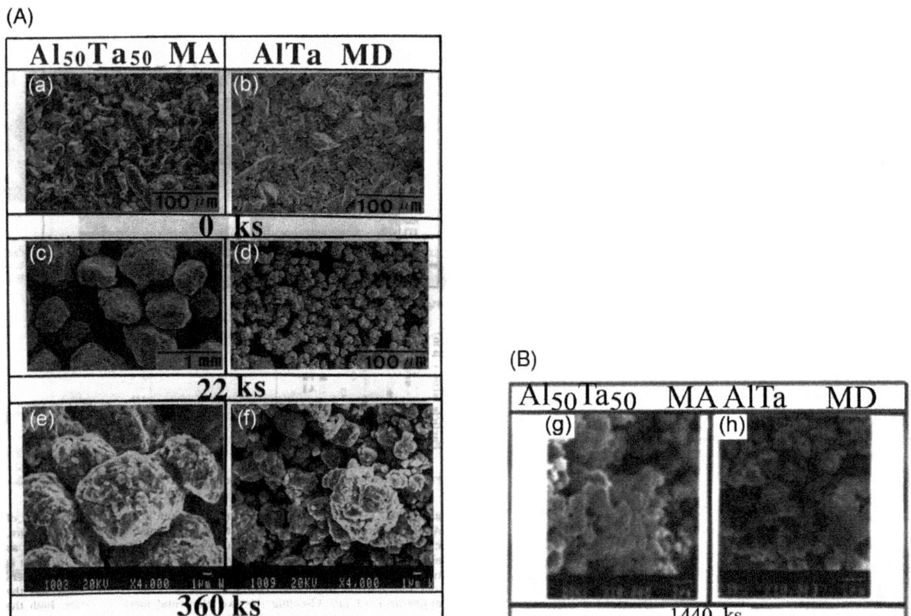

Figure 13.48 Scanning electron micrographs of $Al_{50}Ta_{50}$ alloy powders as a function of MA and MD time.
Source: After El-Eskandarany et al.[49]

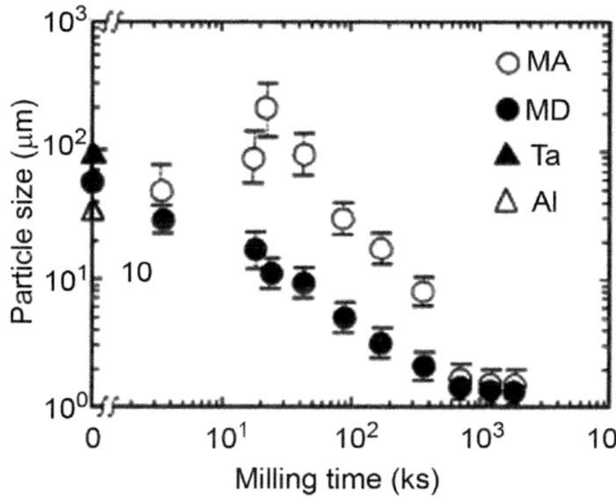

Figure 13.49 Particle size distribution of $Al_{50}Ta_{50}$ alloy powders as a function of MA and MD time.
Source: After El-Eskandarany et al.[49]

less than 20 μm in diameter, and the shape of the starting materials has been changed to a globe-like morphology.

Fig. 13.49 shows the change of particle size for MA and MD $Al_{50}Ta_{50}$ powders as a function of the rod-milling time. The sizes of the powders were determined from the SEM observations. The process for forming alloy powders differs for MA and MD.

The milling process performed by the MA process can be classified into three stages. At the initial stage (4–22 ks), which is called the agglomeration stage, the starting elemental powders of Al and Ta agglomerate during this early stage of milling to form powders of greater diameters; as large as several hundred micrometers peaking at almost 300 μm after 22 ks of milling, as shown in Fig. 13.48C. During this stage of milling, the characteristic structure containing Al and Ta layers appears, as shown in Fig. 13.50A. Using an electron microscope analysis, it was found that the gray parts are Al layers and the dark matrix is Ta rich. In the middle stage corresponding to the MA time from 22 to 360 ks, fragmentation with amorphizing reaction proceeds drastically by the atomic diffusion through the interface between the layers. Moreover, the

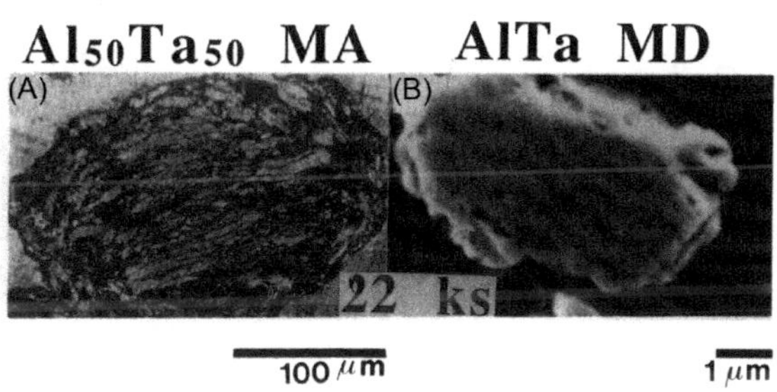

Figure 13.50 Morphology of the powder particles as a function of MA and MD processes after 22 ks of milling (A) optical micrographs of MA $Al_{50}Ta_{50}$ powders and (B) scanning electron micrographs of MD AlTa powders.
Source: After El-Eskandarany et al.[49]

powder particles are subjected to continuous disintegration, until the size of the powders is reduced to less than 10 μm in diameter, as shown in Fig. 13.48E and Fig. 13.49.

The MD process leads to a different behavior for the variation in powder diameters. The starting powders of an AlTa intermetallic compound are continuously reduced in size without any agglomerations or layer-structure morphology, to form fine powders of about 10 μm in diameter after only 22 ks, as shown in Fig. 13.48F. Furthermore, in the MD, however, the size reduction of the particles proceeds at a high rate to provide a narrow size distribution, as shown in Fig. 13.49.

13.8.5 Thermal stability

Fig. 13.51 shows the DTA curves for $Al_{50}Ta_{50}$ alloy powders as a function of the MA time. At 0 ks of milling, a sharp endothermic peak appears at about 930 K, due to the melting of aluminum in the starting material of $Al_{50}Ta_{50}$ powders. Then, this melt reacts with tantalum powders in the mixture, characterized by an exothermic peak, which appears above 1190 K, as shown in Fig. 13.51. During the first few kiloseconds of the MA time (4–22 ks) there is no significant difference in the thermal properties

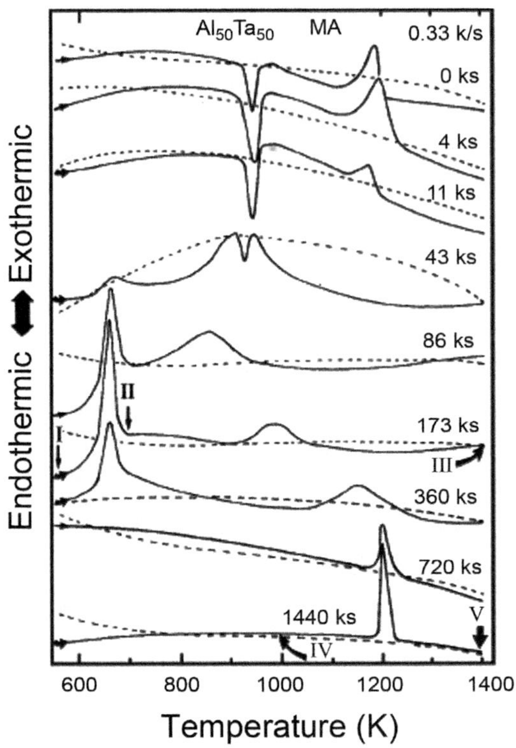

Figure 13.51 DTA curves of $Al_{50}Ta_{50}$ alloy powders as a function of MA time.
Source: After El-Eskandarany et al.[49]

between the alloy powders and the starting materials. Therefore, the process at this stage has the appearance of just blending the two elemental metal powders. Both the exothermic and the endothermic peaks disappear through the second heating runs, as illustrated by the dashed lines in Fig. 13.51. Contrary to this, a single broad exothermic peak appears at around 650 K, and the exothermic reaction peak disappears after 43 ks of MA time. After 86 ks of milling, another broad exothermic peak appears above 800 K, and the first exothermic peak becomes sharper. This second exothermic peak is shifted to an elevated temperature, and becomes sharp toward the end of the MA process (1440 ks). The temperature of the first exothermic peak, however, does not change with the milling time, as shown in Fig. 13.51. These reactions, which take place at the low temperature, occur due to so-called thermally assisted solid-state amorphization reaction[52] that was described in detail previously (Section 13.4.3.1).

In MD alloy powders, however, the exothermic reaction peaks, which are attributed to a solid-state amorphizing reaction at the fresh surfaces of Al/Ta boundaries, are absent, as shown in Fig. 13.52.

Fig. 13.53 shows the correlation between T_a, T_x, and the milling times of MA and MD. The value of T_a for MA alloy powders is independent of the milling time, which suggests that the amorphization by a solid-state reaction occurs simultaneously. Contrary to this, the T_x increases drastically with increasing the MA time in the early and the intermediate stages of milling, indicating a continuous change of the composition of the amorphous phase. The T_x approaches a saturation value (1210 K) during the

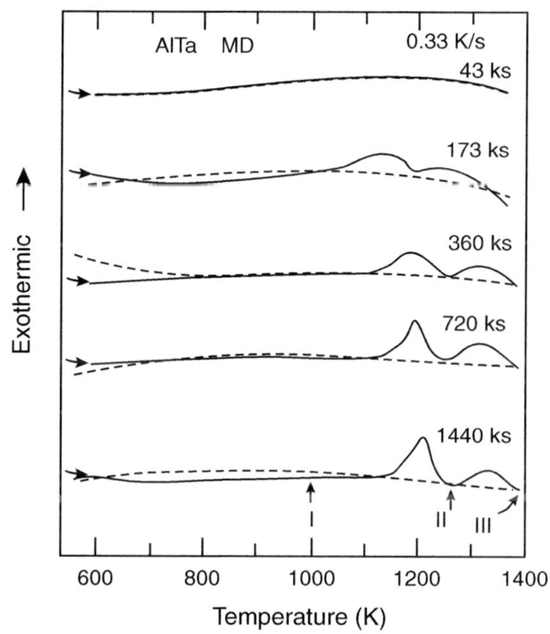

Figure 13.52 DTA curves of $Al_{50}Ta_{50}$ alloy powders as a function of MD time.
Source: After El-Eskandarany et al.[49]

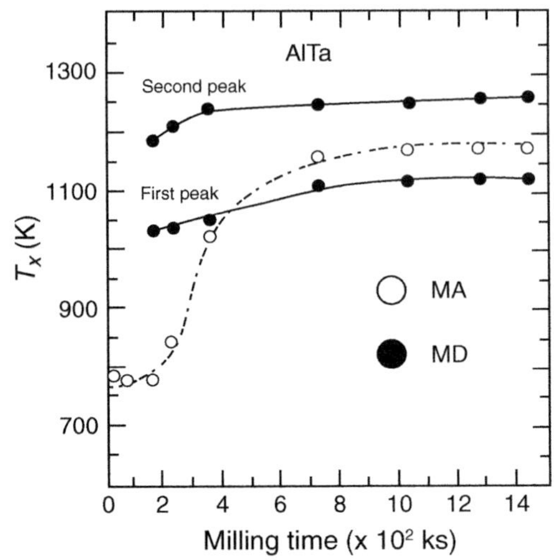

Figure 13.53 Crystallization temperature, T_x, and amorphization temperature, T_a, of $Al_{50}Ta_{50}$ alloy powders as a function of the rod-milling time.
Source: After El-Eskandarany et al.[49]

final stage of the MA time. In MD alloy powders, however, there is no big change in T_x, suggesting that the process occurs homogeneously, without compositional change, as illustrated in Fig. 13.53.

13.8.6 Mechanism of formation of amorphous $Al_{50}Ta_{50}$ via MD method

The MA and MD processes are similar in appearance, but differ completely in their mechanism. The MD process differs from the MA process in that the amorphization of a crystalline AlTa alloy requires an increase in its free energy from a lower level (more stable) to a higher level (less stable). By rod milling, the crystalline compound stores energy in the form of chemical disorder, and point and lattice defects. The crystal-to-amorphous transformation requires releasing of that stored energy to approach a specified value. In addition, the MD process is carried out through only one stage, in which the powder particles are continuously disintegrated without agglomeration or formation of layer-structure morphology of the particles. This is because the starting material here is an intermetallic compound by itself, so the solid-state amorphizing reaction can not be considered in this case.

Moreover, the MD technique differs from the MA technique in the rate of amorphization. The amorphization process by MD occurs fast and much more homogeneously than that by MA. This is demonstrated by the nearly constant values of T_x and ΔH_x at the intermediate and final stages of milling (360–1440 ks).

Furthermore, the DTA curves of mechanically disordered AlTa alloy powders have shown that the exothermic peaks, which are related to a solid-state amorphizing reaction, are absent. However, the XRD patterns of the DTA sample for MD AlTa, which was milled for 1440 ks and then annealed to 1400 K, did not show a clear difference in the Bragg peaks between them and those obtained in the case of the MA process. In addition, there is no significant difference in the Bragg peaks of as-cast AlTa crystalline compounds and as-annealed MA and MD powders at 1400 K, which were milled for 1440 ks. This may suggest that the amorphization by MA and MD leads to the formation of almost the same amorphous phase regardless of the great difference in the mechanism between the two processes. Furthermore, the two exothermic peaks, which appear in the DTA curves of the 1440 ks alloy, show that the crystallization of MD alloy powders occur in two stages. This may be attributed to nucleation and growth morphology of the heated alloy powders.

In addition, the iron contamination content in the MD powders is higher than it is in the MA powders. The higher iron contamination in MD powders can be attributed to the high hardness of the used intermetallic compound powders. Generally, intermetallic compounds are harder than their constituent elements. Therefore, the high iron contamination seems to be a general disadvantage in the MD process.

13.9 Mechanically induced cyclic crystalline-amorphous transformations during mechanical alloying

Since the amorphous phase is a metastable one, it inherently possesses the possibility of transforming into a more stable crystalline phase. This phase transformation usually takes place when an amorphous alloy is heated to a certain temperature (crystallization temperature, T_x), which usually ranges between 0.4 and 0.6 of its melting point, T_m. Trudeau et al.[120] suggested that some iron-based amorphous ribbons crystallize during high-energy ball milling. They mainly attributed this amorphous-crystalline phase transformation to the presence of hcp-Co, which enhances the crystallization reaction. They also discussed the effect of surface oxidation on the crystallization reaction. Eckert et al.[60] attributed the phase changes that occurred upon ball milling elemental $Al_{65}Cu_{20}Mn_{15}$ powder, to the presence of a high local effective temperature during the milling process.

The phase changes (if any) beyond these transformations were not reported until 1997 when El-Eskandarany et al.[121,122] reported the first novel technique for the possibility of a cyclic phase transformation that can occur during the high-energy ball milling of elemental Co and Ti. In addition, they investigated to prove that such cycloidal transformations were not unique to a single system (Co-Ti), and also could be observed during milling an elemental powder mixture of $Al_{50}Zr_{50}$.[123] Moreover, the effect of the milling speed (I) on such cyclic phase transformations has been studied.[123]

13.9.1 Co-Ti binary system

Pure elemental powders (99.9%) of Co (70 μm) and Ti (50 μm) were mixed to give the nominal composition of $Co_{50}Ti_{50}$ or $Co_{75}Ti_{25}$ (at. %) in a glove box under purified argon atmosphere and sealed in a stainless steel vial (SUS 316, 250 mL in volume) together with fifty stainless steel balls (SUS 316, 10 mm in diameter). The ball-to-powder weight ratio was maintained as 17:1. The MA process was performed in a high-energy planetary ball mill (Fritsch P5) at a rotation speed of 4.2 s^{-1}. The temperature of the vial was measured periodically, every 1.8 ks of MA time, and found to be about 320 K. In order to avoid a further temperature increase, the MA experiments were stopped periodically (every 1.8 ks) and resumed when the temperature of the vial decreased to about 300 K. When the vial was charged with the milling media only and operated for 1.8 ks, the vial's temperature reached to about 310 K. The MA experiments were performed under the same experimental conditions 3 times to confirm the reproducibility of the present investigation and to avoid any accidental results.

13.9.1.1 Structural changes with the milling time

Fig. 13.54 shows the XRD patterns of mechanically alloyed $Co_{75}Ti_{25}$ powder after selected milling times. In contrast to the initial mixture of polycrystalline hcp-Co and hcp-Ti [Fig. 13.54A] after 11 ks of MA time [Fig. 13.54B], a broad, diffuse, and smooth halo appears, suggesting the formation of an amorphous phase. The XRD pattern of the sample milled for 11 ks then annealed, well above its crystallization temperature (874 K), reveals an fcc structure corresponding to the equilibrium phase of Co_3Ti,[124] as shown in Fig. 13.55A. This amorphous phase is not stable against the impact and shear forces generated by the milling media (balls) and, surprisingly,

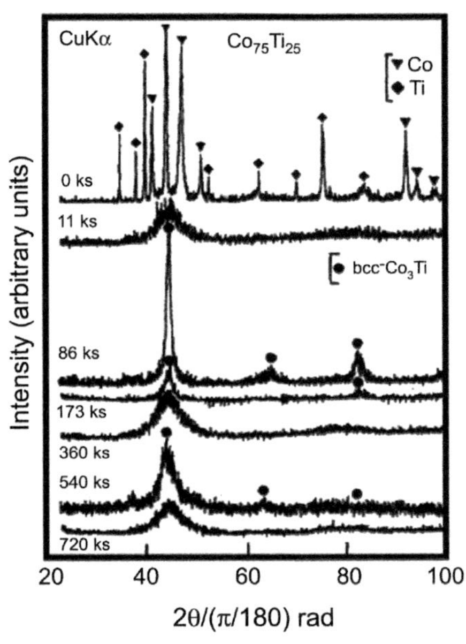

Figure 13.54 XRD patterns of mechanically alloyed $Co_{75}Ti_{25}$ powder milled for (A) 0 ks, (B) 11 ks, (C) 86 ks, (D) 173 ks, (E) 360 ks, (F) 540 ks, and (G) 720 ks of the MA time.
Source: After El-Eskandarany et al.[121]

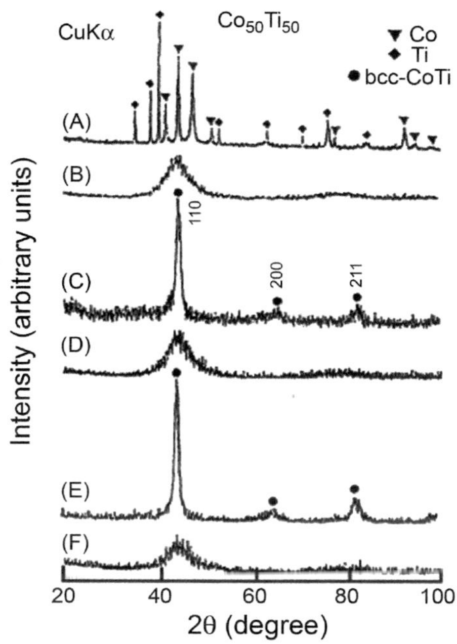

Figure 13.55 XRD patterns of mechanically alloyed $Co_{50}Ti_{50}$ powder after (A) 0 ks, (B) 22 ks, (C) 86 ks, (D) 360 ks, (E) 540 ks, and (F) 720 ks of the MA time.
Source: After El-Eskandarany et al.[122]

transforms into bcc-Co_3Ti upon milling for 86 ks, as shown in Fig. 13.54C. The lattice parameter, a_0, of this new bcc-Co_3Ti phase was calculated to be 0.2855 nm, being smaller than that for the ordered phase of bcc-CoTi (0.2987 nm). After 173 ks of the MA time [Fig. 13.54D] the Bragg peaks for bcc-Co_3Ti become broader, suggesting the existence of an amorphous phase. The XRD pattern of the sample, milled for 173 ks then heated to 1300 K, reveals an fcc structure corresponding to fcc-Co_3Ti coexisting with bcc-Co_3Ti. After 360 ks of the MA time, the bcc-Co_3Ti phase transformed completely to an amorphous $Co_{75}Ti_{25}$, as illustrated in Fig. 13.54E. Increasing the MA time to 540 ks leads again to the formation of nanocrystalline bcc-Co_3Ti coexisting with an amorphous phase, as presented in Fig. 13.54F. This bcc-Co_3Ti returns to the same amorphous phase of $Co_{75}Ti_{25}$ after 720 ks of MA time, as shown in Fig. 13.54G. The XRD pattern of the as-annealed 720 ks alloy at 1300 K reveals an fcc structure corresponding to fcc-Co_3Ti. It is worth noting that such transformations were observed for all the samples, which were milled on three different milling runs.

Fig. 13.55 shows the XRD patterns of mechanically alloyed $Co_{50}Ti_{50}$ powder after selected milling times. A broad, diffuse, and smooth halo appears after 22 ks of MA time [Fig. 13.55B], suggesting the formation of an amorphous phase. The amorphous phase formed at this stage is not stable against the impact and shear forces generated by the milling media (balls) and, surprisingly, crystallizes into bcc-$Co_{50}Ti_{50}$ upon milling for 86 ks, as shown in Fig. 13.55C. The lattice parameter, a_0, of this bcc-$Co_{50}Ti_{50}$

phase was calculated to be 0.2982 nm, being in agreement with the reported value (0.2987 nm). After 360 ks of the MA time, this bcc phase transforms again to an amorphous phase, suggested by the halo that is presented in Fig. 13.55D. This phase transformation occurs due to the accumulation of several lattice imperfections, such as point and lattice defects, which raise the free energy from a stable phase to a less stable phase.

Further milling (540 ks) leads again to crystallization of the amorphous phase into nanocrystalline bcc-CoTi, as presented in Fig. 13.55E. This bcc-$Co_{50}Ti_{50}$ returns to the same amorphous phase of $Co_{50}Ti_{50}$ after 720 ks of MA time, as shown in Fig. 13.55F. These cyclic phase transformations have been observed for all the samples, which were milled for three (and sometimes four) different milling runs. Thus, the sequence of the phase transformation with MA time in ball-milled $Co_{50}Ti_{50}$ is

$$\mathrm{CO + Ti} \xrightarrow{22\,\mathrm{ks}} \mathrm{amorphous} \xrightarrow{86\,\mathrm{ks}} \mathrm{bcc} \xrightarrow{360\,\mathrm{ks}} \mathrm{amorphous} \xrightarrow{540\,\mathrm{ks}} \mathrm{bcc} \xrightarrow{720\,\mathrm{ks}} \mathrm{amorphous}$$

Fig. 13.56 shows the bright field images (BFIs) and the corresponding selected area diffraction patterns (SADPs) of mechanically alloyed $Co_{50}Ti_{50}$ powder after

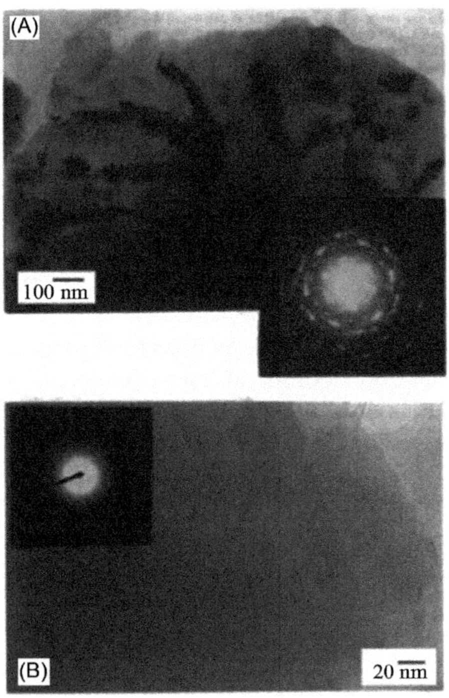

Figure 13.56 BFIs and the corresponding SADPs of mechanically alloyed $Co_{50}Ti_{50}$ powder after (A) 3.6 ks, (B) 22 ks, (C) 86 ks, and (D) 360 ks of the MA time.
Source: After El-Eskandarany et al.[122]

selected MA times. After a few kiloseconds of MA (3.6 ks), the Co powder (dark layers) embedded into the Ti matrix, forming composite Co/Ti particles, as shown in Fig. 13.56A. The SADP [inset of Fig. 13.56A] shows sharp spot patterns (Laue spots) that corresponds to polycrystalline Co and Ti. Fig. 13.56B presents the BFI and the corresponding SADP of a sample milled for 22 ks of the MA time. Overall, the sample appears to have a homogeneous fine structure with no dominant facet structure. Moreover, the SADP shows a typical halo pattern of an amorphous phase in good agreement with the XRD pattern presented in Fig. 13.55B. The BFI of the powder that milled for 86 ks of MA time is presented, together with the corresponding indexed SADP, in Fig. 13.56C. The powder consists of nano-size grains of polycrystalline bcc-$Co_{50}Ti_{50}$, as indicated by the Debye-Scherrer rings. In a cyclic crystalline-amorphous phase transformation, this bcc-$Co_{50}Ti_{50}$ phase returns to an amorphous phase after 360 ks of MA time. The formed amorphous phase has a fine and homogeneous structure with a clear halo diffuse pattern, as presented in Fig. 13.56D.

The magnetization of the milled powder at room temperature is plotted as a function of the MA time in Fig. 13.57. The magnetization of the starting materials (does not appear in Fig. 13.57) is measured to be 74 emu/g. After 22 ks, the magnetization of the alloyed powder decreases dramatically, suggesting a decrease of pure Co particles in the mixture of $Co_{50}Ti_{50}$ powder, and formation of an amorphous phase. During the next stage of MA (43–86 ks), the magnetization of the milled powder decreases, to have a value of less than 2.5 emu/g, suggesting the formation of bcc-$Co_{50}Ti_{50}$. The magnetization of the mechanically alloyed powder tends to increase upon further MA time (173–360 ks), indicating the formation of an amorphous phase. Increasing the MA time (450–540 ks) leads to a remarkable decrease in the magnetization. This is attributed to the crystallization of the amorphous alloy and the formation of crystalline bcc-$Co_{50}Ti_{50}$ phase. Towards the end of the MA time (630–720 ks), crystalline-to-amorphous phase transformation is observed, and the yielded amorphous phase has almost the same magnetic properties of the 22 ks and 360 ks MA amorphous alloys,

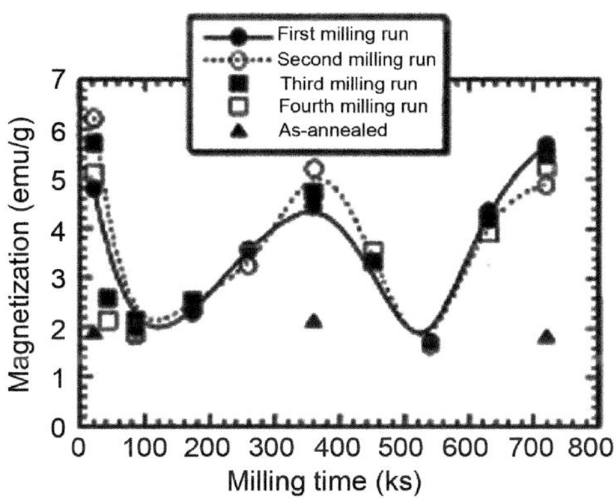

Figure 13.57 Dependence of the magnetization, measured at room temperature of mechanically alloyed $Co_{50}Ti_{50}$ powder, on the MA time.
Source: After El-Eskandarany et al.[122]

indicating cyclic phase transformations. It is worth noting that the magnetization of the amorphous samples that annealed well above their crystallization temperatures (976 K), have almost the same magnetization values of the samples, which milled for 86 and 540 ks. This may prove the similarity of the structure for these samples.

13.9.1.2 Thermal stability

The DTA curves of mechanically alloyed $Co_{50}Ti_{50}$ powder are presented in Fig. 13.58, after selected MA time. The measurements were performed at a constant heating rate of 0.33 K/s under an argon gas atmosphere. All the samples were heated to 1200 K (first run) and cooled to about 400 K. Then, second heating runs *(dashed lines)* were performed in order to get the base line. After 22 ks of the MA time [Fig. 13.58A], a single sharp exothermic peak appears at 968 K. This exothermic reaction is attributed

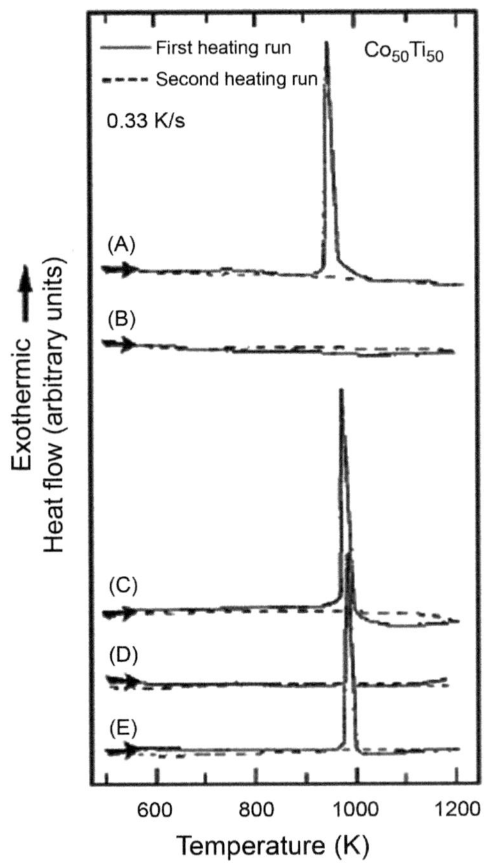

Figure 13.58 DTA curves of mechanically alloyed $Co_{50}Ti_{50}$ powder after (A) 22 ks, (B) 86 ks, (C 360 ks, (D) 540 ks, and (E) 720 ks of the MA time.
Source: After El-Eskandarany et al.[122]

to an amorphous-crystalline phase transformation (crystallization). The XRD patterns of the sample heated up to 1300 K show the formation of an equilibrium bcc-$Co_{50}Ti_{50}$ phase. After 86 ks of MA time, this exothermic reaction disappears and no other reactions can be detected, as illustrated in Fig. 13.58B. This suggests the amorphous-crystalline (bcc-$Co_{50}Ti_{50}$) phase transformation of the as-milled sample. No remarkable changes in the crystallization temperature (the peak temperature of the exothermic peak) can be detected in the sample that was milled for 22 ks. This bcc-$Co_{50}Ti_{50}$phase transforms to a single amorphous phase after 360 ks of MA time, suggested by the sharp exothermic reaction which appears at 976 K in Fig. 13.58C. The crystallized sample reveals a structure of bcc-CoTi phase. After 540 ks of MA time [Fig. 13.58D], no reactions can be noted, suggesting that the milled powder has a stable phase of bcc-$Co_{50}Ti_{50}$. After 720 ks of MA time, the bcc-$Co_{50}Ti_{50}$ phase transforms to an amorphous $Co_{50}Ti_{50}$ powder, which crystallizes through a single exothermic reaction peak [Fig. 13.58E].

13.9.2 Al-Zr binary system

In order to understand the effect of the milling speed on the cyclic phase transformations, several samples of mechanically alloyed $Al_{50}Zr_{50}$ were prepared, using milling speed, 90 rpm ($I = 2$), 180 rpm ($I = 4$), 270 rpm ($I = 6$), and 360 rpm ($I = 8$). In all experiments the temperature of the vial, T_v, was measured every 30 minutes. It is worth noting that when the samples were milled with a low milling speed ($I = 2$ and 4), T_v does not increase remarkably, (307 or less). For those experiments run at $I = 6$ and 8, however, T_v increases rapidly with increasing MA time, and is found to be about 320 K after 30 minutes of MA time. In order to minimize the temperature effect in the study of the cyclic transformations, the ball-milling experiments were stopped periodically (every 30 minutes) and resumed when the temperature of the vial decreased to about 300 K. This on/off mode was also applied to those samples milled at a low milling speed. It is worth noting that when the vial was charged with milling media only and operated at $I = 8$ for 30 minutes, the vial's temperature increased to about 310 K. In order to study the reproducibility of this work, all the milling experiments were repeated, under the same experimental conditions, 3 times.

13.9.2.1 Structural changes with the milling time

Here, unexpected results can be seen in Fig. 13.59, which shows the effect of the MA time on the structural changes of $Al_{50}Zr_{50}$ powder milled at $I = 8$. In contrast to the starting reactant materials of polycrystalline fcc-Al and hcp-Zr [Fig. 13.59A], a nanocrystalline metastable phase of fcc-AlZr solid solution is obtained after 12 h of the MA time [Fig. 13.59B]. Increasing the MA time (24 h) transforms this phase into an amorphous phase of $Al_{50}Zr_{50}$, characterized by a broad, diffuse, and smooth halo, as displayed in Fig. 13.59(*c*). This amorphous phase is not stable against further mechanical deformation. It crystallizes into the most stable phase of orthorhombic-AlZr upon milling for 48 h as in Fig. 13.59(*d*). In a cyclic crystalline-amorphous phase transformation and after 72 h of MA time, the crystalline phase transforms to an amorphous

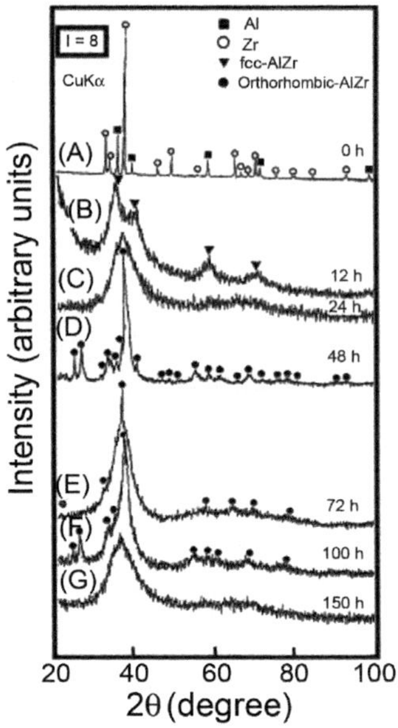

Figure 13.59 XRD patterns of $Al_{50}Zr_{50}$ powder that were ball-milled at milling speed of 8 after (A) 0 h, (B) 12 h, (C) 24 h, (D) 48 h, (E) 72 h, (F) 100 h, and (G) 150 h of the MA time. *Source:* After El-Eskandarany et al.[123]

phase coexisting with nanocrystalline grains of AlZr [Fig. 13.59E]. Presumably, this phase transformation takes place due to the accumulation of several lattice imperfections, such as point and lattice defects, which raise the free energy from a stable phase (orthorhombic-AlZr) to a less stable phase (amorphous-AlZr). Further milling (100 h) leads, again, to crystallization of a large fraction of the amorphous phase, and the product of this stage of milling is a composite nanocrystalline/amorphous powder [Fig. 13.59F]. Finally, at an MA processing time of 150 h, the nanocrystalline phase transforms again to the amorphous phase [Fig. 13.59G]. These cyclic phase transformations have been observed for all the samples milled on three different milling runs, using I = 8. It has been shown in a previous study[19] that when a mixture of $Al_{50}Z$-r_{50}powders were milled for a long-milling run (400 h), using a low-energy ball mill and/or rod mill, no mechanically induced crystallization and/or cyclic phase transformations were observed, and the final-products of those experiments were amorphous $Al_{50}Zr_{50}$ powders. Increasing the low-energy ball milling time[15] does not lead to the crystallization of the amorphous phase, even after a very long milling time (800 h). We can then conclude that the kinetic energy of the ball mill, which is affected by the milling speed, plays an important role in the incidence of cyclic transformations.

In contrast to the results shown in Fig. 13.59, the XRD patterns of the samples milled for different MA times, but at lower milling speed ($I = 4$) (Fig. 13.60), show a slow solid-state amorphization reaction taking place between the diffusion couples of Al and Zr with the absence of the cyclic phase transformations. It is worth noting that, even for samples milled at this low milling speed at longer milling times (400 h, 450 h,

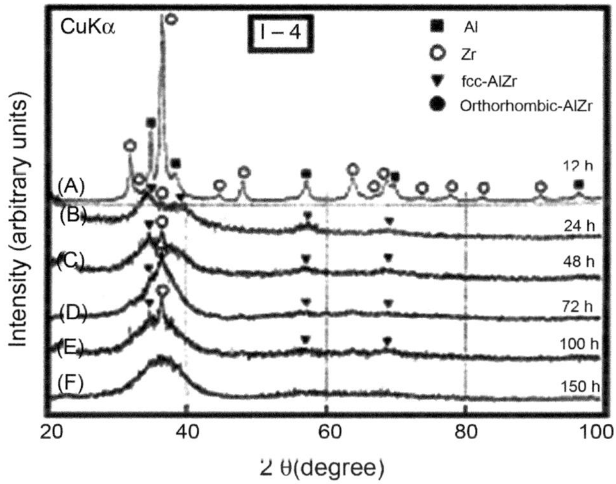

Figure 13.60 XRD patterns of $Al_{50}Zr_{50}$ powder that were ball-milled at milling speed of 4 after (A) 12 h, (B) 24 h, (C) 48 h, (D) 72 h, (E) 100 h, and (F) 150 h.
Source: After El-Eskandarany et al.[123]

500 h, 550 h, and 600 h), the cyclic reactions do not appear. The end-product for each milling run is the same as that shown in Fig. 13.60F.

13.9.2.2 Thermal stability

The crystallization characteristics indexed by the enthalpy change of crystallization, ΔH_x, of amorphous $Al_{50}Zr_{50}$ powder that were milled at different milling speeds, I, are plotted in Fig. 13.61 as a function of the MA time. For I = 8 [Fig. 13.61A], a rapid increase in the value of ΔH_x occurs during the first stage of milling (0–12 h). This indicates a decrease of pure elemental Al and Zr in the milled powder, and the formation of an amorphous phase. After 24 h of MA time, these values increase again, ranging from −36 kJ/mol to −38 kJ/mol. During the next stage of milling (24–48 h), the values of ΔH_x for the as-milled powder decrease remarkably, suggesting mechanically induced amorphous-crystalline phase transformation. After 72 h of milling time, however, the values of ΔH_x for the milled powder tend to increase again, due to the formation of amorphous $Al_{50}Zr_{50}$ alloy. Again, the ΔH_x tends to have lower values during further MA time (72–100 h), indicating a second amorphous-crystalline phase transformation, as shown in Fig. 13.61A. After longer milling (100–150 h), the value of ΔH_x increases to have almost the same values as those milled for 24 h.

The characteristics of the mechanically alloyed powders seem to be very sensitive to the milling speed, I, as suggested by the behavior of ΔH_x with the I and MA time [Fig. 13.61B]. For I = 4 and 6, the value of ΔH_x increases monotonically with increasing MA time, indicating an increasing fraction of the amorphous alloy. At I = 6, this amorphous phase gradually transforms into a stable phase when milled for 72–150 h. Based on the results of the present study, the crystalline phase formed after 150 h of milling is transformed to an amorphous phase after milling for 440 h. No cyclic phase

(A)

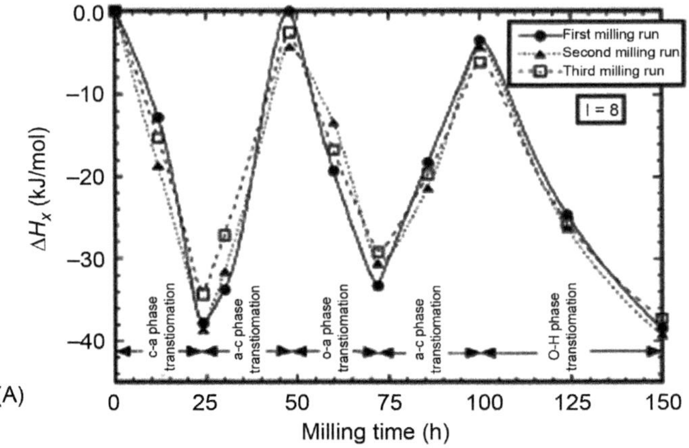

(B)

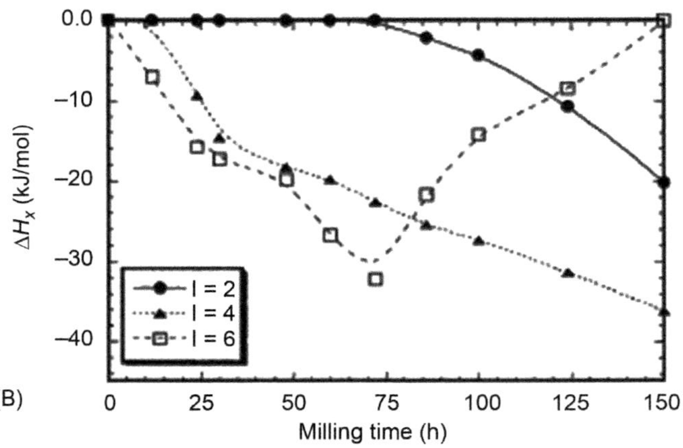

Figure 13.61 Correlation between the enthalpy change of crystallization, ΔH_x, for amorphous $Al_{50}Zr_{50}$ alloy powders that were ball milled at (A) high speed of milling and (B) low speed of milling.
Source: After El-Eskandarany et al.[123]

changes could be observed for those samples milled at I = 4 [Fig. 13.61B], and the end-product is an amorphous phase, even after long milling runs, as long as 600 h. Contrary to the samples milled at $I = 4$, the end-product (150 h) of the samples milled at $I = 2$ are an amorphous phase coexisting with a considerable fraction of elemental reactant powders, as indicated by the low value in ΔH_x [Fig. 13.61B].

13.9.3 Mechanism of amorphous-crystalline-amorphous cyclic phase transformations during ball milling

One possible factor which may lead to such cyclic crystalline-amorphous transformations of Co-Ti or Al-Zr binary systems, is the introduction of contamination in the

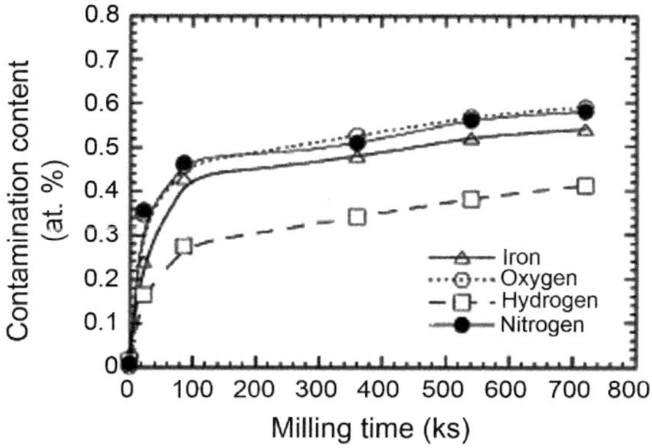

Figure 13.62 Dependence of the contamination content of mechanically alloyed $Co_{50}Ti_{50}$ powder on the MA time.
Source: After El-Eskandarany et al.[122]

ball-milled powder. The iron and gas (oxygen, nitrogen, and hydrogen) contaminating contents of the milled powder are plotted in Fig. 13.62 (for mechanically alloyed $Co_{50}Ti_{50}$) as a function of the milling time. During the first stage of MA, the iron, which comes from the stainless steel milling tool, increases drastically, to about 0.40 at. %. The gas contamination content that may be introduced to the milled powder during the ball-milling process and/or handling of the sample outside the glove box increases during this stage of milling. This is attributed to the formation of active-fresh surfaces of the elemental powders (especially Ti and Zr), which are able to react with the surrounding atmosphere. As the MA time increases, the milling tools are coated with the milled powder, which is wear-resistant to iron. These coated balls play an important role to prevent introduction of further iron contamination to the milled powder. The iron contamination content of the end product is about 0.50 at. %. In addition, no remarkable changes in the gas content can be detected for the samples that milled for longer times. These values of the contaminating contents in the milled powder are almost negligible and cannot be considered to be responsible for the cyclic phase transformations of Co_xTi_{100-x} or $Al_{50}Zr_{50}$ during the ball-milling process.

Another possible factor which can lead to cyclic phase transformations, is excessive temperature during milling. To avoid such a factor, the ball-milling process was interrupted when the temperature of the vial reached about 320 K (almost every 1.8 ks of continuous milling) and then resumed when the temperature decreased to 300 K. In order to realize the effect of the temperature increase on the structure of the powder during milling, the samples that milled for 22 ks (amorphous phase of Co_xTi_{100-x} or AlZr alloys) and 86 ks (bcc-$Co_{50}Ti_{50}$) were heated separately in the DTA to 700 K (well above the measured temperature of the vial). The XRD patterns of these samples prove that no phase transformation takes place, even at this relatively high temperature. However, the amorphous $Co_{50}Ti_{50}$ alloy is transformed to a crystalline bcc-$Co_{50}Ti_{50}$ upon heating to about 968 K. We can then conclude that the temperature, which was recorded during the ball-milling process, is far below the required temperature that can cause structural changes in the milled powder. The same results were obtained from the samples, which were milled separately for the second and third milling runs.

13.10 Consolidation of multicomponent metallic glassy alloy powders into full-dense bulk materials

During the last few years, powder metallurgy, which is considered the most diverse and flexible manufacturing approach for fabricating nearly net-shape high-quality parts with complex shapes, has made remarkable progress in the field of fabrication fully dense metallic glassy alloys.[125] Several interesting and successful methods have been proposed to produce fully dense glassy alloy through the powder metallurgy technique. Of these, the powder atomization method, which involves the formation of liquid metallic droplets from a molten metal mass and subsequent solidification into powder particles, has shown several benefits of introduces large quantity of spherical uniform glassy powders.[126–128]- The obtained glassy powders can be then consolidated into fully dense compacts, using hot-pressing or plasma activated sintering techniques. The high cost of atomization process plus the fact that it is not suitable for systems that have very high melting points are the main drawbacks of this technique. More recently, crushing the rapid-quenched glassy ribbons into coarse particles that are consequently consolidated into different bulk metallic glasses (BMGs) has attracted many research groups due to its simplicity.[129–131] Unfortunately, crushing the sharp-edge glassy ribbons usually leads to contamination of the obtained powders with iron that comes from using the steel crushing tools. In addition, when these metastable ribbons are subjected to severe mechanical deformations during the crushing procedure, their supercooled liquid region disappears and they may transform into different undesired stable and/or metastable phase(s).

13.10.1 Fabrication and consolidation of multicomponent $Zr_{52}Al_6Ni_8Cu_{14}W_{20}$ metallic glassy alloy powders

MA processing route, which is now recognized and respected as a competitive technology and an alternative to casting or conventional glass forming for fabrication of new multicomponent $Zr_{52}Al_6Ni_8Cu_{14}W_{20}$ glassy alloy powders.[125] Selection of this concentration of W comes from the fact that lower atomic percentage of W(5–10%) can be easily obtained by the rapid solidification and/or casting techniques. In contrast, a higher percentage of W (above 30 at.%) does not show the formation of a glassy phase, and the end product is a body-centered cubic (bcc) WZrAlNiCu solid solution that cannot be consolidated into compact sample due to the absence of glass transition temperature (T_g). The obtained glassy powders in this work were consequently consolidated into large compact glassy objects with different shapes using the hot pressing technique.

In the experiments, Pure elemental metallic powders (99.9% or better) of Zr (100 μm), Al (10 μm), Ni (25 μm), Cu (10 μm) and W (50 μm) were balanced to give the desired nominal composition (at.%) of $Zr_{52}Al_6Ni_8Cu_{14}W_{20}$ and mixed in a glove box under an argon (99.999%) atmosphere. The mixed powders of the starting reactant materials were then sealed into a tempered chromium steel vial (1000 ml in volume) together with fifty tempered chromium steel balls (19 mm in diameter) in the glove box. A ball-to-powder weight ratio of 60:1 was chosen. The MA process was performed at room temperature, using a tumbling mill at a rotation speed of 1.1 s^{-1}. The powders that were obtained after 720 ks (final-product) of MA were slightly contaminated with iron (0.28 at.%) and oxygen (0.24 at.%). The glassy powders were subjected to consolidation procedure using uniaxial hot-press equipment under an argon atmosphere. The powders were pressed under a pressure of 936 MPa in tool steel dies of several inner diameters, using tool steel plungers. To minimize the friction between the plungers, punches, and die wall, graphite spray was used to act as a die wall lubricant. No powder binding materials were added to the ball-milled powders. To obtain a correlation between the compaction temperature and some physical and mechanical properties, the die set was heated at heating rate of 0.33 K/s up to several temperatures and then rapidly cooled by flowing an argon gas.

13.10.1.1 Structural change

Fig. 13.63 displays the XRD patterns of ball-milled powders after different MA times. The sample obtained after 173 ks of MA time reveals sharp Bragg-peaks of a bcc phase correlated to a solid solution bcc-W(ZrAlNiCu) metastable phase. The lattice constant a0 of this metastable phase was calculated to be 0.318 nm, larger than the reported value for bcc-W (0.31648 nm). The HRTEM images of the ball-milled powders after 173 ks of MAtime are shown in Fig. 13.64. The powders consist of nanocrystalline fine grains of about 5 nm in diameter [Fig. 13.64A]. Moreover, the image reveals lattice fringe images of the domain structure of solid solution bcc-W(ZrAlNiCu). The measured fringe spacing is 0.225 nm and matches with the interplanar of (110). Considerable broadening and decreasing in the intensity of the major Bragg peaks of solid solution bcc-W(ZrAlNiCu) can be seen in the sample that milled for 259 ks [Fig. 13.63B]. This indicates a crystalline-to-amorphous phase transformation and a decrease in the volume atomic fraction of the bcc-solid solution phase.

The bright-field image (BFI) and the corresponding selected area diffraction pattern (SADP) of the sample that was milled for 259 ks of the MA time are shown together in Fig. 13.65. The BFI image reveals an obvious features of local ordered regions of solid solution bcc-W(ZrAlNiCu) spherical nanocrystalline grains (~18 nm in diameter), which are embedded into the formed amorphous matrix [Fig. 13.65A]. Furthermore, the SADP shows a halo-diffraction coexisted with several spots [Fig. 13.65B], implying a considerable crystalline–amorphous phase transformation.

After 360 ks MA time, the Bragg-peaks that correlated to bcc-W(ZrAlNiCu) solid solution are markedly decreased with obvious broadening, suggesting an increase in the volume fraction of the amorphous phase against the untransformed solid solution phase [Fig. 13.63C]. The Bragg peaks corresponding to the bcc-crystals are hardly visible in the XRD pattern for the sample, which was obtained after 540 ks of MA time [Fig. 13.63D]. Moreover, the primary diffraction maximum tends to shift slightly

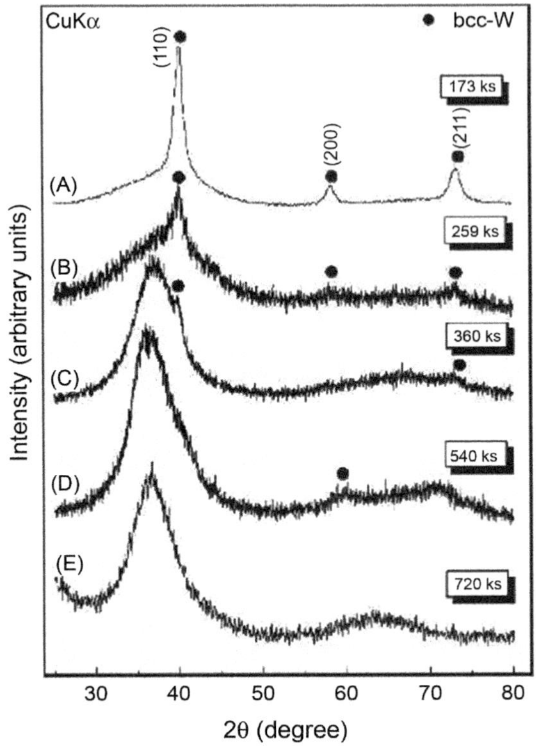

Figure 13.63 XRD patterns of $Zr_{52}Al_6Ni_8Cu_{14}W_{20}$ powders after ball milling for (A) 173 ks, (B) 259 ks, (C) 360 ks, (D) 540 ks, and (E) 720 ks of the MA time.
Source: After El-Eskandarany and Inoue.[125]

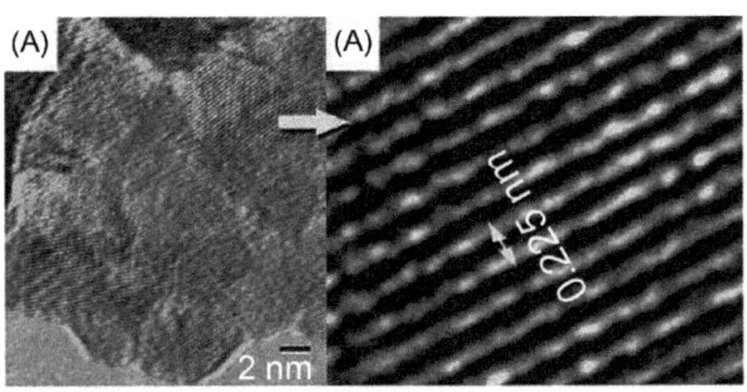

Figure 13.64 HRTEM of ball-milled $Zr_{52}Al_6Ni_8Cu_{14}W_{20}$ powders after 173 ks of the MA time. Part of indexed grain shown in (A) is displayed in (B) at another scale.
Source: After El-Eskandarany and Inoue.[125]

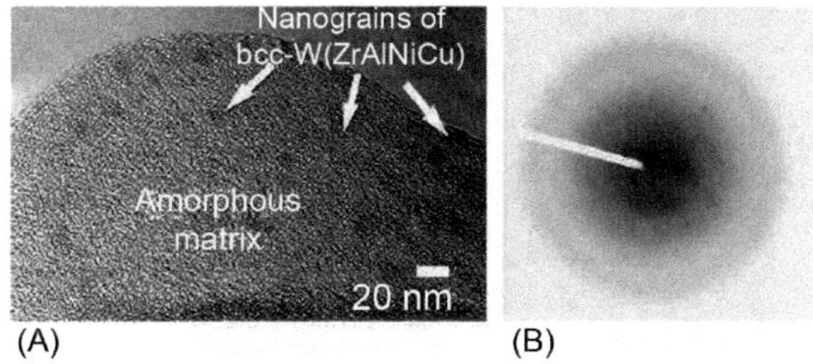

Figure 13.65 TEM micrograph of the (A) BFI and (B) the corresponding SADP of ball-milled $Zr_{52}Al_6Ni_8Cu_{14}W_{20}$ powders after 259 ks of the MA time.
Source: After El-Eskandarany and Inoue.[125]

toward the low angle side, indicating a continuous compositional change of the ball-milled powders. In spite of the samples at the previous stages of MA, the powder of the end product (720 ks) gives broad diffuse smooth halo, suggesting the formation of a single amorphous phase and the completion of a crystalline-amorphous phase transformation [Fig. 13.63E]. The absence of further peak shifting at this stage of milling suggests the formation of a homogeneous $Zr_{52}Al_6Ni_8Cu_{14}W_{20}$ amorphous phase.

The HRTEM image of the synthetic $Zr_{52}Al_6Ni_8Cu_{14}W_{20}$ amorphous powders shows featureless and homogenous internal structure with maze pattern contrast [Fig. 13.66A] with no indication of precipitations of any crystalline phase(s). The EDS analysis of selected regions [indexed by circles in Fig. 13.66A] does not show any serious compositional fluctuations and the average composition is being very close to the starting nominal composition of $Zr_{52}Al_6Ni_8Cu_{14}W_{20}$. In addition, the SADP of this sample displays a typical halo-diffraction of an amorphous phase, as shown in Fig. 13.66B. The end product of amorphous $Zr_{52}Al_6Ni_8Cu_{14}W_{20}$ possesses excellent morphological properties of being spherical ultrafine powders with a diameter ranging from 0.25 to 1 μm [Fig. 13.66C].

13.10.1.2 Thermal stability

The thermal stability of $Zr_{52}Al_6Ni_8Cu_{14}W_{20}$ powders ball-milled for 259 and 720 ks MA time are shown in Fig. 13.67A and B, respectively. After 259 ks [Fig. 13.67A], the scan of the ball-milled powders reveals three opposite reactions. The first reaction, which is endothermic, lies at 785 K and refers to the T_g of the obtained glassy alloy powders. The second and third reactions are overlapped and take place exothermically at 852 and 879 K, respectively, due to the crystallization of the glassy phase. The supercooled liquid region, ΔT_x is calculated from the scan and found to be 67 K [Fig. 13.67A]. Both reactions become more pronounced of the sample that was milled for 720 ks [Fig. 13.67B]. Moreover, the exothermic peak, which refers to the crystallization of the amorphous phase, becomes sharp and single, suggesting the formation of a single homogeneous amorphous phase. The remarked shift in the T_g and T_x to the high temperature side (811 and 884 K, respectively) indicates a compositional change of the obtained glassy phase. Furthermore, ΔT_x shows a larger value of 73 K, suggesting a good glass-forming ability and an excellent thermal stability of

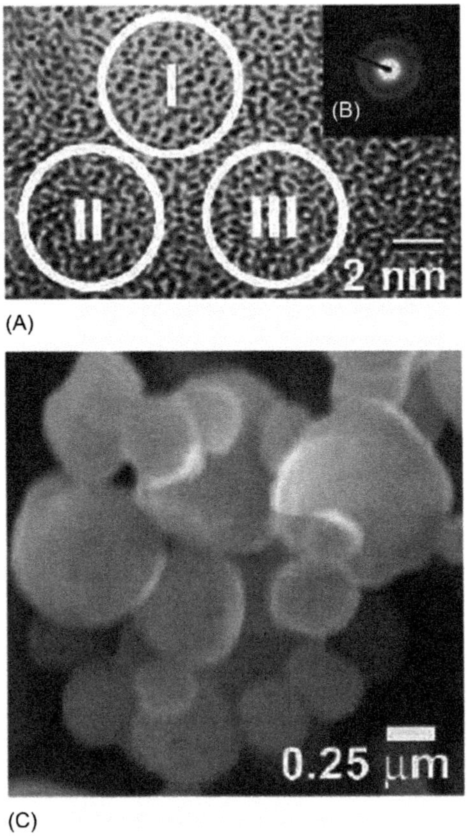

Figure 13.66 (A) HRTEM image, (B) SADP, and (C) SEM micrograph of ball-milled $Zr_{52}Al_6Ni_8Cu_{14}W_{20}$ powders obtained after 720 ks of the MA time.
Source: After El-Eskandarany and Inoue.[125]

glassy $Zr_{52}Al_6Ni_8Cu_{14}W_{20}$ alloy powders that were obtained by MA method. ΔH_x of the obtained glassy phase was measured from the area under the exothermic peak in Fig. 13.67B and found to be –5.48 kJ/mol.

The melting (T_m) and liquidus (T_l) temperatures of the obtained glassy powders were determined using the DTA technique. The DTA curve of glassy $Zr_{52}Al_6Ni_8Cu_{14}W_{20}$ alloy powders obtained after 720 ks of MA time is presented in Fig. 13.68. The displayed single endothermic reaction in this figure indicates that the fabricated glassy alloy possesses very high T_m and T_l, measured to be 1687 and 1768 K, respectively. The reduced glass transition temperature [ratio between T_g and liquidus temperatures, T_l (T_g/T_l)] was then calculated and found to be 0.46.

13.10.1.3 Consolidation

The transverse cross-sectional macroscopic and microscopic views of the sample consolidated at 473 K are shown in Fig. 13.69A and B, respectively. Obviously, densification

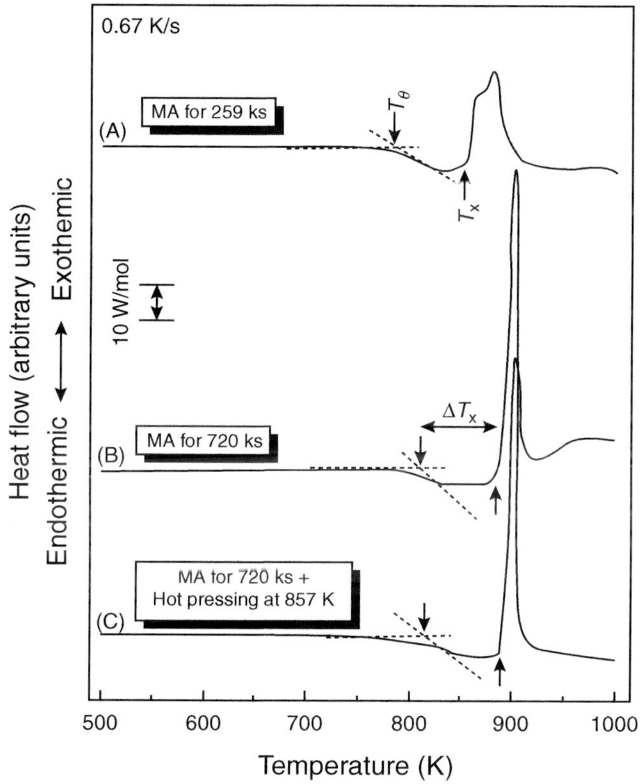

Figure 13.67 DSC curves of ball-milled $Zr_{52}Al_6Ni_8Cu_{14}W_{20}$ powders after ball milling for (A) 259 ks and (B) 720 ks of the MA time (end product). The DSC curve of the sample that was obtained after, hot pressing the powders of the end product at 857 K is displayed in (C). *Source:* After El-Eskandarany and Inoue.[125]

at this temperature does not lead to successful consolidation where a porous green compact is obtained [Fig. 13.69A]. This compacted sample has very poor powder packing [Fig. 13.69B] with a low relative density (74.1%).

The densities and Vickers hardness of the pressed samples were then measured and plotted as a function of the consolidation temperature in Fig. 13.70. It is worth noting that the relative densities of all glassy samples pressed at temperatures below their T_g (solid amorphous) have very low values of less than 76% (Fig. 13.70). When the consolidation takes place at temperature range of 820–857 K (above the T_g and within ΔT_x), the relative density is dramatically increased to approach a saturation value of 99.82% at consolidation temperatures ranging from 832 to 857 K, as shown in Fig. 13.70. The obtained buttons have very high values in the range of 98.60% to 99.85%.

The outer shapes of hot-pressed $Zr_{52}Al_6Ni_8Cu_{14}W_{20}$ glassy alloy buttons with several shapes (cylindrical, button, and cubic) that were consolidated at a temperature of 857 K are shown in Fig. 13.71A–C. The core portions of these buttons are

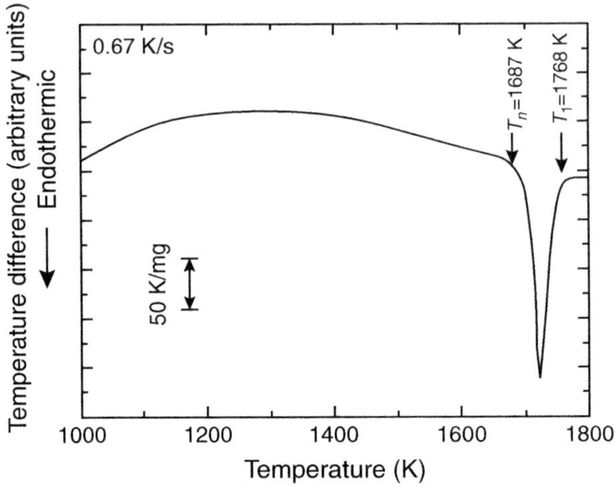

Figure 13.68 DTA curve of ball milled $Zr_{52}Al_6Ni_8Cu_{14}W_{20}$ powders after 720 ks of the MA time.
Source: After El-Eskandarany and Inoue.[125]

presented in Fig. 13.71D–F. In contrast to the sample obtained at 473 K, consolidation the glassy powders at 857 K is capable to fabricate smooth surface and metallic luster bulk metallic glass (BMG) $Zr_{52}Al_6Ni_8Cu_{14}W_{20}$ alloy with different large sizes (ranging from 10 to 30 mm) and various aspect ratios (height /diameter, h/d: 1/2 to 3/2), as shown in Fig. 13.71A–C. The samples that were taken from the core of the consolidated buttons show high dense appearance without any indications of the existence of any cracks or pores [Fig. 13.71D–F]. The average density of the consolidated samples was measured for the all samples and found to be 9.352 Mg/m^3. This value of the synthetic glassy alloy, which is about 99.81% of the calculated theoretical density of $Zr_{52}Al_6Ni_8Cu_{14}W_{20}$ alloy (9.369 Mg/m3), implying the formation of fully dense BMG alloy.

The XRD pattern of as-hot pressed glassy alloy, which reveals broad diffuse haloes of an amorphous phase [Fig. 13.72A], implies that this consolidation step does not lead to any crystallization of the obtained glassy phase. The HRTEM image of the consolidated sample is presented in Fig. 13.72B together with the SADP [Fig. 13.72C]. The indexed circular regions in Fig. 13.72B refer to those regions that were selected for detailed EDS analyses. No features of local ordered regions could be detected and the image shows a homogenous maze contrast of an amorphous phase [Fig. 13.72B]. The SADP displays a typical diffuse halo of an amorphous phase [Fig. 13.72C]. The as-consolidated sample possesses a fair homogeneity in composition, indicated by the close local elemental compositional analysis.

The DSC curve of the same sample reveals the two opposite reactions [Fig. 13.67C] that were previously observed for the glassy powders before consolidation [Fig. 13.67B] with near values of T_g, ΔT_x, and T_x. In addition, the ΔH_x of the as-consolidated glassy sample was measured and found to be -5.34 kJ/mol, being in a fair agreement with the ball-milled powders.

Figure 13.69 (A) Transverse view of consolidated $Zr_{52}Al_6Ni_8Cu_{14}W_{20}$ glassy powders after hot-pressing at 473 K. The corresponding optical microscope of cross-sectional view for this sample is presented in (B).
Source: After El-Eskandarany and Inoue.[125]

We shall return back to Fig. 13.70 to see the effect of the consolidation temperature on the Vickers hardness of hot-pressed $Zr_{52}Al_6Ni_8Cu_{14}W_{20}$ glassy alloy powders. The samples pressed in the temperature region from 373 to 773 K (solid amorphous state) have low hardness values of 4.19–4.31 GPa. The hardness is, however, dramatically increased to be 8.18 GPa upon increasing the consolidation temperature to 820 K (above the T_g). Increasing the densification temperature (857 K) leads to a remarkable increasing in the hardness value, which is saturated at the level of 8.59 to 8.62 GPa after hot pressing at 832 to 857 K, respectively (Fig. 13.70).

Metallographic examinations were achieved on a consolidated cylinder-like glassy sample [Fig. 13.73A] on its longitudinal [Fig. 13.73B] and transverse [Fig. 13.73C]

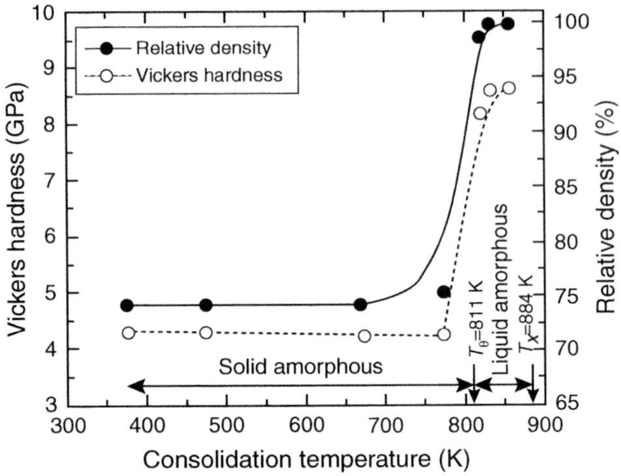

Figure 13.70 Dependence of the Vickers hardness and the relative density of glassy $Zr_{52}Al_6Ni_8Cu_{14}W_{20}$ alloy on the consolidation temperatures.
Source: After El-Eskandarany and Inoue.[125]

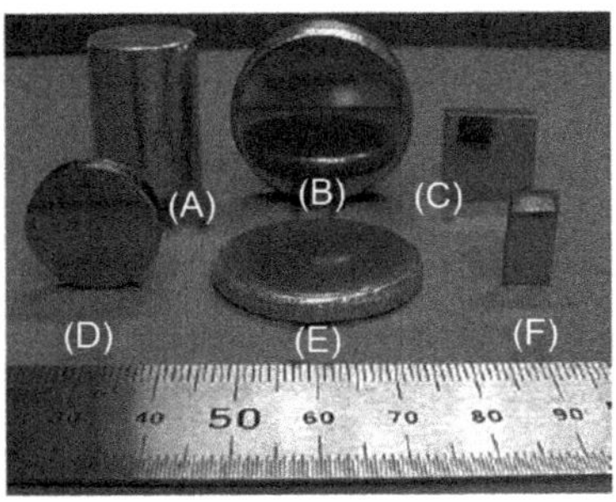

Figure 13.71 (A–C) Outer shape of glassy $Zr_{52}Al_6Ni_8Cu_{14}W_{20}$ compacts obtained with different shapes and size by consolidation the glassy powders at 857 K. The corresponding core portions of the presented samples are shown in (D–F).
Source: After El-Eskandarany and Inoue.[125]

sections at different distances. The indexed arrows in Fig. 13.73 refer to the direction of the consolidated pressure during hot-pressing of the glassy powders. The macroscopic examinations of both sections indicate the absence of any cracks and the metallic luster polished samples are highly compacted [Fig. 13.73B and C]. Figure 13.73D, which displays a SEM micrograph of selected zone for the longitudinal-sectional core of the consolidated sample [Fig. 13.73B], shows a mirror-like morphology with a minimum existence of micro pores (indicated in the figure) and the absence of crystalline

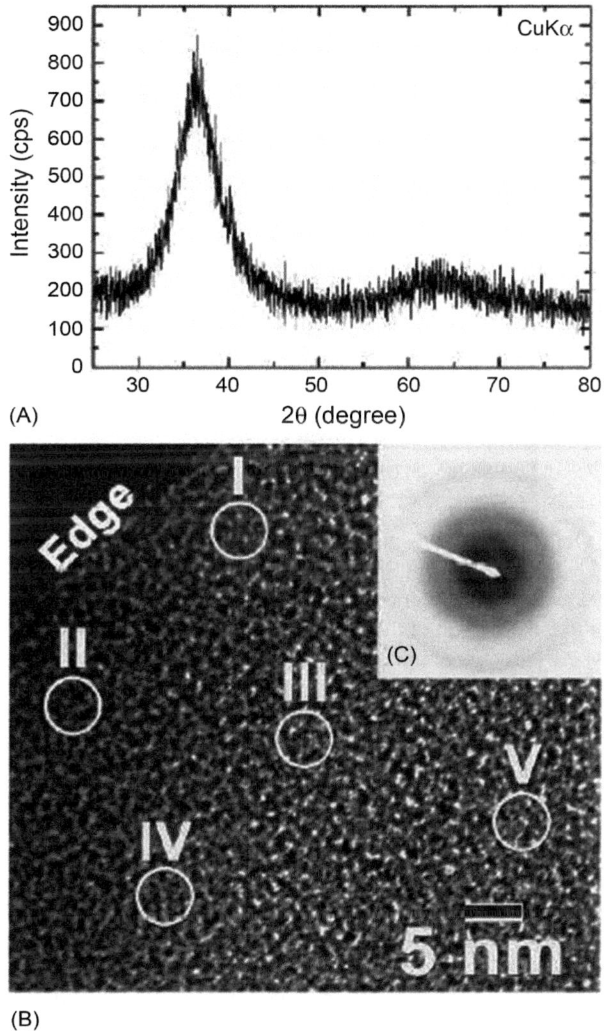

Figure 13.72 (A) XRD pattern of consolidated glassy $Zr_{52}Al_6Ni_8Cu_{14}W_{20}$ alloy after hot pressing at 857 K. The HRTEM image and the corresponding SADP of this sample are shown in (B) and (c), respectively.
Source: After El-Eskandarany and Inoue.[125]

phase. The same SEM observations could be obtained for the transverse-sectional core that also shows a high powder packing of a homogeneous single phase [Fig. 13.73E].

Some useful information of the mechanical properties for hot-pressed $Zr_{52}Al_6Ni_8Cu_{14}W_{20}$ BMG has also been determined, using the room-temperature compression test. This BMG exhibits an extraordinary high value of fracture strength (2.13 GPa) and shows a high Young's modulus of 138 GPa. Neither yielding stress nor plastic strain could be detected; whereas its elastic strain is 1.47%.

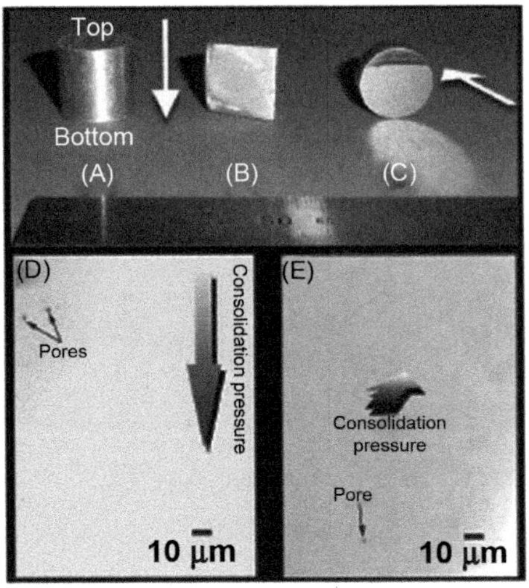

Figure 13.73 (A) Outer macroscopic view of a cylindrical-like glassy $Zr_{52}Al_6Ni_8Cu_{14}W_{20}$ button obtained by hot pressing the glassy powders at 857 K. The longitudinal and transverse cross-sectional macroscopic views for this samples are show in (B) and (C), respectively. The *arrows* denote to the direction of the applied consolidation pressure on the compacted powders. The SEM cross-sectional micrographs taken at the longitudinal and transverse cores of the glassy sample are shown in (D) and (E), respectively.
Source: After El-Eskandarany and Inoue.[125]

13.10.2 Consolidation of mechanically alloyed $Ti_{40.6}Cu_{15.4}Ni_{8.5}Al_{5.5}W_{30}$ metallic glassy alloy powders by SPS

In 2017, El-Eskandarany et al. employed mechanical alloying approach, using a low-energy ball milling technique for synthesizing $Ti_{40.6}Cu_{15.4}Ni_{8.5}Al_{5.5}W_{30}$ metallic glassy alloy powders.[131] The XRD patterns of the powders obtained after different states of MA time are shown in Fig. 13.47A–D. Obviously show, the sample obtained after 50 h of MA time (Fig. 13.74A) reveals crystal structure corresponding to bcc-W metal, as indexed by the (110), (200), and (211) Bragg-peaks. The lattice parameter (a_o) calculated from the major peak (110) of this sample was 0.3169 nm. This value, which is larger than that one reported for pure W metal (0.3165 nm [PDF # 04-806]) indicates a solid-state diffusion of the alloying elements (Ti, Cu, Ni, and Al) into the W lattice to form a solid solution metastable phase. The shifting of W-Bragg peaks to the low angle side seen in the sample milled for 100 h (Fig. 13.74B) suggests a continuous solubility of the alloying elements into the W lattice and further increasing in a_o calculated for W (0.3178 nm). It is worth notifying that the W(TiCuNiAl) solid-solution Bragg peaks were significantly broadened due to the grain refining achieved by ball milling.

The HRTEM micrograph of the powders obtained after 100 h of MA time revealed bcc-W(TiCuNiAl) solid-solution structure possessed Moiré fringes corresponding to

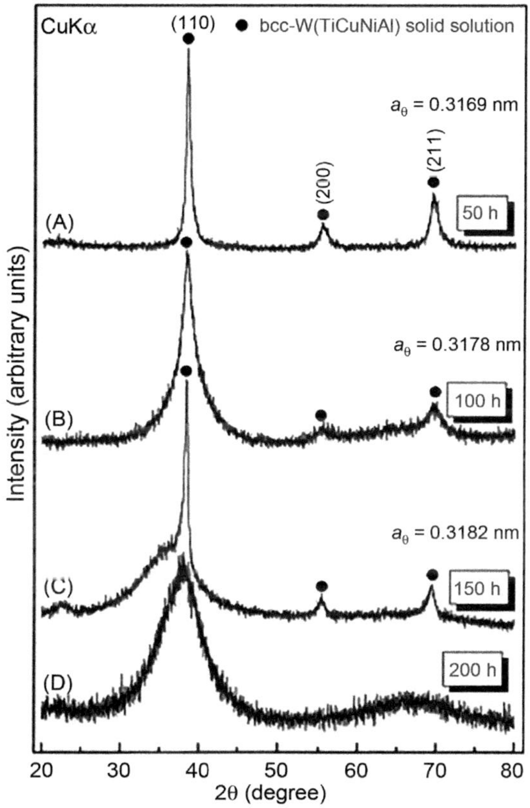

Figure 13.74 XRD patterns ball-milled $(Ti_{58}Cu_{22}Ni_{12}Al_8)_{70}W_{30}$ powders obtained after (A) 50 h, (B) 100 h, (C) 150 h, and (D) 200 h (end product) of MA time.
Source: After El-Eskandarany et al. [131]

W(110), (211) and (200), as shown in Fig. 13.75A. After this stage of milling, the mechanically alloyed powders composed of fine grains with sizes ranging between 8 and 10 nm (Fig. 13.75A). The Fast Fourier Transformation (FFT) spot-diffraction patterns shown in Fig. 13.75B and C are for bcc-W(TiCuNiAl) solid-solution crystals oriented to zone axis [100] and [211], respectively.

Further milling (150 h) led to a slight increasing in a_o (0.3182 nm), where the Bragg peak (110) related to W-solid solution phase was overlapped with a halo-diffuse peak (Fig. 13.74C), indicating existence of a significant volume fractions of an amorphous phase. Formation of such amorphous phase can be attributed to solid solution-to-amorphous phase transformation that was conducted during the milling procedure. Since ball milling technique is an energetic process in which the metastable materials tends to gain more energy upon introducing lattice imperfections (e.g., point defects and dislocations), the authors assumed that W-solid solution phase transformed from a less stable phase to a more stable phase (amorphous) upon further ball milling time.

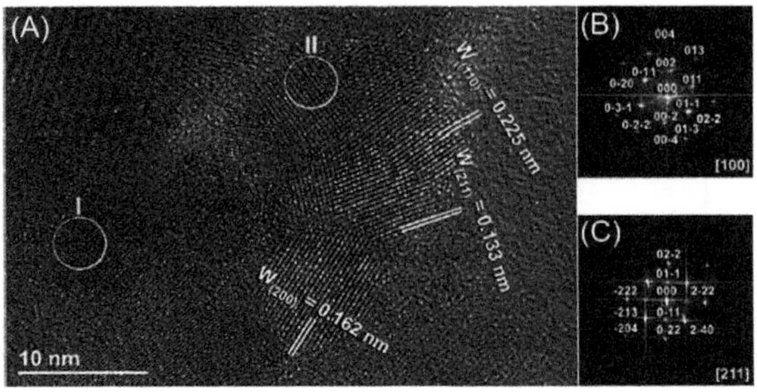

Figure 13.75 (A) HRTEM micrograph and (B and C) the corresponding FFT images of the powders obtained after 100 h the MA time.
Source: After El-Eskandarany et al. [131]

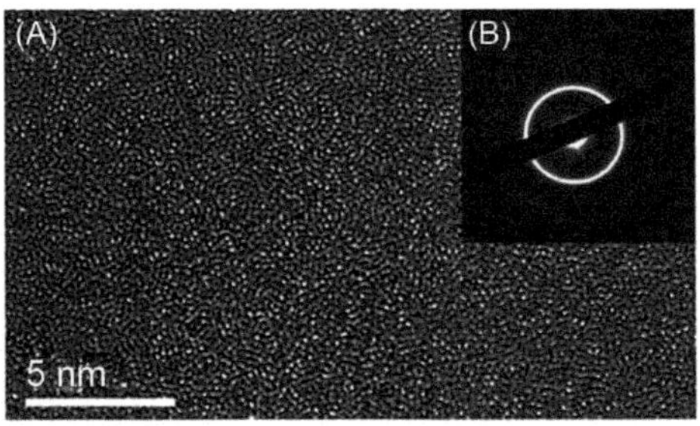

Figure 13.76 The HRTEM micrograph and the corresponding NBDP of the end-product powders (200 h) are shown in (A) and (B), respectively.
Source: After El-Eskandarany et al.[131]

The XRD pattern of the sample obtained after 200 h of MA time revealed a broad halo peak with the absence of the nanocrystalline phase related to W-solid solution structure, as elucidated in Fig. 13.74d. This indicates the completion of the mechanically induced solid solution-amorphous-phase transformation. The HRTEM image of the final-product obtained after 200 h of MA time is displayed in Fig. 13.76A together with the nano beam diffraction pattern, NBDP (Fig. 13.76B). The powders had uniform maze-like morphology with dense random packing structure of an amorphous alloy (Fig. 13.76A).

Based on the target of the present work, the mechanically alloyed $Ti_{58}Cu_{22}Ni_{12}Al_8$ and $(Ti_{58}Cu_{22}Ni_{12}Al_8)_{70}W_{30}$ amorphous alloy powders were consolidated into full dense (above 99.96%) bulk samples, using SPS approach. The densification process was achieved at vacuum under an axial load of 10 MPa. The applied consolidation temperatures were 750 K and 775 for $Ti_{58}Cu_{22}Ni_{12}Al_8$ and $(Ti_{58}Cu_{22}Ni_{12}Al_8)_{70}W_{30}$ samples, respectively. In the SPS process, both die and powder are directly heated by the Joule effect of the direct current (DC). The ON–OFF DC pulse energizing method generates spark plasma, spark impact pressure, joule heating, and an electrical field diffusion effect. Thus, the surfaces of the powders were activated and purified when

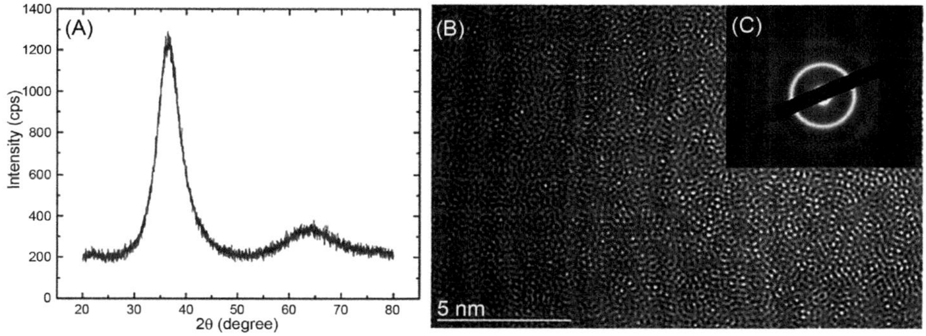

Figure 13.77 (A) XRD pattern, and (B) HRTEM micrograph of as-consolidated $(Ti_{58}Cu_{22}Ni_{12}Al_8)_{70}W_{30}$ bulk metallic glassy system.
Source: After El-Eskandarany et al.[131]

compared with the other traditional sintering methods. Accordingly and based on the formation of fresh metallic glassy powder an easy sintering process was achieved within short processing time sintering temperature being far below the crystallization temperature.

Fig. 13.77A presents the XRD pattern of as-consolidated $(Ti_{58}Cu_{22}Ni1_2Al_8)_{70}W_{30}$ powders. This bulk sample revealed broad halo peaks at the first and second maxima with the absence of crystalline peaks, as displayed in Fig. 13.77A. Useful information related to the structure of the consolidated sample was obtained upon examining the sample structure beyond the atomic level, using HRTEM analysis. The HRTEM micrograph of a polished $(Ti_{58}Cu_{22}Ni_{12}Al_8)_{70}W_{30}$ bulk sample shows a homogeneous random packing structure (Fig. 13.77B) with a clear spot-free halo-diffraction pattern (Fig. 13.77C). The absence of crystalline phase(s) related to order/intermediate structure indicating the formation of a single amorphous phase. The XRD and HRTEM analyses indicate that the consolidation step via SPS approach did not lead to undesirable crystallization and the sample maintained its original structure of short-range order.

Fig. 13.78A and B show the DSC curves of as-consolidated $(Ti_{58}Cu_{22}Ni_{12}Al_8)_{100-x}W_x$ ($0 < x < 30$ at.%) bulk samples. Both samples revealed clear endothermic peaks related to the glass transition temperatures (T_g) followed by sharp pronounced exothermic peaks corresponding to the crystallization (T_x) of the amorphous phase. The regions extended between the endothermic- and exothermic- reactions refer to the supercooled liquid region ($\Delta T_x = T_x - T_g$). The T_g measured for the original $Ti_{58}Cu_{22}Ni_{12}Al_8$ metallic glassy alloy was 727 K, as shown in Fig. 13.78A. This value is higher the one reported for $Ti_{60}Cu_{15}Ni_{15}Al_{10}$ metallic glassy system (~710 K). The T_g tended to increase slightly to 737 K upon alloying with 30 at.% W, as shown in Fig. 13.78B, implying a change in the melting points between the two glassy systems. Bulk metallic glassy $Ti_{58}Cu_{22}Ni_{12}Al_8$ alloy obtained by SPS process crystallized through a single sharp exothermic peak at 776 K (T_x) with a large enthalpy change of crystallization (ΔH_x) of −7.9 kJ/mol. The T_x of this glassy system is far above that one reported for

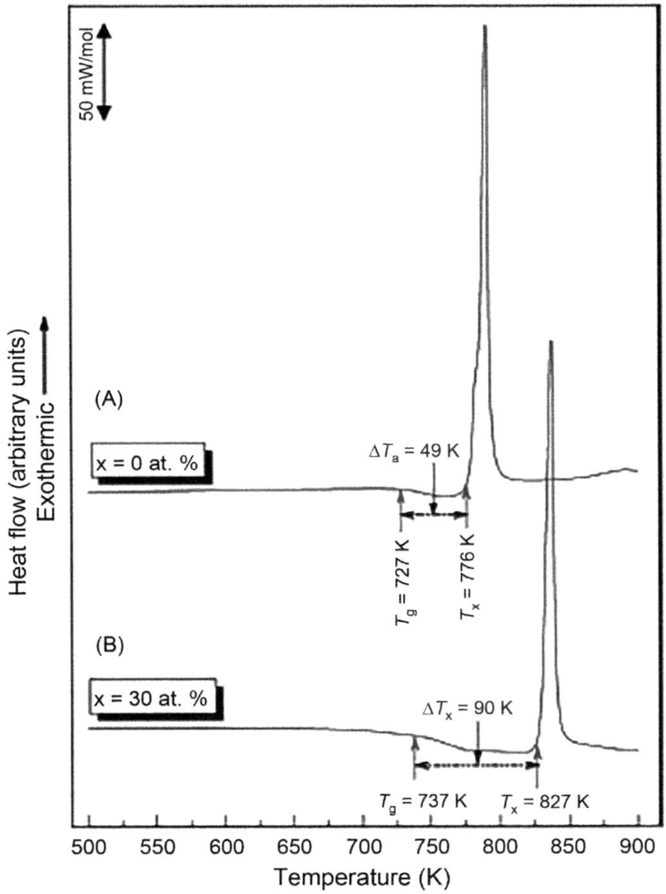

Figure 13.78 DSC traces of as-consolidated (A) $Ti_{58}Cu_{22}Ni_{12}Al_8$, and (B) $(Ti_{58}Cu_{22}Ni_{12}Al_8)90\text{-}W_{10}$ systems.
Source: After El-Eskandarany et al.[131]

$Ti_{60}Cu_{15}Ni_{15}Al_{10}$ system (745 K). Moreover, metallic glassy $Ti_{58}Cu_{22}Ni_{12}Al_8$ system revealed a higher value of ΔT_x (49 K), being larger that the reported value (35 K) for $Ti_{60}Cu_{15}Ni_{15}Al_{10}$ system. The high T_x and relatively large ΔT_x values indicate that metallic glassy $Ti_{58}Cu_{22}Ni_{12}Al_8$ alloy possesses high thermal stability.

The relation between the coefficient of friction (l) behaviors of as-consolidated $(Ti_{58}Cu_{22}Ni_{12}Al_8)_{100-x}W_x$ (x; 0 and 30 at.%) bulk metallic glasses and the W concentrations (x) is presented in Fig. 13.79A. Figure 13.79B and C display the setup of the pin-on-dick machine employed in the present work for studying the wear behavior of the metallic glassy samples, using WC-Ni composite ball (pin) under the application of 200 N, and a rotation speed of 150 rpm. Both metallic glassy samples had almost same behaviors of getting steady-state stages after about 10 m of sliding distance, as shown in Fig. 13.79A. During the second stage (10–48 m) the μ value for 0 at.% W

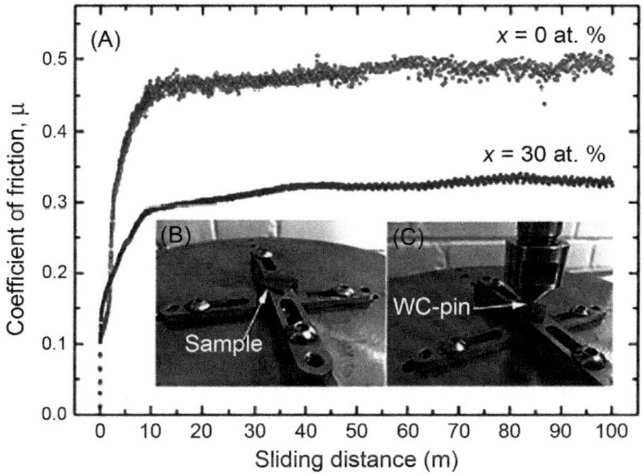

Figure 13.79 (A) Dependence of friction coefficient (*l*) on W concentrations (*x*; 0 and 30 at.%) and sliding distance. The pin-on-disk experimental set-up is shown in (B) and (C). *Source:* After El-Eskandarany et al.[131]

and 30 at.% W samples were slightly increased to be about 0.5 and 0.32, respectively (Fig. 13.79A). No remarkable changes could be seen in μ during the last stage of the test (48–100 m). Since the pin-on-disk test was conducted for both samples (*x*; 0 and 30 at.%) under the sample experimental conditions for the same sliding distance (100 m), the decreasing noted on μ value implies a higher wear resistance of the sample because it required more energy for removing same volume of the materials. The μ value of this present system is less that that one reported for regular Ti-based metallic glassy alloys systems. It can be then concluded that plus the attractive high hardness value demonstrated for metallic glassy $Ti_{40.6}Cu_{15.4}Ni_{8.5}Al_{5.5}W_{30}$ bulk material, this system possesses also superior wear resistance behavior, making it a desirable materials for surface protective coating employed for the cutting tools applications in oil sector.

13.11 Recent studies

Ramasamy et al.[132] studied the effect of milling on the crystallization of amorphous $Fe_{80}Nb_{10}B_{10}$ ribbons prepared by melt spinning technique. Their results have shown that ball milling not only induced the formation of nano-crystalline α-Fe during milling, but also drastically changed the crystallization behavior of the residual amorphous phase. They claimed that nanocrystalline/amorphous composite powders that were obtained had excellent soft magnetic properties. In addition, the as-ball milled nano-crystalline/amorphous powders had a saturation magnetization 2.5–3 times higher than the as-quenched amorphous ribbon along with Curie temperature increase by about 70 K with respect to the parent as-spun material.[132] An interesting study reported the formation of metallic nanoglasses has been just published.[133] The authors have attributed the formation of metallic nanoglasses to inhomogeneous plastic deformation.[133]

Cardinal et al. a new approach aimed to improve the plasticity in a CuZrAl bulk metallic glass.[134] Their approach was based on introducing soft metallic crystalline particles to the glassy matrix of CuZrAl. Tantalum particles with two different geometries were chosen in the present work: (1) irregularly shaped particles of equivalent average diameter of 45 μm and (2) spherical particles of average size 25 μm. When the volume fraction of tantalum goes from 0% to 50%, the modulus increases (from 95-GPa to 140-GPa), the Poisson's ratio decreases only slightly (from 0.37 to 0.35) and the hardness decreases regularly (from 485 HV30 to about 300 HV30), whatever the particles size. They reported that for the highest values of this content, a significant damage of the sample follows this plastic deformation before rupture. But a large dispersion of the values is observed, more particularly for small particles of tantalum. This is due to a problem of heterogeneity of the particles dispersion and the presence of clusters; this difficulty in dispersing the particles is more marked for small spherical particles. An image analysis was performed to characterize this clustering effect. Clusters have a negative effect because they facilitate the propagation of cracks through Ta-Ta weak interface instead of deflecting it. An optimal content to combine a high elastic limit, a good hardness and a high plasticity seems to exist near 30% vol. of crystalline particles.

Qin et al.[135] have recently proposed metallic glasses of $Fe_{78}Si_8B_{14}$ and $Fe_{73.5}Nb_3Cu1Si_{13.5}B_9$ prepared by mechanical alloying to treat coke wastewater. They reported that $Fe_{78}Si_8B_{14}$ metallic glassy powders exhibit superior catalytic capability in wastewater treatment of high chemical oxygen demand removal efficiency with low dosage of H_2O_2.

References

1. Buckel, W., and Hilsch, R., *Z. Phys.*, 138:109 (1954)
2. Duwez, P., Willens, R. H., and Klement, W., Jr., *J. Appl. Phys.*, 31:1136 (1960)
3. Vegard, J., *Phil. Mag.*, 32:20 (1916)
4. Herd, S. R., Tu, K. N., and Ahn, K. Y., *Appl. Phys. Lett.*, 42:597 (1983)
5. Schwarz, R. B., and Johnson, W. L., *Phys. Rev. Lett.*, 51:415 (1983)
6. Blatter, A., and von Allmen, M., *Phys. Rev. Lett.*, 54:2103 (1985)
7. Yeh, X. L., Samwer, K., and Johnson, W. L., *Appl. Phys. Lett.*, 42:242 (1983)
8. Aoki, K., Nagano, M., Yanagitani, A., and Masumoto, T., *J. Appl. Phys.*, 62:3314 (1987)
9. Aoki, K., Yanagitani, A., and Masumoto, T., *Appl. Phys. Lett.*, 52:2212 (1988)
10. Cheng, Y. T., Johnson, W. L., and Nicolet, M. A., *Appl. Phys. Lett.*, 47:800 (1985)
11. Schroder, H., Samwer, K., and Koster, U., *Phys. Rev. Lett.*, 54:197 (1985)
12. Suryanarayana, C., Inoue, A., *Bulk Metallic Glasses*, CRC Press, 1st Edition, Taylor$ Francis Group, New York (2011)
13. White, R. L., *The Use of Mechanical Alloying in the Manufacture of Multifilamentary Superconductor Wire*, Ph.D. Thesis, Stanford University (1979)
14. Koch, C. C., Cavin, O. B., McKamey, C. G., and Scarbrough, J. O., *Appl. Phys. Lett.*, 43:1017 (1983)

15. El-Eskandarany, M. S., Aoki, K., and Suzuki, K., *J. Less-Common Met.*, 167:113 (1990)
16. Schwarz, R. B., Wong, K. L., Johnson, W. L., and Clements, B. M., *J. Non-Cryst. Solids*, 61–62:129 (1984)
17. Schultz, L., *J. Less-Common Met.*, 145:233 (1988)
18. Johnson, W. L., *Prog. Mater. Sci.*, 30:81 (1986)
19. El-Eskandarany, M. S., Zhang, W., and Inoue, A., *Journal of Alloys and Compounds*, 350:232 (2003)
20. R.B. Schwarz, R.R. Petrich, C.K. Saw, *J. Non-Cryst. Solids* 76 (1985) 281
21. Schwarz, R. B., Petrich, R. R., and Saw, C. K., *J. Non-Cryst. Solids*, 76:281 (1985)
22. Schwarz, R. B., and Petrich, R. R., *J. Less-Common Met.*, 140:171 (1988)
23. Weeber, A. W., and Bakker, H., *Physica B*, 153:93 (1988)
24. Weeber, A. W., Haag, W. J., Wester, A. J. H., and Bakker, H., *J. Less-Common Met.*, 140:119 (1988)
25. Weeber, A. W., Bakker, H., Heijligers, H. J. M., and Bastin, G. F., *Europhys. Lett.*, 3:1261 (1987)
26. Eckert, J., Schultz, L., and Urban, K., *J. Mater. Res.*, 6:1874 (1991)
27. Petzoldt, F., Scholz, B., and Kunze, H. D., *Mater. Sci. Eng.*, 97:25 (1988)
28. Hellstern, E., and Schultz, L., *J. Appl. Phys.*, 63:1408 (1988)
29. El-Eskandarany, M. S., Bahgat, A. A., Gomaa, N. S., and Eissa, N. A., *J. Alloys Comp.*, 290:181 (1999)
30. El-Eskandarany, M. S., Aoki, K., and Suzuki, K., *Mater. Sci. Forum*, 88–90:81 (1992)
31. El-Eskandarany, M. S., *J. Alloys Comp.*, 234:67 (1996)
32. El-Eskandarany, M. S., Aoki, K., and Suzuki, K., *Metall. Trans. A*, 23:2131 (1992)
33. El-Eskandarany, M. S., Aoki, K., Masumoto, T., and Suzuki, K., *J. Alloys Comp.*, 209:71 (1994)
34. El-Eskandarany, M. S., *J. Alloys Comp.*, 284:295 (1999)
35. El-Eskandarany, M. S., Aoki, K., and Suzuki, K., *J. Appl. Phys.*, 71:2924 (1992)
36. El-Eskandarany, M. S., Aoki, K., and Suzuki, K., *J. Appl. Phys.*, 72:2665 (1992)
37. El-Eskandarany, M. S., Aoki, A., and Suzuki, K., *J. Jpn Society of Powder and Powder Metallurgy*, 38:59 (1991)
38. El-Eskandarany, M. S., Aoki, K., Itoh, H., and Suzuki, K., *J. Less-Common Met.*, 169:235 (1991)
39. El-Eskandarany, M. S., Suzuki, H., Aoki, K., and Suzuki, K., *J. Jpn Society of Powder and Powder Metallurgy*, 38:96 (1991)
40. El-Eskandarany, M. S., Aoki, K., and Suzuki, K., *J. Non-Cryst. Solids*, 150:472 (1992)
41. El-Eskandarany, M. S., Sumiyama, K., Aoki, K., and Suzuki, K., *J. Jpn Society of Powder and Powder Metallurgy*, 39:836 (1992)
42. El-Eskandarany, M. S., in: *Proceeding of The 6th International Mining, Petroleum and Metallurgical Engineering Conference* (MPM), p. 119, Organized by Faculty of Engineering, Cairo University (Feb. 20–24, 1999)
43. El-Eskandarany, M. S., Aoki, K., and Suzuki, K., *J. Alloys Comp.*, 186:15 (1992)
44. Henaff, M. P., Colinet, C., Pasturel, A., and Buschow, K. H. J., *J. Appl. Phys.*, 56:307 (1984)
45. Miedema, A. R., *J. Less-Common Met.*, 32:117 (1973)
46. Massalski, T., *Binary Alloy Phase Diagrams*, 1st Printing, 1:157, 1st Ed., ASM (1986)
47. Davies, H. A., *Phys. Chem. Glasses*, 17:159 (1976)
48. Johnson, W. L., Poon, S. J., Durand, J., and Duwez, P., *Phys. Rev. B*, 18:201 (1978)
49. Buschow, K. H. J., *J. Phys. F*, 14:593 (1984)
50. El-Eskandarany, M. S., Aoki, K., and Suzuki, K., *J. Alloys Comp.*, 177:229 (1991)

51. Suzuki, K., *J. Non-Cryst. Solids*, 112:23 (1989)
52. El-Eskandarany, M. S., *Metall. Trans. A*, 27:3267 (1996)
53. El-Eskandarany, M. S., Aoki, K., and Suzuki, K., *Scripta Metall.*, 25:1695 (1991)
54. van Rossum, M., Nicolet, M. A., and Johnson, W. L., *Phys. Rev. B*, 29:549 (1984)
55. Cotts, E. J., Wong, G. C., and Johnson, W. L., *Phys. Rev. B*, 37:9049 (1988)
56. Politis, C., and Johnson, W. L., *Appl. Phys. Lett.*, 60:1147 (1986)
57. Hellstern, E., and Schultz, L., *Appl. Phys. Lett.*, 49:1163 (1986)
58. Schultz, L., Hellstern, E., and Thomä, A., *Europhys. Lett.*, 3:921 (1987)
59. Eckert, J., Schultz, L., and Urban, K., *J. Appl. Phys.*, 64:3224 (1988)
60. Eckert, J., Schultz, L., and Urban, K., *J. Less-Common Met.*, 145:283 (1988)
61. Thomä, A., Saemann-Ischenko, G., Schultz, L., and Hellstern, E., *Jpn. J. Appl. Phys.*, 26:977 (1987)
62. Hellstern, E., Schultz, L., and Eckert, J., *J. Less-Common Met.*, 140:93 (1988)
63. Lee, P. Y., and Koch, C. C., *J. Non-Cryst. Solids*, 94:88 (1988)
64. Hellstern, E., Schultz, L., Bormann, R., and Lee, D., *Appl. Phys. Lett.*, 53:1399 (1986)
65. El-Eskandarany, M. S., Aoki, K., and Suzuki, K., *Sci. Rep. Ritu*, Tohoku University, A39:103 (1994)
66. Babic, E., Ristic, R., Miljak, M., Scott, M. G., and Cregan, G., *Solid State Commun.*, 39:139 (1981)
67. Altounian, Z., and Strom-Olsen, J. O., *Phys. Rev. B*, 27:4149 (1983)
68. Karkut, M. G., and Hake, R. R., *Phys. Rev. B*, 28:1396 (1983)
69. El-Eskandarany, M. S., Itoh, F., Aoki, K., and Suzuki, K., *J. Non-Cryst. Solids*, 118:729 (1990)
70. Miedema, A. R., de Châtel, P. F., and de Boer, F. R., *Physica B*, 100:1 (1980)
71. Richards, T. G., and Johari, G. P., *Philos. Mag. B*, 58:445 (1988)
72. Dong, Y. D., Wang, W. H., Liu, L., Xiao, K. Q., Tong, S. H., and He, Y. Z., *Mater. Sci. Eng. A*, 134:867 (1991)
73. Lerf, R., and Morris, D. G., *Mater. Sci. Eng. A*, 128:119 (1990)
74. Fecht, H. J., and Johnson, W. L., *J. Non-Cryst. Solids*, 117–118:704 (1990)
75. El-Eskandarany, M. S., Aoki, K., and Suzuki, K., *Metall. Trans A.*, 23:2131 (1992)
76. Loeff, P. I., and Bakker, H., *Europhys. Lett.*, 8:35 (1989)
77. Radinski, A. P., and Calka, A., *Mater. Sci. Eng. A*, 134:1376 (1991)
78. Kimura, H., Takada, F., and Myung, W. N., *Sci. Eng. A*, 97:125 (1988)
79. Hellstern, E., and Schultz, L., *Mater. Sci. Eng. A*, 93:213 (1987)
80. El-Eskandarany, M. S., Aoki, K., Sumiyama, K., and Suzuki, K., *Scripta Metall.*, 36:1001 (1997)
81. El-Eskandarany, M. S., Aoki, K., Sumiyama, K., and Suzuki, K., *Appl. Phys. Lett.*, 70:1679 (1997)
82. Eckert, J., Schultz, L., and Urban, K., *J. Less-Common Met.*, 166:293 (1990)
83. Kimura, H., Kimura, M., and Takada, F., *J. Less-Common Met.*, 140:113 (1988)
84. Schänzer, M., and Mehrer, H., *J. Phys., Coll. Phys. C4, Supp. 14*, 51:87 (1990)
85. Thompson, J. R., Politis, C., and Kim, T. C., *Mater. Sci. Eng. A*, 97:31 (1988)
86. Fukunaga, T., Nakamura, K., Suzuki, K., and Mizutani, U., *J. Non-Cryst. Solids*, 117–118:700 (1990)
87. Fukunaga, T., Mori, M., Inou, K., and Mizutani, U., *Mater. Sci. Eng. A*, 134:863 (1991)
88. Jang, J. S. C., and Koch, C. C., *Scripta Metall.*, 20:669 (1986)
89. Martin, V. E., Garcia-Escorial, A., Martin, A., Carmona, F., Cebollada, F., Adeva, P., and Gonzalez, J. M., *J. Phys., Coll. Phys. C4, Supp. 14*, 51:198 (1990)
90. Lee, P. Y., Jang, J., and Koch, C. C., *J. Less-Common Met.*, 140:73 (1988)

91. Ennas, G., Magini, M., Boffito, G., and Licheri, G., *J. Mater. Sci.*, 24:3053 (1988)
92. El-Eskandarany, M. S., Sumiyama, K., and Suzuki, K., *Acta Metall.*, 45:1175 (1997)
93. Hellstern, E., and Schultz, L., *Philos. Mag. B*, 56:443 (1987)
94. Lee, P. Y., and Koch, C. C., *J. Non-Cryst. Solids*, 94:88 (1987)
95. Boldrick, M. S., Lee, D., and Wagner, C. N. J., *J. Non-Cryst. Solids*, 106:60 (1988)
96. Fukunaga, T., Homma, Y., Misawa, M., and Mizutani, U., *J. Non-Cryst. Solids*, 117–118:721 (1990)
97. Thompson, J. R., and Politis, C., *Europhys. Lett.*, 3:199 (1985)
98. Loeff, P. I., Spit, F. H., and Bakker, H., *J. Less-Common Met.*, 145:271 (1988)
99. Gaffet, E., and Harmelin, M., *J. Less-Common Met.*, 157:201 (1990)
100. Gaffet, E., and Harmelin, M., *J. Phys., Coll. Phys. C4, Supp. 14*, 51:139 (1990)
101. Weeber, A. W., and Bakker, H., *Z. Phys. Chem. N. F.*, 157:221 (1988)
102. de Boer, F. R., Boom, R., Mattens, W. C. M., Miedema, A. R., and Niessen, A. K., *Cohesion in Metals-Transition Metal Alloys*, North-Holland, Amsterdam, 1st Ed., 1:224, (1988)
103. Wang, R., Merz, M. D., Brimhall, J. L., and Dahlgren, S. D., *Rapidly Quenched Metals III*, Metals Society, London, 1:420 (1978)
104. Maissel, L. I., Mauer, F. A., Piermarini, G. I., and Block, S., *J. Appl. Phys.*, 54:5698 (1983)
105. Sumiyama, K., Hirata, M., and Teshima, W., *Jpn J. of Appl. Phys. A*, 30:2839 (1991)
106. Boldrick, M. S., Yang, E., and Wagner, C. N. J., *J. Non-Cryst. Solids*, 150:478 (1992)
107. Koyano, T., Chatani, K., Fukunaga, T., and Mizutani, U., *Mater. Sci. Eng. A*, 181–182:1277 (1994)
108. Darken, L. S., and Curry, R. W., *Physical Chemistry of Metals*, p. 50, McGraw-Hill, New York (1953)
109. Egami, T., and Waseda, Y., *J. Non-Cryst. Solids*, 64:113 (1984)
110. Güntherod, H. J., Oelhafen, P., Hausera, E., Greuter, F., Lapka, R., Rösel, F., Jacobs, R., d'Albuquerque e Castro, J., Kübler, J., Hague, C. F., Bennemann, K. H., Fairlie, R. H., Temmermann, W. M., and Gyoffy, B. L., *Inst. Phys. Conf. Ser.*, 55:619 (1981)
111. de Boer, F. R., Boom, R., Mattens, W. C. M., Miedema, A. R., and Niessen, A. K., *Cohesion in Metals-Transition Metal Alloys*, 1:224, 1st Ed., North-Holland, Amsterdam (1988)
112. El-Eskandarany, M. S., and Ahmed, H. A., *J. Alloys Comp.*, 216:213 (1994)
113. Abrahamson, E. P., and Lopata, S. L., *Trans.*, AIME, 236:76 (1966)
114. Schultz, L., Hellstern, E., and Zone, G., *Z. Phys. Chem.*, 157:203 (1988)
115. Michaelsen, C., and Schultz, L., *Acta Metall. Mater.*, 39:987 (1991)
116. Ermakov, A. E., Yurchikov, E. E., and Barinov, V. A., *Phys. Met. Metall.*, 52:50 (1981)
117. Jang, J. S. C., and Koch, C. C., *J. Mater. Res.*, 5:498 (1990)
118. Suzuki, K., *J. Non-Cryst. Solids*, 117/118:1 (1990)
119. Suzuki, K., *J. Non-Cryst. Solids*, 112:23 (1989)
120. Trudeau, M. L., Schulz, R., Dussault, D., and Neste, A. V., *Phys. Rev. Lett.*, 64:99 (1990)
121. El-Eskandarany, M. S., Aoki, K., Sumiyama, K., and Suzuki, K., *Appl. Phys. Lett.*, 70:1679 (1997)
122. El-Eskandarany, M. S., Aoki, K., Sumiyama, K., and Suzuki, K., *Scripta Metall.*, 36:1001 (1997)
123. El-Eskandarany, M. S., Aoki, K., Sumiyama, K., and Suzuki, K., *Metall. Trans. A*, 30:1877 (1999)
124. *ASTM* card No. 15-806
125. El-Eskandarany, M. S., and Inoue, A., *J. Mater. Res.*, 21:976 (2006)
126. Shen, B., and Inoue, A., *J. Mater. Res.*, 18:2115 (2003)
127. Kawamura, Y., Shibata, T., Inoue, A., and Masumoto, T., *Scripta Metall.*, 37:431 (1997)
128. Itoi, T., Takamizawa, T., Kawamura, Y., and Inoue, A., *Scripta Metall.*, 45:1131 (2001)

129. Ishihara, S., Zhang, W., and Inoue, A., *Scripta Metall.*, 47:231 (2002)
130. Inoue, A., and Shen, B., *J. Mater. Res.*, 18:2799 (2003)
131. El-Eskandarany M. Sherif, Al-Hajji, L.A., Al-Hazza, Abdulsalam, *Advanced Powder Technology*, 28:814 (2017)
132. Ramasamy, P., Shahid, Nawaz, Scudino, S., Eckert, Jürgen, Stoic, Mihai, *Journal of Alloys and Compounds*, 725: 227 (2017)
133. Shi, Bo, Xu, Yuanli, Jin, Peipeng, *Materialia*, 7: 100390 (2019)
134. Cardinal S., Pelletier, J. M., Xie, G. Q., Mercier, F., Dalmas, F., *Journal of Alloys and Compounds*, 782:59 (2019)
135. Qin, X., Li, Z., Zhu, Z., Fang, D., Zhang, H., *Journal of Physics and Chemistry of Solids*, 133: 85 (2019)

Mechanical alloying for preparing nanocrystalline high-entropy alloys

14

14.1 Introduction

14.1.1 Traditional alloys

Materials, in particular metals and their alloys, have been considered as the backbone of all developments achieved in human civilizations for thousands of years Fig. 14.1. Since the Bronze Age, human has struggled to introduce discover new families of metals and invented different methodologies used to improve and enhance their properties (Fig. 14.2). The idea of alloying base metals (e.g., Cu, Fe, Au, Ag, etc.) with certain molecular fraction(s) of one or two other metallic elements (the so-called alloying elements) was born when human realized the limitations of using a single metal on its pure form in real applications. Thus, he considered that adding such smaller amount of alloying element(s) to the base metal leads to born of new material known as "alloy," having enhanced properties. Binary and ternary metallic of Cu-Zn, Au-Cu, Al-Ti, Fe-Cr-Ni, and Zr-Ni, Mg-Al alloy systems are some typical paradigms of alloys developed throughout the different ages of human civilization.

14.1.2 The birth of high-entropy alloys

The simple paradigm concept of mixing two or more metals to form binary-, ternary-, quarterly-, and multicomponent alloys has been shifted when Cantor et al.[1] designed for the first time in human history multicomponent alloys containing several metallic alloying species with equal atomic proportions to form $Fe_{20}Cr_{20}Mn_{20}Ni_{20}Co_{20}$ alloy system. Very common casting and melt spinning techniques were used to prepare first their alloy, where they examined its properties by means of convenient characterizations approaches of optical microscopy, scanning electron microscopy, electron probe microanalysis, X-ray diffraction, and microhardness measurements.[1] They pointed out that $Fe_{20}Cr_{20}Mn_{20}Ni_{20}Co_{20}$ did not form any metallic glassy phase, where the end product was a single fcc-solid solution phase, which solidifies dendritically.[1] This equiatomic multicomponent alloy is known as "high-entropy alloy (HEA)."[1]

14.1.3 Basic science behind the HEAs

In general, HEAs may be defined as multicomponent alloys composite of five or more alloying elements in equiatomic ratio.[1–3] In contrast to many other multicomponent alloys, HEAs tend to exhibit solid solution structures, instead of complex

Mechanical Alloying. http://dx.doi.org/10.1016/B978-0-12-818180-5.00014-5

Figure 14.1 For more than 5000 years, melting and casting techniques have been used to fabricate wide spectrum of metal alloys, using different approaches. In contrast to this, fabrication of advanced engineering materials via mechanical alloying has been considered as practical solution to obtain metal alloys, which composite of alloying elements with wide gap in their melting points.

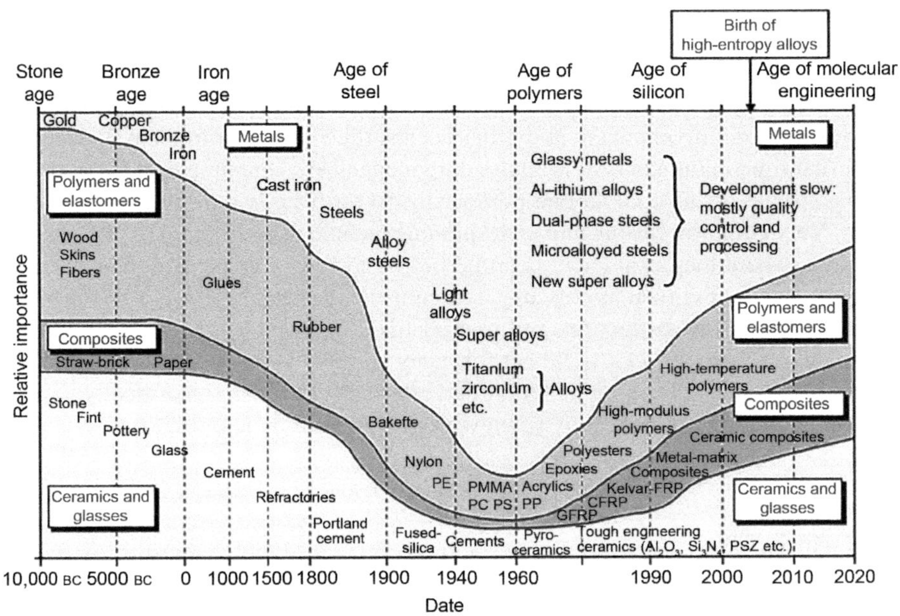

Figure 14.2 Historical evolution of engineering materials—marked with the birth of HEAs. *(After Murty et al.)*[2]

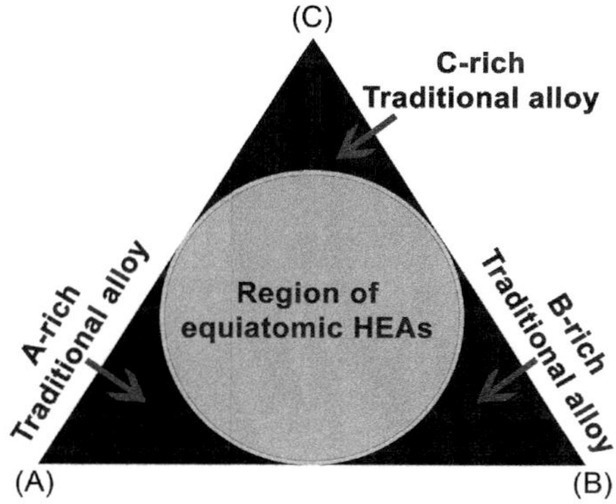

Figure 14.3 Difference between the design of conventional alloys and high-entropy alloys on a ternary plot.
Source: Adopted from Ye et al.[4]

phases, stabilized by their high configurational entropy of mixing.[4,5] According to Yeh et al.,[2] each alloying element should have an atomic percentage between 5% and 35%. They suggested very high configurational entropy of random mixing of the alloying elements in HEAs.[2] Moreover, the atomic percentage of any minor elemental additives should not be more than 5%.[2]

The fundamental science behind the interesting phenomena of HEAs is attributed to the high mixing entropies at high temperatures of solid solution phases that enhance their stability when compared with other alloys such as intermetallic compounds. This leads to a high solubility of the alloying elements easy formation of the end product.[4] Before the birth of HEAs, materials scientists and metallurgists concentrated their efforts to fabricate alloys with concentrations laid on the edge-regions near the corners of phase diagram of traditional alloys (Fig. 14.3). Since 2004 after discovering the HEAs,[1] many researchers focused on that remained zone located in the center part of Fig. 14.3.

For many decades, it is well established that configurational entropy of a system can be calculated from Eq. (14.1):

$$\Delta S_{\text{conf}} = k \ln w \tag{14.1}$$

where k is Boltzmann's constant and w is the number of potential ways in which the available energy can be mixed or shared among the particles of the system. Therefore, the configurational entropy change per mole for the formation of a solid solution from n elements with x_i mole fraction for equiatomic alloy at its regular solid solution state is[3]:

$$\Delta S_{\text{conf}} = -R \ln 1/n = R \ln n \tag{14.2}$$

where R is the gas constant.

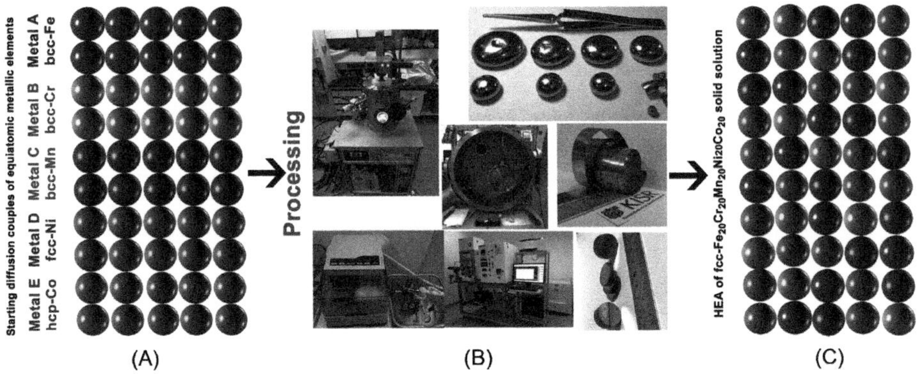

Figure 14.4 (A) Five equiatomic component before mixing, (B) examples of the techniques used for synthesizing HEAs, and (C) multicomponent HEAs after processing.

For example, suppose that we aim to fabricate a quinary alloy system, composing of metals A (bcc-Fe), B (bcc-Cr), C (bcc-Mn), D (fcc-Ni), and E (hcp-Co) with equiatomic concentrations of $Al_{20}Co_{20}Cr_{20}Fe_{20}Ni_{20}$ (Fig. 14.4A), using one or more fabrication technique(s) illustrated in Fig. 14.4B. The ΔS_{conf} of this system required to form a single fcc-solid solution phase (Fig. 14.4C) can be calculated from Eq. (14.2) as $R \ln 5 = 1.61\ R$.[3] This value is larger than that one calculated for nonequiatomic alloy $Al_{1.5}CoCr_{0.5}FeNi_{0.5}$ (1.523 R). Accordingly, it can be realized that the mixing entropy of traditional alloys with nonequiatomic composition is significantly lower than that one for HEAs, as shown in Fig. 14.5.

14.1.4 Advantage and attractive properties of HEAs

14.1.4.1 Preparations

Presumably, the simple way of fabrication, using conventional fabrication processes, such as arc melting, melting and casting, rapid solidification, mechanical alloying (MA), and powder metallurgy (Fig. 14.4B), led many researchers in the world wide to fabricate large number of unique HEAs with different attractive properties. This can be realized when looking at Fig. 14.5A, which presents the number of journal publications reported by Scopus for fabrication of HEAs by different methods (Fig. 14.6A) and via MA (Fig. 14.6B).

14.1.4.2 Properties

Their unique structural characteristics give HEAs unusual properties that cannot be existed in any other multicomponent alloys. High strength,[6,7] high fracture toughness at cryogenic temperatures,[8,9] good thermal stability,[10,11] and excellent corrosion resistance[12,13] are some of attractive properties found in HEAs. In addition, they possess many other excellent behaviors such as wear resistance,[14,15] creep resistance,[16,17] and high-strain-rate superplasticity.[18,19]

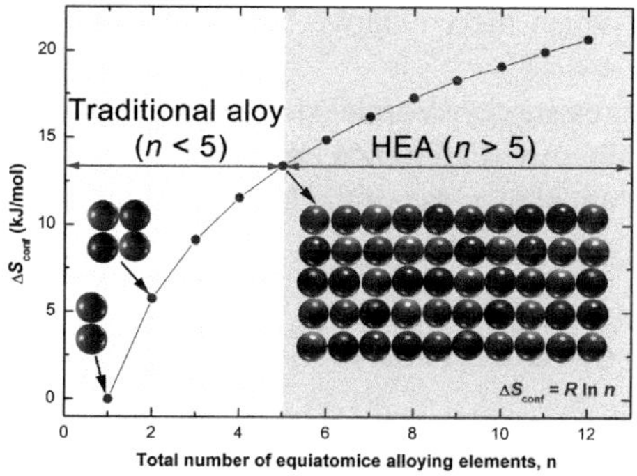

Figure 14.5 Correlation between DS_{conf} and n for traditional alloys (n < 5) and HEAs (n > 5).

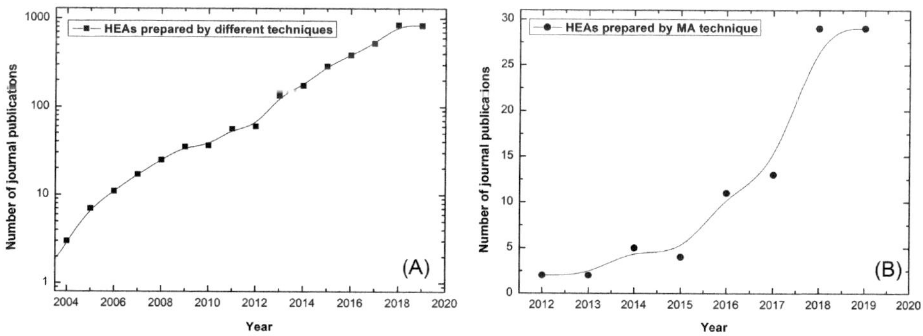

Figure 14.6 Number of journal publications of HEAs, prepared by (A) different approaches, and (B) mechanical alloying technique.

14.2 Preparations of nanocrystalline HEAs by mechanical alloying

Because of its high mixing and alloying capability of various alloying elements with different properties, MA method has been receiving great attention for fabrication of homogeneous multicomponent metastable alloys.[20] $Nb_{50}Zr_{10}Al_{10}Ni_{10}Cu_{20}$,[21] $V_{45}Zr_{20}Ni_{20}Cu_{10}Al_{2.5}Pd_{2.5}$,[22] $Ta_{55}Zr_{10}Ni_{10}Al_{10}Cu_{15}$,[23] $Zr_{52}Al_6Ni_8Cu_{14}W_{20}$,[24] $Ti_{60}Al_{15}Cu_{10}W_{10}Ni_5$,[25] and $Zr_{65}Al_{7.5}Ni_{10}Cu_{12.5}Pd_5$[26] are some of those successfully fabricated multicomponent metastable systems by means of MA technique. In fact, MA process has many obvious advantages when compared with the other routes used for preparing HEAs. MA leads to fabricate large scale of nanocrystalline fine powders that can be used as feedstock materials for cold/thermal spray coating. The as-MA powders can be utilized for preparing nearly full dense bulk HEAs with any desired shapes and sizes, using hot pressing and spark plasma sintering techniques. More importantly, MA leads to extended solid solubility at ambient temperature for those immiscible systems or when the enthalpy change of formation (ΔH_{for}) has zero or negative values.

Based on the literatures survey, Varalakshmi et al.[27] published the first report on fabrication nanocrystalline AlCrCuFeTiZn HEA system by MA technique. Since then, there has been a monotonical increase of the HEAs systems produced by MA, as indicated in Fig. 14.6B.

14.2.1 Examples of recent HEAs systems prepared by mechanical alloying

In this section, selected examples of HEAs, produced recently by MA technique, are demonstrated.

14.2.1.1 Bulk nanocrystalline VNbMoTaW high-entropy alloy

This work studied the microstructure and mechanical properties of equiatomic $V_{0.2}Nb_{0.2}Mo_{0.2}Ta_{0.2}W_{0.2}$ HEA prepared by MA, using a high-energy shaker ball mill (Spex 8000D).[28] The powders obtained after 30 h of milling revealed bcc-structure (Figs. 14.7 and 14.8A) with lattice parameter of 0.3146 nm. This end product consisted of fine grains with average grain size of about 6 nm (Fig. 14.8B and C). The powders obtained after 50 h of MA were then consolidated into rods with a diameter of 10 mm and a length of 10 mm at 10 MPa, using high-pressure/high-temperature (HPHT) technique.

The SAED pattern of as-consolidated VNbMoTaW HEA sintered at 1150°C revealed nanocrystalline structure (Fig. 14.9A) with typical bcc-structure, indexed as (110), (200), (211), and (220), as presented in Fig. 14.9B, where the average grain size was 30 nm, as displayed in Fig. 14.9C. This consolidated alloy had a high density of dislocations, as indicated by the inverse fast Fourier transformed (IFFT) HRTEM image (Fig. 14.9D).[28]

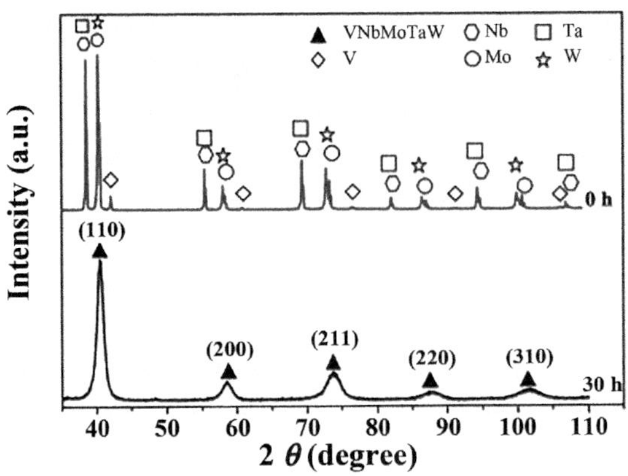

Figure 14.7 XRD patterns of the as-blended (0 h) and as-milled (30 h) VNbMoTaW HEA powders.[28]

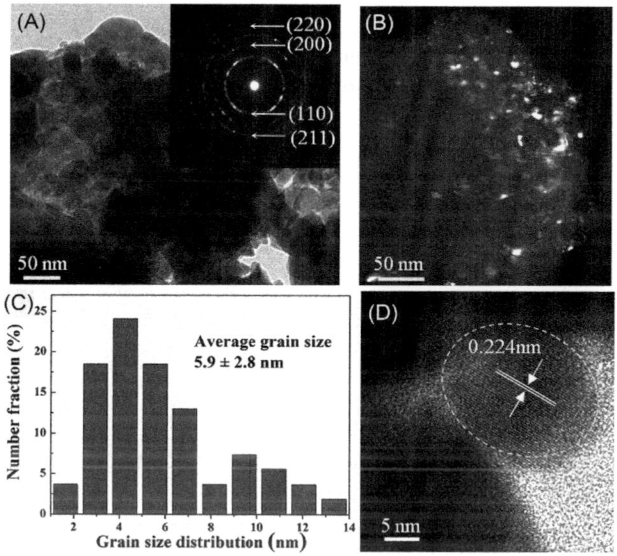

Figure 14.8 TEM images of the NC VNbMoTaW HEA powders: (A) bright-field TEM image with an inset showing the corresponding SAED pattern, (B) dark-field TEM image, (C) grain size distribution, (D) HRTEM image.[28]

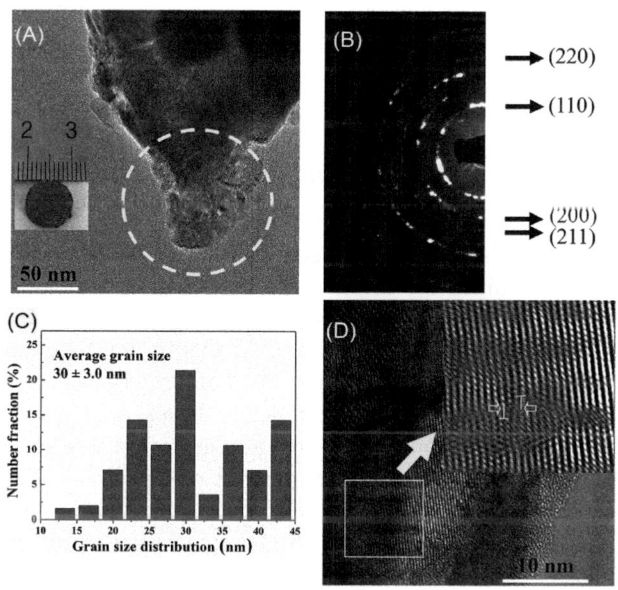

Figure 14.9 TEM images of the bulk NC VNbMoTaW HEA sintered at 1150°C under 5 GPa. (A) Bright-field TEM image with an inset showing the optical image of the bulk specimen in a form of disk with a diameter of 9 mm, (B) SAED pattern corresponding to the area marked with a *yellow* (gray color in print version) circle shown in (A), (C) distribution of grain size, (D) HRTEM image showing dislocations, where the inset is the inverse fast Fourier transform (IFFT) HRTEM image of the area marked with a *yellow* square.[28]

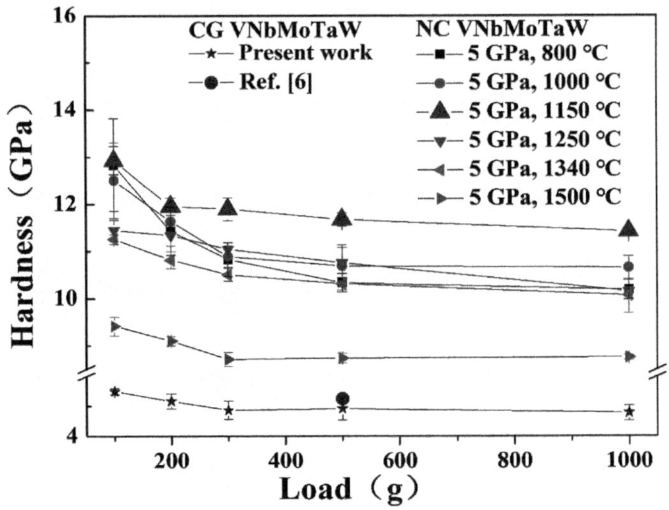

Figure 14.10 Vickers hardness of the bulk NC VNbMoTaW HEA as a function of indentation load.[28]

The microhardness of nanocrystalline bulk VNbMoTaW HEAs was examined as a function of the applied load, and the results are plotted in Fig. 14.10. At lower loads, the hardness decreases with increasing loads; at higher loads, the hardness asymptotes to a value that is dependent on the sintering temperature.[28] When the sintering temperature is 1150°C, the bulk VNbMoTaW HEA has the maximum hardness of 11.4 ± 0.04 GPa.

14.2.1.2 High-entropy multicomponent WMoNbZrV alloy

In this work, the authors used elemental of W, Mo, Nb, Zr, and V powders that were used as feedstock materials with nominal composition of 20 at% of each element.[29] The powders were ball-milled for 100 h, using Fritsch P5 planetary ball mill. XRD technique was used to monitor the progress of solid-state diffusion reaction after different stages of ball-milling time (Fig. 14.11). As the milling time increased (2 and 5 h) reduction in the intensity of diffraction lines of Bragg peaks related to hcp-Zr and bcc-V was gradually decreased and completely disappeared after 20 h of milling (Fig. 14.11).[29] The authors pointed out to the absence of milling effect on the powders obtained after 20 h of milling, which revealed single phase of bcc-solid solution.

14.2.1.3 High-pressure torsion-driven mechanical alloying of CoCrFeMnNi high-entropy alloy

This work presents interesting results obtained upon subjecting mechanically alloyed CoCrFeMnNi (20 at% for each element) powders to high-pressure torsion (HPT).[30] The XRD patterns of as-mechanically alloyed multicomponent powder obtained before and after HPT deformation for selected number of rotations, namely for 1, 10, and 100, are presented in Fig. 14.12. In the lower XRD pattern, all narrow peaks

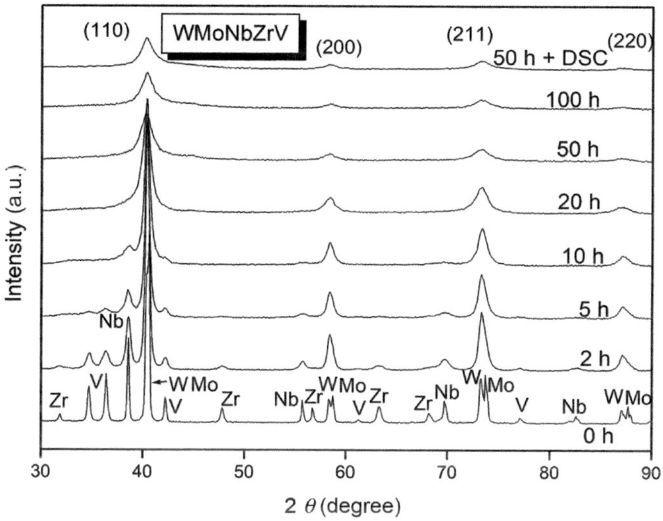

Figure 14.11 XRD patterns for WMoNbZrV powders subjected to milling for different time after heating in the DSC up to 700°C.[29]

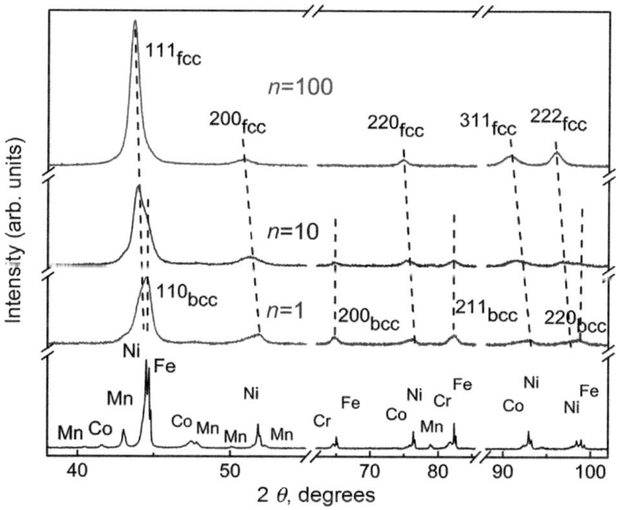

Figure 14.12 X-ray diffraction patterns of "as-mixed" multicomponent powder (bottom curve) and CoCrFeNiMn alloys after HPT for 1, 10, and 100 rotations (from lower to upper curves). *Tilted dashed lines* indicate the changes of peak positions for FCC solid solution.[30]

can be assigned to the five elements present in the initial mixed powder.[30] After HPT at 5 GPa and 1 rotation, XRD peaks appear already as considerably broadened. Subsequent increase in accumulated strain up to 10 rotations results in the formation of two noticeable phases based on fcc- and bcc-types of solid solutions. This implied the HPT-driven dislocation activity and, as a consequence, development of shear deformation texture.[30] The application of HPT for 100 rotations (upper curve in

Fig. 14.12) reveals that only FCC single-phase structure remains with a shift of XRD peaks toward lower diffraction angles.[30]

The authors pointed out to the combination of the hardening mechanisms due to the complex alloy and the nanometer-sized grains, which led to additional precipitation hardening resulted in exceptional values of the hardness of about 6700 MPa. It is speculated that the application of HPT conditions results in accelerated atomic diffusivity, which allows to save both time and energy (lower synthesis temperature) and obtain different types of HEA microstructures for further studies of stability and mechanical properties.[30]

14.2.1.4 Magnetic properties of Co_xCrCuFeMnNi high-entropy alloy powders

HEAs CoxCrCuFeMnNi ($x = 0.5$, 1.0, 1.5, 2.0 mol) powders were synthesized by MA, using a high-energy planetary ball-milled operated for 5, 10, 15, 30, 45, and 50 h at 300 rpm.[31] The powders of the end product (50 h) were then heat treated for 2 h under different temperatures (700, 800, and 900°C) and a vacuum condition of 6×10^{-3} Pa. The XRD patterns of the powders obtained after different milling time are shown in Fig. 14.13. The mixed powders include diffraction patterns of all the pure elements.

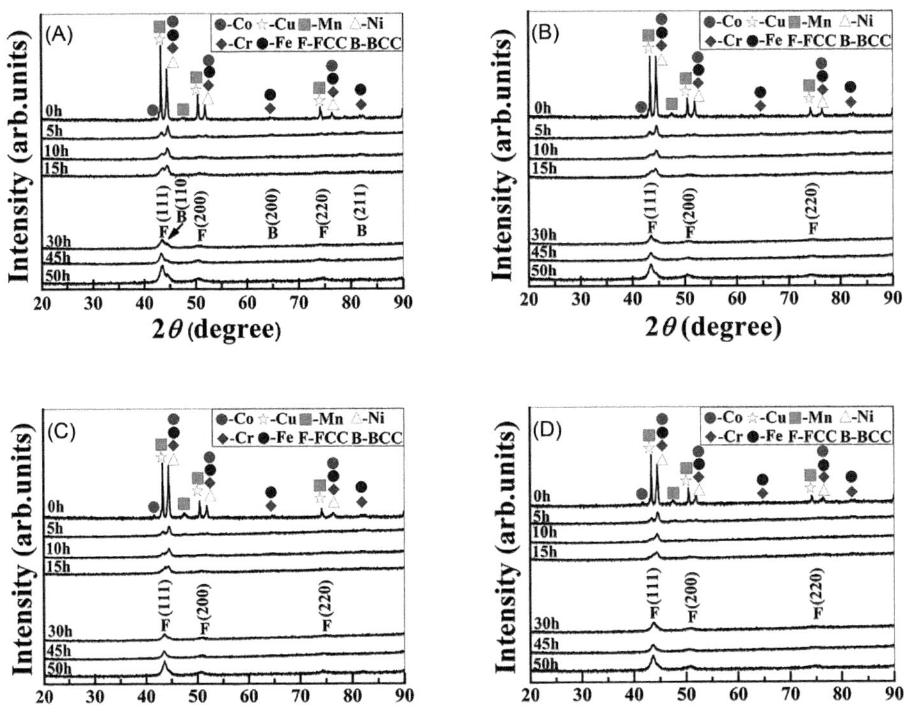

Figure 14.13 XRD patterns of the Co_xCrCuFeMnNi HEA powders prepared under different milling time: (A) $Co_{0.5}$, (B) $Co_{1.0}$, (C) $Co_{1.5}$, (D) $Co_{2.0}$.[31]

As the milling time increased, the constituent elements were gradually dissolved. Both Cu and Mn were dissolved rapidly within 5 h because of their lower melting points in comparison with those of the other elements in the CoxCrCuFeMnNi HEA. In addition, the brittleness of HCP Co also accelerated dissolution.[31] The Bragg peaks corresponding to Ni, Fe, and Cr can still be detected after 15 h of ball milling. As the milling time reached 30 h, two supersaturated solid solution phases with a primary FCC phase and a small amount of BCC phase, respectively, were formed in $Co_{0.5}$CrCuFeMnNi HEA powder. However, only one supersaturated solid solutions phase with FCC structures was detected in other three HEA powders after 30 h of ball milling. As the milling time increased to 45 and 50 h, the diffraction peak of FCC phase was gradually broadening. The diffraction peak broadening can be introduced by crystallite size refinement.

The magnetic hysteresis loops of the 50 h-ball-milled CoxCrCuFeMnNi HEA powders are presented in Fig. 14.14. It can be seen that the Ms, HC, and remanence ratio of $Co_{0.5}$CrCuFeMnNi powder were 21 emu/g, 63 Oe, and 16%, respectively. The Ms gradually increased with the increase of Co content. However, the HC and remanence ratio gradually decreased with the increase of Co content. When the Co content reach 2.0 mol, the HC attained the smallest value of 14 Oe among the four HEA powders.[32] Fig. 14.15 shows the relationship between H_c and crystallite size of

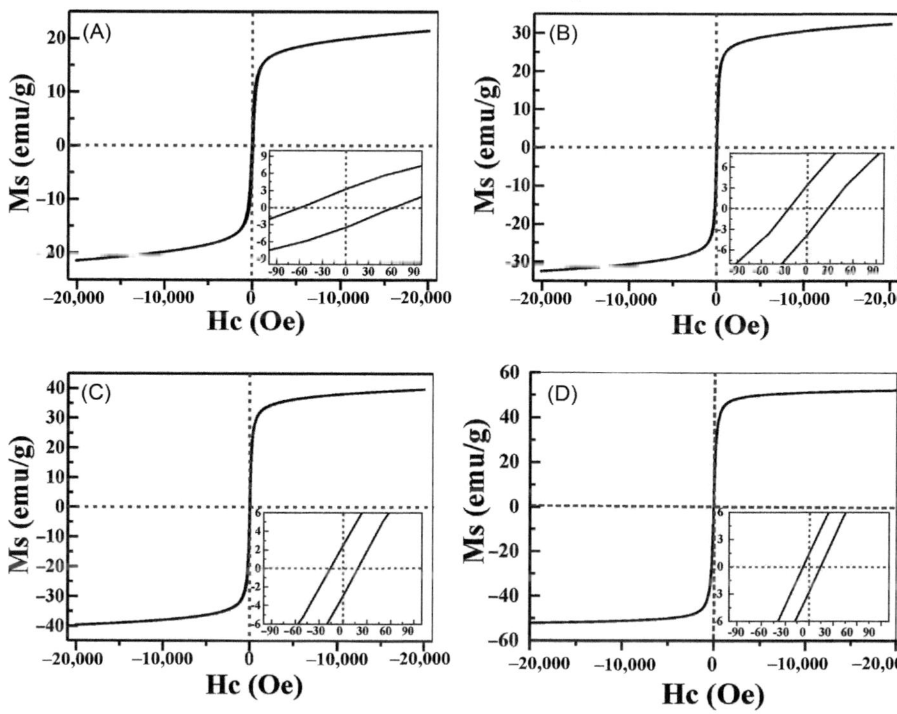

Figure 14.14 Magnetic hysteresis loops of the 50 h-ball-milled Co_xCrCuFeMnNi HEA powders: (A) $Co_{0.5}$ (B) $Co_{1.0}$, (C) $Co_{1.5}$, (D) $Co_{2.0}$.[31]

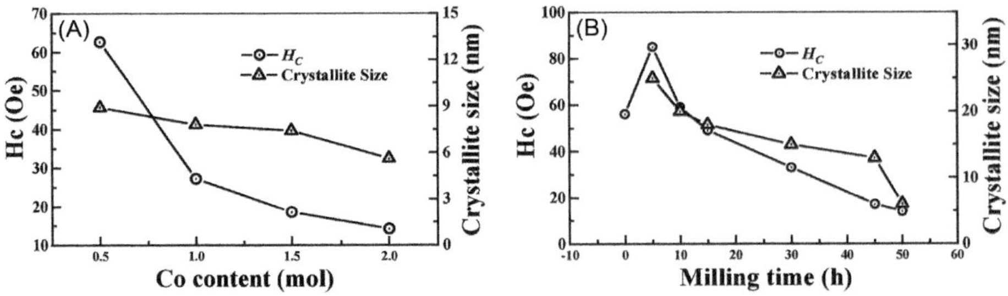

Figure 14.15 Relationship between H_c and crystallite size of the HEA powders: (A) Co_xCrCuFeMnNi powders after 50 h of milling, and (B) $Co_{2.0}$CrCuFeMnNi powder after different milling time.[31]

the HEA powders. H_c decreased with the decrement of crystallite size. The smaller crystallite size, the more crystalline phase boundaries. That could greatly hinder the shift of domain walls.[31]

References

1. Cantor, B., Chang, Knight, I.T.H. P., Vincent, A.J.B., *Mater. Sci. Eng., A* 213: 375 (2004)
2. Yeh, J.W., Chen, S.K., Lin, S.J., Gan, J.Y., Chin, T.S., Shun, T.T., et al., *Adv. Eng. Mater.*, 6: 299 (2004)
3. Murty, B.S., Yeh, J.W., and Ranganathan, S., High-Entropy Alloys, 1st Edition, Elsevier Inc., Butterworth-Heinemann, London, UK, 2014.
4. Ye, Y.F., Wang, Q., Lu, J., Liu, C.T., Yang, Y., *Materials Today*, 19:349 (2016)
5. Vaidya Mayur, Muralikrishna, Garlapati, Murty, Budaraju Srinivasa, *Journal of Materials Research*, 34: 664 (2019)
6. Li, Z. Pradeep, K.G., Deng, Y., Raabe, D., Tasan, C.C., *Nature*, 534: 227 (2016)
7. Zhang, Mengdi, Zhang, Lijun, Fan, Jiantao, Yu, Pengfei, Li, Gong, *Materials Science and Engineering: A*, 764: 138212 (2019)
8. Gludovatz, B., Hohenwarter, A., Catoor, D., Chang, E.H., George, E. P., Ritchie, R.O., *Science*, 345: 1153 (2014)
9. Jo, Yong Hee, Doh, Kyung-Yeon, Kim, Dong Geun, Lee, Kwanho, Kim, Dae Woong, Sung, Hyokyung, et al., *Journal of Alloys and Compounds*, 809: 151864 (2019)
10. Yang, M., Liu, X.J., Ruan, H.H., Wu, Y., Wang, H., Lu, Z.P., *J. Appl. Phys.*, 119: 245112 (2016)
11. Hung, Sheng-Bo, Wang, Chaur-Jeng, Chen, Yen-Yu, Lee, Jyh-Wei, Li, Chia-Lin, *Surface and Coatings Technology*, 375: 802 (2019)
12. Aliyu, Ahmed, Srivastava, Chandan, *Thin Solid Films*, 686: 137437 (2019)
13. Raza, Ahmad, Abdulahad, Syed, Kang, Byungchul, Ryu, Ho Jin, Hong, Soon Hyung, *Applied Surface Science*, 485: 368 (2019)
14. Ji, Xiulin, Ji, Cuicui, Cheng, Jiangbo, Shan, Yiping, Tian, Songya, *Wear*, 398-399: 178 (2018)
15. Kanyane, L. R., Malatji, N., Popoola, A. P. I., O., Fayomi S. I., *Results in Physics*, 14: 102465 (2019)
16. Lee, Kap Ho, Hong, Soon-Ku, Hong, Sun Ig, *Materialia*, 8: 100445 (2019)

17. Xu, Zhenlin, Zhang, Hui, Li, Weihuo, Mao, Aiqin, Wang, Lin, Song, Guangsheng, *Additive Manufacturing*, 28:766 (2019)
18. Li, Jia, Fang, Qihong, Liu, Bin, Liu, Yong, *Acta Materialia*, 147 :35 (2018)
19. Jeong, H. T., Park, H. K., Park, K., Na, T. W., Kim, W. J., et al. *Materials Science and Engineering: A*, 756: 528 (2019)
20. El-Eskandarany, M. Sherif, Matsushita, M. Inoue, A., *J. of Alloys Comp.*, 329: 239 (2001)
21. El-Eskandarany, M. Sherif, Matsushita, M. Inoue, A., *J. Non-Cryst. Solids*, 313: 622 (2002)
22. El-Eskandarany, M. Sherif, Inoue, A., *Materials Transactions, JIM*, 43: 770 (2002)
23. El-Eskandarany, M. Sherif, Zhang, W. Inoue, A., *J. of Alloys Comp.*, 350: 222 (2003)
24. El-Eskandarany, M. Sherif, Inoue, A., *J. Mater. Res.,* 21: 976 (2006)
25. El-Eskandarany, M. Sherif, Inoue, A., *Met. Trans.* A , 37: 2231 (2006)
26. El-Eskandarany, M. Sherif, Saida, J. Inoue, A., *Acta mater.*, 51:4519 (2003)
27. Varalakshmi, S., Kamaraj, M., Murty, B.S., *J. Alloys Compd.*, 460: 253 (2008)
28. Xin, S.W., Zhang, M., Yang, T.T., Zhao, Y.Y., Sun, B.R., Shen, T.D., *J. Alloys Compd.*, 769: 597 (2018)
29. Oleszak, Dariuszk, Antolak-Dudka, Anna, Kulik, Tadeusz, *Materials Letters*, 232: 160 (2018)
30 Kilmametov, A., Kulagin, R., Mazilkin,A., Seils, S., Boll, T., Heilmaier, M., Hahn, H., *Scripta Materialia*, 158: 29 (2019)
31. Zhao, Rui-Feng, Ren, Bo, Zhang, Guo-Peng, Liu, Zhong-Xia, Zhang, Jian-jian, *Journal of Magnetism and Magnetic Materials*, 468: 14 (2018)
32. Hori, Takao, Nagase, Takeshi, Todai, Mitsuharu, Matsugaki, Aira, Nakano, Takayoshi, *Scripta Materialia*, 172: 83 (2019)

Biomedical applications of mechanically alloyed powders

15

Biomaterials are rather new members in functional and advanced materials and rapidly growing area of materials science and engineering. These materials are dedicated either to manufacture biomedical devices to replace a certain function part in the body. Biomaterial may refer to those materials used in medical treatment of diseases. In fact, using of metallic biomaterials dates back to 2000 BC, when old Egyptian employed gold to replace and bound teeth.[1] Since then, gold has been used in wide range of applications in different areas. This chapter presents the applications of mechanical alloying for fabrication of different metallic biomaterial powders, used as feedstock materials in biomedical applications and antimicrobial surface protective coating. The photo presents Ancient Egyptians (2000 BC) replaced and bound teeth using gold wire and plates (source: http://tomboybklyn.com/the-history-of-gold-teeth/) (Fig. 15.1).[1] In the early 1900s bone plates were successfully implemented to stabilize bone fractures and to accelerate their healing. While by the time of the 1950s–60s, blood vessel replacement was in clinical trials and artificial heart valves and hip joints were in development.

15.1 Introduction

In broad definition, biomaterials denoted to a class of materials possessing some favorite characteristics making allowing them to be used for medical treatment (Fig. 15.2).[2] They are nondrug substances suitable for inclusion in systems, which augment or replace the function of bodily tissues or organs. Recently, biomaterials have been receiving great attentions as functional materials dedicated to make devices and components used for replacing a certain part of the body in a safe, reliable, economic, and physiologically acceptable manner.[3] It is necessary for any biomaterials to enjoy excellent biocompatibility with the body. Implant parts must receive acceptance of the body, and they have to be nontoxic, and nonallergenic. Moreover, biomaterials must be chemically stable and inert and do not reveal time-dependent degradation. Based on materials science point of view, biomaterials can be classified into four groups as metal and metal alloys, ceramics, polymers, and composites (Fig. 15.3). More recently, metastable materials such as metallic glassy alloys and nanostructure materials were included in this classification. This chapter will be focused mainly on the metal alloys, metallic glasses, as well as nanostructured biomaterials prepared by mechanical alloying and milling.

Mechanical Alloying. http://dx.doi.org/10.1016/B978-0-12-818180-5.00015-7

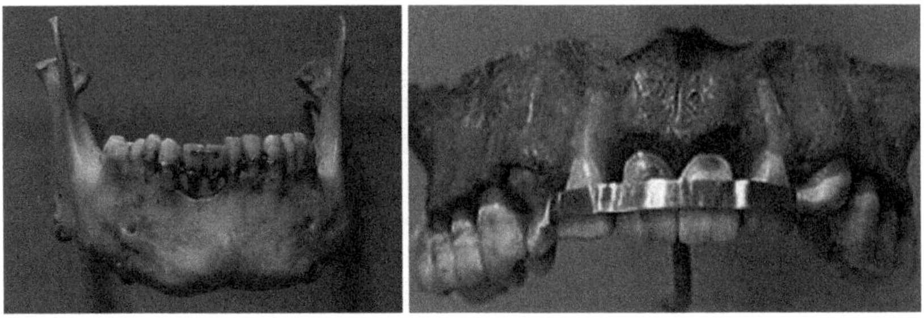

Figure 15.1 Ancient Egyptians (2000 BC) replaced and bound teeth using gold wire and plates.
Source: http://tomboybklyn.com/the-history-of-gold-teeth/.[1]

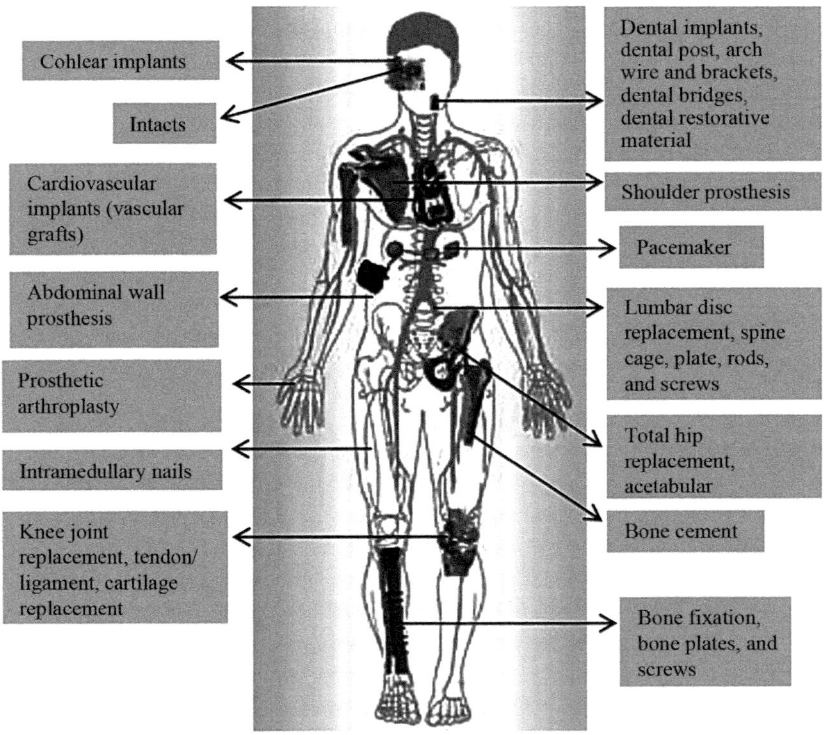

Figure 15.2 An illustration of various implants and devices used to replace and/or enhance the function of diseased or missing tissues and organs. (*After Patel and Gohil.*[2])

15.2 Metallic biomaterials

Need not to be mentioned, materials used for biomedical applications such as orthopedic and cardiac stents can be classified into three major families: polymers, ceramics, and metals. Metallic biomaterials possess superior structural functions when

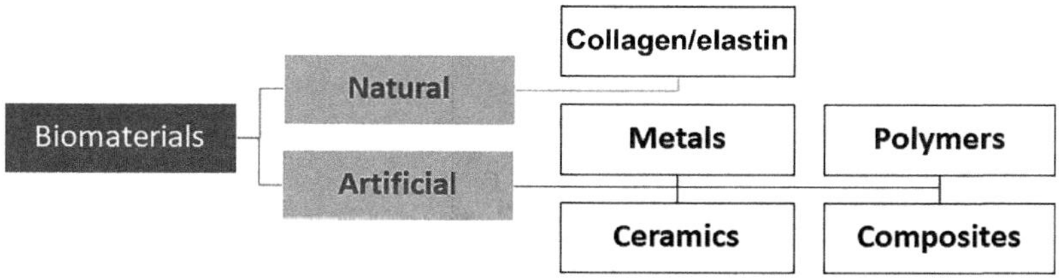

Figure 15.3 General classification of biomaterials.

compared with ceramic and polymeric biomaterials.[4] For example, stainless steel alloys have been used to substitute natural parts and/or to repair different organs in human body for long time. The worldwide interest to use stainless steel biomaterials is attributed to their low cost, good corrosion resistance conferred by the presence of Cr alloying element. Existing of Cr metal ensures the development of a passive and protective oxide layer. However, many of stainless steel alloys show attractive biomedical characteristics, indexed by their excellent fatigue strength, higher ductility, and good machinability,[3] but the high Ni is considered to be the source of allergens.[5,6] Within the last 4 decades, innovated metallic alloys rather than stainless steel have been prepared by different researchers. These advanced metallic biomaterials reveal excellent biocompatibility, high corrosion resistance, good mechanical properties, and high wear resistance.[7,8]

Pure titanium metal and titanium-based alloys, cobalt-based alloys, and tantalum-based alloys are few examples of these new metallic biomaterials that are commonly used nowadays.[9] In addition, magnesium-based alloys have shown excellent biocompatibility for using in orthopedic implants.[10] These light Mg-metal bio-alloys are used in biomedical applications such as bone fixation, cardiovascular stents, hip joints, screws/pins, and dental implants.[11]

Surface modification is considered to be an effective way to enhance the performance and durability of the metallic bioimplants and to modulate the unsafe side effects upon their degradation in human body.[12]

Due to their superior mechanical, stainless steel, Ti- and Co-based alloys are considered to be the best commercial materials used for bone implants. These materials can serve for long term, durable, and load bearing, where they revel outstanding ductility. However, their mechanical properties show significant mismatch in human bones. In contrast to these traditional metallic biomaterials, Mg-based alloys have become desirable bone implants due to its physical and mechanical properties that are very closed with the human bones. In addition, Mg metal and most of its alloying element such as Zn are biodegradable materials.

There are many strategies that are successfully used to modify the surfaces of structural biomaterials, such as physical vapor deposition (PVD),[12] chemical conversion coating,[13] laser surface alloying,[14] thermal spraying,[15] organic coating,[16] and microwave-derived coatings.[17] These techniques offer novel implant surfaces with controlled features at the nanometer scale.[12]

15.3 Mechanical alloying for fabrication of metallic biomaterials

15.3.1 Selected examples

15.3.1.1 Ti-based alloys

15.3.1.1.1 High strength, antibacterial, and biocompatible Ti-5Mo-5Ag alloy

Reference to their high strength, relatively low elastic modulus, enhanced corrosion resistance, and superior biocompatibility of bcc-Ti (β-phase)-based alloys, they have been considered as desirable biomaterials used as dental and orthopedic implant materials.[18,19] However, they are facing challenges from implant-associated infections and the requirements for highly stressed applications. Zhang et al. have recently superseded to employ mechanical alloying process for preparing of fine grains of Ti-5Mo-5Ag alloy, using a high-energy ball mill.[19] The end product obtained after 6 h of milling was consolidated into at 800–900°C, using spark sintering technique.[19] Their results showed that the powders sintered at 900°C revealed a network microstructure consisting of 89% β-phase with average grain size of 8.1 μm as the matrix and 11% α-phase with micron/submicron-scale precipitates at the grain boundaries/triple junctions. They attributed the excellent mechanical properties of this bulk alloy, indexed by compressive yield strength of up to 1694 MPa and fracture strain of 23% to the microstructure structure network, as shown in Fig. 15.4A.[19] In addition, this new alloy possesses enhanced corrosion resistance, as presented in Fig. 15.4B.[19] Moreover, the consolidated Ti-5Mo-5Ag alloy sintered at 900°C reveals superior antibacterial activity, indicated by antibacterial rate up to ~95% against *Staphylococcus aureus*.[19] All this superior characteristics led the author to propose this alloy as a promising candidate for load-bearing implant applications.

15.3.1.1.2 Low-cost Ti-Mn-Nb alloys for biomedical applications

In 2018, new group of Ti-Mn-Nb alloys with different Nb concentrations (0–10 wt.%) were fabricated by mechanical alloying technique.[20] The started elemental powders with nominal composition of Ti-7Mn-xNb ($x = 0$, 3, 7, and 10 wt.%) were mechanically

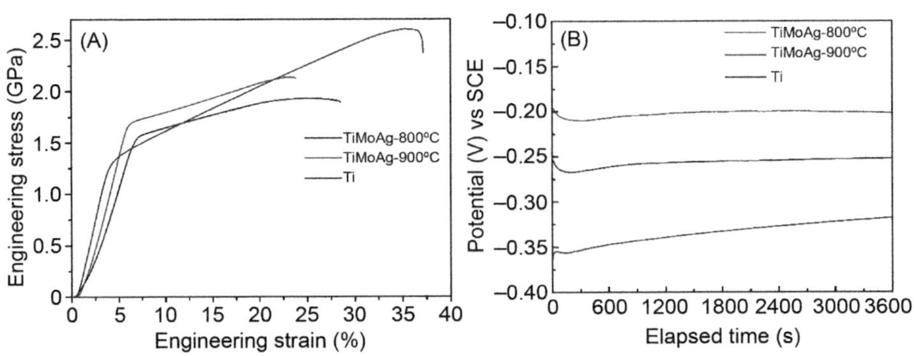

Figure 15.4 (A) Room-temperature compressive engineering stress–strain curves of Ti-5Mo-5Ag samples and pure Ti, and (B) open-circuit potential evolution for Ti-5Mo-5Ag samples and pure Ti during 1 h of immersion in artificial saliva solution at room temperature.[19]

alloyed for 2 h. The blended powders were then cold pressed under 550 MPa to obtain green cylindrical compacts. These green consolidated buttons were subsequently sintered in a high-temperature tubular furnace at 1170°C for 8 h.[20] The authors pointed out that microstructural studies and phase analysis showed a two-phase composed of α- and β-phases in all alloys. They claimed that addition of Nb up to 10 wt.% enhanced the stability of the β-phase hindering the transformation during furnace cooling and increasing the proportion of retained β. In addition, they claimed that sintering and densification of the Ti-7Mn-xNb alloys was enhanced by the fast diffusing Mn and negatively impacted by the slow diffusing Nb due to the reliance of solid sintering processes on diffusion. Based on their results, the compressive strength and hardness of the alloys varied in the range of (1842–2127 MPa) and (341–375 Hv), respectively (Fig. 15.5).[20] The authors

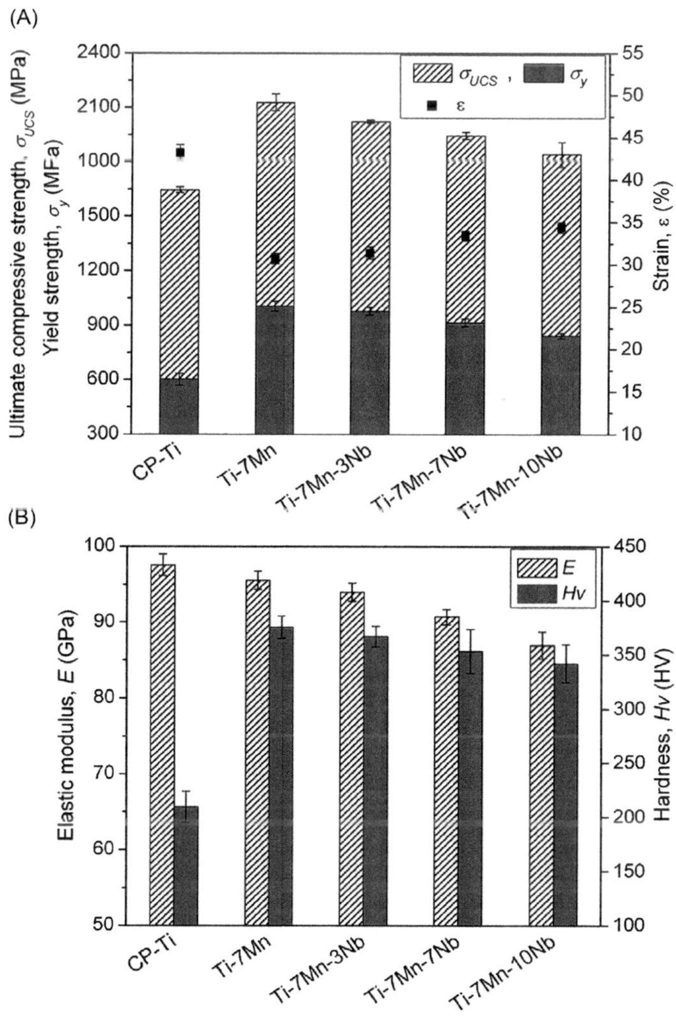

Figure 15.5 Compressive mechanical properties of the sintered Ti-7Mn-xNb and CP-Ti alloys.[20]

have realized that increasing the Nb contents led to decrease the elastic modulus; however, the compressive strain was increased. This is attributed to stabilization of a greater proportion of the β-phase.

15.3.1.1.3 Low modulus titanium-niobium-tantalum-zirconium (TNTZ) alloy

More recently, Mavros et al.[21] have reported the possibility of employing mechanical alloying to fabricated ultrafine grained low modulus TNZT alloys, started from elemental powders. The powders were ball-milled under Ar atmosphere for 60 h, using high-energy planetary ball mill. The powders of the end product (60 h) were pressed under 5 MPa to obtain green compacts, and then sintered by spark plasma sintering at 1373 K for 5 min. The authors pointed out that the microstructures of as fabricated TNZT alloys was composed of large volume fraction of BCC β-phase along with some minor fraction of HCP α-phase.[21] Based on their results, a large volume fraction of the BCC β-phase structure attributed from the use of strong β-phase stabilizing elements Nb, Ta, and Zr. They claimed that the process of mechanical alloying, followed by spark plasma sintering, is an effective and novel procedure for processing low modulus β-phase titanium alloys for potential biomedical applications. They clarify that all alloys exhibited high corrosion resistance.

15.3.1.1.4 β-type Ti-Nb-Ta-Zr-xHaP ($x = 0, 10$) alloy

The formation of nano-porous β-type Ti-Nb-Ta-Zr-xHaP alloys via mechanical alloying was recently reported.[22] The powders of the end product obtained after 8 h of high-energy ball milling were consolidated into compacts, using spark plasma sintering. The results showed that after milling and sintering processes led to the formation of β-type Ti-28Nb-7Ta-5Zr-10HaP. The microhardness of this new alloy was 537 Hv, which is being very high when compared with another Ti-based alloys.

15.3.1.1.5 Ti-13Nb-13Zr alloy with radial porous Ti-HA coatings

Mechanical alloying of elemental Ti, Nb, and Zr powders with nominal compositions of Ti-13Nb-13Zr (wt.%) was employed to obtain homogeneous powder blend. The starting material powders were high-energy ball-milled for 10 h, using planetary ball mill.[23] Then, 10 wt.% of hydroxyapatite (HA) powders were mixed with Ti powders for 5 h. The mixture was firstly pressed into green compacts, using a hydraulic press and then finally sintered at 1100°C, using spark plasma sintering. The as-consolidated button was radial porous composite, consisting of an inner compact Ti-13Nb-13Zr alloy and an outer porous Ti-HA composite (Fig. 15.6). This fabricated member revealed low elastic modulus (14.9 GPa), but it showed high compressive strength of 1289.7 MPa.

15.3.1.2 Mg-based alloys

15.3.1.2.1 High-performance MgFe biodegradable alloy

Fabrication of $Mg_{30}Fe_{70}$ alloys powders by a high-energy ball milling of the elemental powders was reported by Xie et al. in 2016.[24] In their work, the powders obtained after 37.5 h were consolidated into full-dense compacts, using spark plasma sintering technique. The results have shown that consolidated $Mg_{30}Fe_{70}$ alloy button sintered

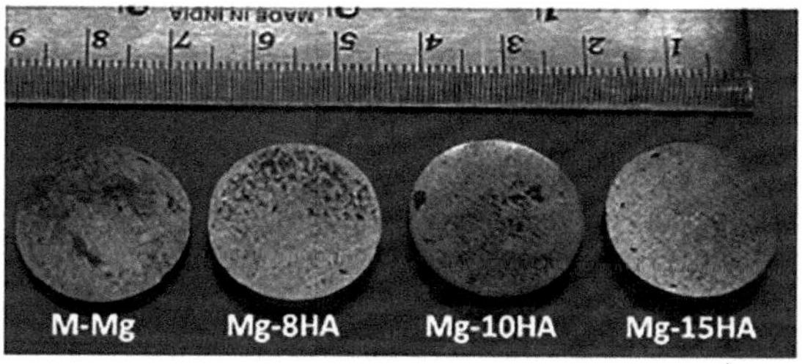

Figure 15.6 Macroscopic patterns (A) and SEM photographs of sintered TNZ/TH composites along longitudinal (B) and transverse directions (C), and XRD patterns of powders and composites (D): (1) as-milled Ti-10HA mixture, (2) as-sintered outer Ti-10HA part, (3) as-milled Ti-13Nb-13Zr mixture, and (4) as-sintered inner Ti-13Nb-13Zr part.[21]

at 773 K exhibited ultrahigh strength of 1380 MPa, and had high hardness and high densification, as well as had improved biodegradation rate.[24]

15.3.1.2.2 Biodegradable Mg-Zn/HA composite

Mg-x_1Zn/x_2HA (x_1 = 5.6 and x_2 = 5, 10, 15, 20 wt.%) nanocomposite powders were fabricated by means of mechanical alloying, using ball-milling technique.[25] The microhardness of the sintered powders containing 10 wt.% HA was 131 Hv, where the maximum compressive strength was ~292 MPa. This gives about 17% increase in the value of compressive strength when compared with pure Mg-5.6Zn alloy at 10 wt.% HA.

15.3.1.2.3 Nanocrystalline AZ31 magnesium alloy with titanium additive

Mechanical milling was used to prepare nanocrystalline AZ31 magnesium (Mg) alloy with titanium (Ti) addition.[26] During milling, the crystallite size of the AZ31 Mg matrix decreased steadily with the increase of milling time. After milling for 110 h, the crystallite size of the Mg phase of the alloy with 18 wt.% and 27 wt.% Ti additions was refined to 86 and 66 nm, respectively, leading to the formation of a nanocrystalline Mg matrix microstructure. Meanwhile, the size of Ti particles added in the alloy also decreased gradually with the increase of milling time, and the Ti dispersions presented a positive effect on the refinement of the Mg matrix grains. By milling for 110 h, the average size of Ti particles was refined to less than 1 μm in the AZ31 Mg alloy with 18 wt.% Ti addition. After mechanical milling, the as-milled powders were cold pressed into green compacts, and the mechanical properties were characterized by Vickers' hardness and uniaxial compressive tests. It was found that the microhardness of mechanically milled AZ31 Mg alloy with 27 wt.% Ti addition reached about 147 Hv, which was almost 3 times larger than that of the as-received AZ31 Mg alloy. Meanwhile, the yield strength of mechanically milled Ti-added AZ31 Mg alloy was seen to be as high as 293 MPa at room temperature and 60 MPa at 300°C, respectively.[26]

15.3.1.2.4 Lamellar structured degradable magnesium–hydroxyapatite implants

Pure Mg powders and nano-hydroxyapatite (HA) particles (32 nm in diameter) were used as feedstock materials to prepare three nanocomposite batches with Mg/HA with 8, 10, and 15 wt.% HA.[27] The desired nanocomposite samples were prepared by

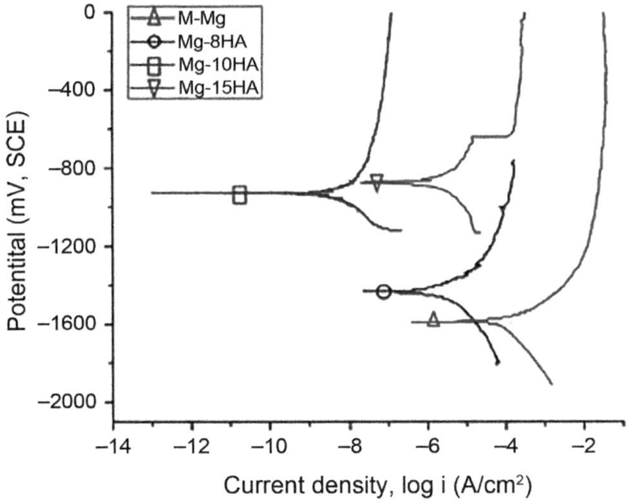

Figure 15.7 Image of the sintered compacts of pure Mg, Mg/*x* wt.% HA (*x*; 8, 10, 15).[27]

mechanical alloying approach, using a high-energy ball-milling technique. The mixed powders obtained after 20 h of milling were then charged into graphite die under uniaxial pressure of 50 MPa in vacuum at 450°C for 10 min. These green compacts were consolidated into full-dense buttons (Fig. 15.7), using spark plasma sintering. The consolidated samples revealed lamellar structure of the composites (Fig. 15.8), where

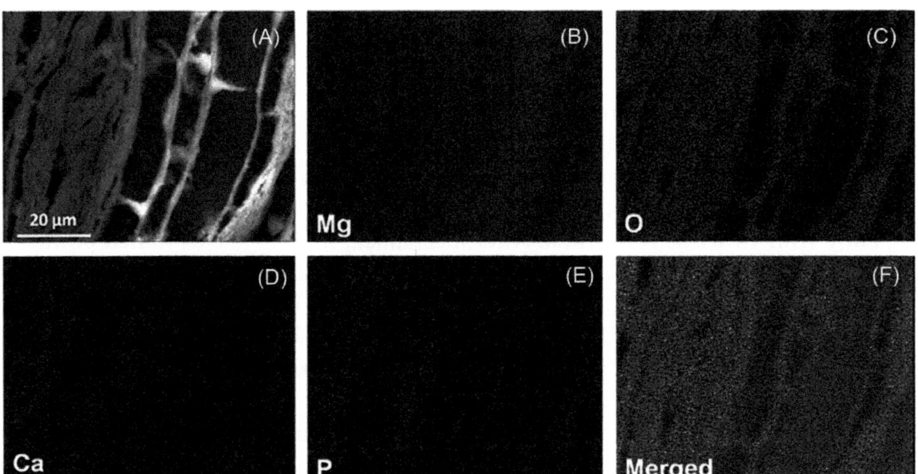

Figure 15.8 SEM observation (2000x magnification) and EDS analysis of Mg-8 wt.% HA composite: (A) image in secondary electron mode and corresponding EDS mapping showing (B) magnesium, (C) oxygen, (D) calcium, (E) phosphorous, and (F) merged image after overlapping the individual elements.[27]

the individual Mg flakes coated with HA were bonded together during the sintering process, as shown in Fig. 15.8. Based on the results of this work, the corrosion behavior of the composites investigated by potentiodynamic polarization shows a reduction in the interlamellar corrosion with increase in HA content, where the best corrosion resistance is found for the Mg-10% HA composite.[27] The results indicated that this composite exhibited maximum Vickers hardness; however, the Young's modulus and fracture toughness measured were higher for the Mg/8% HA composite.

References

1. http://tomboybklyn.com/the-history-of-gold-teeth/
2. Patel, Nitesh R., Gohil, Piyush P., *International Journal of Emerging Technology and Advanced Engineering*, 2: 92-101 (2012)
3. Park, Joon, Lakes, R.S., Biomaterials: An Introduction, 3rd Edition, Springer Science+Business Media, LLC, 2007, , Cham/Heidelberg, New York/Dordrecht, London
4. Su, Y., Luo, C., Zhang, Z., Hermawan, H., Zhu, D., Huang, J., et al. *Journal of the Mechanical Behavior of Biomedical Materials*, 77: 90-105 (2018)
5. Singh R, Dahotre N. *Journal of Materials Science: Materials in Medicine*; 18: 725-751 (2007)
6. Chen Qizhi, Thouas, George A. *Materials Science and Engineering R*; 87: 1-57 (2015)
7. Zindani, D., Kumar K., Davim, J. Mechanical Behavior of Biomaterials, Woodhead Publishing Series in Biomaterials, ELSEVIER, 2019, 83-99
8. Wilson, J. Metallic biomaterials: State of the art and new challenges, Woodhead Publishing Series in Biomaterials, ELSEVIER, 2018, 1-33
9. Ping DH. Acta Metall Sin Engl Lett.; 27:1 (2014)
10. Zeng, R. C., Dietzel, W., Witte, F., Hort, N., Blawert, C. *Adv. Eng. Mater.*; 10: B3-B14 (2008)
11. Ali, M., Hussein, M.A., Al-Aqeeli, N. *Journal of Alloys and Compounds*; 792: 1162-1190 (2019)
12. Kumara, S., Nehra, M., Kedia, D., Dilbaghia, N., Tankeshwar,K., Kim, Ki-Hyun, *Materials Science & Engineering C*; 106: 110154 (2020)
13. Rojaee, R., Fathi, M., Raeissi, K., *Appl. Surf. Sci.*; 285: 664-673 (2013)
14. Vorobyev, A.Y. Guo, C. *Laser Photon. Rev.*; 7: 385-407 (2013)
15. Bolelli, G., Bellucci, D., Cannillo, V., Lusvarghi, L., Sola, A., Stiegler, N., et al. *Mater. Sci. Eng. C*; 34: 287-303 (2014)
16. Sima, F., Davidson, P.M., Dentzer, J., Gadiou, R., Pauthe, E., Gallet, O., et al. *ACS Appl. Mater. Interfaces*; 7: 911-920 (2014)
17. Ren, Y., Babaie, E., Bhaduri, S.B., *Prog. Org. Coat.*; 118: 1-8 (2018)
18. 18. Geetha, M., Singh, A.K., Asokamani, R., Gogia, A.K., *Prog. Mater. Sci.*, 54: 397-425 (2009)
19. Zhang, Y., Chu, K., He, S., Wang, B., Zhu, W., Ren, F., *Materials Science & Engineering C*, 106: 110165, (2020)
20. Ehtemam-Haghighi, S., Attar, H., Dargusch, M. S., Kent, D., *Journal of Alloys and Compounds*, 787: 570-577 (2019)
21. Mavros, N., Larimian, T., Esqivel J., Gupta, R. K., Contieri, R., Borkar, T., *Materials & Design*; 183: 108163 (2019)

22. Bhushan, B., Singh, A., Singh, R., Mehta, J. S., Gupta, A., Prakash, C., *Materials Today: Proceedings*; 5: 27749-27756 (2018)
23. He, Y., Zhang, Y., Zhang J., Jiang, Y., Zhou, R., *Materials Letters*; 209: 543-546 (2017)
24. Xie, G., Takada, H., Kanetaka, H., *Materials Science & Engineering A*; 671: 48-53 (2016)
25. Salleh, E.M. Zuhailawati, H., Ramakrishnan, S. Dhindaw, B.K., *Metall. Mater. Trans. A*, 48: 2519-2528 (2017)
26. Lianxi, H. Z., Sun, H. Y., Zhang, H., Congwen Yu, D. H., *Materials Characterization*; 113:108-116 (2016)
27. Sunil B. R., Ganapathy, C., Kumarn, T.S. S., Chakkingal, U., *Journal of the Mechanical Behavior of Biomedical Materials*; 40: 178-189 (2014)

Index

Note: Page numbers followed by "f " indicate figures and "t" indicate tables

A

Advanced materials, 1. *See also specific names and techniques*
  chemically assisted approach, 2
  defined, 1
  fabrications, strategies, 1. *See also specific entries*
  high-energy-assisted approach, 2
  liquid-phase fabrication approach, 2
  lithographic approach, 2
  mechanically assisted approach, 2, 4. *See also specific entries*
    ball-milling technique, 4
    classification, 2
    mechanical alloying (MA), 5, 6f
    powder metallurgy (P/M), 4, 5f
    severe plastic deformation, 7f
  mechanically induced solid-state reaction (MISSR) approach, 2
  preparation methods, 4
  processing/structure/properties/performance interrelationship of, 2, 3f
  thermal approach, 2, 8
    droplet method (gas/water atomization), 8
    rapid solidification, 8, 9f
    thermal plasma processing, 9
    vapor deposition, 10
  uses, 1
  vapor deposition approach, 2
AFM. *See* Atomic force microscopy (AFM)
Agglomerates coarse composite SiC_p/Al powder particles
  disintegration, 206
  formation, 205
Agitation, 32
Al-CNT, 330
AlCrCuFeTiZn HEA system, 422
ALD. *See* Atomic layer deposition (ALD)
$Al_{93}Fe_3Ti_2Cr_2$ powders, 83
Al-Hf, 361
$Al_{50}Hf_{50}$ alloy
  BFI images, 350, 351f
  particle size distribution of, 348f
  SADPs, 350, 351f
  XRD patterns, 343f
Al-Mg alloys, 26f
  binary system, 28
Al-Nb, 361
Al/SiC MMNCs, fabrication, 196. *See also* SiC/Al MMNCs; fabrication
Al-Ta, 361
$Al_{50}Ta_{50}$ amorphization
  reaction in rod mill (MA *vs.* MD), 377, 384
    background, 377, 378f
    morphological changes with milling time, 380f, 381f
    procedure, 378
    structural changes with milling time, 379f
    thermal stability, 382f
  XRD patterns, 83, 86f, 87f
$Al_{30}Ta_{70}$ powders, 337f, 338
Al-Ti alloy, 361, 417
Al5356/TiN composites powders, 321
$Al_xHf_{100\text{-}x}$ powders
  crystallization, 356f, 357f
  mechanism, 358
  X-ray analysis, 340, 342f, 343f
Al_xTa_{1-x}, 361
Al-Zr binary system, 73, 391
  crystallization, 393, 394f
  structural changes with milling time, 391–393
  thermal stability, 392f–394f
  XRD patterns, 391–393
$Al_{50}Zr_{50}$ powder, 73–75
Ammonia gas (NH_3), 147
Amorphization, solid-state. *See* Solid-state amorphization

Amorphous and metallic glassy alloys, 1, 3f, 19. *See also* Glass(es)
background, 335, 336
consolidation into full-dense bulk materials, 396
$Ti_{40.6}Cu_{15.4}Ni_{8.5}Al_{5.5}W_{30}$ metallic glassy alloy powders, by SPS, 407f–411f, 406
$Zr_{52}Al_6Ni_8Cu_{14}W_{20}$ glassy alloy powders, 396
fabrication by MA, 336, 337f, 338f
MA method for, 28
melt spinning technique, 80
preparation methods, 336
Amorphous austenitic stainless steel ($Fe_{74}Cr_{18}Ni_8$), 373f, 374f
DTA curve, 374f
rod-milling time, 373f
XRD patterns, 374f
Amorphous-crystalline-amorphous cyclic phase transformations, during ball milling, 394, 395f
Amorphous phases, crystallization of, 99
Ancient Egyptians, gold wire and plates uses for teeth replacement, 431, 432f
Apparent activation energy, MgH_2 powders, 250, 251f
Arc-discharge technique, 4, 102
Archimedes' principle, 139
Arc-melting method, 8
Argon sparging, 19
Atmosphere, milling, 80
Atomic disordering of crystalline lattices, 335
Atomic force microscope (AFM), 96t
Atomic force microscope (AFM) nanostencil, 4
Atomic force microscopy (AFM), 17, 18f
WC-5Co-1Cr-3MgO-0.7VC-0.3Cr_3C_2 nanocomposite, 216, 218f, 219f
WC powders, 175, 178f
Atomic layer deposition (ALD), 4, 93
Atomization, 4, 93
defined, 8
gas/water (droplet method), 8
Attritor/attrition ball mill, 32, 33f, 55, 57t
Au-Cu alloy, 417
AZ31 magnesium alloy with titanium additive (nanocrystalline), 437

B

Back electron scattering (BES) FE-SEM micrograph, MgH_2/10 wt.% (8 Nb_2O_5/2 wt.% Ni) system, 295f, 296
Backscattered electron (BSE) micrographs, MgH_2 powders, 246, 247f
Ball-milled powders, for surface protective coating
cold gas-dynamic spray technology, 310
as feedstock materials for HVOF, 312
overview, 309
thermal spraying, 310, 312f
cold spray, 315–317
combustion-based processes, 311
high-feed-milled HVOF sprayed WC-Co coatings, 315
high velocity oxygen thermal spraying (HVOF), 311, 313f
HVOF of nanocrystalline iron aluminide, 315
HVOF of nanostructured Cr_3C_2-Ni20Cr coatings, 314f
HVOF reactive spraying of mechanically alloyed Ni-Ti-C powders, 314
HVOF sprayed diamond reinforced bronze coatings, 315
types of, 312f
Ball-milling (BM), 4, 19, 93. *See also* Rod milling (RM)
amorphization during, 342. *See also* Solid-state amorphization
amorphous-crystalline-amorphous cyclic phase transformations during, 394, 395f
atmosphere, 80
background, 5
combined with cold rolling, MgH_2 powders and, 286f
dry mode, 83
of elemental Ni and Nb powders, 27
environment, 84f–85f, 83
history, 19, 20, 27
homogeneous nanocomposite materials, 28
metal matrix/CNTs, fabrication, 225
metal nitrides preparation, 28
methods, 4

milling media
balls sizes, 63, 64, 65f
brittleness, 63t
Cu-15 vol.% Nb powders, media dynamics study, 64f, 65f
Cylpebs (Doering International), 60t, 66, 67f, 69f
hardness, 63t
influences of, 62, 63t
Millpebs (Wheelabrator Allevard Enterprise), 66, 67f
Powerpebs (Donhad), 66, 67f
requirements of, 62
shape of, 66f–69f, 60
specific gravity, 63t
milling speed, 67
balls and particle motions, tumbling mill, 70f
balls motion during milling at different, 73f
defined, 67
Fe-7 mass% C-6 mass% Mn, 75, 77f
HfN system, 74, 76f, 77f
selection, 67
theoretical investigations, 71, 72f
trajectory of milling media, 70, 71f
XRD patterns ($Al_{50}Zr_{50}$ powder), 73–75
milling time, 75
characterizations techniques, 78
crystallite size and lattice strain, 78, 79f
DSC thermograms, 80f
HRTEM image, 78, 79f
lattice constant, 78f
SEM images, 80–82
milling tools, impurities and, 23
gas contamination, 60, 61f
organic contaminations, 59f
PAC, contamination by, 60f
powder contaminations, 61
solid contamination, 61
milling vials shape, effects, 14, 58f
nanocrystalline materials formation, 28, 99
effects on CNTs structure, 102, 104f, 105f
first stage mechanism, 100
mechanical alloying, 99
mechanism(s), 100
methods, 99, 100f
second stage mechanism, 101
third stage mechanism, 101
overview, 30
oxygen content, increased, 81
reactive. *See* Reactive ball milling (RBM)
temperature, 88–90
TiC powders, 157f
BFI images, 160f
Bragg peaks, 158
consolidation techniques, 162
EDS, 161f
HAADF, 161f
HRTEM, 156, 161f
SADP, 160f
SEM, 156, 162f
STEM, 156, 160f, 161f
XRD pattern, 156, 158, 159f
wet mode, 83
Ball mill(s)
cataracting zone, 66f
inactive zone, 66f
shearing zone, 66f
types, 31
attritor/attrition ball mill, 32, 33f
bead mills, 31, 55
drum ball mills, 31, 55
effects on MA/MD/MM techniques, 50, 57t
high-energy ball mills, 32, 55
horizontal rotary ball mills, 31
jet ball mills, 31, 55
low-energy tumbling mill, 44, 55
Planetary Ball Mills, 31, 37–39, 55
powder contaminations and, 62
Retsch Mixer CryoMill, 36, 37f
Retsch Mixer Mills (MM 200 and MM 400), 35f
roller ball mills, 55
selection, 31
shaker mills, 33–35
SPEX Freezer/Mill-6970EFM, 36, 37f
Super Misuni, 36f
tumbler ball mill, 44, 45f
tumbler rod mill, 45, 46f
uni-ball mill, 42f, 43f
vibration ball mills, 31, 55
Ball-powder-ball collision, 48f
Balls size(s), milling media, 63, 64, 65f
Ball-to-powder weight ratio (BPR), 86f–88f, 83, 88

bcc-Nb powders, 150, 151f
Bead mills, 31, 55
Benjamin, J.S, 19
BET. *See* Brunauer, Emmett, and Teller Approach (BET)
BFIs. *See* Bright-field images (BFIs)
Biomaterials, 1, 3f, 431
  classification, 431, 433f
  defined, 431, 432f
  gold wire and plates in teeth replacement in Ancient Egyptians, 431, 432f
  history, 431, 432f
  mechanical alloying for fabrication, 434
  metallic, 432
  Mg-based alloys, 436
    high-performance MgFe biodegradable alloy, 436
    lamellar structured degradable magnesium-hydroxyapatite implants, 437, 438f
    Mg-Zn/HA composite, biodegradable, 437
    nanocrystalline AZ31 magnesium alloy with titanium additive, 437
  techniques, 433
  Ti-based alloys, 434
    β-type Ti-Nb-Ta-Zr-xHaP (x = 0, 10) alloy, 436
    low modulus titanium-niobium-tantalum-zirconium (TNTZ) alloy, 436
    Ti-Mn-Nb alloys, low-cost, 434, 435f
    Ti-5Mo-5Ag alloy, high strength, antibacterial, and biocompatible, 434f
    Ti-13Nb-13Zr alloy with radial porous Ti-HA coatings, 436, 437f
Biomedical applications, of mechanically alloyed powders. *See also* Biomaterials
  metallic biomaterials, 432
  Mg-based alloys, 436
    high-performance MgFe biodegradable alloy, 436
    lamellar structured degradable magnesium-hydroxyapatite implants, 437, 438f
    Mg-Zn/HA composite, biodegradable, 437
    nanocrystalline AZ31 magnesium alloy with titanium additive, 437
  overview, 431
  Ti-based alloys, 434
    low modulus titanium-niobium-tantalum-zirconium (TNTZ) alloy, 436
    Ti-Mn-Nb alloys, low-cost, 434, 435f
    Ti-5Mo-5Ag alloy, high strength, antibacterial, and biocompatible, 434f
    β-type Ti-Nb-Ta-Zr-xHaP (x = 0, 10) alloy, 436
    Ti-13Nb-13Zr alloy with radial porous Ti-HA coatings, 436, 437f
BMGs. *See* Bulk metallic glasses (BMGs)
Born nitride (BN)
  applications, 147
  BN nanotubes, by RBM, 148f–150f, 147
    SEM image, 149f
    TEM image, 149, 150f
    XRD patterns, 148, 149f
  cubic phase (c) of. *See* c-BN
  properties, 147
Bottom-up approach, nanofabrication methods, 93
BPR. *See* Ball-to-powder weight ratio (BPR)
Bragg peaks
  MgH_2 powders, 241, 243
  $SiC_{10}/A1_{90}$ powder, 198
  β-SiC powders, 167
  TiC powders, 158
  TiN powders, 182, 183f
  WC/Al_2O_3 nanocomposites, 207, 208f
  WC powders, 170, 173, 174
Brazing, 309
Bright-field images (BFIs)
  $Al_{50}Hf_{50}$ powders, 350, 351f
  amorphization by MA, 344f
  Co-Ti binary system, 388f
  Cu_2O reduction with Ti, 132f–134f, 132
  low magnification BFI, MWCNTs, 104f
  MgH2/5Ni/5Nb2O5 powders, 263f
  MgH_2 powders, 244f
  $SiC_{10}/A1_{90}$ powder, 200f
  SWCNTs, 104, 105f
  TiC powders, 160f, 162, 163f
  TiN powders, 185, 186f
  WC/Al_2O_3 nanocomposites, 207, 210f, 213f

WC and WC-MgO fabrication by MSSR method, 140–141, 141f, 142f, 143f
WC-5Co-1Cr-3MgO-0.7VC-0.3Cr_3C_2 nanocomposite, 215f
WC powders, 170, 172f
$Zr_{52}Al_6Ni_8Cu_{14}W_{20}$ glassy alloy powders, 397, 399f
Brittleness, milling media, 63t
Bronze Age, 417, 418f
Brunauer, Emmett, and Teller Approach (BET), 97t
Bucky balls, 93
Bulk metallic glasses (BMGs), 396
Bulk modulus, $SiC_{10}/A1_{90}$ powder, 204, 205f

C

Carbon dioxide (CO_2) emissions, impacts, 293
Carbon fiber reinforced polymer (CFRP), 234
Carbon nanotubes (CNTs), 1, 19. *See also* Multiwall carbon nanotubes (MWCNTs); Single-wall carbon nanotubes (SWCNTs)
ball-milling effect on structure, 102, 104f, 105f
cold spraying of metal matrix reinforced with, 330, 332f
fabrication, by mechanical alloying, 225
fabrication, by milling, 114
discussion, 123, 124f
materials and equipment, 115
method, 115
results, 116f–123f, 116
MWCNTs, 226, 235
reinforced ceramics nanocomposites, 195
SWCNTs, 235
Cataracting zone, ball mill, 66f
c-BN, 147. *See also* Born nitride (BN)
Ceramics, 432
Ceramics matrix composites (CMC), 194
Chemical conversion coating, 433
Chemically assisted approach, 2
Chemical vapor deposition (CVD), 4, 10, 93, 309, 310f
β-SiC powders, 166
TiN powders, 180
Chemisorption, 235
Classic powder metallurgy, 106f, 107f
CNT-MMC-based materials, 330
CNTs. *See* Carbon nanotubes (CNTs)
Coating. *See also* Surface protective coating
approaches, 309
correlation between processes, 309, 311f
defined, 309
dry coating approach, 309, 310f
process classification, 309
uses, 309
wet coating approach, 309
Co-based alloy powders, 333
Co-based alloys, 433
CoCrFeMnNi HEA, high-pressure torsion-driven MA of, 424, 425f
Coefficient of friction, TiN powders, 188, 189f
Cold densification, 109
cold isostatic pressing, 109
cold uniaxial pressing, 109
Cold gas-dynamic spray technology, 310, 315–317
background, 315
equipment, robotic system, 310f
technique, 309
Cold isostatic pressing, 109
Cold rolling, combined with ball milling, MgH_2 powders and, 286f
Cold spray. *See* Cold gas-dynamic spray technology
Cold spray coating feedstock powders
applications, 332
Co-based alloy powders, 333
Cu-based alloy powders, 333
Ni-based alloy powders, 333
Ti- and hard Cr-based alloy powders, 333
Cold uniaxial pressing, 109
Cold-welding effect, PCA in controlling, 83
Combustion-based processes, thermal spraying
high velocity oxygen thermal spraying (HVOF), 311, 313f
advantage of, 312
gun for, 313f
high-feed-milled HVOF sprayed WC-Co coatings, 315
HVOF of nanocrystalline iron aluminide, 315
HVOF of nanostructured Cr_3C_2-Ni20Cr coatings, 314f

Combustion-based processes, thermal spraying (*cont.*)
HVOF reactive spraying of mechanically alloyed Ni-Ti-C powders, 314
HVOF sprayed diamond reinforced bronze coatings, 315
uses, 312, 313f
utilization of powders as feedstock for HVOF, 312
Configurational entropy, 419. *See also* High-entropy alloys (HEAs)
Copper mold casting, 8
Co-Ti binary system, 385
BFIs of, 388f
SADPs of, 388f
structural changes with milling time, 386f
thermal stability, 390f
XRD patterns, 386f, 387f
$Co_xCrCuFeMnNi$ HEA powders, 426f, 427f
Co-Zr, 361
Cr_3C_2-Ni20Cr coatings (nanostructured), HVOF of, 314f
CRing, 286
Cryo-compressed hydrogen storage (CcH2), 234, 235
CryoMill, 36, 37f
Crystallization
of amorphous phases, 99
thermal stability, of amorphization by MA, 356f–358f, 356, 358t
Crystal structure, nanocrystalline materials, 94, 95t
Crystal-to-glass transition, 339f
metastable phase diagram, 339, 340f, 341f
Cu-based alloy(s). *See also specific entries*
nanoparticle, STEM-BF image, 19, 20f
powders, 333
Cu-CNT-AlN, 330
Cu-CNT-SiC, 330
Cu_2O reduction with Ti
BFIs, 132f–134f, 132
DTA measurements, 134, 136f
MSSR, mechanism, 136
by room temperature rod-milling technique, 129
SADPs, 132f–134f, 132
structural changes with milling time, 130–132
TEM analyses, 132f–134f, 132
XRD patterns, 130, 131f
$Cu_{30}Ta_{70}$ powders, 89f
Cu-Ti, 361
$Cu_{50}Ti_{20}Ni_{30}$ alloy powders, 322
SEM micrograph, 322
XRD patterns, 322, 326f
Cu-15 vol.% Nb powders, media dynamics study, 64f, 65f
Cu-Zn alloy, 417
CuZrAl bulk metallic glass, 412
CVD. *See* Chemical vapor deposition (CVD)
Cyclic crystalline-amorphous transformations, during MA, 385
Al-Zr binary system, 73, 391
crystallization, 393, 394f
structural changes with milling time, 391–393
thermal stability, 392f–394f
XRD patterns, 391–393
amorphous-crystalline-amorphous cyclic phase transformations during ball milling, 394, 395f
Co-Ti binary system, 385
BFIs of, 388f
SADPs of, 388f
structural changes with milling time, 386f
thermal stability, 390f
XRD patterns, 386f, 387f
Cylpebs (Doering International), 66, 67f
milling action of linear contact, 66, 69f
physical properties, comparative data, 66, 67t
Cyro-melting process, 7

D

Dark-field image (DFI)
MgH_2 powders, 244f, 245f
β-SiC powders, 168f
TiN powders, 184f
WC and WC-MgO fabrication by MSSR method, 140, 141f
WC powders, 174, 175f
Debye-Scherrer rings, 160, 162
Dehydrogenation reaction, defined, 236
Densification
cold, 109
equal channel angular press (ECAP), 109

magnetic pulse compaction (MPC), 109
powder, 109
rapid omnicompaction (ROC), 109
Detonation technique, 114
DFI. *See* Dark-field image (DFI)
Diamond-Bronze composite coatings, HVOF sprayed, 315
Diamond-reinforced MMCs, cold spraying, 331
Differential scanning calorimetry (DSC), 17, 18f, 78, 80f, 96t
MgH_2 powders, 248f
MgH_2/10 wt.% (8 Nb_2O_5/2 wt.% Ni) system and, 296, 297f
$Ni_{60}Nb_{40}$ amorphous alloy, 295f, 337
$Ti_{40.6}Cu_{15.4}Ni_{8.5}Al_{5.5}W_{30}$ metallic glassy alloy powders, 409, 410f
Dip-pen methods (DPNL), 93
Doping, of MgH_2 with catalysts, 266
in-situ catalyzation, 269
intermetallic catalytic agents, 271–274, 276t, 278f, 279t
metal and metal alloys, 266–268
with metal/metal oxide nanocomposite powders, 278, 280f
with pure metals, 269f
with titanium carbide nanopowders, 280, 281f
Droplet method (gas/water atomization), 8
Drum ball mills, 31, 55
Dry coating approach, 309, 310f
Dry pressing, 106
DSC. *See* Differential scanning calorimetry (DSC)
DTA measurements, 352
Al_50Hf_{50} powders, 350f
$Al_{50}Ta_{50}$ alloy powders, 382f, 383f
$Al_{30}Ta_{70}$ powders, 335f, 338
amorphization reactions and, 350f
Cu_2O reduction with Ti, 134, 136f
$Fe_{74}Cr_{18}Ni_8$, 374f
NbN alloy powders, 150, 151f
Duwez, Pol, 8

E

EBPVD. *See* Electron beam physical vapor deposition (EBPVD)
ECAP. *See* Equal channel angular press(ing) (ECAP)
Elastic modulus
SiC_{10}/Al_{90} powder, 204, 205f
$Ti_{55}C_{45}$ samples, 164, 166f
TiN powders, 188, 189f
Electric discharge-assisted mechanical milling, 4
Electrodeposition, 99
Electron beam physical vapor deposition (EBPVD), 4, 10, 309, 310f
Electron probe micro-analyzer (EPMA), 96t
Electrospinning, 93
Energy dispersive X-ray fluorescence (XRF), 15, 16f
Energy-dispersive X-ray spectroscopy (EDS)
MgH_2 powders, 246, 247f
TiC powders, 161f
TiN powders, 184f, 185
ENG. *See* Expanded natural graphite (ENG)
Engineering materials, historical evolution, 418f
Entropy, configurational, 419. *See also* High-entropy alloys (HEAs)
Environment, milling, 84f–85f, 83
EPMA. *See* Electron probe micro-analyzer (EPMA)
Equal channel angular press(ing) (ECAP), 109, 286
Ethanol, 80, 83
Ethyl alcohol, 80
Evaporation, 93
Evico-magnetics, 42, 156
Expanded natural graphite (ENG), 294
Extrusion process, plastic deformation, 7f

F

FAST. *See* Field-assisted sintering technique (FAST)
Fast Fourier transformations (FFT)
HRTEM image, VNbMoTaW HEA, 422, 423f
TiN powders, 184f
fcc-NbN alloy powders, 150, 151f
Fe-base high-temperature alloys, 22f
Fe-7C-6Mn, milling speed, 75, 77f
FeCoCrMoCB, XRD patterns, 67, 323f
$Fe_{20}Cr_{20}Mn_{20}Ni_{20}Co_{20}$ alloy system, 417
$Fe_{74}Cr_{18}Ni_8$. *See* Amorphous austenitic stainless steel ($Fe_{74}Cr_{18}Ni_8$)
Fe-Cr-Ni alloy, 417

Feedstock-HVOF materials, ball-milled powders as, 312
FE-HRTEM. *See* Field emission high-resolution transmission electron microscope (FE-HRTEM)
FE-HRTEM/EDS. *See* Field emission high-resolution transmission electron microscope/energy dispersive X-ray spectroscopy (FE-HRTEM/EDS)
200 kV FE-HRTEM/EDS/STEM (JEOL-Japan) equipment, 17f
$Fe_{52}Nb_{48}$ amorphous alloy, 375f
Fe-Ni alloy, 28
FE-SEM/EDS. *See* Field emission scanning electron microscope/EDS (FE-SEM/EDS)
FE-SEM micrographs
  TiN powders, 185f, 187f, 188
  WC-5Co-1Cr-3MgO-0.7VC-0.3Cr_3C_2 nanocomposite, 221f, 225f
  WC powders, 173f, 174, 176f
FE-TEM. *See* Field emission transmission electron microscope (FE-TEM)
$Fe_{50}W_{50}$ binary system
  amorphous FeW formation stage, 372
  composite FeW powder particles formation stage, 371
  FeW solid solution formation stage, 372
  solid-state amorphization by MA when ΔH^{for} equal to zero, 366
    magnetic studies, 368, 369f
    mechanism, 371
    structural changes with milling time, 366–368
    TEM/EDS technique, 367
    thermal stability, 369, 370f, 371f
    XRD patterns, 366, 367f
Fe-Zr, 361
Fe-Zr-B system, 376f, 377f
FFT. *See* Fast Fourier transformations (FFT)
Field-assisted sintering technique (FAST), 110, 113
  TiN powders, 185
Field emission high-resolution transmission electron microscope (FE-HRTEM)
  MgH_2/10 wt.% (8 Nb_2O_5/2 wt.% Ni) system, 295f, 297f
  WC-5Co-1Cr-3MgO-0.7VC-0.3Cr_3C_2 nanocomposite, micrographs, 221f
Field emission high-resolution transmission electron microscope/energy dispersive X-ray spectroscopy (FE-HRTEM/EDS), 15, 17f
  crystal structure and nanoelemental composition of nanocrystalline materials, 95t
Field emission scanning electron microscope/EDS (FE-SEM/EDS), 15, 16f, 96t
Field emission transmission electron microscope (FE-TEM)
  MgH_2/10 wt.% (8 Nb_2O_5/2 wt.% Ni) system, 295f, 296
Filamentary reinforcements, 15
Fourier-transform infrared spectroscopy (FT-IR), 13, 15f, 96t
Fracture toughness, WC powders, 178, 179f
Freezer/Mill, 36, 37f
Fritsch GmbH, 38, 39f, 42
Fritsch P6 high-energy, 138
Fritsch Pulverisette P5, 38, 39f
Fritsch Pulverisette P6, 38, 39f
FT-IR. *See* Fourier-transform infrared spectroscopy (FT-IR)
FTIR-8400S, 15f
Fuel cell applications, MgH_2-based nanocomposite for
  developments in, 294
  hydrogen reactors, 294
    background, 294
    MgH_2/10 wt.% (8 Nb_2O_5/2 wt.% Ni) system (. *See* MgH_2/10 wt.% (8 Nb_2O_5/2 wt.% Ni) system)
  overview, 293
  as solid-state hydrogen source, 293
Fullerenes, 156
Full width at half maximum (FWHN) values, 158

G

Gangue minerals, 5
Gas atomization, 8
Gas contamination, milling tools, 60, 61f
Gaseous storage method, hydrogen energy, 234
Gas phase condensation thermolysis, 93
Gas-temperature monitoring (GTM) system, 42, 156, 239

Glass(es). *See also* Amorphous and metallic glassy alloys
history, 335
metallic, 335
oxide glasses, 335
Glass-forming range, amorphization, 362f–364f, 361, 365t
Gold wire and plates, for teeth replacement in Ancient Egyptians, 431, 432f
Graphene, 65
Graphite, 65
expanded natural graphite (ENG), 294
Graphite aluminum alloys, 19
Graphite powders
FE-SEM micrograph, 73f, 117
XRD patterns, 72f, 116
Gravimetric hydrogen storage, 232, 233f

H

Hall-Petch equation, 98
Hall-Petch relationship, 163, 179
Hardness, milling media, 63t
HEAs. *See* High-entropy alloys (HEAs)
Helium carrier fusion-thermal conductivity method, SiC_p/Al MMNCs, fabrication, 197
HfN system, milling speed, 74, 76f, 77f
Hiden isotherma, MgH_2 powders, 251
High-angle annular dark field (HAADF)
MgH_2 powders, 244, 245f
TiC powders, 161f
High-energy-assisted approach, 2
High-energy ball milling (HEBM), 103, 286
High-energy ball mills, 32, 55
attritor/attrition ball mill, 32, 33f
Freezer/Mill, 36, 37f
Planetary Ball Mills, 31, 37–39
Retsch Mixer Mills (MM 200 and MM 400), 35f
shaker mills, 33–35
Super Misuni, 36f
uni-ball mill, 42f, 43f
High-energy sonication, 93
High-entropy alloys (HEAs). *See also* Traditional alloys
basic science behind, 417, 419f, 420f
birth of, 417, 418f
concept of, 417
defined, 417
nanocrystalline HEAs, preparations by MA, 421
bulk nanocrystalline VNbMoTaW HEA, 422f–424f, 422
CoCrFeMnNi HEA, high-pressure torsion-driven MA of, 424, 425f
Co_xCrCuFeMnNi HEA powders, 426f
high-entropy multicomponent WMoNbZrV alloy, 424, 425f
preparations, 420, 421f
properties, 420
vs. conventional alloys, 419f
High-feed-milled HVOF sprayed WC-Co coatings, 315
High frequency induction heated sintering, 109
High pressure casting, 8
High-pressure/high-temperature (HPHT) technique, 422
High-pressure torsion (HPT), 286
driven MA, of CoCrFeMnNi HEA, 424, 425f
High-resolution transmission electron microscopy (HRTEM), 78, 79f, 101, 118f, 138
MgH_2 powders, 244–246
TiC powders, 156, 161f
$Ti_{40.6}Cu_{15.4}Ni_{8.5}Al_{5.5}W_{30}$ metallic glassy alloy powders, 406, 408f
TiN powders, 182, 183f, 184f
WC/Al_2O_3 nanocomposites, 208, 211f
WC-5Co-1Cr-3MgO-0.7VC-0.3Cr_3C_2 nanocomposite, 221f
WC powders, 170, 172f, 173, 174f
$Zr_{52}Al_6Ni_8Cu_{14}W_{20}$ glassy alloy powders, 397–399
"High-tech" materials. *See* Advanced materials
High velocity oxygen thermal spraying (HVOF), 311, 313f
advantage of, 312
ball-milled powders as feedstock materials for, 312
gun for, 313f
high-feed-milled HVOF sprayed WC-Co coatings, 315
nanocrystalline iron aluminide, 315
nanostructured Cr_3C_2-Ni20Cr coatings, 314f

High velocity oxygen thermal spraying (HVOF) (*cont.*)
  reactive spraying of mechanically alloyed Ni-Ti-C powders, 314
  sprayed diamond reinforced bronze coatings, 315
  uses, 312, 313f
HIP. *See* Hot isostatic pressing (HIP)
HIPing technique, 108, 373
Horizontal rotary ball mills, 31
Hot extrusion, 108
Hot isostatic pressing (HIP), 110
  metal matrix/CNTs, fabrication, 225
  WC powders, 169, 170
Hot press (HP) technique, 108, 373
  TiC powders, 156, 162, 163f, 164, 166f
  WC powders, 169, 170
HPT. *See* High-pressure torsion (HPT)
HRTEM. *See* High-resolution transmission electron microscopy (HRTEM)
HVOF. *See* High velocity oxygen thermal spraying (HVOF)
Hydrogenation/dehydrogenation kinetics, MgH_2 powders, 251, 264
  absorption and consequent desorption curves, 258, 259f
  doping effect
    of intermetallic compounds and alloys, 268t
    metal oxides, 275, 276t
  experimental procedure, 255–259
  hiden isotherma, 251
  PCT, 251, 254, 257f
  pure metal additives effect on, 267t
Hydrogenation reaction, defined, 236
Hydrogen energy, 231, 293
  economy, 232
  properties, 231, 232t
  storage, 232
    comparison of various fuels, 233f
    gaseous storage method, 234
    liquid storage method, 234
    volumetric and gravimetric hydrogen storage, 232, 233f
Hydrogen fuel cells, 294. *See also* Fuel cell applications
Hydrogen gas
  conversion into liquid hydrogen, 293
  storage considerations, 293. *See also* MgH_2-based nanocomposite; fuel cell applications
Hydrogen micro-fuel cells, 294
Hydrogen reactors, 294
  background, 294
  MgH_2/10 wt.% (8 Nb_2O_5/2 wt.% Ni) system, 295f–302f, 295
    back electron scattering (BES) FE-SEM micrograph, 295f, 296
    β-to-γ phase transformation, 295
    continuous cyclic hydrogenation/dehydrogenation processes, 301, 302f
    dehydrogenation, kinetics of, 300f, 301
    drawbacks, 304
    DSC experiments, 296, 297f
    FE-HRTEM image, 295f, 297f
    FE-TEM image, 295f, 296
    hydrogenation, kinetics of, 299–300, 300f
    implementation of, 302, 303f, 305f–307f
    pressure-composition-temperature (PCT) curves, 298f
    SADP, 295f
    solid-hydrogen Ti-tank integrated with, 304, 305f
    van't Hoff approach, 298, 299
    van't Hoff plot, 298, 299f
    XRD pattern, 295f, 296, 302f
Hydrogen storage nanomaterials, 229
  energy sector, 231
    comparison of various fuels, 233f
    economy, 232
    gaseous storage method, 234
    liquid storage method, 234
    storage, 232
    volumetric and gravimetric hydrogen storage, 232, 233f
  overview, 229
  solid-state, 235
    hosting hydrogen, 235
    magnesium hydride, 237
    metal hydrides, 235, 237t
  world fuel consumption, 230f
  world production of crude oil, 230f
Hydrometallurgy, 1

I

ICP-MS. *See* Inductively coupled plasma mass spectrometry (ICP-MS)
Ignition surface coating, 19
Impellers, 32
Inactive zone, ball mill, 66f
INCO. *See* The International Nickel Company (INCO)
INCONEL MA 754, 22, 23f
 stress rupture properties, 23, 24f
INCONEL MA 956, 22f, 24–27
INCONEL MA 6000, 22, 23, 25f
Indium (In), 265
Induction coupled plasma emission, SiC_p/Al MMNCs, fabrication, 197
Inductively coupled plasma mass spectrometry (ICP-MS), 13
Inert gas condensation, 99
Infrared transmission, $Si_{50}C_{50}$ powders, 167f
In-situ catalyzation, 269
Institute of Theoretical and Applied Mechanics, of Russian Academy of Sciences in Novosibirsk, 315
Intelligent materials, 1. *See also* Advanced materials
Intermetallic catalytic agents, doping of MgH_2, 271–274, 276t, 278f, 279t
Internal oxidation, 19
The International Nickel Company (INCO), 19
Iron aluminide (nanocrystalline), HVOF of, 315
Iron-niobium alloys, 375

J

Jet ball mills, 31, 55

K

Kuwait Institute for scientific Research (KISR), Kuwait, 269, 309
 Nanotechnology Laboratory, Energy and Building Research Center (EBRC), 309, 310f

L

LA. *See* Laser ablation (LA)
Lamellar structured degradable magnesium-hydroxyapatite implants, 437, 438f
$La_{1.5}Mg_{0.5}Ni_{6.5}Co_{0.5}$ alloy, 265
Laminate nanocomposites, 195
$LaNi_5$ alloy, 294
Laser ablation (LA), 4, 93, 114
Laser diffraction particle size analyzer, 97t
Laser surface alloying, 433
Liquid hydrogen, 293
 hydrogen gas conversion into, 293
Liquid metallurgy, 19, 28
Liquid-phase fabrication approach, 2
Liquid storage method, hydrogen energy, 234
Lithographic approach, 2
Longitudinal optical (LO)-like phonon mode, 167
Low-energy ball mill(s), 44, 55
 tumbler ball mill, 44, 45f
 tumbler rod mill, 45, 46f

M

MA. *See* Mechanical alloying (MA)
MacQueen, Ewan C, 19
Magnesium hydride, solid-state hydrogen storage material, 237
 RBM-MgH_2 powders, characterization, 240
  absorption and consequent desorption curves, 258, 259f
  apparent activation energy, 250, 251f
  endothermic peaks, 250
  experimental procedure, 255–259
  hiden isotherma, 251
  hydrogenation/dehydrogenation behavior, 251
  morphological changes of Mg powders, 246
  PCT, 251, 254, 257f
  structural change of Mg powders, 241
  thermal stability, 248
 synthesizing, 237
  by RBM technique, 238, 239f
Magnesium-hydroxyapatite implants, 437, 438f
Magnetic pulse compaction (MPC), 109
Magnetic studies, $Fe_{50}W_{50}$ binary system, 368, 369f
MA/MD/MM techniques, factors affecting, 56f, 57
 ball mills selection and, 45, 57t
 ball-to-powder weight ratio, 86f–88f, 83

MA/MD/MM techniques, factors affecting (*cont.*)
gas contamination, 60, 61f
impurities and milling tools, 50
milling atmosphere, 80
milling environment, 84f–85f, 83
milling media, 62, 63t
milling speed, 70f–77f, 67
milling temperature, 88–90
milling time, 78f–82f, 75
milling vials, shape of, 57, 58f
organic contaminations, 59f
PAC, contamination by, 60f
solid contamination, 61
Materials. *See also* Advanced materials
from Stone Age to Nanotechnology Age, 2f
traditional categories, 1
Matrix materials
CMC, 194
defined, 193
fabrication, by mechanical alloying, 225
MMC, 194
PMC, 194
types, 194
MBE. *See* Molecular beam epitaxy (MBE)
MD. *See* Mechanical disordering (MD)
MDSSA. See Mechanically driven solid-state amorphization *(MDSSA)*
$\Delta H_a MDSSA$, 353
Mechanical alloying (MA), 4–6, 265, 377. *See also* MA/MD/MM techniques
advantages, 335
amorphization by, mechanism, 342
morphology and metallography changes with milling time, 347f–349f, 347
structural changes with milling time, 343
TEM observations, 344f–346f, 344
thermal stability, 349
X-ray analysis, 342f, 343f
amorphous alloys, fabrication, 336, 337f, 338f
ball mills, types, 31
attritor/attrition ball mill, 32, 33f
high-energy ball mills, 32
low-energy tumbling mill, 44
Planetary Ball Mills, 37–39
Retsch Mixer CryoMill, 36, 37f
Retsch Mixer Mills (MM 200 and MM 400), 35f
shaker mills, 33–35
SPEX Freezer/Mill-6970EFM, 36, 37f
Super Misuni, 36f
tumbler ball mill, 44, 45f
tumbler rod mill, 45, 46f
uni-ball mill, 42f, 43f
cold spraying of metastable powders, 321f–331f, 320
cyclic crystalline-amorphous transformations. *See* Cyclic crystalline-amorphous transformations, during MA
free-energy (ΔG) diagram for phases in, 30, 31f
history, 19, 336
ball-milling, 19, 20
ignition surface coating, 19
internal oxidation, 19
process, defined, 30
selective reduction, 19
simple mixing, 19
time-line presentation of process development during last 6 decades, 27, 29f
mechanism, 47f
ball-powder-ball collision, 48f
as MISSR approach, 5
MSSR via, 137
nanocrystalline HEAs by, 421
bulk nanocrystalline VNbMoTaW HEA, 422f–424f, 422
CoCrFeMnNi HEA, high-pressure torsion-driven MA of, 424, 425f
$Co_xCrCuFeMnNi$ HEA powders, 426f
high-entropy multicomponent WMoNbZrV alloy, 424, 425f
nanocrystalline materials, 99
necessity of, 48, 49f
ODS alloys, fabrications, 21
Fe-base high-temperature alloys, 22f
ODS Ni-base superalloys, 22f
process control agent, decomposition, 26
processes in, 31
thermodynamical processes, 377, 378f
vs. MD, in $Al_{50}Ta_{50}$ amorphization reaction in rod mill, 377, 384
background, 377, 378f

morphological changes with milling time, 380f, 381f
procedure, 378
structural changes with milling time, 379f
thermal stability, 382f
Mechanical disordering (MD), 4, 30, 31f, 377. *See also* MA/MD/MM techniques
vs. MA, in $Al_{50}Ta_{50}$ amorphization reaction in rod mill, 377, 384
background, 377, 378f
morphological changes with milling time, 380f, 381f
procedure, 378
structural changes with milling time, 379f
thermal stability, 382f
Mechanical grinding (MG), 30, 377
thermodynamical processes, 377, 378t
Mechanically alloyed powders
biomedical applications, 431. *See also* Biomaterials
metallic biomaterials, 432
Mg-based alloys, 436
overview, 431
Ti-based alloys, 434
characterization techniques, 13
energy dispersive X-ray fluorescence (XRF), 15, 16f
field emission high-resolution transmission electron microscope/ energy dispersive X-ray spectroscopy (FE-HRTEM/EDS), 15, 17f
field emission scanning electron microscope/EDS (FE-SEM/EDS), 15, 16f
Fourier-transform infrared spectroscopy (FT-IR), 13, 15f
inductively coupled plasma mass spectrometry (ICP-MS), 13
photon probe methods, 13, 14f, 15f, 16f, 17f
scanning probe methods, 17, 18f
small angle X-ray scattering (SAXS), 13
thermodynamic methods, 17, 18f
X-ray diffraction (XRD), 13, 14f
X-ray photoelectron spectroscopy (XPS), 13
extrinsic and intrinsic properties, 13, 14f
overview, 13
Mechanically assisted approach, advanced materials, 2, 4. *See also specific techniques*
ball milling, 4
classification, 2
mechanical alloying (MA), 5, 6f
powder metallurgy (P/M), 4, 5f
severe plastic deformation, 7f
Mechanically driven solid-state amorphization *(MDSSA)*, 371
amorphization via, 360
thermal stability, amorphization by MA, 360
Mechanically induced self-propagating reactions (MSR), 157, 158
Mechanically induced solid-state amorphization. *See* Solid-state amorphization, mechanically induced
Mechanically induced solid-state mixing (MISSM), 4
Mechanically induced solid-state reaction (MISSR) approach, 2, 4, 55, 78, 88
MA as, 5
Mechanical milling (MM), 30, 377. *See also* MA/MD/MM techniques
nanocrystalline WC powders, 169
WC-5Co-1Cr-3MgO-0.7VC-0.3Cr_3C_2 nanocomposite, 222
WC powders, 169, 170, 172f–174f, 178, 179t
Mechanical properties, nanocrystalline materials, 98t
strengthening by grain size reduction, 94, 99f
Mechanical solid-state mixing, SiC_p/Al MMNCs fabrication, 197
properties, 198
Mechanical solid-state reduction (MSSR), 129
Cu_2O reduction with Ti, 130, 131f
mechanism, 136
by rod-milling technique, 137
via MA, 137
WC/Al_2O_3 nanocomposites, 207, 208f, 210
WC and WC-MgO, fabrication, 138
ball-milled powders, properties, 139

Mechanical solid-state reduction (MSSR) (*cont.*)
 hardness, toughness, and elastic moduli, 146f, 147t
 structural changes with milling time, 140f–143f, 139
 temperature change with milling time, 144f, 145f
Mechanochemistry approaches, 4
Media, milling. *See* Milling media
Melt spinning technique, 8, 9f, 80
Metal alloy(s). *See also specific names*
 doping of MgH_2, 266–268
 history, 335
Metal hydrides, 235
 high-temperature, 236
 hydrogen capacity per weight and cost, 237t
 low-temperature, 236
Metallic biomaterials, 432. *See also specific entries*
Metallic glass, 335. *See also* Glass(es)
 properties, 335
Metallic glassy alloy powders, multicomponent
 consolidation into full-dense bulk materials, 396
  $Ti_{40.6}Cu_{15.4}Ni_{8.5}Al_{5.5}W_{30}$ metallic glassy alloy powders, by SPS, 407f–411f, 406
  $Zr_{52}Al_6Ni_8Cu_{14}W_{20}$ glassy alloy powders, 396
Metallographical examinations, amorphization by MA, 347f, 348f
Metal matrix composite (MMC), 194
Metal-matrix nanocomposites (MMNCs), 195
 particulate, fabrications methods, 195
  SiC/Al MMNCs, 196
Metal organic framework (MOFs) systems, 235
Metal oxide(s)
 catalytic effect, 275
 catalyzation with, 278, 280f
 hydrogenation/ dehydrogenation kinetics MgH_2 powders, 275, 276t
Metal(s), 432. *See also specific names*
 catalyzation with, 278, 280f
 doping of MgH_2, 266–268
Metastable phase diagram, crystal-to-glass transition, 339, 340f, 341f
Methanol, 83
MG. *See* Mechanical grinding (MG)
Mg-Al alloy, 417
Mg-based alloys, 433, 436. *See also specific entries*
 high-performance MgFe biodegradable alloy, 436
 lamellar structured degradable magnesium-hydroxyapatite implants, 437, 438f
 Mg-Zn/HA composite, biodegradable, 437
 nanocrystalline AZ31 magnesium alloy with titanium additive, 437
Mg_2Cu eutectic alloy, 266
$Mg_{30}Fe_{70}$ alloy, 436
MgFe biodegradable alloy, 436
MgH_2, 263
 catalyzation, with metastable phases of Zr-based nanopowders, 281–284
 characterization techniques, 263
 cold rolling and ball milling, combination of, 286f
 doping, with catalysts, 266
  in-situ catalyzation, 269
  metal and metal alloys, 266–268
  with metal/metal oxide nanocomposite powders, 278, 280f
  new intermetallic catalytic agents, 271–274, 276t, 278f, 279t
  with pure metals, 269f
  with titanium carbide nanopowders, 280, 281f
 hydrogenation/dehydrogenation kinetics, 264
  doping effect of intermetallic compounds and alloys, 268t
  pure metal additives effect on, 267t
 improvement scenarios, 264
 limitations, 240
 lowering stability of, 265
 ß-MgH_2, micrograph, 94f
 properties, 240
MgH_2-based nanocomposite, fuel cell applications
 developments in, 294
 hydrogen reactors, 294
  background, 294
  MgH_2/10 wt.% (8 Nb_2O_5/2 wt.% Ni) system. *See* MgH_2/10 wt.% (8 Nb_2O_5/2 wt.% Ni) system

overview, 293
as solid-state hydrogen source, 293
MgH_2/7 $Mn_{3.6}Ti_{2.4}$ nanocomposite, 271, 272f
MgH2/5Ni/5Nb2O5 powders, 263f
MgH_2/10 wt.% (8 Nb_2O_5/2 wt.% Ni) system, 295f–302f, 295
back electron scattering (BES) FE-SEM micrograph, 295f, 296
β-to-γ phase transformation, 295
continuous cyclic hydrogenation/ dehydrogenation processes, 301, 302f
dehydrogenation, kinetics of, 300f, 301
drawbacks, 304
DSC experiments, 296, 297f
FE-HRTEM image, 295f, 297f
FE-TEM image, 295f, 296
hydrogenation, kinetics of, 299–300, 300f
implementation of, 302, 303f, 305f–307f
pressure-composition-temperature (PCT) curves, 298f
SADP, 295f
solid-hydrogen Ti-tank integrated with, 304, 305f
van't Hoff approach, 298, 299
van't Hoff plot, 298, 299f
XRD pattern, 295f, 296, 302f
MgH_2/10 wt.% $ZrNi_5$ powders, 273, 274f
$Mg_{85}In_5Al_5Ti_5$ alloy, 265
Mg-In-H system, 265
$Mg_{0.95}In_{0.05}$ solid solution, 265
Mg-Nd system, 264
Mg-5Ni-3La (at.%) alloys, 265
$Mg_{89}Pd_7Nd_4$ alloy, 264
Mg-Pd system, 264
$Mg_{0.95}Sn_{0.05}$ solid solution, 265
Mg-Zn/HA composite, biodegradable, 437
Microhardness, $Ti_{55}C_{45}$ samples, 163–165
Microwave-derived coatings, 433
Microwave sintering, 109
Milling, defined, 5. *See also* Ball-milling (BM)
Milling atmosphere, 80
Milling environment, 84f–85f, 83
Milling media
balls sizes, 63, 64, 65f
brittleness, 63t
Cu-15 vol.% Nb powders, media dynamics study, 64f, 65f
Cylpebs (Doering International), 66, 67f, 69f
hardness, 63t
influences of, 62
Millpebs (Wheelabrator Allevard Enterprise), 66, 67f
Powerpebs (Donhad), 66, 67f
requirements of, 62
shape of, 66f–69f, 70
specific gravity, 63t
trajectory of, in tumbling mill, 70, 71f
Milling speed, 67
balls and particle motions, tumbling mill, 70f
balls motion during milling at different, 73f
defined, 67
Fe-7 mass% C-6 mass% Mn, 75, 77f
HfN system, 74, 76f, 77f
selection, 67
theoretical investigations, 71, 72f
trajectory of milling media, 70, 71f
XRD patterns ($Al_{50}Zr_{50}$ powder), 73–75
Milling temperature, 88–90
Milling time, 75
$Al_{50}Ta_{50}$ amorphization
morphological changes, 380f, 381f
structural changes with, 379f
amorphization by MA
morphology and metallography changes, 347f–349f, 347
SEM observations, 347f, 348f
structural changes, 343
TEM observations, 344f–346f, 344
X-ray analysis, 342f, 343f
characterizations techniques, 78
Co-Ti binary system, structural changes, 386f
crystallite size and lattice strain, 78, 79f
Cu_2O reduction with Ti, structural changes, 130–132
DSC thermograms, 80f
$Fe_{50}W_{50}$ binary system, structural changes, 366–368
HRTEM image, 78, 79f
lattice constant, 78f
SEM images, 80–82
WC and WC-MgO fabrication by MSSR method
structural changes, 140f–143f, 139
temperature change, 144f, 145f

Milling tools, impurities and, 23
gas contamination, 60, 61f
organic contaminations, 59f
PAC, contamination by, 60f
powder contaminations, 61
solid contamination, 61
Milling vials, shape, effects, 58f, 59
Millpebs (Wheelabrator Allevard Enterprise), 66, 67f
MISSM. *See* Mechanically induced solid-state mixing (MISSM)
MISSR. *See* Mechanically induced solid-state reaction (MISSR) approach
MM. *See* Mechanical milling (MM)
$Mn_{3.6}Ti_{2.4}$ alloys, 271, 272
Molecular beam epitaxy (MBE), 93
MPC. *See* Magnetic pulse compaction (MPC)
Multicomponent metallic glassy alloy powders
consolidation into full-dense bulk materials, 396
$Ti_{40.6}Cu_{15.4}Ni_{8.5}Al_{5.5}W_{30}$ metallic glassy alloy powders, by SPS, 407f–411f, 406
$Zr_{52}Al_6Ni_8Cu_{14}W_{20}$ glassy alloy powders, 396
Multiwall carbon nanotubes (MWCNTs), 102, 103, 226, 235. *See also* Single-wall carbon nanotubes (SWCNTs)
low magnification BFI of, 104f
XRD patterns, 104f
MWCNTs. *See* Multiwall carbon nanotubes (MWCNTs)

N

Nano-beam diffraction (NBD), MgH_2 powders, 244, 245f
Nanocomposite(s)/nanocomposite materials, 1, 3f, 5, 19, 193. *See also specific entries*
homogeneous, 28
metal matrix/carbon nanotubes, 225
MMNCs, 195
overview, 193, 194f
SiC/Al MMNCs, fabrication, 196
SiC_p/Al powder particles
consolidation, 206
formation, 206
WC-based nanocomposites, 207
WC/Al_2O_3, 207
WC-5Co-1Cr-3MgO-0.7VC-0.3Cr_3C_2 nanocomposite, 214, 219
Nanocrystalline aluminum alloy, 7
Nanocrystalline fcc metals, formation of, 101, 103f
Nanocrystalline materials, 5, 93
ball-milling technique for, 28, 99
effects on carbon nanotubes structure, 102, 104f, 105f
first stage mechanism, 100
mechanical alloying, 99
mechanism(s), 100
methods, 99, 100f
second stage mechanism, 101
third stage mechanism, 101
characterization techniques, 94
crystal structure and, 94, 95t
mechanical properties, 94, 98t
morphology, topological, and surface properties, 94, 96t, 97t
thermal stability and phase transformations and, 94, 96t
defined, 94
examples, 101
nanocrystalline fcc metals, 101, 103f
Ni_xMo_{100-x} (x = 60 and 85 at.%) powders, 101, 102f
mechanical properties, 98t
strengthening by grain size reduction, 94, 99f
nanocrystalline powders, consolidation of, 108
of ball-milled powders, 109
spark plasma sintering (SPS), 110
nanodiamonds and CNTs fabrication by milling, 114
discussion, 123, 124f
materials and equipment, 115
method, 115
nanodiamonds syntheses, 115
results, 116f–123f, 116
pressing and sintering of powders materials, 105
classic powder metallurgy, 106f, 107f
Nanocrystalline nickel aluminide (Ni_3Al), 7
Nanocrystalline nickel metal, 7
Nanocrystalline powders, 93

average particle size, surface area, and pore size distributions, 97t
consolidation of, 108
ball-milled powders, 109
cold densification, 109
powder densification and sintering, 109
pressure-assisted sintering, 109, 110
problems-related to, 108
spark plasma sintering, 110
crystal structure and nanoelemental composition, 95t
morphological, topological, and surface characterizations, 96t
thermal stability and phase transformation temperatures, 96t
Nanocrystalline refractory materials, 153
overview, 153
preparation challenges and difficulties, 154
powder metallurgy, technique, 154, 155t
SHS, 154
β-SiC powders, 166
TiC powders, 155
ball milling. *See* Ball-milling (BM)
consolidation techniques, 162
mechanical properties, 163
TiN powders, 180
consolidation, 182
preparation, 181f
results, 182
WC powders, 169
ZrC powders, 180
Nanocrystals, 93
Nanodiamonds (NDs)
cold spraying, 331
fabrication by milling, 114
background, 114
discussion, 123, 124f
materials and equipment, 115
method, 115
results, 116f–123f, 116
syntheses, 115
Nanohardness *(NH)*, WC powders, 175, 177f–179f
Nanoimprint lithography (NIL), 4
Nanoindentation, 98t
Nanolithography, 93
Nanomaterial(s). *See also* Nanocomposite(s)/ nanocomposite materials; Nanocrystalline materials; Nanocrystalline refractory materials
one-dimensional (1D), 93
preparation methods, 93
bottom-up approach, 93
top-down approach, 93
three-dimensional (3D), 93
two-dimensional (2D), 93
types of, 1, 3f, 93
Nanoparticle(s), 1, 5, 19, 93
Nanopowder(s), 93. *See also* Nanocrystalline powders; *specific entries*
Nanorods, 93
Nanostructured materials, 1, 19, 94
Nanotechnology Age. *See also* Advanced materials
materials form Stone Age to, 2f
Nanotubes, 93
Nanowires, 93
National Defense Academy of Japan, 32
NbN alloy powders, 150, 151f
DTA measurements, 150, 151f
XRD patterns, 150, 151f
Nb_2O_5/Ni system, 304. *See also* MgH_2/10 wt.% (8 Nb_2O_5/2 wt.% Ni) system
$Nb_{50}Zr_{10}Al_{10}Ni_{10}Cu_{20}$, 421
Negative milling, 22
$Ni_{33}Al_{67}$ powders, 23f
Ni-based alloy powders, 333
Ni-based superalloys, 19, 21, 22f
Nickel, 21
Nickel-chromium alloy, 19
Nickel titanium alloy powders, 322
NIL. *See* Nanoimprint lithography (NIL)
Ni-Mo alloy, 28
Ni-Nb, 361
$Ni_{60}Nb_{40}$ amorphous alloy, 24, 27
DSC plots, 24f, 337
XRD patterns, 330f, 336
$Ni_{60}Nb_{40}$ powders, 80, 81
Niobium, 375
Ni-Ti, 361
Ni-Ti-C powders, HVOF reactive spraying, 314
Ni_xMo_{100-x} (x = 60 and 85 at.%) powders (nanocrystalline), formation of, 101, 102f
Ni-Zr, 361

O

Obsidian, 335
ODS alloys, fabrications, 21, 21. *See also* Mechanical alloying (MA)
  examples, 21t
  Fe-base high-temperature alloys, 22f
  INCONEL MA 754, 23f, 24f
  INCONEL MA 956, 24–27
  INCONEL MA 6000, 23, 25f
  ODS Ni-base superalloys, 22f
One-dimensional (1D) nanomaterials, 93
ON-OFF DC pulse energizing method, 113, 114f
Organic coating, 433
Organic contaminations, milling tools, 59f
Oxide glasses, 335. *See also* Glass(es)

P

PAC, contamination, milling tools, 60f
Particle size distribution, $Al_{50}Hf_{50}$ alloy, 348f
Particulate composites, 194
PCA. *See* Processing control agent (PCA)
PECVD. *See* Plasma-enhanced chemical vapor deposition (PECVD)
PEM fuel cell system. *See* Proton exchange membrane (PEM)-fuel cell system
Phase transformations, nanocrystalline materials, 94, 96t
Photolithography, 4
Photon probe methods, 13, 14f, 15f, 16f, 17f
  energy dispersive X-ray fluorescence (XRF), 15, 16f
  field emission high-resolution transmission electron microscope/energy dispersive X-ray spectroscopy (FE-HRTEM/EDS), 15, 17f
  field emission scanning electron microscope/EDS (FE-SEM/EDS), 15, 16f
  Fourier-transform infrared spectroscopy (FT-IR), 13, 15f
  inductively coupled plasma mass spectrometry (ICP-MS), 13
  X-ray diffraction (XRD), 13, 14f
  X-ray photoelectron spectroscopy (XPS), 13
Physical vapor deposition (PVD), 4, 10, 309, 433
  TiN powders, 180
Pin-on-disk, 98t
Planetary Ball Mill PM 100, 40, 41f
Planetary Ball Mill PM 200, 40, 41f
Planetary Ball Mill PM 400, 40, 41f
Planetary Ball Mills, 31, 37–39, 55, 57t
Planetary Micro Mill PULVERISETTE 7, 38, 40f
Plasma-activated sintering (PAS) method, 138, 139
  SiC_p/Al MMNCs, fabrication, 197, 201, 206
Plasma-enhanced chemical vapor deposition (PECVD), 4, 114
Plastic deformation
  extrusion, 7
  severe, 7f
P/M. *See* Powder metallurgy (P/M)
Polymer electrolyte membrane (PEM), hydrogen-PEM fuel cells, 232
Polymer matrix composites (PMC), 194
Polymers, 432
Powder densification and sintering, 109
Powder metallurgy (P/M), 1, 4, 5f
  classic, 106f, 107f
Powder-milling process, MA/MD/MM, factors, 56f, 57
  ball mills selection and, 56, 57t
  ball-to-powder weight ratio, 86f–88f, 83
  gas contamination, 60, 61f
  impurities and milling tools, 75
  milling atmosphere, 80
  milling environment, 84f–85f, 83
  milling media, 62, 63t
  milling speed, 70f–77f, 67
  milling temperature, 88–90
  milling time, 78f–82f, 75
  milling vials, shape of, 58f, 59
  organic contaminations, 59f
  PAC, contamination by, 60f
  solid contamination, 61
Powders materials. *See also* Nanocrystalline powders
  consolidation, 105
  pressing and sintering of, 105
    classic powder metallurgy, 106f, 107f
  production, 105
Powerpebs (Donhad), 66, 67f
PPS. *See* Pulse plasma sintering (PPS)

Pressing and sintering, powders materials, 105. *See also* Nanocrystalline powders
classic powder metallurgy, 106f, 107f
Pressure-assisted sintering, 109, 110
hot isostatic pressing (HIP), 110
pulse plasma sintering (PPS), 110
spark plasma sintering (SPS), 110
Pressure-composition isotherm (PCI), MgH_2 powders, 251
Pressure-composition-temperature (PCT)
MgH_2 powders, 251, 254, 257f
Pressure-composition-temperature (PCT) curves, 264
MgH_2/10 wt.% (8 Nb_2O_5/2 wt.% Ni) system, 298f
Processing control agent (PCA), 47
advantages, 83
cold-welding effect, controlling, 83
in ductile materials milling, 59
examples, 60
MA, 26
Proton exchange membrane (PEM)-fuel cell system, 293, 294f, 304–306
Pt-Co alloy, 28
Pulse plasma sintering (PPS), 110
PVD. *See* Physical vapor deposition (PVD)
Pyrometallurgy, 1

Q

Quantum dots (QDs), 93

R

Raney-Ni alloy, 28
Rapid omnicompaction (ROC), 109
Rapid solidifications (RSs), 4, 8, 9f, 99, 265, 335
categories, 8
droplet method, 8
melt spinning, 8, 9f
surface melting, 8
RBM. *See* Reactive ball milling (RBM)
Reactive ball milling (RBM), 4, 28, 32, 82. *See also* Ball-milling (BM)
BN nanotubes synthesis, 148f–150f, 147
SEM image, 149f
TEM image, 149, 150f
XRD patterns, 148, 149f
doping MgH2 with pure metals, 269f
MgH_2 powders, characterization, 240
absorption and consequent desorption curves, 258, 259f
apparent activation energy, 250, 251f
endothermic peaks, 250
experimental procedure, 255–259
hiden isotherma, 251
hydrogenation/dehydrogenation behavior, 251
morphological changes of Mg powders, 246
PCT, 251, 254, 257f
structural change of Mg powders, 241
thermal stability, 248
nanocrystalline MgH_2 powders, synthesizing, 238
flowchart diagram, 239f
high-energy, 238
low-energy, 240–242
TiN powders, 180
BFI and corresponding SADP, 185, 186f
bulk densities, 187f
coefficient of friction, 188, 189f
consolidation, 182
FE-SEM micrographs, 185f
HRTEM micrograph, 183, 184
Vickers hardness, 187f, 188
XRD pattern, 182, 183f
Young's and shear modulus values, 188, 189f
Reducing agent (RA), 129
Reinforcement materials
defined, 193
particulate composites, 194
Renewable energy system, 293. *See also* Hydrogen energy
Retsch, 38, 40, 41f, 42
Retsch Mixer Mills
CryoMill, 36, 37f
MM 200 and MM 400, 35f
RFDA HTVP1600, 156
RF sputtering, 93
Ribbons, 8
RM200-ZOZ GmbH, Germany, 181
ROC. *See* Rapid omnicompaction (ROC)

Rod milling (RM), 30, 66. *See also* Ball-milling (BM)
  advantages, 66
  amorphization during, 342. *See also* Solid-state amorphization
  atmosphere, 80
  MSSR by, 137
  room temperature, Cu_2O reduction with Ti by, 129
  tumbler rod mill, 45, 46f
Roller ball mills, 55, 57t
  balls and particle motions, 70f
Rolling, 93
Room temperature rod-milling technique, Cu_2O reduction with Ti by, 129
RSs. *See* Rapid solidifications (RSs)
Run-of-mine, 5
Russian Academy of Sciences in Novosibirsk, Institute of Theoretical and Applied Mechanics of, 315
Rutherford back-scattering spectroscopic (RBS) marker experiments, 336

S

SADPs. *See* Selected area diffraction patterns (SADPs)
SAP. *See* Sintered aluminum particles (SAP)
SARR. *See* Solid-state amorphization reaction (SARR)
SAXS. *See* Small angle X-ray scattering (SAXS)
Scanning electron microscope (SEM), 78, 80–82, 138
Scanning electron microscopy (SEM)
  $Al_{50}Ta_{50}$ alloy powders, 380f
  BN nanotubes synthesis, 149f
  $Cu_{50}Ti_{20}Ni_{30}$ alloy powders, 322
  MgH_2 powders, 246, 247f
  observations, amorphization by MA, 347f, 348f
  SiC_{10}/Al_{90} powder, 197, 198, 202, 203f
  TiC powders, 156, 162f
  TiN powders, 185f, 187f, 188
  WC/Al_2O_3 nanocomposites, 207–209, 211f, 212f
  WC-5Co-1Cr-3MgO-0.7VC-0.3Cr_3C_2 nanocomposite, 214f, 220f, 221f, 225f
  WC powders, 173f, 174, 176f
Scanning probe methods, 17, 18f
Scanning transmission electron microscopy (STEM)
  crystal structure and nanoelemental composition of nanocrystalline materials, 95t
  MgH_2 powders, 244, 245f
  TiC powders, 156, 160f, 161f
  TiN powders, 184f
  WC powders, 174, 175f, 179
Selected area diffraction patterns (SADPs)
  $Al_{50}Hf_{50}$ powders, 350, 351f
  amorphization by MA, 344f
  Co-Ti binary system, 388f
  Cu_2O reduction with Ti, 132f–134f, 132
  MgH_2 powders, 244f
  MgH_2/10 wt.% (8 Nb_2O_5/2 wt.% Ni) system, 295f
  SADP of, 397, 399f
  SiC_{10}/Al_{90} powder, 200f
  β-SiC powders, 168f
  TiC powders, 160f, 162, 163f
  TiN powders, 185, 186f
  WC/Al_2O_3 nanocomposites, 208, 211f
  WC and WC-MgO fabrication by MSSR method, 140–142
Selective reduction technique, 19
Self-propagating combustion method, TiN powders, 180
Self-propagating high-temperature synthesis (SHS), 154, 158, 312, 314, 322
SEM. *See* Scanning electron microscope (SEM)
Semiconductors, 1, 3f
Severe plastic deformation (SPD), 7f, 286
  extrusion process, 7f
Shaker mills, 33–35
  Freezer/Mill, 36, 37f
  Retsch Mixer Mills (MM 200 and MM 400), 35f
  Spex, 33
  Spex 8000D, 33, 34f
  Spex 8000M, 33, 34f
Shape
  milling media, 66f–69f, 67
  milling vials, effects, 58f, 67
Shearing zone, ball mill, 66f
Shear modulus

SiC_{10}/Al_{90} powder, 204, 205f
$Ti_{55}C_{45}$ samples, 164, 166f
TiN powders, 188, 189f
WC/Al_2O_3 nanocomposites, 212
SHS. *See* Self-propagating high-temperature synthesis (SHS)
SiC/Al MMNCs, fabrication, 196
by mechanical solid-state mixing, 197
BFIs, 200f
Bragg peaks, 198
bulk density, 203f
elastic modulus, 204, 205f
helium carrier fusion-thermal conductivity method, 197
induction coupled plasma emission, 197
optical micrograph, 199f
PAS method, 197, 201
properties, 198
SADP, 200f
SEM, 197, 198, 202, 203f
TEM, 197, 200, 201
Vickers hardness, 204f
XRD patterns, 197, 198f, 201, 202f
mechanism, 204
agglomerates coarse composite SiC_p/Al powder particles, 205, 206
nanocomposite SiC_p/Al powder particles, 206
β-SiC powders, 166
infrared transmission, 167f
XRD pattern, 167f
Sievert-type apparatus, 251, 252f, 255
$Si_{80}Ge_{20}$ alloy powders, 58f
Sigma-Aldrich-GF14536537, 181
Simple mixing, 19
Single-wall carbon nanotubes (SWCNTs), 28, 102, 235. *See also* Multiwall carbon nanotubes (MWCNTs)
BFI images of, 104, 105f
fabrication by milling, 117, 118f
Sintered Al powder (SAP), 24
room temperature tensile strength *vs.* dispersoid content, 25f
Sintered aluminum particles (SAP), 19
Sintering
consolidation of nanocrystalline powders, 108
high frequency induction heated, 109
microwave, 109
powder densification and, 109
powders materials, 105, 108. *See also* Nanocrystalline powders
classic powder metallurgy, 106f, 107f
pressure-assisted, 109, 110
vacuum, 109
Sintering forging, 108
Sliding distance, TiN powders, 188, 189f
Small angle X-ray scattering (SAXS), 13
crystal structure and nanoelemental composition of nanocrystalline materials, 95t
9kW-SmartLab-XRD/SAXS (Rigaku-Japan) equipment, 14f
Smart materials, 3. *See also* Advanced materials
Solgel, 4
Sol-gel approach, TiN powders, 180
Solid contamination, milling tools, 61
Solid-hydrogen Ti-tank, integrated with mgH_2/10 wt.% (8 Nb_2O_5/2 wt.% Ni) system, 304, 305f
Solidification approach, particulate MMNCs, fabrications method, 195
Solid-state amorphization, mechanically induced
$Al_{50}Ta_{50}$ amorphization reaction in rod mill, MA *vs.* MD in, 377, 384
background, 377, 378f
morphological changes with milling time, 380f, 381f
procedure, 378
structural changes with milling time, 379f
thermal stability, 382f
amorphous alloys, fabrication by MA, 336, 337f, 338f
crystal-to-glass transition, 339f
metastable phase diagram, 339, 340f, 341f
cyclic crystalline-amorphous transformations during MA, 385
Al-Zr binary system, 391
amorphous-crystalline-amorphous cyclic phase transformations during ball milling, 394, 395f
Co-Ti binary system, 385
of $Fe_{50}W_{50}$ binary system (by MA when ΔH^{for} equal to zero), 366

Solid-state amorphization, mechanically induced (*cont.*)
magnetic studies, 368, 369f
mechanism, 371
structural changes with milling time, 366–368
thermal stability, 369, 370f, 371f
glass-forming range, 362f–364f, 361, 365t
history, 336
mechanism, MA process, 342
morphology and metallography changes with milling time, 347f–349f, 347
structural changes with milling time, 343
TEM observations, 344f–346f, 344
thermal stability, 349
X-ray analysis, 342f, 343f
multicomponent metallic glassy alloy powders, consolidation into full-dense bulk materials, 396
$Ti_{40.6}Cu_{15.4}Ni_{8.5}Al_{5.5}W_{30}$ metallic glassy alloy powders, by SPS, 407f–411f, 406
$Zr_{52}Al_6Ni_8Cu_{14}W_{20}$ glassy alloy powders, 396
overview, 335
recent studies, 411
requirements for, 336
special systems and applications
amorphous austenitic stainless steel, 373f, 374f
$Fe_{52}Nb_{48}$ special steel, 375f
Fe-Zr-B system, 376f, 377f
Solid-state amorphization reaction (SARR), 339
Solid-state hydrogen storage, 235
hosting hydrogen, nanomaterials for, 235
magnesium hydride, 237
synthesizing, 237, 238
metal hydrides, 235, 237t
Solid-state reaction (SSR)
rates, 129
WC/Al_2O_3 nanocomposites, 207, 208f, 210
Spark plasma sintering (SPS) method, 110, 302
ball-milled nanocrystalline powders consolidation, 110
components and system configurations, 110, 111f
mechanism, 113, 114f
powder specimen filling procedure, 110, 112f
procedure, 113
TiC powders, 156, 162, 163f
$Ti_{40.6}Cu_{15.4}Ni_{8.5}Al_{5.5}W_{30}$ metallic glassy alloy powders, 407f–411f, 406
TiN powders, 182, 185, 187
WC-5Co-1Cr-3MgO-0.7VC-0.3Cr_3C_2 nanocomposite, 222, 223
WC powders, fabrication, 169
SPD. *See* Severe plastic deformation (SPD)
Specific gravity, milling media, 63t
Speed, milling. *See* Milling speed
Spex, 33
Spex 8000D, 33, 34f
SPEX 8000D Mixer/Mill, 56, 58
SPEX Freezer/Mill-6970EFM, 36, 37f
Spex 8000M, 33, 34f
SPEX Mixer/Mill, 62
"Splats", 312
"Sponge iron", 4
SPS. *See* Spark plasma sintering (SPS) method
Sputtering, 4, 99
Stainless steel alloys, 432
Startack Co. Ltd, 309, 310f
Stearic acid, 83
Steel, mechanical properties, 375
STEM. *See* Scanning transmission electron microscopy (STEM)
STEM-BF image, of Cu-based alloy nanoparticle, 19, 20f
Stirred ball mill, 32
Stone Age to Nanotechnology Age, materials, 2. *See also* Advanced materials
Super Misuni, 36f
Surface engineering, 309
Surface melting, 8
Surface modification, 433
Surface protective coating
ball-milled powders for, 309. *See also* Ball-milled powders; for surface protective coating
correlation between processes, 309, 311f
defined, 309
dry coating approach, 309, 310f
process classification, 309
thermal spraying, 310, 312f

cold spray, 310, 315–317
combustion-based processes, 311
high-feed-milled HVOF sprayed WC-Co coatings, 315
high velocity oxygen thermal spraying (HVOF), 311, 313f
HVOF of nanocrystalline iron aluminide, 315
HVOF of nanostructured Cr_3C_2-Ni20Cr coatings, 314f
HVOF reactive spraying of mechanically alloyed Ni-Ti-C powders, 314
HVOF sprayed diamond reinforced bronze coatings, 315
types of, 312f
utilization of powders as feedstock for HVOF, 312
uses, 309
wet coating approach, 309
SUS304, 62, 286
SWCNTs. *See* Single-wall carbon nanotubes (SWCNTs)
Szegvari attritor grinding mill, 32, 33f

T

Tantalum-based alloys, 433
TASSA. *See* Thermally assisted solid-state amorphization *(TASSA)*
ΔH_a*TASSA*, 352
$Ta_{55}Zr_{10}Ni_{10}Al_{10}Cu_{15}$, 421
TEM. *See* Transmission electron microscopy (TEM)
Temperature, milling, 88–90
TG/DTA. *See* Thermogravimetry/differential thermal analysis (TG/DTA)
Thermal approach, advanced materials, 2, 8
droplet method (gas/water atomization), 8
rapid solidification, 8, 9f
thermal plasma processing, 9
vapor deposition, 10
Thermally assisted solid-state amorphization *(TASSA)*, 4
amorphization via, 359, 360
ΔH_a*TASSA*, 352
thermal stability, of amorphization by MA, 359, 360
Thermal mechanical analysis (TMA), 18, 96t
Thermal plasma processing, 9
Thermal spraying, 309, 310, 312f, 433
cold spray, 315–317
advantages, 317, 318f, 319f
background, 315
Co-based alloy powders, 333
Cu-based alloy powders, 333
feedstock powders, 332
indoor process, 315, 317f
mechanism, 320
of metal matrix reinforced with CNTs, 330, 332f
of metal matrix reinforced with diamond powders, 331
metal matrix reinforced with tungsten carbide, 332
of metastable powders obtained by mechanical alloying, 321f–331f, 320
Ni-based alloy powders, 333
process representation, 315, 316f
robot system, 317f, 318f
scientific publications, 316, 318f
Ti- and hard Cr-based alloy powders, 333
combustion-based processes, 311
high-feed-milled HVOF sprayed WC-Co coatings, 315
high velocity oxygen thermal spraying (HVOF), 311, 313f
HVOF of nanocrystalline iron aluminide, 315
HVOF of nanostructured Cr_3C_2-Ni20Cr coatings, 314f
HVOF reactive spraying of mechanically alloyed Ni-Ti-C powders, 314
HVOF sprayed diamond reinforced bronze coatings, 315
utilization of powders as feedstock for HVOF, 312
metallic and nonmetallic groups, 310
types of, 312f
Thermal stability, 349
$Al_{50}Ta_{50}$ amorphization, 382f
amorphization by MA
amorphization process, 350f–355f, 349, 356t
crystallization process, 356f–358f, 356, 358t

Thermal stability (*cont.*)
MDSSA process, 360
mechanism, 358
TASSA process, 359, 360
Co-Ti binary system, 390f
$Fe_{50}W_{50}$ binary system, solid-state amorphization by MA when ΔH^{for} equal to zero, 369, 370f, 371f
nanocrystalline materials, 94, 96t
$Zr_{52}Al_6Ni_8Cu_{14}W_{20}$ glassy alloy powders, 399, 401f, 402f
Thermodynamic methods, 17, 18f
Thermogravimetry/differential thermal analysis (TG/DTA), 17, 18f, 96t
ThO_2-dispersed nickel, 19
Three-dimensional (3D) nanomaterials, 93
mechanochimical process for fabrication
c-BN, 147
DTA measurements, 134, 136f
metallography, 134, 135f
MSSR, mechanism of, 136
NbN, 150, 151f
overview, 129, 130f
properties, 130
room temperature rod-milling technique (Cu_2O reduction with Ti), 129
structural changes with milling time, 131f–134f, 130
WC alloys, 138
WC-MgO, nanocomposite, 138
Ti-Al coatings, Wb:Wp ratio, 87f
$Ti_{60}Al_{15}Cu_{10}W_{10}Ni_5$, 421
Ti- and hard Cr-based alloy powders, 333
Ti-based alloys, 433, 434. *See also specific entries*
biomedical applications, 434
low modulus titanium-niobium-tantalum-zirconium (TNTZ) alloy, 436
Ti-Mn-Nb alloys, low-cost, 434, 435f
Ti-5Mo-5Ag alloy, high strength, antibacterial, and biocompatible, 434f
β-type Ti-Nb-Ta-Zr-xHaP (x = 0, 10) alloy, 436
Ti-13Nb-13Zr alloy with radial porous Ti-HA coatings, 436, 437f
TiC powders, synthesizing and properties, 155
ball milling, 157f
BFI images, 160f
Bragg peaks, 158
EDS, 161f
HAADF, 161f
HRTEM, 156, 161f
SADP, 160f
SEM, 156, 162f
STEM, 156, 160f, 161f
XRD pattern, 156, 158, 159f
BFI image, 160f
Bragg peaks, 158
consolidation techniques, 162
EDS, 161f
FWHN values, 158
HAADF, 161f
HRTEM, 156, 161f
mechanical properties, 163
elastic moduli, 164, 166f
microhardness, 163–165
MSR, 157, 158
SADP, 160f
SEM, 156, 162f
STEM, 156, 160f, 161f
XRD technique, 156, 158, 159f
$Ti_{40.6}Cu_{15.4}Ni_{8.5}Al_{5.5}W_{30}$ metallic glassy alloy powders
DSC curves, 409, 410f
friction coefficient, 410, 411f
HRTEM micrograph, 406, 408f
by SPS, 407f–411f, 406
XRD patterns, 406, 407f
Ti-HA coatings, Ti-13Nb-13Zr alloy with, 436, 437f
Time, milling. *See* Milling time
Ti-Mn-Nb alloys, low-cost, 434, 435f
Ti-5Mo-5Ag alloy (high strength, antibacterial, and biocompatible), 434f
β-type Ti-Nb-Ta-Zr-xHaP (x = 0, 10) alloy, 436
Ti-13Nb-13Zr alloy, with radial porous Ti-HA coatings, 436, 437f
TiN powders, 180
chemical analyses, 190t
consolidation, 182
preparation, 181f
results, 182
Ti powders, Cu_2O reduction with
BFIs, 132f–134f, 132
DTA measurements, 134, 136f
MSSR, mechanism, 136

by room temperature rod-milling technique, 129
SADPs, 132f–134f, 132
structural changes with milling time, 130–132
TEM analyses, 132f–134f, 132
XRD patterns, 130, 131f
Titanium carbide nanopowders, catalyzation with, 280, 281f
Titanium-niobium-tantalum-zirconium (TNTZ) alloy, low modulus, 436
TMA. *See* Thermal mechanical analysis (TMA)
Top-down approach, 5
nanofabrication methods, 93
Traditional alloys, 417, 418f. *See also* High-entropy alloys (HEAs)
vs. HEAs, 419f
Transition metals (TM). *See also specific names*
doping of MgH_2, 266–268
Transmission electron microscopy (TEM), 78, 101, 129, 138
amorphization by MA, 344f–346f, 344
BN nanotubes synthesis, 149, 150f
Cu_2O reduction with Ti, 132f–134f, 132
$SiC_{10}/A1_{90}$ powder, 197, 200, 201
VNbMoTaW HEA, 422, 423f
WC/Al_2O_3 nanocomposites, 210, 213f
WC powders, 170, 172f, 174
$Zr_{52}Al_6Ni_8Cu_{14}W_{20}$ glassy alloy powders, 397, 398f
Transverse optical (TO)-like phonon mode, 167
Tumbler ball mill, 44, 45f
Tumbler rod mill, 45, 46f
Tumbling mill(s), 44, 240
balls and particle motions, 70f
defined, 44
trajectory of milling media in, 70, 71f
Tungsten carbide (WC), 20
cold spraying of metal matrix reinforced, 332
Two-dimensional (2D) nanomaterials, 93

U

Uni-ball mill, 42f, 43f
UNILAB Pro Glove Box Workstation, 181
Union Process, 32

V

Vacuum sintering, 109
van der Waals-Zeeman Laboratory, 34
vibratory ball mill at, 34, 35f
van't Hoff approach, 298, 299
van't Hoff equation, 253, 298
van't Hoff plot, 298, 299f
Vapor deposition, 2, 10, 309
CVD. *See* Chemical vapor deposition (CVD)
EBPVD, 10
PVD. *See* Physical vapor deposition (PVD)
techniques, 10
Vapor-transfer processes, 10
Vials, milling, shape of, 58f, 59
Vibratory ball mill(s), 31, 55, 57t
Super Misuni, 36f
at van der Waals-Zeeman Laboratory, 34, 35f
Vickers hardness
$SiC_{10}/A1_{90}$ powder, 204f
$Ti_{55}C_{45}$ samples, 163–165
TiN powders, 187f, 188
values, 63t
WC/Al_2O_3 nanocomposites, 211, 214f
WC-5Co-1Cr-3MgO-0.7VC-0.3Cr_3C_2 nanocomposite, 215, 216f
Vickers indenter, 156
VNbMoTaW HEA, 422f–424f, 422
hardness, 424f
IFFT HRTEM image, 422, 423f
TEM images, 422, 423f
XRD patterns, 422f
Volumetric hydrogen storage, 232, 233f
$V_{45}Zr_{20}Ni_{20}Cu_{10}Al_{2.5}Pd_{2.5}$, 421

W

Water atomization, 8
Water quenching, 8
Wb:Wp ratio, Ti-Al coatings, 87f
WC/Al_2O_3 nanocomposites, 207
BFIs, 207, 210f, 213f
Bragg peaks, 207, 208f
HRTEM, 208, 211f
MSSR, 207, 208f
SADP, 208, 211f
SEM, 207–209, 211f, 212f
TEM, 210, 213f

WC/Al_2O_3 nanocomposites (*cont.*)
vial's temperature and the milling time, 210, 213f
Vickers hardness, 211, 214f
XRD patterns, 207, 208f
Young's and shear moduli, 212
WC and WC-MgO fabrication, MSSR method, 138
ball-milled powders, properties, 139
BFI of, 140–142, 143f
DFI of, 140, 141f
hardness, toughness, and elastic moduli, 146f, 147t
SADP, 140–142
structural changes with milling time, 140f–143f, 139
temperature change with milling time, 144f, 145f
XRD patterns, 139–141
WC and WC-MgO refractory materials, wide industrial applications, 138
WC-based nanocomposites, 207. *See also specific entries*
WC/Al_2O_3, 207
WC-5Co-1Cr-3MgO-0.7VC-0.3Cr_3C_2 nanocomposite, 214, 219
WC-5Co-1Cr-3MgO-0.7VC-0.3Cr_3C_2 nanocomposite, 214, 219
AFM micrographs, 216, 218f, 219f
BFIs, 215f
FE-SEM micrograph, 221f, 225f
MM approach, 222
optical micrographs, 226f
SEM micrograph, 214f, 220f
SPS technique, 222, 223
XRD patterns, 219, 220f, 222, 224f
Young's modulus, 216, 223
WC/Co powders, 62
WC powders, nanocrystalline, 169
properties, 169, 178, 179f
spark plasma sintering for, 169
synthesis, 169
Weld overlays, 309
Wet coating approach, 309
Wet milling technique, 83
Wilson Vickers Hardness Tester, Germany, 156
WMoNbZrV alloy, 424, 425f
WO3/3Mg/C powders, 130f

X

XPS. *See* X-ray photoelectron spectroscopy (XPS)
X-ray analysis, amorphization by MA, 342f, 343f
X-ray diffraction (XRD), 13, 14f, 58f, 78, 101, 129
$Al_{50}Ta_{50}$ alloy powders, 379f
$Al_{50}Ta_{50}$ powders, 83, 86f, 87f
Al_xHf_{100-x} powders, 342f–344f
Al_xTa_{1-x}, 361
Al-Zr binary system, 391–393
$Al_{50}Zr_{50}$ powder, milling speed, 73–75
bcc-Nb powders, 150, 151f
BN, 148, 149f
CoCrFeNiMn alloys, 424, 425f
Co-Ti binary system, 386f, 387f
CRed MgH_2 powders, 287
crystal structure and nanoelemental composition of nanocrystalline materials, 95t
$Cu_{50}Ti_{20}Ni_{30}$ alloy powders, 322, 326f
fcc-NbN alloy powders, 150, 151f
FeCoCrMoCB, 67, 323f
$Fe_{74}Cr_{18}Ni_8$, 374f
$Fe_{50}W_{50}$ powders, 366, 367f
graphite powders, 72f, 116
MgH_2 powders, 241, 243f, 249f
MgH_2/10 wt.% (8 Nb_2O_5/2 wt.% Ni) system, 295f, 296, 302f
MWCNTs, 104f
$Ni_{60}Nb_{40}$ amorphous alloy, 335f, 336
Pd powders, 102, 103f
SiC_{10}/Al_{90} powder, 197, 198f, 201, 202f
β-SiC powders, 167f
TiC powders, 156, 158, 159f
$Ti_{40.6}Cu_{15.4}Ni_{8.5}Al_{5.5}W_{30}$ metallic glassy alloy powders, 406, 407f
TiN powders, 182, 183f
VNbMoTaW HEA powders, 422f
WC/Al_2O_3 nanocomposites, 207, 208f
WC and WC-MgO fabrication, 139–141
WC-5Co-1Cr-3MgO-0.7VC-0.3Cr_3C_2 nanocomposite, 219, 220f, 222, 224f
WC powders, 170, 171f, 174
WMoNbZrV powders, 425f
$Zr_{52}Al_6Ni_8Cu_{14}W_{20}$ glassy alloy powders, 397, 398f

X-ray line broadening, 98
X-ray photoelectron spectroscopy (XPS), 13
XRD. *See* X-ray diffraction (XRD)
XRF. *See* Energy dispersive X-ray fluorescence (XRF)
XRF (S8 TIGER 4 kW) system, 16f

Y

Young's modulus
  mechanically mixed SiC_{10}/Al_{90} powder, 204, 205f
  $Ti_{55}C_{45}$ samples, 164, 166f
  TiN powders, 188, 189f
  WC/Al_2O_3 nanocomposites, 212
  WC-5Co-1Cr-3MgO-0.7VC-0.3Cr_3C_2 nanocomposite, 216, 223
  WC powders, 175, 177f–179f
Yttrium-aluminum garnet $5Al_2O_3$ $3Y_2O_3$ (YAG), 23

Z

Zirconium oxide, 20
Zoz GmbH, 32
Zoz Simoloter mill, 56
$Zr_{65}Al_{7.5}Ni_{10}Cu_{12.5}Pd_5$, 421
$Zr_{52}Al_6Ni_8Cu_{14}W_{20}$ glassy alloy powders, 396, 421
  BFIs of, 397, 399f
  consolidation, 403f–406f, 400
  HRTEM images, 397, 398f
  SADP of, 397, 399f
  structural change, 397, 398f, 399f
  TEM observations, 397, 399f
  thermal stability, 399, 401f, 402f
  XRD patterns, 397, 398f
Zr-based nanopowders
  metastable phases, MgH_2 catalyzation with, 281–284
ZrC powders, fabrication, 180
Zr-Ni alloy, 417
$Zr_{100-x}Ni_x$ alloy powders, 267

9780128181805